Himalaya: Mountains of Destiny

Suresh Chand Rai
Editor

Himalaya: Mountains of Destiny

Springer

Editor
Suresh Chand Rai
Department of Geography
Delhi School of Economics
University of Delhi
Delhi, India

ISBN 978-3-031-94904-3 ISBN 978-3-031-94905-0 (eBook)
https://doi.org/10.1007/978-3-031-94905-0

This Springer imprint is published by the registered company Springer Nature Switzerland AG
The registered company address is: Gewerbestrasse 11, 6330 Cham, Switzerland

Preface

The Himalayas are vital to the life support system of millions of people in the Indian subcontinent's highlands and many more in its lowlands and eventually to the entire world. Among the major issues that the area and its residents must address sustainably are climate change, conflicts, and other risks. The vast majority of the Indian subcontinent depends heavily on the ecosystem services provided by this enormous mountain chain. All of this has contributed to the region's international significance, along with its riches and distinctive biophysical and socio-cultural diversity.

Although the Himalaya has a complex and significant ecology, little is understood about it, its ecosystems, or its people—especially in light of the region's rapid development. Over the last few decades, there has been a surge in research on the region, but the data collected is often scattered and only reaches a limited audience. This book has several justifications. The foremost is about making the knowledge base more accessible. Gathering those involved in creating knowledge about the Himalayas has been incredibly beneficial in compiling what is already known. The evaluation also incorporates information from many viewpoints on the mountains by collaborating across disciplines. Global evaluations and programs, such as the Intergovernmental Panel on Climate Change (IPCC), can now benefit from a valuable source of information about this region, and the book is a valuable resource for educating people around the world.

However, the primary motivation of this book is to extend beyond the compiling of information. Its purpose is to address a variety of policy-related issues that we all face. Therefore, providing decision-makers with the finest available research and knowledge is the book's primary goal. The release of the Himalaya: Mountains of Destiny marks a significant step in a broader effort to bring together researchers, policymakers, and the public in successfully managing the HKH region for the benefit of all individuals, including women and children, to enhance their overall well-being in a sustainable mountain environment.

This book includes an introduction with four parts. The introduction covers the conceptual framework, mountains of the world, the Indian Himalayan region, multiple threats, challenges, and future directions, as well as the SDG's and Himalaya to protect natural resources, so that readers can better comprehend the book. The first part mainly focuses on the geo-physical environment of the Indian

Himalayan region. The second part mainly focuses on ecosystems, resources, and development. The third part is about geo-environmental hazards. The fourth part is focused on governance: key for environmental sustainability. All the contributors are young and middle-aged researchers from diverse fields, along with accomplished earth scientists. The book is meant for professionals, researchers, and undergraduate and graduate students in this case. At various levels, case studies were used to convey the research papers in this book. We can easily achieve the Sustainable Development Goals (SDGs) if we combine all of these factors, according to the editor.

This book would not be possible without the kind assistance of many researchers. The editor would like to thank several professionals for their insightful recommendations. The editor would like to thank the chapter author or authors for contributing original content to this book.

Delhi, India Suresh Chand Rai

About This Book

This book explains the significance of the Himalayas to the millions of people living in the highlands and many more in the lowlands of the Indian subcontinent. The region and its people face a range of new challenges moving forward with climate change, globalization, migration, conflicts, and environmental degradation. Despite its significance, comparatively little is understood about the Himalayas, its ecosystems, and its inhabitants—especially in light of the region's rapid transformation. Natural resources are degrading at an alarming rate these days as a result of poor conservation techniques, changes in land-use/cover, human pressure, infrastructure expansion, etc. The entire book, which is busted up into four parts, covers different aspects of the Himalaya's ecosystem, resources, development, and many threats. Physical aspects are reflected in the first part, ecosystem, resource, and development are in the second part, geo-environmental risks are covered in the third, and the governance: key for environmental sustainability is covered in the fourth part. This book explains and critically examines the Himalaya's significance in managing natural resources. This book will be the ideal source for researchers, undergraduate, and postgraduate students in India and outside. It also assists legislators, planners, and administrators in comprehending problems and effectively managing the Himalayan environment.

About This Book

[illegible]

Contents

About the Editor

Prof. Suresh Chand Rai is Professor and Former Chair of the Department of Geography, Delhi School of Economics, University of Delhi. He did his Master's in Geography (1979) and Doctor of Philosophy (1984) from Banaras Hindu University, Varanasi. His research expertise includes watershed management, mountain hydrology, ecotourism, biodiversity conservation, wetland hydrology, etc. His significant contributions are in the field of land and water resource management. Professor Rai has received international recognition for his pioneering studies on various topics in the Eastern Himalayas. Professor Rai won in the category of "*Scientific Achievement*" for the prestigious international award known as "Crystal Compass" which was given by the Russian Geographical Society, Russia, 2024. The Soil and Water Conservation Society of the USA awarded him in 1999 Award in recognition of his outstanding work in the Sikkim Himalaya's Intergrated Watershed Management. The extensive studies made by Prof. Rai on various aspects of mountain hydrology have not only helped in understanding the structure and function of the systems but also in finding out means of how to make the system sustainable. He has published more than 150 research papers in various impact factor/peer-reviewed journals and books (*Scopus h-index-24*). He served as Secretary General of the National Association of Geographers, India (NAGI), the largest society of Geographers in India, from 2010 to 2022. Prof. Rai is currently serving as the Vice-President of the Asian Geographers Association, Beijing, P.R. China. Prof. Rai has been Editor of the *Journal of the National Association of Geographers, India* since 2023.

Chapter 1
Himalaya: Mountains of Destiny

Suresh Chand Rai

Abstract The Himalayan mountain range, encompassing the Hindu Kush and the Tibet Autonomous Region of China, spans over eight countries—from Afghanistan in the west to Myanmar in the east, including Pakistan, India, China, Nepal, Bhutan, and Bangladesh. This vast range stretches over 3000 km, rising from low-lying plains to towering heights of approximately 8848 m above sea level. Within the global mountain ecosystem, the Indian Himalayan region occupies a distinct and vital position. The majority of the people are still marginalized and subsistence farmers despite the abundance of natural resources. The Himalayas are recognized for several features, including their religious shrines and monasteries, breathtaking scenic beauties, diversity of human races, religions, and cultures, high level of geological fragility, enormous landscape variability, rich forests and biodiversity, glaciers, perennial rivers, lakes, and wetlands. The Himalayan ecosystem provides services to millions of people across a broad range of climate zones and habitats. Natural resource overuse without scientific justification exacerbates the effects of natural risks and degrades the ecosystem. For instance, the severe floods that recently occurred in the Indian Himalayan region were probably made worse by climate change. The dynamics of snow, ice, and glaciers in the Himalayas are not well understood. Politically and socially, the area is plagued by border conflicts. The Himalayas present an ideal setting for interdisciplinary research, yet changes in this region and their consequences must be closely monitored. Sustainable management of the Himalayas demands transboundary cooperation, integrating research findings into policy and practice. A new paradigm is needed that balances ecological priorities with economic interests while respecting socio-cultural traditions.

Keywords Climate change · Glacial Lake Outburst Flood (GLOF) · Multi-hazard risks transboundary cooperation · Himalayan ecosystem services

---

S. C. Rai (✉)
Department of Geography, Delhi School of Economics, University of Delhi, Delhi, India
e-mail: raisc1958du@gmail.com

S. C. Rai (ed.), *Himalaya: Mountains of Destiny*,
https://doi.org/10.1007/978-3-031-94905-0_1

1.1 The Context

The Himalayas, often referred to as the "Mountains of Destiny," stand as a formidable and awe-inspiring range that not only shapes the geographical and ecological landscape of South Asia but also profoundly influences the cultural, spiritual, and socio-economic life of the millions who live within its shadow. Spanning over 3000 km across eight countries, the Himalayas are home to the world's highest peaks, including Mount Everest, and serve as the source of some of the planet's major rivers, such as the Ganges, Brahmaputra, and Indus. The region's unique position and unparalleled biodiversity make it a focal point for scientific research, environmental conservation, and cultural preservation.

1.2 Mountains of the World: Global Priorities

Mountainous regions comprise 22% of the planet's total area and are inhabited by around 13% of its population (FAO 2015). Mountains are home to a significant portion of the world's biodiversity, including many endemic species. They are also vital sources of freshwater, supplying nearly half of the global population with water for drinking, agriculture, and industry. Within the mountainous regions, where there are around 915 million people, less than 150 million reside over 2500 m above sea level and barely 20–30 million dwell above 3000 m above sea level. About half of the world's biodiversity hotspots are found in mountains, which contribute 25% of the terrestrial biodiversity worldwide. From water to hydroelectricity, from forest and lumber resources to mineral resources and leisure, mountains offer goods and services that are vital to the entire world (Schild and Sharma 2011; Molden and Sharma 2013). The mountains are a site rich in traditional wisdom and ethnic variety. People in mountainous areas are typically more vulnerable to poverty and risk than those in other areas. A recent analysis by the FAO found that in developing nations, 39% of the mountain population (rural and urban combined) was vulnerable to food insecurity in 2012. This was a 30% rise compared to 12 years prior (FAO 2015).

Mountains have extremely fragile geological structures, and the forces driving environmental and human change are rapidly degrading mountain ecosystems. Mountain regions are significant to those living elsewhere as well as to the local population from an ecological, aesthetic, and socio-economic perspective. According to Nogues-Bravo et al. (2008), mountains offer distinctive environments for identifying and evaluating the effects of climate change. With increasing awareness of the effects of climate change on mountains, mountain ecosystems, and mountain communities, mountains have been brought to the forefront of international discussions at conferences such as the 1992 United Nations Conference on Environment and Development in Rio de Janeiro, the 2015 Sendai Framework for Disaster Risk Reduction, the 2015 United Nations Framework Convention on Climate Change Paris Agreement, and the 2030 Sustainable Development Goals (SDGs). Thus, more efforts should be made to ensure that mountain viewpoints are included in any conversations regarding the strategy for the sustainable development of mountainous areas.

One of the planet Earth's most recent and complex mountain systems is the Himalayas. An area of more than 3.44 million km^2 is distributed throughout Afghanistan, Pakistan, India, Bhutan, Myanmar, Nepal, and Bangladesh and makes up the Himalayan region, which includes the Hindu Kush Mountains and the Tibet Autonomous Region of China. Stretching over 3000 km, it rises to an altitude of about 8848 m above sea level from low-lying valleys (GBPIHED 2013). The Himalayas are known for their high level of geological fragility, tremendous landscape variability, rich forests and biodiversity, glaciers, perennial rivers, lakes, and wetlands, as well as their religious shrines and monasteries, breathtakingly beautiful locations, and diversity of human races, religions, and cultures. The Himalayas are a result of the ongoing tectonic collision between the Indian Plate and the Eurasian Plate, a process that began approximately 50 million years ago during the Eocene epoch. This continental collision not only led to the uplift of the Himalayas but also continues to drive seismic activity in the region today. The Himalayan orogeny is a classic example of plate tectonics in action, showcasing processes such as thrust faulting, folding, and metamorphism on a grand scale (Gansser 1964).

The Himalayas are one of the world's most biodiverse regions, encompassing a wide range of ecosystems from tropical forests at lower elevations to alpine meadows and tundra at higher altitudes. This biodiversity hotspot is home to numerous endemic species, including the snow leopard (*Panthera uncia*), Himalayan blue sheep (*Pseudois nayaur*), and the red panda (*Ailurus fulgens*). The region's flora is equally diverse, with over 10,000 species of plants, many of which have medicinal properties and are used in traditional medicine (Mittermeier et al. 1998).

The complex topography and varying climatic conditions across the Himalayas create a mosaic of microhabitats, supporting a rich array of life forms. However, the region is also highly vulnerable to the impacts of climate change, such as rising temperatures, changing precipitation patterns, and glacial retreats. These changes threaten not only the unique biodiversity of the Himalayas but also the livelihoods of millions of people who depend on the region's natural resources.

The Himalayas are strategically important because it is a natural border between nations and because the Himalayan environment greatly enhances the ecological and economic security of the nations it crosses. As the water tower and climate regulator of Asia, it has regional significance. The Himalayan landmass's orography interacts with the East Asian and Indian summer monsoons to produce a massive amount of precipitation. Ten of the world's major rivers originate and drain through the Himalayas, and the basins of these rivers support around 1.35 billion people (Bandyopadhyay and Gyawali 1994). It is also the world's largest reservoir and storage of snow and ice outside of the two poles. More than three billion people benefit from the water supply, food, and energy produced with the water from these river basins (www.icimod.org). The Himalayan region owing to its unique biogeography has a peculiar identity in the world (GBPIHED 1992). It supplies a large number of ecosystems (forest ecosystem, grassland ecosystem, desert ecosystem, wetland ecosystem, etc.). Not only do the Himalayan ecosystems support the survival of the resident communities, but they also play a major role in the establishment of downstream civilizations and economies, biodiversity, water supplies, aesthetic and cultural significance, and forests.

The ecosystems of the Himalayas are a treasure trove of plants and animals, and a portion of the area is known as a hotspot for biodiversity worldwide. A wide variety of cultures, customs, and knowledge systems have emerged in the area as a result of this biodiversity. The Himalayan forests support the communities that live there and supply raw materials for industry. The rivers that flow through them carry rich soil and energy, which boosts the productivity of agriculture in areas downstream and creates potential for the development of hydropower. A significant contributing cause to their degradation is inadequate or nonexistent environmental governance. Encouraging the management of the Himalayan ecosystems requires a multi-faceted, integrated approach that takes into account the interconnectivity between different sectors and involves all stakeholders.

The Himalayas are not only a physical barrier but also a cultural and spiritual haven. The mountains have been revered as sacred in Hinduism, Buddhism, Jainism, and Sikhism for centuries. Mount Kailash, for instance, is considered the abode of Lord Shiva in Hinduism, while Buddhists believe it to be the home of the Buddha Demchok. The region is dotted with ancient monasteries, temples, and pilgrimage sites, attracting millions of devotees and spiritual seekers from around the world. The spiritual heritage of the Himalayas is deeply intertwined with the daily lives of its inhabitants, influencing their customs, traditions, and worldviews. The concept of living in harmony with nature, which is central to many Himalayan cultures, is increasingly recognized as a valuable paradigm in the face of global environmental challenges.

In addition, the strategic location of the Himalayas has made the region a focal point of geopolitical interest for centuries. The mountains form a natural border between India and China, two of the world's most populous and powerful nations. The ongoing border disputes and military presence in the region highlight the Himalayas' role as a critical buffer zone. Moreover, the Himalayas are a key source of water for several countries, leading to complex transboundary water management issues. The potential for conflict over water resources, especially in the context of climate change and growing demand, underscores the need for regional cooperation and sustainable management practices.

1.3 The Indian Himalayan Region

Situated between 21° 57′–37° 5′ N and 72° 40′–97° 25′ E, the Indian Himalayan region spans 5.37 lakh km^2 and accounts for approximately 16.2% of the country's total land area. Jammu and Kashmir, Himachal Pradesh, Uttarakhand, Sikkim, Arunachal Pradesh, Nagaland, Manipur, Mizoram, Tripura, and Meghalaya are the 10 states that make up the administratively recognized Indian Himalayan Region (Box 1.1). By contrast, only the hill districts of West Bengal and Assam are included, making them just partially represented. Table 1.1 presents the detailed statistical information.

Box 1.1: Indian Himalayan Region (IHR)

Geographical Coverage:

States Fully Covered

Jammu and Kashmir, Himachal Pradesh, Uttarakhand, Sikkim, Tripura, Meghalaya, Mizoram, Nagaland, Manipur, Arunachal Pradesh

States Partially Covered

Assam—two districts (Karbi Anglong, Dima Hassao)

West Bengal—one district (Darjeeling)

Total Geographical Area

5,33,604 km^2

Total Population (2011)

4,85,98,561 persons (48.5% Females)

About 17% of the Indian Himalayan Region is permanently covered in snow and glaciers, while the remaining 30–40% is covered in snow during the winter, which supplies water for residential and industrial use to the region's residents as well as to people and economies downstream. These areas feed North India's major rivers.

Table 1.1 Important features of states falling within the Indian Himalayan Region

S. No	States/ regions	Geographical area (km^2)	Decennial population (2011)	Sex ratio (2011)	Literacy rate (%)	Decadal growth rate (%) (2001–2011)
1	J&K	222,236	12,540	889	68.74	23.64
2	H. P	55,673	6864	972	83.78	12.94
3	Uttarakhand	53,483	10,117	963	79.63	18.81
4	Sikkim	7096	608	890	82.2	12.89
5	Arunachal Pradesh	83,743	1383	938	66.95	26.03
6	Nagaland	16,579	1980	931	80.11	-0.58
7	Manipur	22,327	2722	992	79.85	24.50
8	Mizoram	21,081	1091	976	91.58	23.48
9	Tripura	10,486	3671	960	87.75	14.84
10	Meghalaya	22,429	2964	989	75.48	27.95
11	Assam Hills	19,153	5517	954		NA
12	W. B. Hills	3149	1847	970		14.77
	India	32,87,263	12,10,193	943	74.04	

Source Economic Survey, 2012–13, http://indiabudget.nic.in

The Himalayan Rivers carry about 1,200,000 million cubic meters of water annually (Singh 2006).

With about 41.5% of its land covered by forests, the Indian Himalayan Region boasts a rich forest cover. This accounts for one-third of India's total forest cover and over half (47%) of its very good quality forest cover. Along with acting as a sink for CO_2, these forests offer a vast range of commodities and services that are important to the entire world (Singh 2002). The Indian Himalayan Region encompasses a significant portion of the Himalayan biodiversity hotspot. Its flora comprises approximately 8000 endemic species of angiosperms (40% endemic), 44 endemic gymnosperms (16% endemic), 600 endemic pteridophytes (25% endemic), 1737 endemic bryophyte (33% endemic), 1159 lichen species (11% endemic), and 6900 species of fungi (27% endemic). These species represent 1748 species of medicinal plants with both traditional and modern therapeutic uses, 675 species of wild edible plants, 118 species of essential oil-producing medicinal plants, 279 varieties of fodder, and 155 sacred plants.

There are a lot of tourist attractions and shrines in the Indian Himalayan Region. Approximately 4% of the nation's population lives there and depends on the area for their livelihood. The people who live in the Himalayan region as well as those who live downstream, both inside and outside the nation, value the numerous ecological services that the region provides, including fresh water, food, energy, life-saving medications, and biodiversity.

The fragile and vulnerable Himalayan ecosystems are currently threatened by unscientific development and climate change, which shows up as widespread environmental degradation, glacier melting, an increase in the intensity of extreme events, deforestation, biodiversity loss, etc. These trends imply a decreased supply of raw materials and resources for economic growth and subsistence, as well as potential threats to development and human well-being. Additionally harming the development and interests of ecosystem protection is the hill-specific characteristics, non-remunerative agriculture, and outmigration. Recently, there has been a persistent threat to the economies and civilizations of the occupants as well as those downstream due to the recurrent floods, landslides, and forest fires that cause enormous losses in terms of life, property, and natural resources.

Because of their importance to life, their wealth of resources, and the diverse range of service flows they produce, Himalayan ecosystems must be maintained. Furthermore, it holds significance beyond national borders and establishes connections to mitigate climate risk and provide additional services. The goals of this mission are as follows: (i) to demonstrate the Himalayas' global significance; (ii) to study hydro-geomorphological aspects of Himalaya; (iii) to value and conserve existing ecosystems, culture, societies, knowledge, and distinctive Himalayan solutions that are valuable to the rest of the world; (iv) to address current policy issues and influence policy processes with strong evidence for sustainable mountain development.

1.4 The Himalayas Are Facing Multiple Threats

The Himalayas are confronting a range of significant and interrelated threats that endanger their ecological integrity, cultural heritage, and the livelihoods of millions of people. Several regions of the Himalayas have reported seeing an elevation-dependent shift. As a result, over the twenty-first century, the amount of snow and glaciers as well as their covered area will decline (IPCC 2021). The decrease of glaciers in the Himalayas is the most commonly cited hazard to the region, as it will have dire implications for future water supply downstream. The Himalayas face similar challenges to other ecosystems, such as deforestation, habitat loss, fragmentation, and degradation, overexploitation of resources, illegal hunting, intensified agriculture, loss of genetic resources, conflict between humans and wildlife, invasive species, and air pollution. In the Himalayas, changes in land use and cover are the primary factors contributing to soil degradation and modifying ecosystem services, which impede ecosystems' capacity to sustain human demands. There is a dearth of comprehensive studies on changes in land cover and use in the Himalayas. Many rangelands, forests, croplands, and wetlands are supported by the Himalayan Mountains, but changes in land use and cover are affecting the quantity and quality of these ecosystems. A major threat to the Himalayan ecosystems is overexploitation of resources, which includes illegal or unsustainable mining, overharvesting of non-timber forest products, and destruction of forests and shrub lands for fuelwood or commercial timber.

The Himalayan ecosystems are also under threat from overgrazing and wildlife hunting. Overgrazing occurs frequently in the Himalayan rangelands. Throughout the whole Himalayan region, it is illegal to hunt red pandas, pangolins, tigers, musk deer, and rhinoceroses. Loss of vegetation, soil erosion, and disturbance of wildlife have all resulted from uncontrolled mining. The area has seen a sharp increase in unplanned development, including mining activities, the haphazard building of hydroelectric dams, and the badly designed and built roadways. The annual mean surface temperature of the Himalayan region increased by around 0.11 °C per decade between 1901 and 2014, indicating a notable increase in warming in recent times (Ren et al. 2017). The Himalaya is destined for this. Tracking the trends and impacts of climate change is one of the most challenging issues the Himalayan countries are currently confronting.

1.5 The Conceptual Framework

The Himalayan Conceptual Framework is an integrative approach that seeks to understand and address the complex and interconnected challenges faced by the Himalayan region. This framework recognizes the Himalayas as a dynamic and multi-faceted system where environmental, social, economic, and cultural factors are deeply intertwined. It emphasizes the need for a holistic perspective that goes beyond isolated disciplines to capture the full complexity of the region.

The conceptual framework was rationally created step-by-step by posing important questions, articulating important challenges, envisioning the exercise, and selecting objectives that may help meet the Sustainable Development Goals by 2030. The important key questions are:

(i) Which factors are changing the Himalayan region the most, what role does climate change play, and what are the possible effects on resources, livelihood, ecosystem services, biodiversity, and ecosystems?
(ii) How does the cryosphere currently stand, what modifications are most likely to occur, and what effects might those modifications have?
(iii) How much water is there now and how much is likely to be there in future in the three main river basins of the Indian Himalayan region? What are the possible effects of changing the water availability, and how may adverse effects be lessened?
(iv) What are the most effective ways to lessen the likelihood of floods and droughts, and how can they be implemented on different scales, such as Transboundary Rivers?
(v) In the diverse contexts present in the Himalayan region, how can ecosystems be managed to ensure the conservation of soil and biodiversity as well as enhanced livelihoods?
(vi) What ecological services can mountains offer, and how may these services be managed and supplied in exchange for payment?

In essence, the Himalayan Conceptual Framework serves as a comprehensive guide for understanding the region's challenges and opportunities, promoting an integrated approach to development and conservation that respects the Himalayas' unique ecological, cultural, and socio-economic context. By embracing this holistic perspective, the framework aims to secure a sustainable future for the Himalayas and the millions of people who depend on them.

1.6 Challenges and the Ways Forward

The Himalayas, a region of immense ecological, cultural, and geopolitical significance, face numerous challenges that threaten their sustainability and the well-being of the communities that depend on them. Climate change is a major concern, as rising temperatures lead to the accelerated melting of glaciers, altering water availability and increasing the risk of glacial lake outburst floods (GLOFs). Deforestation and land degradation, driven by unsustainable agricultural practices, infrastructure development, and illegal logging, further exacerbate soil erosion and biodiversity loss. Additionally, the region grapples with socio-economic challenges, including poverty, limited access to education and healthcare, and the impacts of outmigration, which weaken traditional knowledge systems and community resilience.

In the context of climate change and glaciers, monitoring the trends and impacts of climate change is one of the most challenging issues the Himalayan nations are now dealing with. The melting glaciers in the Himalayas have garnered the majority of discussion; nevertheless, due to the sparse hydrological and meteorological stations and difficult terrain, systematic monitoring is absent. The lack of sufficient spatial density and/or temporal extent of observations affects the accuracy of climate change predictions; therefore, there are still uncertainties in the identification and attribution of atmospheric drivers. The 2013 outburst of the Chorabari glacier in Uttarakhand, which claimed more than 6000 lives, is one example of the Glacial Lake Outburst Floods (GLOFs) caused by the diminishing cryosphere. The area is vulnerable to a lack of knowledge regarding the effects on ecosystems and biodiversity of development initiatives like infrastructure, hydropower projects, tourism, and transportation. Due to numerous boundary conflicts in the area, countries are unable to exchange information about the state of their resources.

There are possibilities to be gained from the Himalaya's environmental challenges and data gaps. Particular consideration must be given to the governance frameworks and policy execution. These voids might be filled by transboundary research conducted in conjunction with foreign partners. The eco-social and economic systems that make up the Himalayan landscapes are intricate. The landscape approach is the most effective way to study socio-ecological systems to comprehend both natural and anthropogenic ecological changes (Sayer 2009). In the Himalayas, a transdisciplinary landscape approach has promise for optimizing synergy and ensuring integrated activities by various stakeholders. Further benefits will come from incorporating indigenous knowledge and citizen science into the Himalayan climate change monitoring system. To increase the breadth of scientific knowledge and incorporate traditional knowledge and citizen science into policy and practice, initiatives like the National Mission for Himalayan Studies and the National Mission for Sustaining Himalayan Ecosystem (NMSHE) have been established (GOI 2016, 2017). Throughout the Himalayan region, these kinds of programs are required.

Himalaya can be sustained with cross-calibration and long-term monitoring along with ground validation. There is some hope that citizen science may transcend politics and give scientists and environmentalists a global platform to collaborate. For practical planning, more reliable data from higher elevations are desperately needed. To track the risk of a natural disaster, the Himalayan countries must establish an international network including regional expertise and logistical support from richer nations.

Addressing these challenges requires a multi-faceted approach that includes enhancing regional cooperation on transboundary water management, promoting sustainable land-use practices, and strengthening disaster risk reduction strategies. Conservation efforts must focus on protecting the unique biodiversity of the Himalayas, while climate adaptation measures should prioritize the resilience of local communities. Empowering these communities through education, capacity building, and the preservation of indigenous knowledge is essential for sustainable development. Moreover, fostering international collaboration and integrating traditional wisdom with modern scientific approaches will be crucial in securing the future of the Himalayas and ensuring that they continue to provide vital resources for the region and beyond.

1.7 SDGs and Himalayas

The Sustainable Development Goals (SDGs) and the Himalayas are closely interconnected, as the region plays a crucial role in achieving several of these global targets. The Himalayas, home to unique ecosystems and diverse cultures, are vital for regional and global sustainability. However, the region faces numerous challenges that align with various SDGs, necessitating focused efforts to ensure that development in the Himalayas is sustainable and inclusive.

1. **SDG 1**: **No Poverty**—The Himalayan region, despite its rich natural resources, is home to many marginalized communities that rely on subsistence agriculture and have limited access to basic services. Efforts to reduce poverty in the Himalayas must address issues such as access to education, health care, and economic opportunities, while also promoting sustainable agricultural practices and rural development.
2. **SDG 2**: **Zero Hunger**—Food security in the Himalayas is challenged by harsh environmental conditions, soil erosion, and climate change. Promoting sustainable agriculture, improving access to nutritious food, and enhancing local food systems are essential to achieving food security in this region.
3. **SDG 6**: **Clean Water and Sanitation**—The Himalayas are often referred to as the "Water Towers of Asia," providing fresh water to millions of people across South Asia. Ensuring the sustainable management of water resources, protecting watersheds, and improving water and sanitation infrastructure are critical for this region.
4. **SDG 7**: **Affordable and Clean Energy**—Many Himalayan communities still lack access to modern energy services. Expanding access to clean and renewable energy sources, such as hydropower, solar, and wind energy, is crucial for sustainable development and reducing reliance on traditional biomass, which can lead to deforestation.
5. **SDG 13**: **Climate Action**—The Himalayas are highly vulnerable to climate change, with rising temperatures causing glacier melt, shifting weather patterns, and increasing the frequency of natural disasters. Climate adaptation and resilience-building measures are essential for protecting the region's ecosystems and communities.
6. **SDG 15**: **Life on Land**—The Himalayas are a biodiversity hotspot, home to many endemic species and vital ecosystems. Protecting and restoring these ecosystems, preventing deforestation, and conserving biodiversity are critical to maintaining the ecological balance and achieving sustainable development in the region.
7. **SDG 16**: **Peace, Justice, and Strong Institutions**—The geopolitical significance of the Himalayas, marked by border disputes and resource conflicts, highlights the need for strong institutions and regional cooperation. Ensuring peace, stability, and effective governance in the region is essential for achieving sustainable development.

8. **SDG 17**: **Partnerships for the Goals**—The challenges facing the Himalayas require collaborative efforts across borders, involving governments, local communities, international organizations, and civil society. Strengthening partnerships and fostering regional cooperation are key to addressing the complex issues of the Himalayan region and achieving the SDGs.

In conclusion, the Himalayas are central to achieving several of the SDGs, but doing so requires a coordinated and sustainable approach that addresses the unique challenges of the region. Integrating the SDGs into development strategies for the Himalayas will help ensure that the region's natural resources and cultural heritage are preserved for future generations while improving the quality of life for its inhabitants.

The Himalayas, truly the "Mountains of Destiny," are a region of unparalleled natural beauty, ecological diversity, and cultural richness. They are central to the lives of millions of people, providing essential resources, spiritual inspiration, and a natural barrier between nations. However, the Himalayas also face significant challenges, from environmental degradation and climate change to geopolitical tensions. Preserving this unique and vital region requires concerted efforts at local, national, and international levels, as well as a deep respect for the traditional knowledge and practices of the Himalayan communities.

References

Bandyopadhyay J, Gyawali D (1994) Himalayan water resources: ecological and political aspects of management. Mt Res Dev 14(1):1–24

Barnett TP, Adam JC, Lettenmaier DP (2005) Potential impacts of a warming climate on water availability in snow-dominated regions. Nature 438(7066):303–309

Bernbaum E (2006) Sacred mountains: themes and teachings. Mt Res Dev 26(4):304–309

Chhetri N, Chaudhary P, Tiwari PC, Yadav RB (2017) Promoting sustainable development in the Indian Himalayan region: challenges and opportunities. In: Singh TB, Choudhary RK (eds) Climate change and sustainable development: global perspective, Springer, Singapore

FAO (2015) Mapping the vulnerability of mountain peoples in food insecurity, Rome, Italy. Food and Agriculture Organization of the United Nations

Gansser A (1964) Geology of the Himalayas. Interscience Publishers

GBPIHED (1992) II Pt. Govind Ballabh Pant Memorial lecture, Gangtok, Sikkim, 14 Sept 2002

GBPIHED (2013) 19th Pt. Govind Ballabh Pant Memorial Lecture, Koshi-Katarmal, Almora

GOI (2016) National mission for sustaining the Himalayan ecosystem, Government of India, New Delhi

GOI (2017) National mission on Himalayan studies. Government of India, New Delhi

International Centre for Integrated Mountain Development (ICIMOD) (2020) The Hindu Kush Himalaya assessment: mountains, climate change, sustainability and people

IPCC (2021) Climate change 2021: the physical science base. In: Masson-Delmotte V, Zhai P, Pirani A, Connors SL, Pean C, Berger S, Caud N, Chen Y, Goldfarb L, Gomis MI, Huang (eds) Contribution of working group I to the sixth assessment report of the IPCC, Cambridge University Press, Cambridge, U.K.

Karki M, Adhikari A (2019) Himalayas: water towers of Asia in a changing climate. In: Mukherjee A, Das B, Sen D, Choudhury A (eds) Climate change and water resources in South Asia. Springer, Singapore

Mittermeier RA, Myers N, Thomsen JB, Da Fonseca GAB, Olivieri S (1998) Biodiversity hotspots and major tropical wilderness areas: approaches to setting conservation priorities. Conserv Biol 12(3):516–520

Molden D, Sharma E (2013) ICIMOD's strategy for delivering high-quality research and achieving impact for sustainable mountain development. Mt Res Dev 33(2):179–183

Nogues-Bravo D, Arujo MB, Romdel T, Rahbek C (2008) Scale effects and human impact on the elevational species richness gradients. Nature 453(8):216–220

Ren YY, Ren GY, Sun XB, Shrestha AB, You QL, Zhan YJ, Rajbhandari R, Zhang PF, Wen KM (2017) Observed changes in surface air temperature and precipitation in the Hindu-Kush Himalayan Region during 1901–2014. Adv Chim Change Res 8:148–156

Sayer J (2009) Reconciling conservation and development: are landscape the answer? Biotropica 41:649–652

Schild A, Sharma E (2011) Sustainable mountain development revisited. Mt Res Dev 31(3):237–241

Shrestha AB, Aryal R (2011) Climate change in Nepal and its impact on Himalayan glaciers. Reg Environ Change 11(1):65–77

Singh SP (2002) Balancing the approaches of environmental conservation by considering ecosystem services as well as biodiversity. Curr Sci 82(11):1331–1335

Singh JS (2006) Sustainable development of the Indian Himalayan region: linking ecological and economic concerns. Curr Sci 90(6):784–788

Singh SP, Thadani R (2015) Complexities and challenges in Himalayan research: the context of climate change and sustainable development. Mt Res Dev 35(3):237–246

Swain A (2011) Challenges for water sharing in the Nile basin: changing geopolitics and changing climate. Hydrol Sci J 56(4):687–702

Wymann von Dach S, Bachmann F, Adler C (2018) High mountain ecosystems and societies: addressing the challenges of sustainable development in the Andes and the Hindu Kush Himalaya. Sustainable Development in Mountain Regions, Springer, Cham

Xu B, Cao J, Hansen J, Yao T, Joswia D, Wang N, He J (2009) Black soot and the survival of Tibetan glaciers. Proc Natl Acad Sci 106(52):22114–22118

Part I
Geo-physical Environment

Chapter 2
An Overview of Himalayan Geology

Vimal Singh, Tanya, and Parv Kasana

Abstract The Himalayas, 50 million years old, the result of the collision between the Indian and Eurasian plates, exemplify plate tectonics and regional and global climates. Models such as channel flow and critical taper wedge theories explain the evolution of these distinct geological units and fault systems. Shaped by tectonics, glaciers, rivers, and climate, the region hosts a variety of landforms such as towering peaks, high-altitude lakes, and intermontane valleys. The Himalayas are the most prone to the natural catastrophes like earthquakes, landslides, flash floods, and avalanches. They indeed reflect the intricate interaction between geological and climatic forces.

Keywords Channel flow · Critical taper wedge theory · Catastrophes · Global climate

2.1 Introduction

One of the most prominent examples of plate tectonics is the Himalayas. Around 50 million years ago, the Indian and Eurasian continents collided, triggering the development of the Himalayas (Searle 2013). Since then, the rocks that once formed the ocean floor in the Tethys Ocean have risen to the highest elevation on Earth. The Himalayas also play an important role in shaping regional and global climates. They act as a barrier to the monsoons, forcing moist winds to rise, cool, and condense, resulting in heavy rainfall on the southern slopes while creating arid conditions in the Tibetan Plateau to the north. The Himalaya is an open laboratory for scientists from various disciplines due to its diverse topography, climate, biota, and conflict between nature and humans in recent times.

V. Singh (✉) · Tanya · P. Kasana
Department of Geology, University of Delhi, Delhi, India
e-mail: vimalgeo@gmail.com

S. C. Rai (ed.), *Himalaya: Mountains of Destiny*,
https://doi.org/10.1007/978-3-031-94905-0_2

2.2 Division of the Himalayas

Several divisions of the Himalayas exist in literature based on political boundaries, structural boundaries, stratigraphic boundaries, and geographic boundaries (Yin 2006). One of the widely accepted divisions from south to north is the Siwaliks, or Sub-Himalaya, Lesser Himalaya, Greater or Higher Himalaya, Tethyan Himalaya, and the Trans-Himalaya (Fig. 2.1). Along the length, the Himalaya is divided into the Western Himalaya, the Central Himalaya, and the Eastern Himalaya (Yin 2006).

2.2.1 *Trans-Himalaya*

This region lies north of the plate boundary between the Indian and Eurasian plates. It extends to around 1000 km in an east–west direction and includes important ranges like Kailash and Karakoram. Granitic and metamorphic rocks dominate its composition.

2.2.2 *Tethyan Himalaya*

These comprise sequences varying in age from 1840 to 40 Ma (Yin 2006). The rocks mostly consist of siliciclastic and carbonate interbedded with volcanics. The

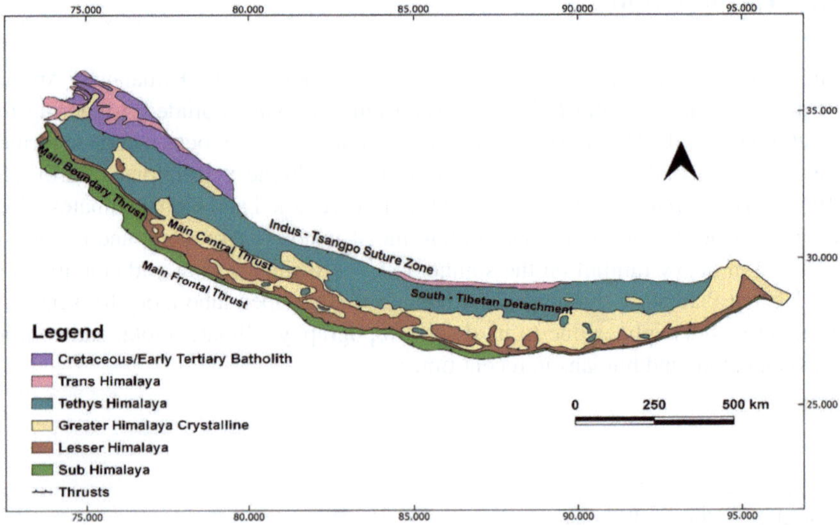

Fig. 2.1 Map showing regional geology of the Himalayas (Modified after Yin 2006)

topmost sequence of the Tethyan Himalaya was deposited during the collision of the Indian and Eurasian plates, whereas the bottom sequences consist of Proterozoic to Cretaceous sequences (Fig. 2.1) (Yin 2006). The elevation of the Tethyan Himalayas varies between 4500 and 6500 m.

2.2.3 *Greater/Higher Himalaya*

These generally comprise high-grade metamorphic rocks occurring as a continuous belt along the length of the Himalayas. They also occur as isolated patches in a few places, both in the Tethyan and Lesser Himalayas. In general, the elevation of the Higher Himalayas ranges between 3500 and 8848 m. The age of the Greater Himalayan rocks varies from Neoproterozoic to Ordovician (Fig. 2.1) (Yin 2006). In the Greater Himalayan rocks, we observe an intriguing phenomenon known as inverted metamorphism, where the metamorphism grade first increases upwards in the lower part and then decreases from the middle to the upper part toward the South Tibetan Detachment (STD). In some places, such as Himachal Pradesh, we observe this inverted metamorphism across the entire area between the Main Central Thrust (MCT) and the STD (Searle 2013).

2.2.4 *Lesser Himalaya*

These comprise low-grade metasedimentary rocks. Though these rocks are difficult to date, the radiometric ages of interlayered gneisses and detrital zircons suggest that the age of the rocks comprising the Lesser Himalaya is between 1870 and 520 Ma (Fig. 2.1) (Yin 2006). These are overlain by the Permian to Cretaceous Gondwana sequence. The main lithology in the Lesser Himalayan zone is siliciclastics, carbonates, phyllites, quartzites, and metavolcanics (Yin 2006). These vary in elevation from 1500 to 3500 m.

2.2.5 *Sub-Himalaya*

These are 10–25 km wide belts of Neogene sediments (Fig. 2.1). These are dominated by the Siwalik Group, comprising fluvial sediments deposited from the Middle Miocene to the Pleistocene (~20–0.5 Ma) (Yin 2006). The Siwalik Group represents a foreland basin system that was formed due to crustal shortening caused by the collision of the Indian and Eurasian plates. This zone is geologically significant as it represents the youngest phase of Himalayan mountain-building processes (Thakur et al. 2020). The elevation of these mountains varies from 300 to 1500 m.

2.3 Major Structures of the Himalayas and their Ages

The structural geology of the Himalaya is defined by five major structures: South Tibetan Detachment (STD), Main Central Thrust (MCT), Ramgarh Thrust (RT), Main Boundary Thrust (MBT), and the Main Frontal Thrust (MFT; also known as Himalayan Frontal Thrust, HFT) (Fig. 2.1). We can trace these five major structures along the length of the Himalaya, despite the presence of several other important structures. The Karakoram fault in the Trans-Himalaya is not considered here. Studies reveal that the major structures of the Himalayas converge into a low-angle fault system known as the Main Detachment Fault/Main Himalayan Thrust (MHT), or décollement (Schelling and Arita 1991). The activity along these structures varies in age, but we have a general idea of the fault's age based on several studies.

2.3.1 South Tibet Detachment

This is a normal fault along which unmetamorphosed or low-grade Tethyan rock is positioned adjacent to the high-grade Greater Himalayan rocks. However, details of its character along the length of the Himalayas are not yet known. Radiometric ages suggest that this fault was active in the Zanskar region between 23 and 19 Ma, with its activity ceasing between 11 and 9 Ma (Yin 2006). The initiation of the fault is not known. The estimated slip on the fault in the Zanskar region ranges from 25 to 35 km (Yin 2006).

2.3.2 Main Central Thrust

This fault puts high-grade metamorphic rocks (gneisses and schists) over the low-grade metamorphic rocks. It has been identified based on the contact between the Greater Himalayan and Lesser Himalayan rocks or by a sudden change in the metamorphic grade. This occurs as a several-kilometer-thick, broad shear zone. This structure is characterized by a sudden increase in topographic elevation. Studies suggest that the MCT was active around 20 Ma (Yin 2006). The activity on the MCT continued up to 14–10 Ma. Several parts of the Himalayas have reported out-of-sequence thrusting around the MCT between 6 and 2 Ma (Yin 2006). The slip on the MCT varies in different regions; for example, in Sikkim and Bhutan, it is around 75 km, whereas in Arunachal it is over 195 km (Yin 2006).

2.3.3 *Main Boundary Thrust*

This fault places Lesser Himalayan rocks over Sub-Himalayan sedimentary strata. If we define the Main Boundary Thrust based on the age of the rocks in the hanging wall and the footwall, we encounter issues, as certain areas in the hanging wall contain Cenozoic sequences. This fault is suggested to have developed around 10 Ma, based on various datasets (Meigs et al. 1995).

2.3.4 *Main Frontal Thrust*

This thrust is contact between the Siwalik sediments and the Indo-Ganga–Brahmaputra sediments. The Siwalik sediments in the MFT's hanging wall typically exhibit folds. MFT has been suggested to be blind in some parts along the length of the Himalayas (Yeats and Lillie 1991). The age of the MFT varies as it has developed in various segments; it is suggested to be between 2.47 and 0.63 Ma (Rao 1993).

2.4 Formation of the Himalayas

Driven by plate tectonics, the Indian plate traveled a distance of over 8000 km in 120 million years and moved from the southern hemisphere to the northern hemisphere. Its velocity during this period reached up to 20 cm/year; however, in general, it moved northwards with a velocity of 5 cm/year (van der Wiel et al. 2024). At around 50 Ma, the Indian plate collided with the Eurasian plate, and its velocity slowed down. This meeting of the two plates led to the closure of the Tethys Sea. The heavier Indian plate subducted beneath the Eurasian plate, causing the rocks on its upper part to fold and fault.

During the Miocene epoch, the rate of uplift in the Himalayas significantly increased with the breakup of the Main Central Thrust (MCT) (Yin 2006). This bounds the Higher Himalayas, which feature the high peaks we see today. A new structure, the Main Boundary Thrust, emerged after this, bounding the Lesser Himalaya. We traverse this structure on our way to Mussoorie from Dehradun or Haldwani to Nainital. Geologists have mapped both these structures, demonstrating their presence throughout the entire Himalayan region (Fig. 2.1) (Yin 2006). Another major structure in the Himalayas is the Main Frontal Thrust (MFT), which separates the vast Indo-Ganga plains from the Himalayas and marks the southernmost boundary of the Siwaliks—the frontal Himalayas. This is considered to be the youngest and the most active major structure, and it is still developing in the eastern Himalayas; these are ruptured during large earthquakes. Paleoseismologists have identified several large earthquakes along the MFT in the past 2000 years. Geologists have also shown that this frontal Himalaya takes up nearly 50% of the total convergence between the Indian and Eurasian plates (Lavé and Avouac 2000).

Here we describe two popular models suggested for the evolution of the Himalayas. Though there are several models of evolution given by researchers, each has certain advantages and disadvantages.

2.4.1 Channel Flow Model

Geological mapping, geothermobarometry, metamorphic modeling, and U–Pb geochronology helps us understand the structural and thermal evolution of the Himalayan belt. Geophysical data have shown that the Moho discontinuity deepens beneath the Himalayas and Tibet; the depth reaches 75 km beneath Tibet (Searle 2013). The presence of a mid-crustal layer is noticed between 15 and 30 km and has been called the zone of partial melting. Their continuity has been traced up to the South Tibetan Detachment (STD) and Main Central Thrust (MCT) shear zones that bound the Greater Himalayas (Fig. 2.2a). It has been suggested that pockets of liquid must have been present beneath the Greater Himalayas around 20–17 Ma, similar to the one recognized beneath Tibet today (Searle et al. 2003, 2006; Gaillard et al. 2004). The melting is believed to result from thermal relaxation caused by crustal thickening and elevated internal heat production within the Proterozoic source rocks located in the middle crust (Searle 2013). As melting occurred, southward extrusion of the ductile mid-crustal layer took place between two deep-seated faults, viz., MCT and STD; this is also known as the channel flow model (Fig. 2.2a) (Searle 2013). This model fits all the geological and geophysical observations; for example, it helps in explaining the inverse metamorphism (lower-grade metamorphism at the bottom and the higher-grade metamorphism at the top) problem in the Greater Himalayas. Furthermore, models have shown it to be a viable process.

2.4.2 Critical Taper Wedge Model

The critical taper wedge model explains how mountains and other geological structures are formed when one tectonic plate pushes against another (Boyer and Elliott 1982; Davis et al. 1983). To make it simple, imagine a pile of sand being pushed from one side. When you push a pile of sand, it forms a wedge-like shape with a steep front and flat base. The Earth's crust undergoes the same process during mountain-building, but instead of sand, rock layers form the "wedge." The wedge maintains a "critical angle" or slope. If it's too steep, materials from the top slide off, and if it's too flat, the base pushes out more material to adjust its slope. The system balances itself to stay stable. This model elucidates the formation process of large systems such as the Himalayas.

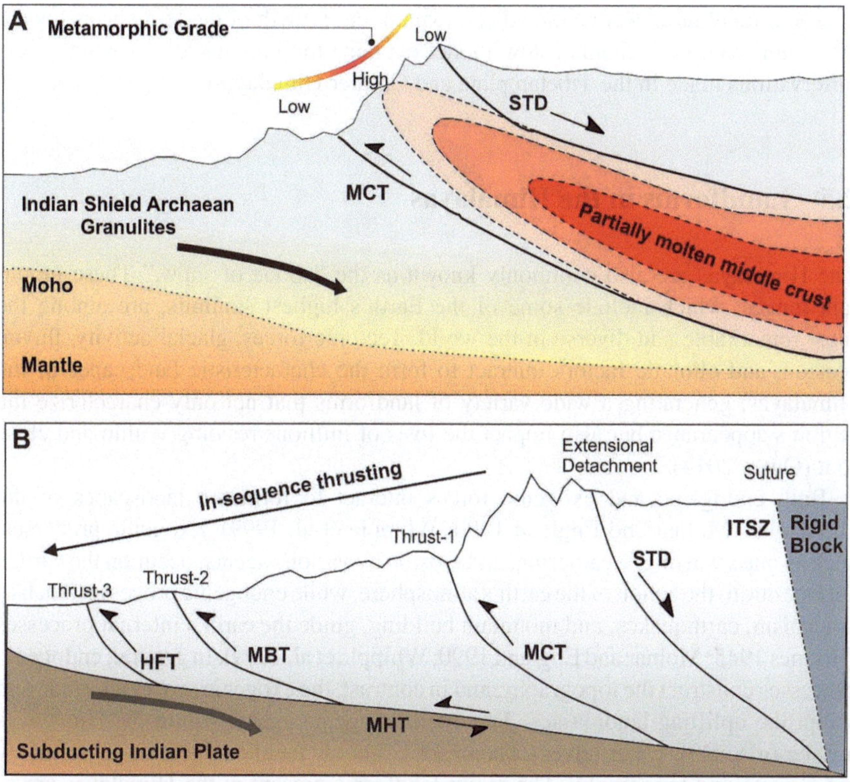

Fig. 2.2 Diagrams of (**a**) channel flow model, which explains inverted metamorphism, and (**b**) critical taper wedge model, showing three accretionary prisms (Modified after Searle 2013)

The Indian tectonic plate pushes under the Eurasian plate, creating a wedge of rocks and sediment that forms the mountains. The Indus-Tsangpo Suture zone delineates the boundary between the Indian and Eurasian plates. To the south of the ITSZ is the Himalayan Fold-Thrust Belt. The subducting Indian plate led to the accumulation of a sedimentary prism on its northern margin. This sedimentary prism is detached from the underlying basement and is divided into several south-vergent thrust sheets. Over time, the Greater Himalayas emerged as the initial wedge. As the MCT, along which the Greater Himalayan rocks were pushed southward, grew steeper and reached the critical angle, a new fault, the MBT, emerged, resulting in the formation of the Lesser Himalaya wedge. A new wedge Sub-Himalaya (pushed along MFT) developed in front of the Lesser Himalaya once it achieved the critical angle (e.g., DeCelles et al. 2001; Kapp and DeCelles 2019). Currently, the Sub-Himalayan wedge is undergoing folding and faulting to attain the critical angle. Now, any movement along a fault to the north of the MFT is termed an out-of-sequence thrust.

The critical taper wedge model can explain the growth of the Himalayas toward the south, while the channel flow model explains the geological and geophysical observations made in the Tibetan plate and Greater Himalayas.

2.5 Landforms in the Himalayas

The Himalayas are also commonly known as the "abode of snow." These mountain ranges, which include some of the Earth's highest summits, are among the most remarkable and diverse in the world. Tectonic forces, glacial activity, fluvial erosion, and climatic factors interact to form the characteristic landscapes of the Himalayas, generating a wide variety of landforms that not only characterize the region's appearance but also impact the lives of millions residing within and close to it (Owen 2014).

Both endogenic and exogenic forces interact to form the landscapes of the Himalayas (Molnar and England 1990; Whipple et al. 1999). Exogenic processes, such as mass wasting, weathering, and erosion by various agents, occur on the earth's surface due to the action of the earth's atmosphere, while endogenic processes, such as volcanism, earthquakes, and mountain building, guide the earth's internal processes (Holmes 1945; Molnar and England 1990; Whipple et al. 1999). In general, endogenic processes construct the topography, and in contrast, the exogenic processes erode and sculpt the uplifting landmasses. The major exogenic agents within the Himalayas are ice (glaciers), water (rivers, lakes, etc.), and air (aeolian agents), governed by the climate and topography. The major landforms present in the Himalayas are as follows:

2.5.1 Mountain Peaks and Summits

The Himalayan region houses the world's largest mountain ranges and the tallest peaks. Mount Everest (8848 m), the highest peak on the Earth, is the most renowned of these summits. Other notable summits in the range are K2 (8611 m), Kangchenjunga (8586 m), Lhotse (8516 m), and Makalu (8485 m). The Himalayan orogeny has created not only summits but also a complex network of ridges, horns, and minor mountain ranges throughout the region (Yin and Harrison 2000 and references within).

The significant elevation of the Himalayas affects the climate, leading to altitudinal variations that establish distinct ecological zones from tropical to alpine habitats. The considerable elevation of these summits has created extensive valleys and canyons between the mountains, increasing the breathtaking vista (Fig. 2.3).

Fig. 2.3 Photograph showing the mountain peaks and spires (Pinnacles) in the Himalayan range

2.5.2 Glaciers and Glaciogenic Landforms

The glaciers are the large ice masses that slowly creep down a slope or a valley. During this process, large ice masses grind the bedrock, sculpt the valleys, and deposit large amounts of eroded materials (Holmes 1945; Benn and Owen 2002). The glacial systems in the Himalayas range from cirques to valley glaciers in the monsoon-affected areas of the Higher Himalayas (Owen 2014).

Known as the third pole, the Himalayas house some of the largest glaciers outside the Polar Regions (Fig. 2.4a). These extensive glacier systems carve large U-shaped valleys inside the rocky terrains of the Himalayas (Fig. 2.4b). The **Siachen Glacier**, located in the Karakoram range, is one of the longest glaciers (outside the polar regions) in the world, while others, such as the **Gangotri Glacier** in Uttarakhand, serve as the source of the holy **Ganga River** (Bolch et al. 2012).

Glaciers erode and transport substantial quantities of debris. Glaciers, upon advancing or retreating, deposit **moraines**—collections of soil and rock debris—that frequently create natural dams and barriers, therefore further shaping the landscape (Benn and Owen 2002; Fig. 2.4c, d). The Himalayan glaciers are significant not just for their geological influence but also as vital freshwater supplies that nourish the principal river systems supporting millions of individuals (Owen 2014).

2.5.3 Hydrological Systems and Landforms

The river systems in the Himalayas mainly receive water from the monsoon, groundwater, and glacial and snowmelt (Owen 2014). As we move from the west to the east in the Himalayas, the contribution from glacial and snow melt in the major rivers decreases (Dortch et al. 2011; Owen 2014). The large volume of sediments transported by the Indus and Ganga–Brahmaputra River systems feeds the alluvial plains and the deltas.

Fig. 2.4 Photographs showing (**a**) valley glacier (near Kunzum pass, Spiti), (**b**) large U-shaped valley created by the glacial erosion (near Khardung-La, Ladakh), (**c**) and (**d**) lateral moraines in the Kedarnath valley (Photograph courtesy: Dr. Naresh Rana)

Throughout their journey, river systems erode and sculpt bedrock, depositing sediments to create a variety of landforms (Gilbert 1877; Whipple 2004). The rivers in their upper reaches have higher gradients and carry large volumes of debris (Gilbert 1877). The powerful rivers, along with the sediments, act as very efficient tools for bedrock erosion and cut through their respective beds to form deep gorges (Holmes 1945; Whipple 2004). Rapid riverbed erosion forms the narrow, steep-sided (almost 90 degrees) valleys known as **gorges** (Fig. 2.5a). Occasionally, these gorges contain **strath terraces.** Strath terraces are flat or gently sloping landforms formed by the downcutting and erosion of the bedrock by rivers (Hancock and Anderson 2002). These provide important information about the timing of river incision and uplift.

Bedrock erosion and sustained uplift in the Himalayas force the river systems to undergo vertical erosion. This causes the rivers to develop V-shaped river valleys (Fig. 2.5b). As the river system flows further downslope, the valleys generally widen, and the widening of the valley provides ample accommodation space for the rivers to deposit sediments (Vandenberghe 2015). This sediment deposition leads to the development of thick valley-fill deposits and fluvial terraces (Fig. 2.5c).

(a) (b) (c)

Fig. 2.5 Photographs of (**a**) Gorge in the Ganga River valley, (**b**) fluvial terraces in the Srinagar Valley, Uttarakhand, along the Alaknanda River, and (**c**) alluvial fans in the Spiti Valley. (Photograph courtesy a and b: Dr. Rahul Devrani)

Lakes

There are numerous lakes in the Himalayas. It has been suggested that the lakes in the Himalayas in general have originated due to tectonic processes and are found in thrust-bounded basins along the southern margin of the Himalayas (e.g., the Kashmir Basin, Fig. 2.6b) and along strike-slip faults (e.g., Pangong Tso, Fig. 2.6a) (Owen et al. 1996). Lakes also form in glacial environments behind ice dams, moraines (e.g., along the Gangotri Glacier), and within eroded bedrock depressions (e.g., Chandra Tal, Fig. 2.6c) (Owen et al. 1996). Furthermore, lakes may form behind obstructions created by debris flows, rockfalls, and landslides (e.g., Naini Lake, Fig. 2.6d). The lakes sometimes burst/overflow and create havoc in the downstream settlements located within a basin. Glacial Lake Outburst Floods (**GLOFs**) are sudden and rapid releases of water from glacial lakes, which occur when a dam, typically composed of ice, debris, or moraine (accumulated sediments), fails (Emmer 2017).

High-altitude lakes are another striking feature of the Himalayan landscape. These lakes, often formed by the retreat of glaciers or tectonic activity, are not only visually stunning but also ecologically significant (Zutshi 1991). Located near Mount Kailash in Tibet at an altitude of 4,556 m, **Manasarovar Lake** stands as one of the world's highest freshwater lakes.

Fig. 2.6 Photographs of lakes in the Himalayas—**a** Pangong Lake in Leh, Ladakh, **b** Dal Lake in Srinagar, Kashmir, **c** Chandratal Lake in the Chandra valley, Spiti, **d** Naini Lake in the Nainital, Uttarakhand

Intermontane Valleys

Large spacing between thrusts in the frontal part of the Himalayas led to the development of a structural landform known as intermontane valleys, locally referred to as "dun" or "doon" (Tandon and Singh 2014). Some segments of the Himalayas contain these intermontane valleys. Some of the popularly known duns are Soan dun, Pinjaur dun, Dehra dun, Deokhury dun, and Chitwan dun. These are important for the Himalayan region as they provide a vast, gently sloping surface where developmental activities occur along with agriculture; also, they store a large volume of sediments, which are important for understanding the long-term tectonic and climatic shifts (Singh and Tandon 2008; Sinha and Sinha 2016; Sharma et al. 2019; Divyadarshini et al. 2020). The intermontane valleys are also traversed by structures parallel to the axis of the valley. These intermontane valleys have several fluvial landforms like alluvial fans, surfaces, terraces, and isolated hills (Tandon and Singh 2014).

Alluvial Fans

When a sediment-laden stream enters from a steep segment to a gentle segment and changes from a confined valley system to an unconfined river system, its carrying capacity drops, and as a result, it starts depositing sediments. These sediments get deposited as a gently sloping cone called an **alluvial fan** (Bull 1968; Blair and McPherson 2009). In the proximal part of the alluvial fan, depositional characteristics is debris flow, whereas in the middle part, it shows intercalation of clasts and mud. In the distal part, the fine-grain sediments are deposited. In arid or semi-arid settings (e.g., the Ladakh region), most rainfall infiltrates the permeable deposit, reducing the load and creating an alluvial cone (Owen 2014). Gradations vary from coarse debris cones to extensive sloping sheets of fine alluvium. Where closely spaced streams discharge from a mountainous region, the coalescence of multiple alluvial fans results in the formation of **piedmont alluvial plains**.

The Kosi River, which drains the Himalayas, is responsible for creating the Kosi **megafan** in the Ganga plains. Similar fans have also been suggested to be made by other Himalayan rivers like the Gandak and Teesta.

2.5.4 Aeolian Landforms

Semi-arid to tropical climatic conditions primarily dominate the Himalayan region, but arid conditions also exist in rain shadow zones (i.e., the Tethyan and Trans-Himalayas) and higher altitudes. These regions include aeolian landforms like dunes and dune fields. The junction of the Nubra and Shyok Rivers in Ladakh demonstrates the development of dunes and dune fields (Fig. 2.7a). Features such as sand ramps can be observed in the Indus valleys in the Ladakh region (Fig. 2.7b).

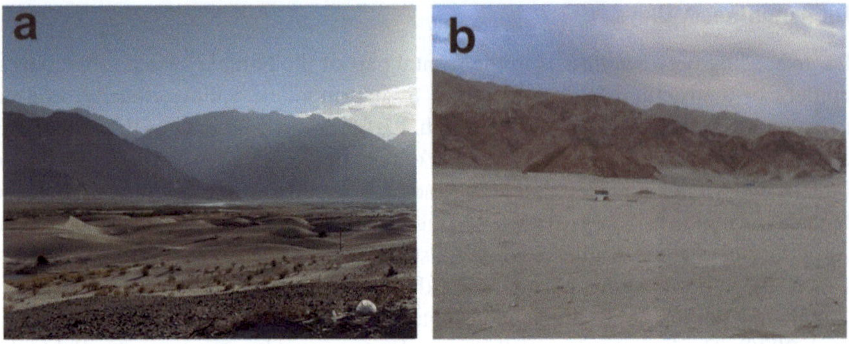

Fig. 2.7 Photograph of (**a**) sand dunes present near the Shyok and Nubra River valleys confluence in Ladakh, and (**b**) Sand ramps near Leh city, Ladakh

2.5.5 Hills and Slopes

Lush forests and terraced fields cover the hills and slopes that dominate the lower elevations of the Himalayas. The **foothills** are part of the **Siwalik range**, which lies at the outermost edge of the Himalayas and marks the southernmost limit of the Himalayas (Yin and Harrison 2000). The Siwalik Hills exhibit significant variations in elevation, from gentle slopes to steep, rugged terrain. This area is often characterized by thick forests, rich biodiversity, and small settlements. However, for the most part, these are under forest cover and preserved as national forests.

2.5.6 Plateau

The Tibetan Plateau, in the Trans-Himalaya, is one of the world's most remarkable landforms. Often referred to as the "Roof of the World," this vast, high-altitude plateau covers an area of over 2.5 million square kilometers and sits at an average elevation of around 4500 m (Royden et al. 2008). The towering peaks of the Himalayas to the south and the Kunlun Mountains to the north define its northern and southern limits. Despite its harsh, cold desert environment and sparse population, the Tibetan Plateau holds significant geological and cultural importance. The Tibetan Plateau has been instrumental in shaping the weather patterns of the surrounding regions, contributing to the monsoon climate that affects much of South Asia (Wu et al. 2015). Vast, flat expanses, deep canyons, and lakes like Lake Manasarovar, sacred to both Buddhists and Hindus, characterize the terrain here.

2.6 Natural Hazards

The active tectonic setting of the Himalayas makes it susceptible to natural hazards. Himalayan States account for about 35 percent of total disasters in India *(EM-DAT International Disaster Database)*. These hazards, including earthquakes, landslides, flash floods, and avalanches, are more pronounced during the monsoon period.

2.6.1 Earthquakes

Nearly 50% of the convergence between the Indian and Eurasian plates is distributed across the Himalayas. This convergence is responsible for the seismicity in the Himalayas, and as a result, entire Himalayas lie in the high seismic zone. The rate of convergence varies across the Himalayas: ~ 12 mm/year in the west, which increases in the Central Himalayas (~20 mm/year) and again reduces toward the East (Jade et al. 2017). The earthquakes in the Himalayas are shallow and rarely originate deeper than 40 kms (Bilham 2019).

These earthquakes are concentrated along the deécollement fault of the Himalayas (i.e., the Main Himalayan Thrust, MHT). Based on movement or the slip along the fault, MHT is divided into different zones; locked, creeping, and partially locked segments (Fig. 2.8) (Avouac et al. 2015). Over time, the locked segments of the MHT accumulate stress, which, when it surpasses the rock's strength, releases an earthquake (e.g., Avouac et al. 2015).

Historical records and paleoseismological studies have identified large earthquake events in the past (Kumar et al. 2010; Bilham 2019; Wesnousky 2020). Recent examples of such events are the 2015 Gorkha Earthquake (M 7.8), the 2005 Kashmir Earthquake (M 7.6), 1950 Assam (M 8.7), 1934 Bihar (M 8.4), and 1905 Kangra (M 7.8).

Certain parts along the Main Himalayan Thrust (MHT) have been devoid of seismic activity for a long period. Such regions, known as seismic gaps, are areas where the probability of a potential earthquake is high due to the accumulation of unreleased stress over time. The presence of seismic gaps indicates a possibility of large-magnitude events occurring in future, and one such seismic gap has been identified in western Nepal and adjoining region by the paleoseismologists.

2.6.2 Landslides

Landslides are one of the most frequent hazards in the Himalayas. These events occur by a combination of factors, including regional geology, hydrology, and precipitation, and are often enhanced by human activities. Broadly, landslides can be defined as the downslope movement of soil, rock, or debris under the influence of gravity.

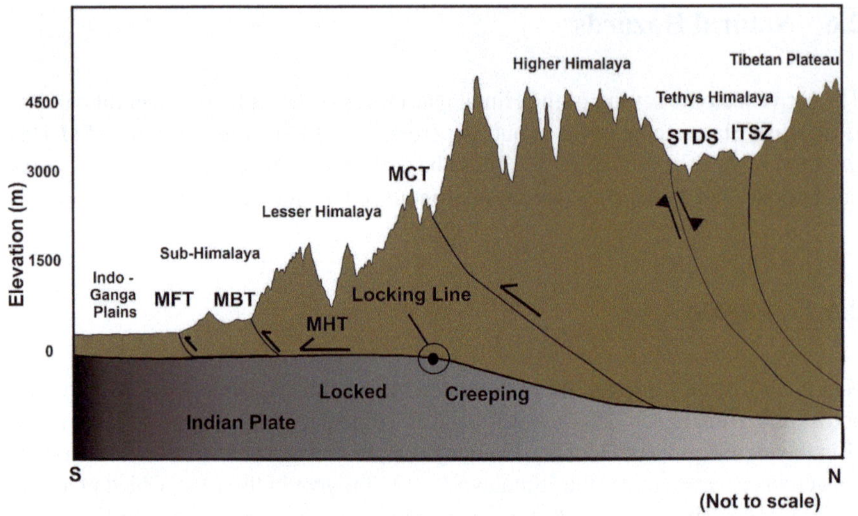

Fig. 2.8 Cross-section of the Himalaya. All major thrusts merge with the Main Himalayan Thrust (MHT) at the base. The locking line marks the boundary between the locked segments of the MHT and the underlyig creeping section, which generates seismicity in the region. Modified after Bilham et al. (2001)

Landslides occur when shear stress acting on a slope becomes higher than the shear strength of the rock, due to which the slope becomes unstable. Landslides can be translational or rotational, depending upon the shear plane involved (planar and concave shear plane, respectively). Examples of translational slides include debris slides and rock slides. In contrast, rotational slides-typically occurring in conditions of low to moderate water content-comprise rock slumps and debris slumps.

In the Himalayas, landslides are especially prevalent during the monsoon when high rainfall saturates the soil and rock layers increasing the pore water pressure on the slopes. This destabilizes the slope by decreasing the cohesive strength of slope material, and as the cohesive strength becomes less than the shear stress, landslides occur. For example, a series of landslides occurred in August 2023 in Mandi (Himachal Pradesh), triggered by heavy continuous rain in the region. Landslides are more common on barren or deforested slopes, as vegetation plays a critical role in stabilizing materials through root reinforcement and moisture regulation. Earthquakes can also trigger landslides by shaking the surface, and liquefaction can also destabilize the slope at times. Uttarakhand has the highest landslide susceptibility zones, followed by Jammu and Kashmir. District-wise, Rudraprayag (Uttarakhand) has both the highest landslide density and the highest risk to population (Fig. 2.9).

Fig. 2.9 Photograph of Kaliasaur landslide in Rudraprayag district of Uttarakhand

2.6.3 Flash Floods

The sudden rise in the river stage over a short period causes flash floods (USGS). These floods are common in mountainous settings because of the terrain, which allows water to flow overland rather than infiltrate into the ground. In the Himalayas, a flash flood occurs due to three phenomena (or their combination):

- **Cloudburst**: Cloudburst events in the Himalayas are concentrated high-intensity precipitation events (> 100 mm in an hour or 300 mm in 24 h, as defined by the Indian Meteorological Department) over a small region, causing flash floods and associated landslides in the region. The frequency of such events has increased since 1970 (Kumar et al. 2018). Rainfall is much higher on the southern slopes of the Himalayas as compared to its northern counterparts due to the direction of monsoon, and therefore, extreme precipitation events are more common over the southern slopes (Kumar et al. 2018).
- **Glacial Lake Outburst Flood (GLOF)** occurs when the glacial lakes are breached, leading to sudden outbursts of large volumes of water stored in the lake downstream at high velocity, and causing floods. Glacial Lakes form when glaciers retreat, which leaves depressions that get filled with meltwater. At times, moraine sediments temporarily dam the melt water leading to the development of lakes. It takes events like earthquakes, avalanches, and high rainfall to destabilize this dam, leading to a GLOF event. Areas with a high density of glacial lakes (Sikkim, Uttarakhand) are highly vulnerable to GLOF events. GLOFs carry high sediment load along with high water discharge and therefore can be extremely devastating.

Flash flood events caused by cloudbursts and GLOFs are distinct events but are not mutually exclusive. High rainfall by a cloudburst can trigger GLOF as happened during the 2013 Uttarakhand floods in Kedarnath (Fig. 2.10). These floods (16–17 June) were initiated by a cloudburst event (315 mm rainfall in 24 hours), which led to the lake outburst at Chorabari Glacier. This flood alone caused the widening of the valley by 120 m with sediment deposition in some reaches reaching more than

Fig. 2.10 Photographs (a–d) from the Mandakini Valley, a year after the 2013 floods; (**a**) shows the landslides induced by the floods, changing the geomorphology of the area; (**b**) showing abrupt aggradation caused by the floods; (**c**) and (**d**) highlight the damage to infrastructure by the floods

25 m (Devrani et al. 2015), leading to a huge loss of lives and property. We expect a rise in the frequency and intensity of flash floods due to climate change.

- **Landslide Lake Outburst Flood (LLOF)** occurs by the rupturing of a lake formed by a river blocked due to a landslide. Along many narrow valleys, large landslides can block the flow of the river. The Gauna (Gohna) lake outburst in Uttarakhand serves as an example of this phenomenon. In 1893 Gauna Lake was formed due to the blocking of the Birahi Ganga River by the landslide (Bhat 2024). The lake formed was around 50 km^2 in area. It breached partially in 1894 and fully in 1970. It caused a major flood in 1970 leading to huge loss of lives and properties.

2.6.4 *Avalanches*

An avalanche, a type of mass movement, is a rapid downslope flow of a large mass of snow, ice, or debris, under the influence of gravity. Avalanches form above the tree line, which is usually at elevations higher than 4000 m above mean sea level in the Himalayas. These are some of the least understood hazards, mainly because of inaccessible high-altitude regions where they occur, and their complex relationship with changing climate. Studies indicate that in the Himalayas, avalanches occur at a slope angle between 25° and 40° (Acharya et al. 2023 and references therein). Avalanches are considered one of the main sources of supraglacial debris in the Himalayas, controlling the glacial dynamics in the region (Benn and Evans 2014; Iturrizaga 2011). However, avalanches are one of the leading causes of fatalities among mountaineers. Around 952 avalanches related deaths in India and 508 in Nepal have been documented.

The nature of the Himalayan glaciers varies from west to east, depending upon the regional climatic variability. Western Himalayas are influenced more by the western disturbance, while the central and eastern glaciers accumulate during the summer monsoon. This also impacts the pattern of avalanches across the Himalayas. Avalanches are more common during winters in the western Himalayas and during summers in the Central Himalayas. Since hazards in the Himalayan region are often interrelated, seismic activities can also trigger avalanches. For example, the Gorkha earthquake in 2015 initiated two avalanches: one in the Langtang Valley and another in the Khumbu region (Kargel et al. 2016; Bartelt et al. 2016).

2.7 Conclusion

The Indian plate moved over 8000 km in 120 million years, colliding with the Eurasian plate about 50 million years ago, which initiated the formation of Himalayas. They act like a monsoonal dam and receive immense precipitation on their southern slopes, while it remains arid towards the Tethyan and Trans-Himalayas.

The Himalayas are divided into the Siwaliks, Lesser Himalaya, Greater Himalaya, Tethyan Himalaya, and Trans-Himalaya. The South Tibetan Detachment (STD), the Main Central Thrust (MCT), the Main Boundary Thrust (MBT), and the Main Frontal Thrust (MFT) make up their structural framework. The MFT is the youngest structure and is found at the southern edge of the Siwaliks.

Studies using geophysics show that the Moho discontinuity under the Himalayas and Tibet is getting deeper. They also found a 15–30 km wide area where some rocks are melting because the crust is getting thicker and heat is being produced. The channel flow model accounts for inverse metamorphism in the Greater Himalaya, whereas the critical taper wedge concept describes mountain building to the south of ITSZ.

Landforms in the Himalayas are formed by glaciers, rivers, tectonics, and wind. This tectonically active region is vulnerable to natural hazards, including earthquakes, landslides, flash floods, and avalanches.

References

Acharya A, Steiner JF, Walizada KM, Zakir ZH, Ali S, Caiserman A, Watanabe T (2023) Snow and ice avalanches in high mountain Asia–scientific, local and indigenous knowledge. Nat Hazards Earth Syst Sci Dis 2023:1–35

Avouac JP, Meng L, Wei S, Wang T, Ampuero JP (2015) Lower edge of locked Main Himalayan Thrust unzipped by the 2015 Gorkha earthquake. Nat Geosci 8(9):708–711

Bartelt P, Buser O, Valero CV, Bühler Y (2016) Configurational energy and the formation of mixed flowing/powder snow and ice avalanches. Ann Glaciol 57(71):179–188

Benn D, Evans DJ (2014) Glaciers and glaciation. Routledge, London

Benn DI, Owen LA (2002) Himalayan glacial sedimentary environments: a framework for reconstructing and dating former glacial extents in high mountain regions. Quat Int 97–98:3–26

Bhat GM (2024) Landslide scenario in India with special reference to the Himalaya and community resilience to hazards. J Geol Soc India 100(11):1507–1510. https://doi.org/10.17491/jgsi/2024/174010

Bilham R (2019) Himalayan earthquakes: a review of historical seismicity and early 21st century slip potential. Geol Soc, London, Spec Publ 483(1):423–482

Bilham R, Gaur VK, Molnar P (2001) Himalayan seismic hazard. Science 293(5534):1442–1444

Blair TC, McPherson JG (2009) Processes and forms of alluvial fans. In: Geomorphology of desert environments. Dordrecht, Springer Netherlands, pp 413–467

Bolch T, Kulkarni A, Kääb A, Huggel C, Paul F, Cogley JG, Frey H, Kargel JS, Fujita K, Scheel M, Bajracharya S, Stoffel M (2012) The state and fate of Himalayan glaciers. Science 336(6079):310–314

Boyer SE, Elliott D (1982) Thrust systems. Am Assoc Petrol Geol 66(9):1196–1230

Bull WB (1968) Alluvial fans. J Geol Educ 16(3):101–106

Davis D, Suppe J, Dahlen FA (1983) Mechanics of fold-and-thrust belts and accretionary wedges. J Geophys Res 88:1153–1172

DeCelles PG, Robinson DM, Quade J, Ojha TP, Garzione CN, Copeland P, Upreti BN (2001) Stratigraphy, structure, and tectonic evolution of the Himalayan fold-thrust belt in western Nepal. Tectonics 20:487–509

Devrani R, Singh V, Mudd SM, Sinclair HD (2015) Prediction of flash flood hazard impact from Himalayan river profiles. Geophys Res Lett 42(14):5888–5894

Divyadarshini A, Singh V, Jaiswal MK, Rawat M (2020) Exploring the roles of climate and tectonics in the geomorphic evolution of the Chitwan Intermontane valley, Central Himalaya. Geomorphology 367:107298

Dortch JM, Owen LA, Caffee MW, Kamp U (2011) Catastrophic partial drainage of Pangong Tso, Northern India and Tibet. Geomorphology 125:109–121

Emmer A (2017) Glacier retreat and glacial lake outburst floods (GLOFs). In: Oxford Research Encyclopedia of Natural Hazard Science

Gaillard F, Scaillet B, Pichavent M (2004) Evidence for present-day leucogranite pluton growth in Tibet. Geology 32:801–804

Gilbert GK (1877) Geology of the Henry mountains. United States Geological Survey, pp 1–160

Hancock GS, Anderson RS (2002) Numerical modelling of fluvial strath-terrace formation in response to oscillating climate. Geol Soc Am Bull 114(9):1131–1142

Holmes A (1945) Principles of physical geology. Geologiska Föreningen i Stockholm Förhandlingar 67(1):115–116

Iturrizaga L (2011) Trends in 20th century and recent glacier fluctuations in the Karakoram Mountains. Zeitscrift Fuer Geomorphologie, Supplementary Issues 55(3):205–231

Jade S, Shrungeshwara TS, Kumar K, Choudhury P, Dumka RK, Bhu H (2017) India plate angular velocity and contemporary deformation rates from continuous GPS measurements from 1996 to 2015. Sci Rep 7(1):11439

Kapp P, DeCelles PG (2019) Mesozoic-cenozoic geological evolution of the himalayan-tibetan orogen and working tectonic hypotheses. Am J Sci 319:159–254

Kargel JS, Leonard GJ, Shugar DH, Haritashya UK, Bevington A, Fielding EJ, Young N (2016) Geomorphic and geologic controls of geohazards induced by Nepal's 2015 Gorkha earthquake. Science 351(6269):aac8353

Kumar A, Gupta AK, Bhambri R, Verma A, Tiwari SK, Asthana AKL (2018) Assessment and review of hydrometeorological aspects for cloudburst and flash flood events in the third pole region (Indian Himalaya). Polar Sci 18:5–20

Kumar S, Wesnousky SG, Jayangondaperumal R, Nakata T, Kumahara Y, Singh V (2010) Paleoseismological evidence of surface faulting along the northeastern Himalayan front, India: timing, size, and spatial extent of great earthquakes. J Geophys Res: Solid Earth 115(B12)

Lavé J, Avouac JP (2000) Active folding of fluvial terraces across the Siwalik Hills, Himalayas of Central Nepal. J Geophys Res, 105(B3):5735–5770

Meigs AJ, Burbank DW, Beck RA (1995) Middle–late Miocene (N10 Ma) formation of the Main Boundary Thrust in the western Himalaya. Geology 23:423–426

Molnar P, England P (1990) Late Cenozoic uplift of mountain ranges and global climate change: chicken or egg? Nature 346(6279):29–34

Owen LA, Dortch JM (2014) Quaternary glaciation of the Himalayan Tibetan orogen. Quatern Sci Rev 88:14–54

Owen LA (2014) Himalayan Landscapes of India. In: Vishwas SK (ed) Landscapes and landforms of India. Springer Dordrecht, pp 135–143

Owen LA, Sharma M, Bigwood R (1996) Landscape modification and geomorphological consequences of the 20 October 1991 earthquake and the July–August 1992 monsoon in the Garhwal Himalaya. Z Geomorphologie 103:359–372

Rao AR (1993) Magnetic-polarity stratigraphy of Upper Siwalik of north-western Himalayan foothills. Curr Sci 64(11–12):863–873

Royden LH, Burchfiel BC, van der Hilst RD (2008) The geological evolution of the Tibetan Plateau. Science 321(5892):1054–1058

Schelling D, Arita K (1991) Thrust tectonics, crustal shortening, and the structure of the far-eastern Nepal, Himalaya. Tectonics 10:851–862

Searle MP (2013) Colliding continents: a geological exploration of the Himalaya, Karakoram, and Tibet. Oxford University Press, USA

Searle MP, Law RD, Jessup MJ (2006) Crustal structure, restoration and evolution of the Greater Himalaya in Nepal–South Tibet: implications for channel flow and ductile extrusion of the middle crust. In: Law RD, Searle MP, Godin L (eds) Channel flow, ductile extrusion and exhumation in continental collision zones. Geological Society, London, Special Publications, vol 268, pp 355–378

Searle MP, Simpson RL, Law RD, Parrish RR, Waters DJ (2003) The structural geometry, metamorphic and magmatic evolution of the Everest massif, high Himalaya of Nepal. J Geol Soc, London 160:345–366

Sharma M, Divyadarshini A, Singh V (2019) Morphotectonic evolution of the Siwalik Hills between the Yamuna and the Markanda river exits, NW Himalaya. J Geol Soc India 94:453–463

Singh V, Tandon SK (2008) The Pinjaur dun (intermontane longitudinal valley) and associated active mountain fronts, NW Himalaya: tectonic geomorphology and morphotectonic evolution. Geomorphology 102(3–4):376–394

Sinha S, Sinha R (2016) Geomorphic evolution of Dehra Dun, NW Himalaya: tectonics and climatic coupling. Geomorphology 266:20–32

Tandon SK, Singh V (2014) Duns: Intermontane basins in the Himalayan Frontal Zone. In: Kale VS (ed) Landscapes and landforms of India. Springer, Dordrecht, pp 135–143

Thakur VC, Joshi M, Jayangondaperumal R (2020) Active tectonics of Himalayan frontal fault zone in the Sub-Himalaya. In: Gupta N, Tandon S (eds) Geodynamics of the Indian plate. Springer Geology. Springer, Cham. https://doi.org/10.1007/978-3-030-15989-4_12

van der Wiel E, Pokorný J, Čížková H, Spakman W, van den Berg AP, van Hinsbergen DJJ (2024) Slab buckling as a driver for rapid oscillations in Indian plate motion and subduction rate. Commun Earth Environ 5:316. https://doi.org/10.1038/s43247-024-01472-x

Vandenberghe J (2015) River terraces as a response to climatic forcing: formation processes, sedimentary characteristics and sites for human occupation. Quat Int 370:3–11

Wesnousky SG (2020) Great pending Himalaya earthquakes. Seismological Soc Ame 91(6):3334–3342

Whipple KX (2004) Bedrock rivers and the geomorphology of active orogens. Annu Rev Earth Planet Sci 32(1):151–185. https://doi.org/10.1146/annurev.earth.32.101802.120356

Whipple KX, Kirby E, Brocklehurst SH (1999) Geomorphic limits to climate-induced increases in topographic relief. Nature 401(6748):39–43

Wu G, Duan A, Liu Y, Mao J, Ren R, Bao Q, He B, Liu B, Hu W (2015) Tibetan Plateau climate dynamics: recent research progress and outlook. Natl Sci Rev 2(1):100–116. https://doi.org/10.1093/nsr/nwu045

Yeats RS, Lillie RJ (1991) Contemporary tectonics of the Himalayan frontal fault system: folds, blind thrusts and the 1905 Kangra earthquake. J Struct Geol 13(2):215–225

Yin A (2006) Cenozoic tectonic evolution of the Himalayan orogen as constrained by along-strike variation of structural geometry, exhumation history, and foreland sedimentation. Earth Sci Rev 76:1–131

Yin A, Harrison TM (2000) Geologic evolution of the Himalayan-Tibetan orogen. Annu Rev Earth Planet Sci 28(1):211–280

Zutshi DP (1991) Limnology of high altitude lakes of Himalayan region. Internationale Vereinigung für theoretische und angewandte Limnologie: Verhandlungen 24(2):1077–1080

Chapter 3
Impact of Land Use Land Cover and Precipitation Change on Surface and Sub-surface Discharge in Central Himalaya

Ayushi Vijhani and Vinay Shankar Prasad Sinha

Abstract The study addresses the critical interplay between land use and land cover (LULC) changes and future precipitation patterns, emphasizing their combined impact on surface discharge dynamics. Across diverse ecological and geographical contexts, alterations in LULC have profound implications for hydrological processes, influencing factors such as surface and sub-surface discharge. In this study, the innovative trend analysis (ITA) and the additive data decomposition method were employed to assess the presence of trends in surface and sub-surface discharge. We analyze how future precipitation scenarios, driven by climate change, may exacerbate the effects of LULC changes on surface and sub-surface discharge. This study highlights the necessity of integrated watershed management strategies that account for both anthropogenic influences and climate variability. Our findings underscore the need for adaptive management practices that can enhance resilience in water resource systems while providing a framework for policymakers to mitigate negative impacts. Furthermore, this research contributes to a deeper understanding of hydrological responses to synergistic environmental changes, offering insights for sustainable land and water management in the face of future climate uncertainties.

Keyword LULC dynamics · Innovative trend analysis · Representative concentration pathway (RCP) · Uttarakhand

A. Vijhani
TERI School of Advanced Studies, New Delhi, India
e-mail: ayushi.vijhani@terisas.ac.in

V. S. P. Sinha (✉)
Centre for the Study of Regional Development, Jawaharlal Nehru University, New Delhi, India
e-mail: sinha_vinay@yahoo.co.uk; vsinha@jnu.ac.in

S. C. Rai (ed.), *Himalaya: Mountains of Destiny*,
https://doi.org/10.1007/978-3-031-94905-0_3

3.1 Introduction

Water, one of the most important elements in the ecosystem, is the primary medium through which mountain communities predominantly feel the consequences of climate change. The Central Himalayan region is critical in terms of water resources, but the scarcity of long-term hydro-meteorological and cryospheric data has impeded our knowledge of the high-elevation catchments' response to climate change (Cogley 2011; Kargel et al. 2011). Watersheds fed by snow are considered to be much more vulnerable (Immerzeel et al. 2013). Although precipitation and flow datasets are available for some basins, the spatio-temporal variability of precipitation in mountains and adjacent basins can result in a very different precipitation pattern (Chalise et al. 2003). Furthermore, the complex inter-relationship between various hydrological processes such as precipitation, evapotranspiration, infiltration, and runoff makes it increasingly difficult to anticipate the hydrological response of high-elevation catchments (Dwarakish and Ganasri 2015). Falkenmark and Rockström (2004) state that hydrological modeling is important for assessing water availability and making policy decisions in water resource management. In addition to climate change, the Central Himalayan region faces challenges of increasing water demand and water scarcity; therefore, it becomes imperative to predict spatial and temporal hydrological processes using physically based distributed hydrological models to assess the water budget of the mountain catchment (Shukla et al. 2018).

Since rainfall is the most critical input for hydrological modeling, having reliable and accurate data that reflect the spatio-temporal variability of precipitation is essential (Oudin et al. 2006; Mul et al. 2009; Tobin and Bennett 2009). Neglecting spatial variability of precipitation can lead to substantial mistakes in model outcomes (Andréassian et al. 2001; Bárdossy and Das 2008). Moreover, the change in precipitation and the flow volume of mountain tributaries have severe repercussions on adjoining plain areas eventually affecting the seasonal water availability. Hannah et al. (2005) observed that the seasonal variability of flow has significant impacts on the livelihood of people (Viviroli et al. 2007) and hazards (Cenderelli and Wohl 2003). Due to complex topography and inaccessibility, most of the Himalayan basins are ungauged or poorly gauged (Han et al. 2019). As a result, the most fundamental strategy in hydrologic modeling is to calculate flow using meteorological data and catchment properties of a basin or the neighboring basin (Lørup et al. 1998; Ye et al. 2013).

Land cover characteristics are important to hydrological processes under investigation and must be described in hydrologic models by parameters related to land activities (Twine et al. 2004; Patidar et al. 2022). As a result, distributed hydrologic models are being utilized more frequently to explore the hydrological implications of changes in land use land cover (Legesse et al. 2003; Krysanova et al. 2005). The type and intensity of vegetative land cover are two major elements that influence surface runoff and evapotranspiration. It has been established that forests have higher rates of evapotranspiration and interception than grasslands and scrublands, which has an impact on the available water for drainage into streams or recharge of aquifers

(Farley et al. 2005). The influence of changing climate on water resources has been estimated using hydrological models. The flow volumes of the Kosi River Basin in Nepal fluctuate with seasons and sub-basins (Bharati et al. 2014). With the Representative Concentration Pathway (RCP) scenario in the Himalayas, Soncini et al. (2015) found that increases in winter precipitation result in increased stream flow. In the Himalayan region, Ragettli et al. (2013) found that the decadal mean runoff is largely steady with the future projections up to 2050. Modeling the hydrology of Himalayan River basins is challenging due to their high climatic variability and the strong influence of extreme events. Narula and Bhadwal (2003) found out that the sub-basin would face significant water stress due to a 20–30% reduction in monthly flows in a small-scale catchment of the Upper Yamuna sub-basin in Uttarakhand. The infiltration excess and saturation overland flow contribute to runoff generation in Nepal (Collins et al. 1998). To investigate the future impact of climatic change on water resources, it is essential to benchmark the current status of water resources in this region as a baseline for future projections.

The land use/cover (LULC) has changed spatially and temporally as a result of agricultural growth, urbanization, and deforestation, impacting water flow routes and water balance (Rawat and Kumar 2015). Changes in land use have a significant impact on the quantity of surface water and groundwater (Dwarakish and Ganasri 2015). Surface runoff, evapotranspiration (ET), sub-surface flow, infiltration, and other elements of the hydrologic cycle are all affected by climate change and land use dynamics (Xu and Singh 1998; DeFries and Eshleman 2004; McColl and Aggett 2007). These interlinked and complex processes control the water budget on the surface and in underground aquifers and thus control the water dynamics of the catchment (Choi and Deal 2008; Baker and Miller 2013; Chapuis et al. 2020). Furthermore, the studies focused on investigating the impact of land use change on mean annual runoff (Brown et al. 2005). Determining how climate change and LULC might affect hydrological components regionally and locally is a challenge, especially in the data-scarce region of Uttarakhand in the Central Himalayan region.

3.2 Study Area

The Himalayas are categorized into four physiographic zones, namely Terai and Bhabar region (175–600 m), Shivaliks (600–1200 m), Middle or Lesser Himalayas (1200–3000 m), and Great Himalayas or Himadri (3000–7000 m) (Kumar and Chopra 2009; Basistha et al. 2008).

Uttarakhand is the source of several major rivers and encompasses four major river catchments of Yamuna (upper Yamuna basin), Ganga (upper Ganga basins), Ramganga, and Ghaghar (Sharda) which supply fresh water to millions of people in downstream regions (Fig. 3.1). With over 12,000 glaciers and eight major river catchments, Uttarakhand is well-endowed in terms of water resources, providing a lifeline for the hydrological system of the Gangetic Plain. The state, on the other hand, recognizes the impending threat to the Himalayan catchments posed by deforestation,

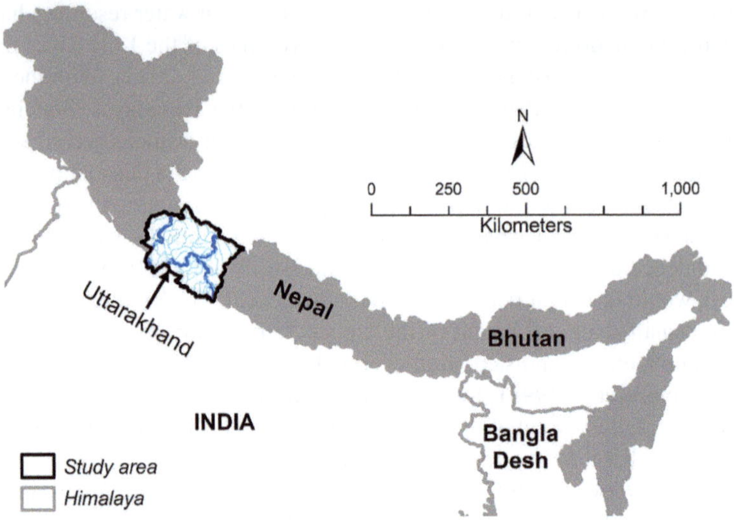

Fig. 3.1 Uttarakhand, Central Himalaya, India

improper agronomic practices, hydrologic fluctuations, and natural disasters. Most water supply schemes that have been laid in Uttarakhand have their origins in a spring (NITI Aayog 2018). Some natural springs have dried up completely and become seasonal; therefore, the issue of dying springs and rivulets is being progressively felt among the mountain communities (Tiwari et al. 2000).

The Gangetic alluvial plain and the Himalayan Mountain belt are the two hydrogeological regimes that make up the hydrogeological setting of Uttarakhand. The Gangetic alluvial plain is a prospective zone for groundwater development since it is covered by a wide tract of alluvium and unconsolidated sedimentary material of diverse size fractions ranging from boulders to clay. Groundwater exploitation is significantly less likely in the Himalayan Mountain belt. In the hilly region, groundwater is typically found in fissures/fractures and emerges as springs.

3.3 Dataset Used and Method

3.3.1 Dataset Used

Land use and land cover (LULC) is one of the most crucial parameters affecting the surface and sub-surface hydrology of a catchment. The forest classes were obtained from a forest class and forest type dataset prepared by the Forest Survey of India (FSI) for the year 2019. Besides forest classes, the other LULC classes were created with the help of Landsat-8 OLI satellite data for the year 2019. The LULC map for the year 2019 was prepared using a hybrid technique that involves digital image classification and visual image interpretation. To strengthen the quality of visual interpretation, referenced and published land use/land cover maps on a 1:50,000 scale for the years 2005 and 2012 by the National Remote Sensing Centre (NRSC), GoI, were used. Moreover, a clear long-term coherence (since 2005) of individual classes over a particular pixel/area served as one of the key elements of visual image interpretation and for defining classes in digital image classification. Finally, 21 LULC classes were prepared for the year 2019.

Daily gridded data of meteorological parameters (maximum/minimum temperature) were procured at 1 × 1 degree and daily precipitation at 0.25 × 0.25 degrees from the Indian Meteorological Department (IMD) Government of India (GoI) for 1983–2019. Other parameters such as solar radiation, relative humidity, and wind speed were downloaded from the NASA POWER websites (https://power.larc.nasa.gov/data.access.viewer/).

The dynamic downscaling of climate data was carried out using the Weather Research and Forecasting model (WRF) (Bonekamp et al. 2018; Maussion et al. 2011; Collier and Immerzeel 2015) regional climate model (RCM) to capture the regional climate effects. Future simulation for ten years (2020–2030) was conducted using an NCAR coupled model known as Community Climate System Model Version4 (CCSM4) (http://www.cesm.ucar.edu/models/ccsm4.0/) RCP 4.5 scenario initial and boundary conditions to force a regional climate model with 25 km horizontal resolution. In the Himalayan region, the RCP 4.5 scenario is widely used by researchers for assessing the impact of climatic change on water resources such as estimating streamflow/discharge (Shrestha et al. 2016) and for analyzing the temperature, precipitation, and runoff linkage (Jasrotia et al. 2021).

3.3.2 *Method*

3.3.2.1 Hydrological Modeling

The fully dynamic MIKE Hydro river model simulates the unsteady flow in rivers. The surface flow, interflow, and baseflow were evaluated using the rainfall-runoff (NAM) Mike model. The model was supplied with time series meteorological data (precipitation and evapotranspiration) and discharge data to calculate the surface flow, interflow, and baseflow (Vijhani et al. 2022). The semi-distributed Soil Water Assessment Tool (SWAT) was used to calculate the evapotranspiration and discharge data of Uttarakhand's ungauged river system. The basin-scale hydrological model SWAT is time-continuous and operates on a daily time step. It is frequently used to study how basin hydrology is affected by land use and cover (Bonekamp et al. 2018) and climate change (Dahal et al. 2020). The watershed's LULC, soil profile features, maximum and minimum temperatures, precipitation, solar radiation, relative humidity, and wind speed are all needed by the model. To calculate the effect of LULC on surface and surface flow, the model was given three alternative time

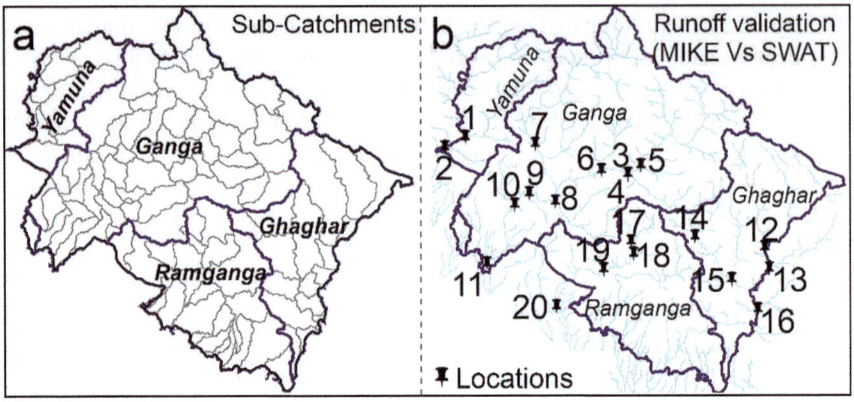

Fig. 3.2 **a** Major catchments of Uttarakhand, and **b** validation location of surface runoff

LULCs (2005, 2012, and 2019). In a similar vein, to illustrate how climate change affects basin hydrology, future daily precipitation was also included in the model under RCP 4.5 scenarios. This was done by calculating changes in surface and sub-surface flow. The future scenario considered the land cover and land use for the most current year, 2019.

Since the study area is an ungauged river system, it was calibrated using a recommended set of parameters based on prior research conducted by multiple researchers (Kushwaha and Jain 2013). The model was then validated using the MIKE Hydro NAM model. To validate the baseflow of the MIKE NAM model, a survey of spring discharge data was done at 200 places, as the majority of the river system is either ungauged or has inadequate gauged stations. Following the validation of the MIKE NAM model's based flow, the SWAT surface flow was verified using the MIKE NAM surface flow. Finally, surface and sub-surface (groundwater/springs) discharge was computed at 20 locations in all four river catchments: Yamuna, Ganga, Ramganga, and Ghaghar (Fig. 3.2). These data were further analyzed using Innovative Trend Analysis (ITA) to highlight the impact of LULC and climate change.

3.3.2.2 Innovative Trend Analysis (ITA)

Testing the trend hypothesis on the time series is one of the most common techniques documented in the hydro-meteorological literature to detect the implications of human activities on climate and hydrological processes. The innovative trend analysis (ITA) (Sen 2012) divides any given hydro-meteorological time series into two equal sections from the beginning to the end date. Both the sub-series are arranged in ascending order. The first half of the series is arranged on the x-axis and the second half is arranged at the y-axis of the Cartesian coordinate system. There is no trend in the data if the data points are collected on the 1:1 line. A positive or a negative trend is indicated by data points that fall above or below the 1:1 line, respectively

(Sen 2014). The ITA method does not make any assumptions about serial correlation, normality, or record length.

3.3.2.3 Data Decomposition Methods

To examine how these sub-components affect the data in the series, the time series data can be decomposed into numerous sub-components. Seasonal and trend patterns constitute the majority of the time series data. There are two types of decomposition methods: additive and multiplicative (Prema and Rao 2015; Jebb et al. 2015), which are represented in Eqs. 3.1 and 3.2, respectively.

$$Y_t = T_t + S_t + R_t, \tag{3.1}$$

$$Y_t = T_t \times S_t \times R_t, \tag{3.2}$$

where Y_t is the time series data, T_t is trend-cycle component, S_t is the seasonal component, and R_t is a random (or remainder) component at period t

According to additive decomposition, time series data are a function of the sum of its components. The method is useful when the seasonal variation is steady across time. Time series data, according to the multiplicative decomposition, are a function of the product of its components. They are useful when the seasonal variation increases over time. In this study, the additive decomposition method was used for surface and sub-surface discharge assuming that there is no increase in seasonal variation over time.

The ITA method and the additive data decomposition method were used to assess the presence of trends in surface discharge based on LULC for the years 2005, 2012, and 2019, and the future (2030) precipitation under the RCP 4.5 scenario at 20 locations of Uttarakhand to spatially cover the entire area. The first three simulations were based on three different land use changes during 2005, 2012, and 2019, and the fourth simulation for the year 2030 was based on 2019 land use land cover practices with future precipitation patterns. The study investigated a two-dimensional approach to analyze the trends of surface runoff and the associated impact of land practices in response to climate change. Emphasizing the need to monitor the trend in surface runoff, these simulations were carried out under dynamic changes in land practices and changes in future precipitation. Further, the study investigates the changes in the spatio-temporal trend, which highlights the cause of the change in surface runoff over the current and future climate under the RCP 4.5 scenario. Each catchment was separately investigated to identify the spatio-temporal trend of surface runoff.

3.4 Results and Discussion

The result summarized the impact of land use land cover and future precipitation trends at basin hydrology on two aspects, i.e., (i) impact on surface discharge and (ii) impact on base flow (ground water discharge) at catchment scale from upstream to downstream.

3.4.1 Impact on Surface Discharge

3.4.1.1 Yamuna Catchment

In the Yamuna catchment, two sites were selected to investigate the surface runoff trend analysis. The surface runoff at Kalsi and Assan Barrage Bird Sanctuary indicated a positive trend over the period. Simulations based on dynamic land practices (2005, 2012, and 2019) have a coinciding trend line (Fig. 3.3) through the ITA method that indicates there is no significant change in surface runoff due to change in land practices rather it is due to current precipitation patterns.

Future climate-based simulation indicated a negative trend for both locations, which could be due to the underestimated values of future precipitation by CCSM4. Overall, the current surface flow revealed a relatively increasing trend. This provides important evidence that the trend is attributable to precipitation patterns rather than changes in land use behaviors. The study shows that the current surface flow in the Yamuna basin has positive trends, but the future surface flow has decreasing trends due to underestimation of precipitation value by the forecasted model.

3.4.1.2 Ganga Catchment

In the Upper Ganga Basin, nine locations on different river systems such as Alaknanda (Karnaprayag, Nandaprayag); Pindar (Galnau); downstream of Alaknanda (Rudraprayag); Bhagirathi (Tehri Dam Reservoir, Matoli); Nayar (Shivpuri); Ganga (Rishikesh, Kalsia) were evaluated. There was no significant trend in surface runoff in the Alaknanda River at Karnaprayag from 2005 to 2019, despite changes in land practices and predicted precipitation behavior captured by climate models (Fig. 3.4). On the other hand, next to Karnaprayag at Nandaprayag in the Alaknanda River, the surface runoff showed a positive trend between 2005 and 2012 and a decreasing trend in 2019 due to changing land use practices. In future, it is expected to have high surface runoff with an increasing trend. A similar condition exists at Galnau in the Pindar watershed (a tributary of the Alaknanda River). The Alaknanda River meets the Pindar River flows south and reaches Rudraprayag. There is a positive trend in surface runoff which increased comparatively higher during 2012 and 2019 land

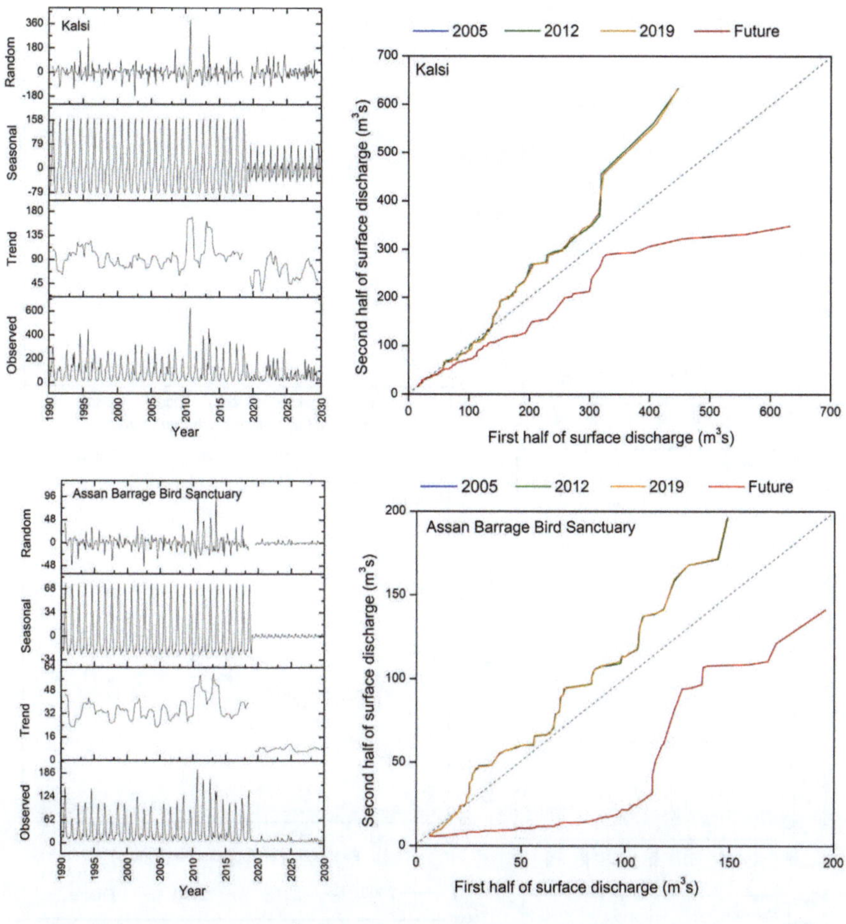

Fig. 3.3 Impact of LULC and future precipitation on surface discharge in Yamuna catchment

use concerning 2005 due to changes in land use practices. The future precipitation behavior shows a significant upward trend.

The next sub-catchment is associated with the Bhagirathi River system. The trend analysis shows the impact of land practices and future climate on the Upper Bhagirathi River at the end of the Tehri Dam. Under the current land use practices (2019), surface runoff has shown decreasing trends. However, under the 2005 and 2012 land use practices, there was an upward trend, and the future trend of surface runoff would continue to increase in a positive direction. Further, Bhagirathi reaches downstream at Matoli, where surface runoff shows an increasing trend over time. This indicates the impact of land practices on runoff and presumes that similar conditions will occur in future precipitation.

In the sub-catchment of the Nayar River (a tributary of Ganga) at Shivpuri, the surface runoff shows a decreasing trend and there is no evidence of the impact of

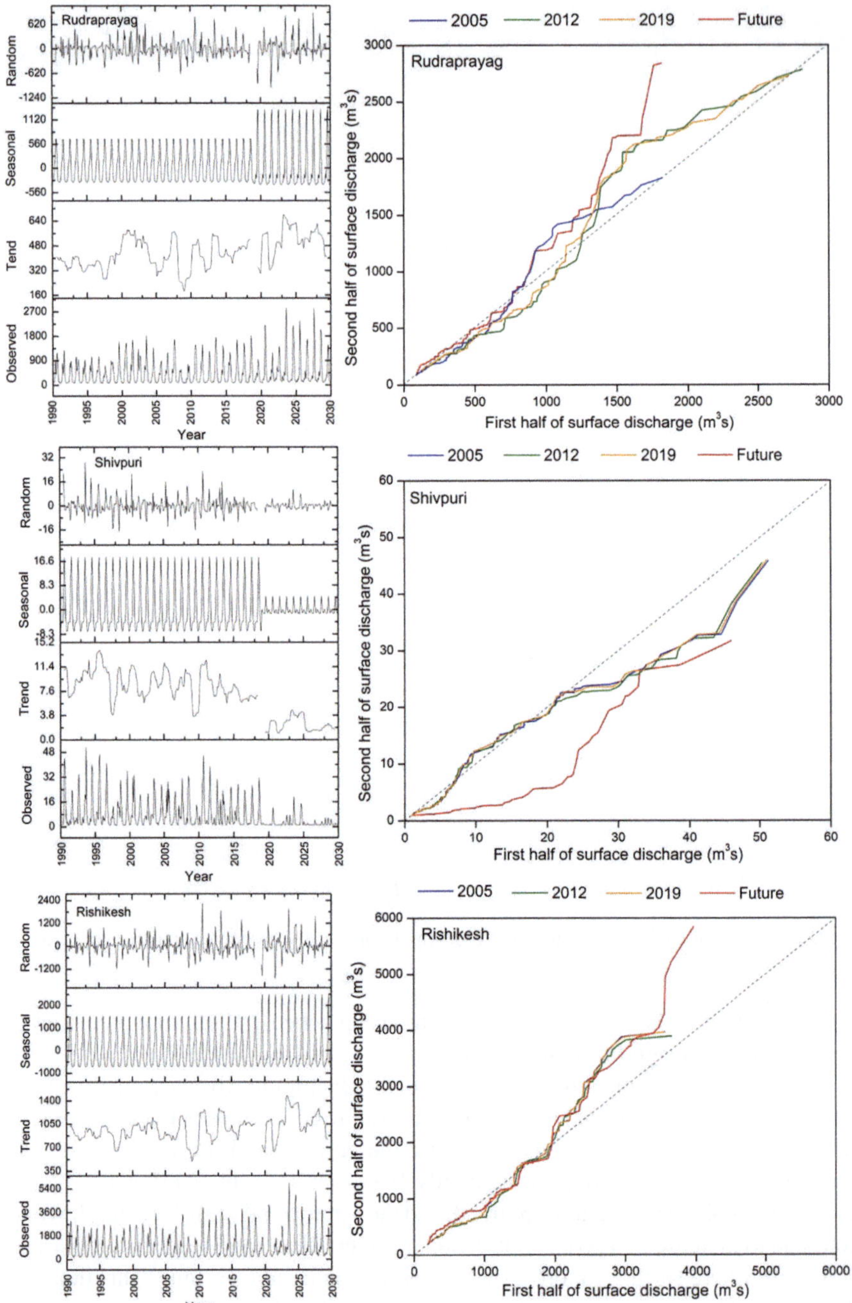

Fig. 3.4 Impact of LULC and future precipitation on surface discharge in the Ganga catchment

land practices which is due to the current precipitation trend. The climate model (2020–30) predicted low precipitation concerning current precipitation trends from 1991 to 2019 and shows a relatively decreasing trend due to the impact of climate change. The Bhagirathi River meets the Alaknanda River at Devprayag further south and is then known as the Ganga River. Two locations such as Rishikesh and Kalsia showed an increasing trend, but there is no significant evidence of the impact of land use practices on surface runoff while future precipitation indicates the impact of surface runoff due to climate change.

3.4.1.3 Ghaghar Catchment

Three river systems were considered under the Ghaghar catchment. First, the Kali River system (Jauljibi and Materi), second, the Saryu River system (Kaphalkhet and Saryu Bridge), and third Sharda River system (Naag) were considered. The Saryu River system meets the Kali river at Pancheshwar which further flows down as Sharda river. In the entire catchment, a negative trend was observed in surface flow and there is no evidence of the impact of land use on surface runoff. Furthermore, a similar trend of surface runoff is expected in future in the Ghaghar catchment (Fig. 3.5).

3.4.1.4 Ramganga Catchment

In the Ramganga catchment, four locations from top to down (Chachroti, Garhkhet, Kalakhand, and Laidarpur) were selected. No significant trend was observed from 2005 to 2019 in surface runoff, and the same has been revealed in the future trend of surface runoff at Chachroti and Kalakhand. A positive trend of surface runoff has been observed in Garhkhet whereas a negative trend was observed at Laidarpur in the plain area of Ramganga catchment. There is no significant impact identified on surface runoff has due to land use dynamics and future precipitation in this catchment (Fig. 3.6).

3.4.2 Estimating The Trend of Baseflow

The agricultural practices of Uttarakhand are not only dependent on surface runoff; rather most of the agricultural practices use spring water already documented in the introduction chapter. Spring flow is a function of the interaction between surface and sub-surface water. It is important to understand the trend of baseflow of sub-surface aquifers. In this study, baseflow was simulated using the MIKE Hydro model and validated for the year 2019. The trend analysis was performed based on land use land cover for the year 2019. The ITA and decomposition method was applied to monthly simulated baseflow from 1991 to 2019.

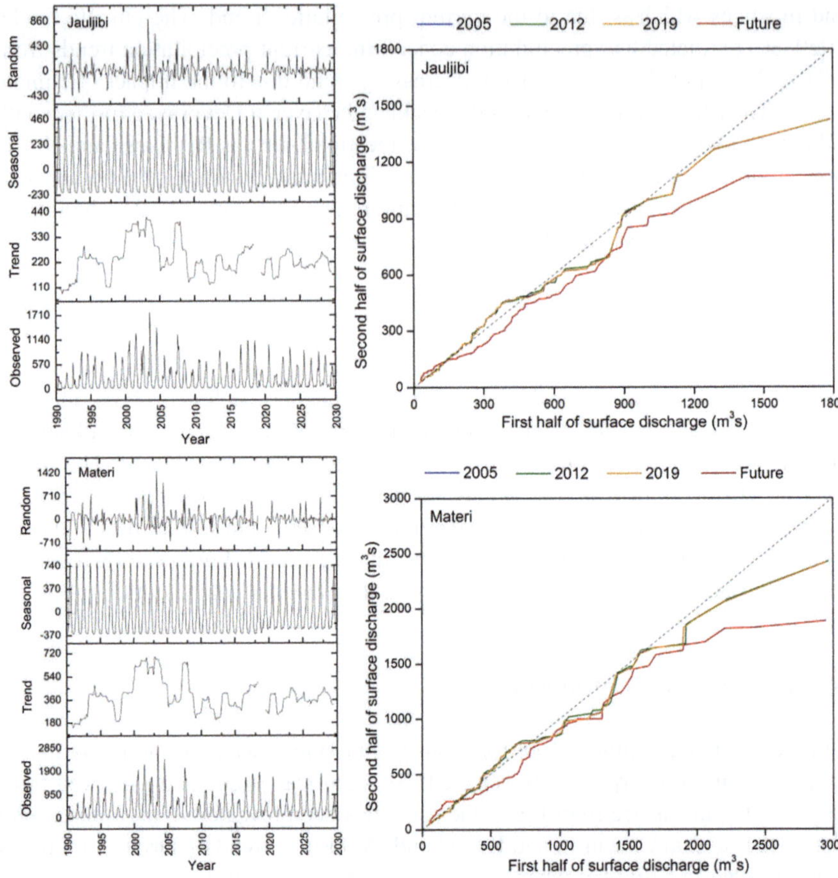

Fig. 3.5 Impact of LULC and future precipitation on surface discharge in Ghaghar catchment

The study result reveals that there is a positive trend in baseflow at Kalsi and Assan Barrage Bird Sanctuary in the Yamuna catchment (Fig. 3.7). Similarly, in the Alaknanda catchment, two locations, namely Karnaprayag and Rudraprayag, observed a positive trend of baseflow, whereas a negative trend of baseflow was observed at three locations, namely Shivpuri in the Nayar river system and Rishikesh and Kalsia in the Ganga River system. A negative trend of baseflow was observed at Nandaprayag in the Alaknanda catchment, whereas Galnau in Pindar catchment initially showed a positive trend which was later converted into a negative trend. In the Bhagirathi catchment, Tehri Dam reservoir and Matoli showed a negative trend in baseflow (Fig. 3.8).

The majority of the locations in the Ghaghar catchment have observed a negative trend of baseflow except at the Saryu bridge, where a positive trend of baseflow was reported (Fig. 3.9). In the Ramganga catchment, no significant trend of baseflow

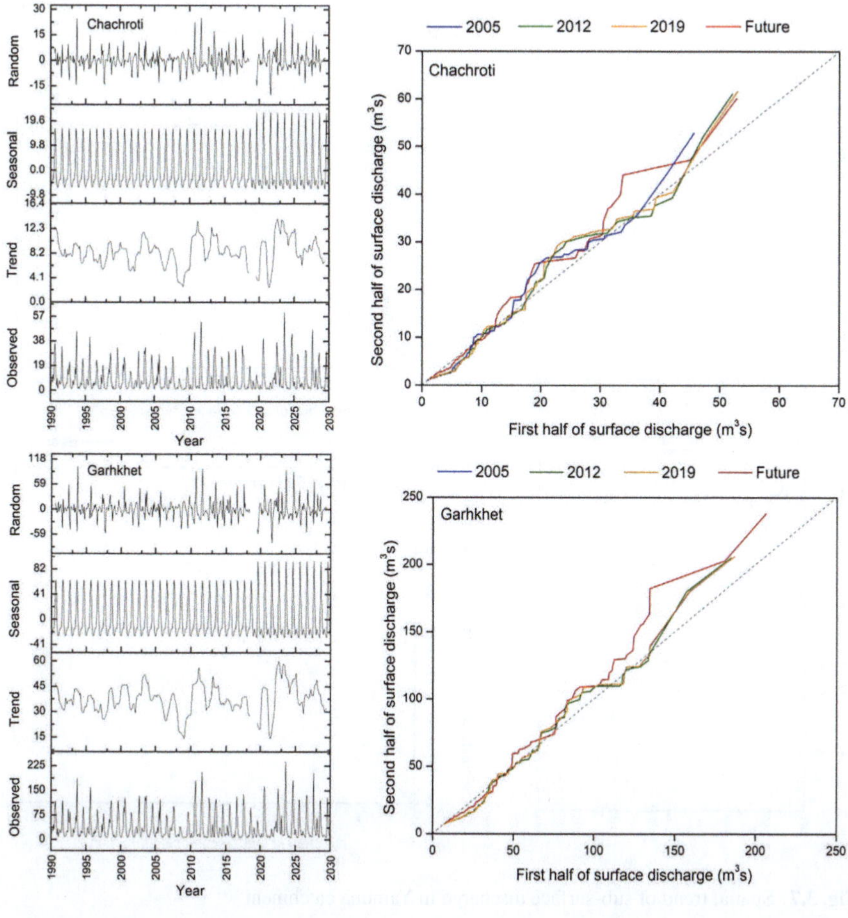

Fig. 3.6 Impact of LULC and future precipitation on surface discharge in Ramganga catchment

was observed except for Laidarpur which showed a positive trend in the plain area (Fig. 3.10).

The results of the study indicate diminishing discharge of sub-surface aquifer in most of the watersheds at the micro-level which is also indicated in most of the research work conducted by researchers in Uttarakhand. The impact of LULC and future precipitation on the surface and sub-surface discharge was analyzed in all 20 locations (Table 3.1).

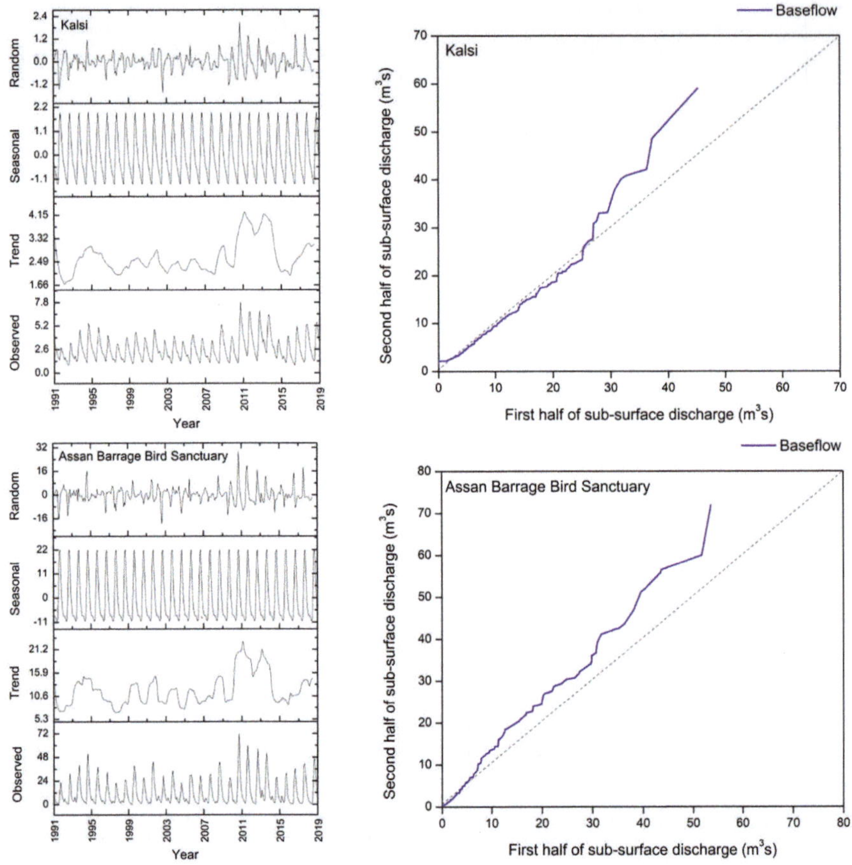

Fig. 3.7 Spatial trend of sub-surface discharge in Yamuna catchment

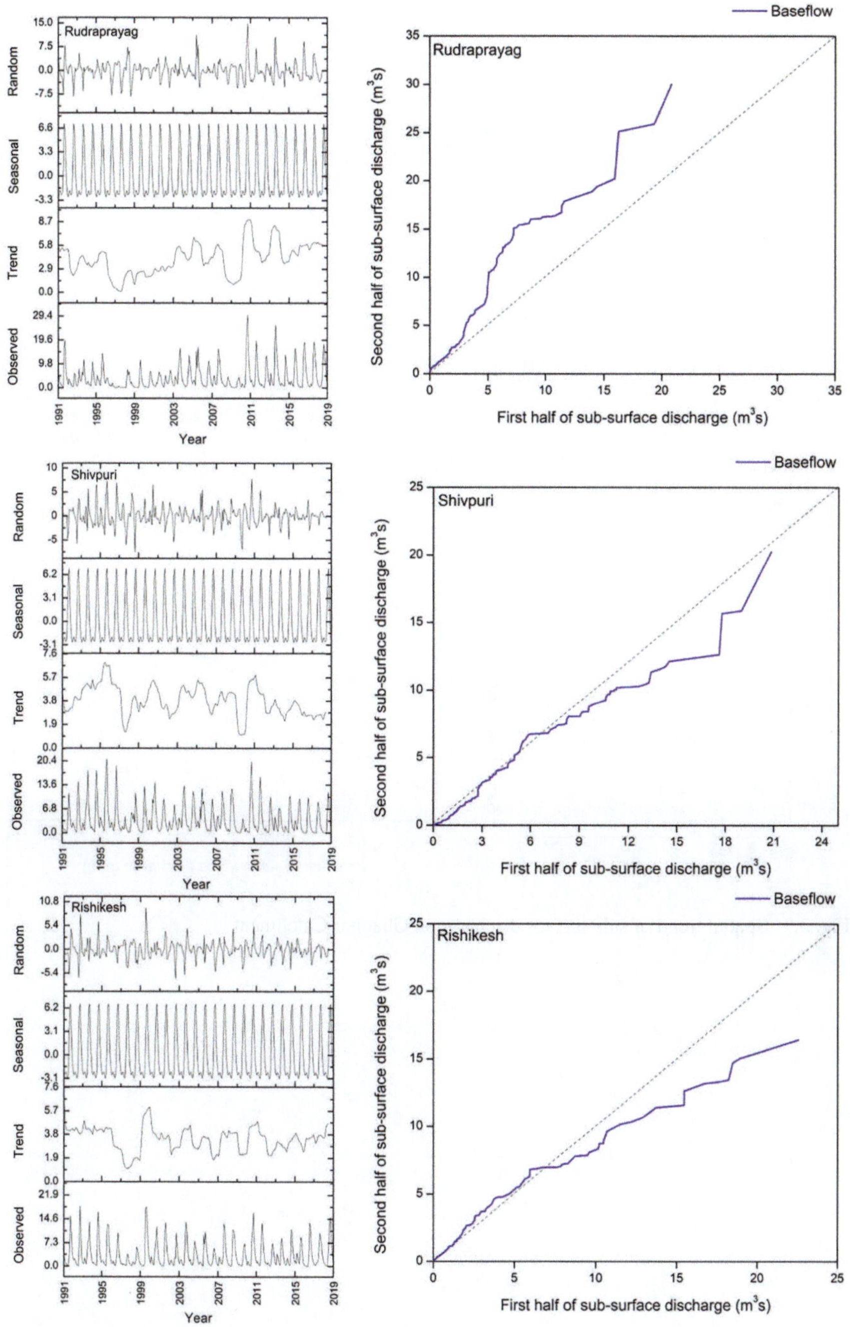

Fig. 3.8 Spatial trend of sub-surface discharge in Ganga catchment

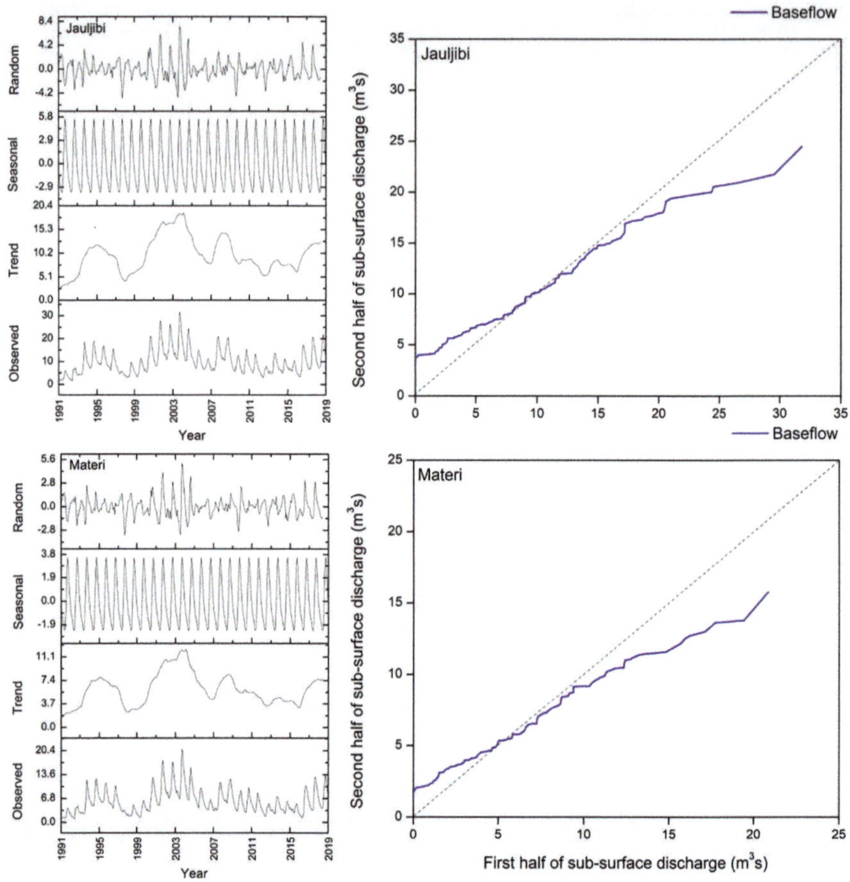

Fig. 3.9 Spatial trend of sub-surface discharge in Ghaghar Catchment

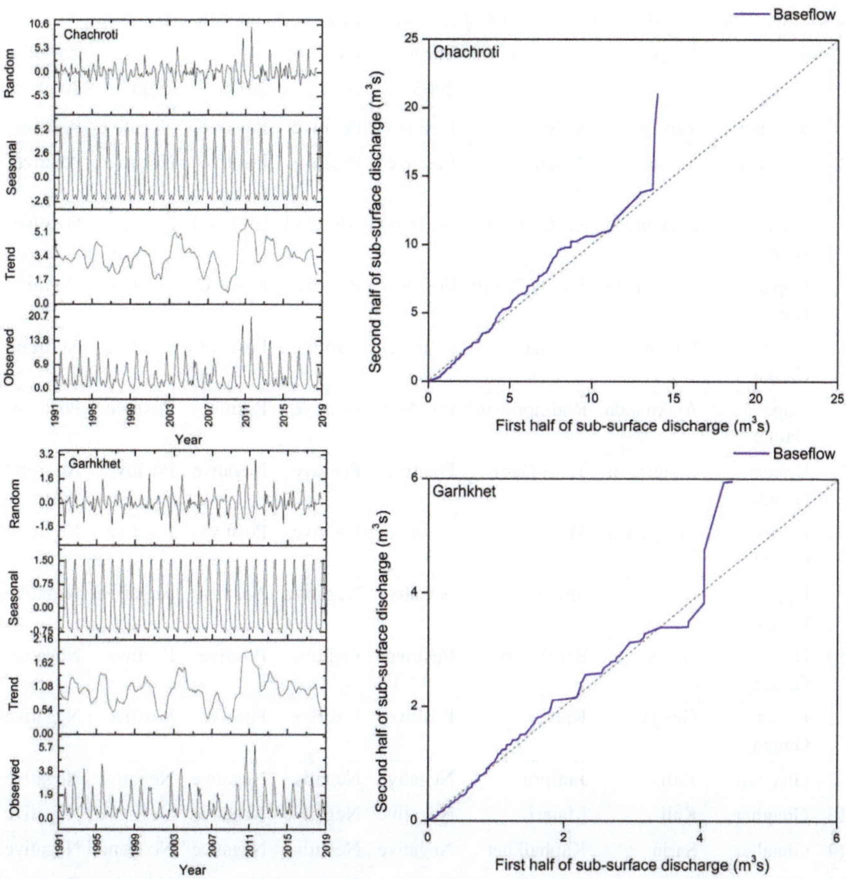

Fig. 3.10 Spatial trend of sub-surface discharge in Ramganga Catchment

Table 3.1 Analyzing the impact of LULC and future precipitation on water resources

ID	Catchment	River	Location	Surface runoff				Baseflow
				2005	2012	2019	2030	2019
1	Yamuna	Yamuna	Kalsi	Positive	Positive	Positive	Negative	Positive
2	Yamuna	Yamuna	Assan Barrage	Positive	Positive	Positive	Negative	Positive
5	Upper Ganga	Alaknanda	Karnaprayag	No trend	No trend	No trend	No trend	Positive
3	Upper Ganga	Alaknanda	Nandaprayag	Positive	Positive	Positive	Positive	No trend
4	Upper Ganga	Pindar	Galnau	Positive	Positive	Positive	Positive	No trend
6	Upper Ganga	Alaknanda	Rudraprayag	Positive	Positive	Positive	Positive	Positive
7	Upper Ganga	Bhagirathi	Tehri Dam	Positive	Positive	Negative	Positive	No trend
8	Upper Ganga	Bhagirathi	Matoli	Positive	Positive	Positive	Positive	No trend
9	Upper Ganga	Nayar	Shivpuri	Negative	Negative	Negative	Negative	Negative
10	Upper Ganga	Ganga	Rishikesh	Positive	Positive	Positive	Positive	Negative
11	Upper Ganga	Ganga	Kalsia	Positive	Positive	Positive	Positive	Negative
12	Ghaghar	Kali	Jauljibi	Negative	Negative	Negative	Negative	Negative
13	Ghaghar	Kali	Materi	Negative	Negative	Negative	Negative	Negative
14	Ghaghar	Sarju	Kaphalkhet	Negative	Negative	Negative	No trend	Negative
15	Ghaghar	Sarju	Saryu Bridge	Negative	Negative	Negative	Negative	Positive
16	Ghaghar	Sharda	Naag	Negative	Negative	Negative	Negative	Negative
17	Ramganga	Ramganga	Chachroti	No trend	No trend	No trend	No trend	Positive
18	Ramganga	Ramganga	Garhkhet	Positive	Positive	Positive	Positive	No trend
19	Ramganga	Ramganga	Kalakhand	No trend	No trend	No trend	No trend	No trend
20	Ramganga	Ramganga	Laidarpur	Negative	Negative	Negative	Negative	Positive

3.5 Conclusion

Hydrological modeling is a prerequisite for water resource management and planning in agriculture. The discharge data for the Himalayan Rivers are classified in nature and are not available in the public domain due to transboundary issues and the data dissemination policy by GoI. Therefore, the hydro-meteorological dataset from the CWC was not used in the study. Moreover, the networks of CWC discharge locations are inadequate throughout the Himalayan locale due to topographic and climatic barriers. Further, the remoteness and physical barriers also restrict the installation of gauging stations in high-altitude regions. Hence, it was not feasible to use the observed streamflow discharge at the micro-watershed level. Considering all the constraints, the research work focuses on integrating two rainfall-runoff models: SWAT and the MIKE Hydro model with limited field measurements for spring discharge. The findings of the study indicate significant trends in the spatio-temporal domain due to changes in land practices and future precipitation behavior at the watershed or sub-catchment scale. The study provides evidence of diminishing discharge of baseflow which impacts the agricultural practices relying on spring-fed water in Uttarakhand. The results also highlight that the surface and the sub-surface flow of aquifer systems are impacted by the future precipitation behavior as well as land use land cover dynamics. This thereby impacts the agricultural practices and livelihood of the locals. Therefore, it is necessary to understand and assess the future agricultural land distribution and its patterns through hydrological-based drivers.

Acknowledgements The authors are thankful to Uttarakhand Jal Sansthan (UJS) for providing support facilities for the study under the "National Mission on Himalayan Studies (NMHS)" project titled "Water Resource Management through Spring and Catchment Rejuvenation in Uttarakhand for Improving Water Security" funded by the Ministry of Environment, Forest & Climate Change (MoEF&CC), GoI (Ref. No.: GBPNI/NMHS-2018-19/MG2).

References

Andréassian V, Perrin C, Michel C, Usart-Sanchez I, Lavabre J (2001) Impact of imperfect rainfall knowledge on the efficiency and the parameters of watershed models. J Hydrol 250(1–4):206–223

Baker TJ, Miller SN (2013) Using the soil and water assessment tool (SWAT) to assess land use impact on water resources in an East African watershed. J Hydrol 486:100–111

Bárdossy A, Das T (2008) Influence of rainfall observation network on model calibration and application. Hydrol Earth Syst Sci 12(1):77–89

Basistha A, Arya DS, Goel NK (2008) Spatial distribution of rainfall in Indian Himalayas–a case study of Uttarakhand region. Water Resour Manage 22(10):1325–1346

Bharati L, Gurung P, Jayakody P, Smakhtin V, Bhattarai U (2014) The projected impact of climate change on water availability and development in the Koshi Basin Nepal. Mountain Res Dev 34(2):118–130

Bonekamp PNJ, Collier E, Immerzeel WW (2018) The impact of spatial resolution, land use, and spinup time on resolving spatial precipitation patterns in the Himalayas. J Hydrometeorol 19(10):1565–1581

Brown AE, Zhang L, McMahon TA, Western AW, Vertessy RA (2005) A review of paired catchment studies for determining changes in water yield resulting from alterations in vegetation. J Hydrol 310(1–4):28–61

Cenderelli DA, Wohl EE (2003) Flow hydraulics and geomorphic effects of glacial-lake outburst floods in the Mount Everest region Nepal. Earth Surf Proc Land 28(4):385–407

Chalise SR, Kansakar SR, Rees G, Croker K, Zaidman M (2003) Management of water resources and low flow estimation for the Himalayan basins of Nepal. J Hydrol 282(1–4):25–35

Chapuis H, Ré-Bahuaud J, Jolivet J, Paran F, Graillot D (2020) Karst–River interaction, elaboration of an indicator of the karst hydrological conditions applied to the Ceze River (Gard, France). In: Eurokarst 2018, Besançon. Springer, Cham, pp 77–81

Choi W, Deal BM (2008) Assessing hydrological impact of potential land use change through hydrological and land use change modeling for the Kishwaukee River basin (USA). J Environ Manage 88(4):1119–1130

Cogley JG (2011) Present and future states of Himalaya and Karakoram glaciers. Ann Glaciol 52(59):69–73

Collier E, Immerzeel WW (2015) High-resolution modeling of atmospheric dynamics in the Nepalese Himalaya. J Geophys Res: Atmospheres 120(19):9882–9896

Collins R, Dickinson A, Johnson R (1998) The use of GIS and remote sensing techniques to predict erosion in the Nepal middle hills. Soil Erosion at multiple scales: principles and methods for assessing causes and impacts, pp 303–315

Dahal P, Shrestha ML, Panthi J, Pradhananga D (2020) Modeling the future impacts of climate change on water availability in the Karnali River Basin of Nepal Himalaya. Environ Res 185:109430

DeFries R, Eshleman KN (2004) Land-use change and hydrologic processes: a major focus for the future. Hydrol Process 18(11):2183–2186

Dwarakish GS, Ganasri BP (2015) Impact of land use change on hydrological systems: a review of current modeling approaches. Cogent Geosc 1(1):1115691

Falkenmark M, Rockström J (2004) Balancing water for humans and nature: the new approach in ecohydrology. Earthscan

Farley KA, Jobbágy EG, Jackson RB (2005) Effects of afforestation on water yield: a global synthesis with implications for policy. Glob Change Biol 11(10):1565–1576

Han P, Long D, Han Z, Du M, Dai L, Hao X (2019) Improved understanding of snowmelt runoff from the headwaters of China's Yangtze River using remotely sensed snow products and hydrological modeling. Remote Sens Environ 224:44–59

Hannah DM, Kansakar SR, Gerrard AJ, Rees G (2005) Flow regimes of Himalayan rivers of Nepal: nature and spatial patterns. J Hydrol 308(1–4):18–32

Immerzeel WW, Pellicciotti F, Bierkens MFP (2013) Rising river flows throughout the twenty-first century in two Himalayan glacierized watersheds. Nat Geosci 6(9):742–745

Jasrotia AS, Baru D, Kour R, Ahmad S, Kour K (2021) Hydrological modeling to simulate stream flow under changing climate conditions in Jhelum catchment, western Himalaya. J Hydrol 593:125887

Jebb AT, Tay L, Wang W, Huang Q (2015) Time series analysis for psychological research: examining and forecasting change. Front Psychol 6:727

Kargel JS, Cogley JG, Leonard GJ, Haritashya U, Byers A (2011) Himalayan glaciers: the big picture is a montage. Proc Natl Acad Sci 108(36):14709–14710

Krysanova V, Hattermann F, Habeck A (2005) Expected changes in water resources availability and water quality with respect to climate change in the Elbe River basin (Germany). Hydrol Res 36(4–5):321–333

Kumar V, Chopra AK (2009) Impact of climate change on biodiversity of India with special reference to Himalayan region-An overview. J Appl Nat Sci 1(1):117–122

Kushwaha A, Jain MK (2013) Hydrological simulation in a forest dominated watershed in Himalayan region using SWAT model. Water Resour Manage 27:3005–3023

Legesse D, Vallet-Coulomb C, Gasse F (2003) Hydrological response of a catchment to climate and land use changes in Tropical Africa: case study South Central Ethiopia. J Hydrol 275(1–2):67–85

Lørup JK, Refsgaard JC, Mazvimavi D (1998) Assessing the effect of land use change on catchment runoff by combined use of statistical tests and hydrological modelling: case studies from Zimbabwe. J Hydrol 205(3–4):147–163

Maussion F, Scherer D, Finkelnburg R, Richters J, Yang W, Yao T (2011) WRF simulation of a precipitation event over the Tibetan Plateau, China–an assessment using remote sensing and ground observations. Hydrol Earth Syst Sci 15(6):1795–1817

McColl C, Aggett G (2007) Land-use forecasting and hydrologic model integration for improved land-use decision support. J Environ Manage 84(4):494–512

Mul ML, Savenije HHG, Uhlenbrook S (2009) Spatial rainfall variability and runoff response during an extreme event in a semi-arid catchment in the South Pare Mountains Tanzania. Hydrology Earth Syst Sci 13(9):1659–1670

Narula KK, Suruchi B (2003) Impact of climate change on hydrology for better decision-making at a river basin level in India: a case study. Environmental threats, vulnerability and adaptation: case studies from India, pp 53–77

NITI Aayog (2018) Report of working group i inventory and revival of springs in the Himalayas for Water Security. Available at: http://niti.gov.in/writereaddata/files/document_publication/doc1.pdf

Oudin L, Perrin C, Mathevet T, Andréassian V, Michel C (2006) Impact of biased and randomly corrupted inputs on the efficiency and the parameters of watershed models. J Hydrol 320(1–2):62–83

Patidar N, Das P, Tripathi P, Behera MD (2022) Covariation between LULC change and hydrological balance in River Basin Scale. In: Geospatial technologies for land and water resources management. Springer, Cham, pp 279–294

Prema V, Rao KU(2015) Time series decomposition model for accurate wind speed forecast. Renewables: Wind, Water, and Solar 2(1):1–11

Ragettli S, Pellicciotti F, Bordoy R, Immerzeel WW (2013) Sources of uncertainty in modeling the glacio-hydrological response of a Karakoram watershed to climate change. Water Resour Res 49(9):6048–6066

Rawat JS, Kumar M (2015) Monitoring land use/cover change using remote sensing and GIS techniques: a case study of Hawalbagh block, district Almora, Uttarakhand, India. The Egyptian J Remote Sens Space Sci 18(1):77–84

Sen Z (2012) Innovative trend analysis methodology. J Hydrol Eng 17(9):1042–1046

Sen Z (2014) Trend identification simulation and application. J Hydrol Eng 19(3):635–642

Shrestha S, Shrestha M, Babel MS (2016) Modeling the potential impacts of climate change on hydrology and water resources in the Indrawati River Basin Nepal. Environ Earth Sci 75(4):1–13

Shukla AK, Pathak S, Pal L, Ojha CSP, Mijic A, Garg RD (2018) Spatio-temporal assessment of annual water balance models for upper Ganga Basin. Hydrol Earth Syst Sci 22(10):5357–5371

Soncini A, Bocchiola D, Confortola G, Bianchi A, Rosso R, Mayer C, Lambrecht A, Palazzi E, Smiraglia C, Diolaiuti G (2015) Future hydrological regimes in the upper Indus basin: a case study from a high-altitude glacierized catchment. J Hydrometeorol 16(1):306–326

Tiwari PC (2000) Land-use changes in Himalaya and their impact on the plains ecosystem: need for sustainable land use. Land Use Policy 17(2):101–111

Tobin KJ, Bennett ME (2009) Using SWAT to model streamflow in two river basins with ground and satellite precipitation data 1. JAWRA J Am Water Res Assoc 45(1):253–271

Twine TE, Kucharik CJ, Foley JA (2004) Effects of land cover change on the energy and water balance of the Mississippi River basin. J Hydrometeorol 5:640–655

Vijhani A, Sinha VSP, Vishwakarma CA, Singh P, Sharma SK (2022) Assessment of diminishing discharge of springs in Central Himalayan region India. Hydrol Process 36(5):e14582

Viviroli D, Dürr HH, Messerli B, Meybeck M, Weingartner R (2007) Mountains of the world, water towers for humanity: typology, mapping, and global significance. Water Res Res 43(7)

Xu CY, Singh VP (1998) A review of monthly water balance models for water resources investigations. Water Resour Manage 12(1):20–50

Ye X, Zhang Q, Liu J, Li X, Xu CY (2013) Distinguishing the relative impacts of climate change and human activities on variation of streamflow in the Poyang Lake catchment, China. J Hydrol 494:83–95

Chapter 4
Spatio-temporal Evolution of Glacial Lakes in the Upper Ganga Basin, Central Himalayas

Atul Kumar, Suraj Mal, and Udo Schickhoff

Abstract Glacial lakes in the central Himalayas are expanding at an unprecedented rate due to the rapid melting and thinning of glaciers. The continuous growth of glacial lakes increases the availability of freshwater for the downstream communities and escalates the risk of future Glacial Lake Outburst Floods (GLOFs) in the upper Ganga basin. In the present study, glacial lake inventories for the years 1990, 2000, 2010, and 2020 were prepared at the sub-basin level to understand the evolution of glacial lakes in the upper Ganga basin at the micro-regional scale. We found that, between 1990 and 2020, the total number of glacial lakes increased by 564 (22.08%) and the total area increased by 40.71 km^2 (23.93%). We found that glacial lakes were present in the 28 sub-basins of the upper Ganga basin. Among different types of glacial lakes, end-moraine-dammed (M(e)) (n: 2413 and a: 178.89 km^2) had the highest proportion both in terms of number and area of different types of glacial lakes. We, furthermore, estimated the volume and mean depth of glacial lakes using (Huggel et al. 2002) empirical equations. The total estimated volume of glacial lakes was 5.548 km^3 (2020), and the average estimated mean depth was 8.26 m (2020). Between 1990 and 2020, the estimated volume of glacial lakes increased by 1.46 km^3 and the average estimated mean depth decreased by −0.20 m.

Keywords Central Himalayas · Glacial lakes · Ganga basin · Inventory

---

A. Kumar · U. Schickhoff
CEN Center for Earth System Research and Sustainability, Institute of Geography, University of Hamburg, Hamburg, Germany
e-mail: udo.schickhoff@uni-hamburg.de

S. Mal (✉)
Jawaharlal Nehru University, Delhi, India
e-mail: surajmal@jnu.ac.in

S. C. Rai (ed.), *Himalaya: Mountains of Destiny*,
https://doi.org/10.1007/978-3-031-94905-0_4

4.1 Introduction

Glacial lakes are water bodies mainly of melted glaciated water deposited in the depressions created by glacial processes. They are a vital component of watersheds in the cryospheric environment of the Indian sub-continent. Several of the glacial lakes in the Himalayan region are a source of the largest rivers in the Indian sub-continent. Ganga and its tributaries receive large amounts of freshwater from these glacial lakes and feed the downstream region before discharging the water into the Bay of Bengal. In the recent past, several studies have highlighted the expansion of glacial lakes in the Hindukush-Karakoram-Himalayas (HKH) (Zhang et al. 2015; Nie et al. 2017; Maharjan et al. 2018; Shugar et al. 2020). Zhang et al. (2015) showed that the number of glacial lakes increased by 23.88% and the area increased of those glacial lakes by 23.19% between 1990 and 2010 in the High Mountain Asia (HMA). Similarly, Nie et al. (2017) found that the number of glacial lakes increased by 8.81% and the area increased by 14.16% between 1990 and 2015 in the Himalayan region. Nie et al. (2013) in another study found that the number of glacial lakes increased by 10.32% and the area increased by 17.10% in the central Himalayan region. The increase in the number and area of glacial lakes not only increases the availability of freshwater resources but also causes potential Glacial Lake Outburst Floods (GLOFs) risk (Maharjan et al. 2018). The upper Ganga river basin, which covers the central Himalayan region, observed the highest expansion of glacial among major river basins in the Indian sub-continent (Zhang et al. 2015; Nie et al. 2017; Li et al. 2022). Zhang et al. (2015) found that the number of glacial lakes increased by 23.8% and the area of glacial lakes increased by 51.1% between 1990 and 2010 in the Ganga basin. Nie et al. (2017) found that glacial lakes are expanding at a rate of 23% in the central Himalayan region. Khadka et al. (2018) found 511 newly emerged glacial lakes between 1987 and 2017 in the Nepal Himalayas. Therefore, it is vital to monitor the evolution of glacial lakes in the upper Ganga basin continuously.

Several studies have been conducted to map the glacial lakes in the upper Ganga basin or parts of it (Nie et al. 2013; Zhang et al. 2015; Khadka et al. 2018; Maharjan et al. 2018; Pandey et al. 2021; Rao et al. 2021; Li et al. 2022). Zhang et al. (2015) mapped 4082 (>0.003 km^2) glacial lakes with a total area of 208.59 km^2 in the Ganga basin using the Landsat (TM & ETM+) missions. Rao et al. (2021) mapped 4707 (> 0.0025 km^2) with a total area of 206.85 km^2 in the Ganga basin using the Resourcesat-2 satellite mission. Pandey et al. (2021) mapped 1353 (0.0005 km^2) glacial lakes with an area of 7.96 km^2 in the Uttarakhand state of India (northern part of the Ganga basin). Khadka et al. (2018) mapped 1541 ($\geq$ 0.0036 km^2) glacial lakes with a total area of 80.95 $\pm$ 15.25 km^2 using the Landsat (MSS, TM & OLI) mission. In previous studies, it has been found that the majority of the glacial lakes are end-moraine type (Nie et al. 2013; Maharjan et al. 2018).

Furthermore, end-moraine-dammed lakes experienced the highest growth rates among different types of glacial lakes in the upper Ganga basin (Nie et al. 2017; Pandey et al. 2021; Li et al. 2022). These lakes are generally connected to the glaciers. The rapid melting of glaciers in the upper Ganga basin is the prime reason for the exponential growth of these glacial lakes (Zhang et al. 2023). End-moraine lakes are dammed by terminal moraine, which are highly unstable. In scenarios of dam failure, end-moraine lakes can cause huge GLOF events in the downstream region (Aggarwal et al. 2017; Mal et al. 2021). Therefore, it becomes more important to continuously monitor the growth of these glacial lakes.

Glacial lake volume is an important lake characteristic that helps in identifying potentially hazardous glacial lakes (Fujita et al. 2013; Mal et al. 2021). However, it is extremely difficult to calculate the volume of each glacial lake because of the limited availability of bathymetric surveys in the upper Ganga basin (Zhang et al. 2023). This is one of the major reasons why none of the existing inventories attempted to calculate or estimate the volume of glacial lakes in the upper Ganga basin. However, there are several existing empirical equations developed by researchers that can estimate the volume and mean depth of glacial lakes (Huggel et al. 2002; Cook and Quincey 2015). These equations can help researchers estimate the volume and mean depth of glacial lakes, which plays a crucial role in identifying potentially hazardous glacial lakes.

Discrepancies in the results of different glacial lake inventories mainly because of different data sources employed and methods used make it extremely complicated to understand the evolution and expansion of glacial lakes between the past and present. Therefore, the main objectives of the present study are: (i) to generate an updated glacial lake inventory for the upper Ganga basin; (ii) to understand the evolution of glacial lakes between 1990 and 2020 at the sub-basin level; and (iii) to generate information about the estimated volume and mean depth of glacial lakes.

4.2 Study Area

The upper Ganga river system is the largest in the central Himalayas. The river system is spread across three countries (China, India, and Nepal), forming a trans-boundary river system. Geographically, the study domain lies between 28°29′36″N and 31°28′12″N (latitudes); and 77°35′27″E and 88°56′36″E (longitudes). The total area of the study domain is 162792.81 km^2, out of which, 59.14% lies in Nepal, followed by China (20.63%), and India (20.22%) (Fig. 4.1). The study domain has 32 sub-basins, out of which the Arun sub-basin (30,144.81 km^2) is the largest.

The physiography of the study domain is mainly dominated by Mountains and valleys. The elevation ranges between 0 and 8848 m asl, including the highest peak in the world (Mt. Everest). The climate of the study domain is dominated by the Indian summer Monsoon, maximum temperature reaches ~ 40 °C (June), whereas in winter temperature drops lower than ~ −10 °C. The study domain receives ~ 80% of its precipitation in July, August, and September (liquid precipitation), whereas ~ 20% is in solid form (winter months) (Krishnan and Indu 2023).

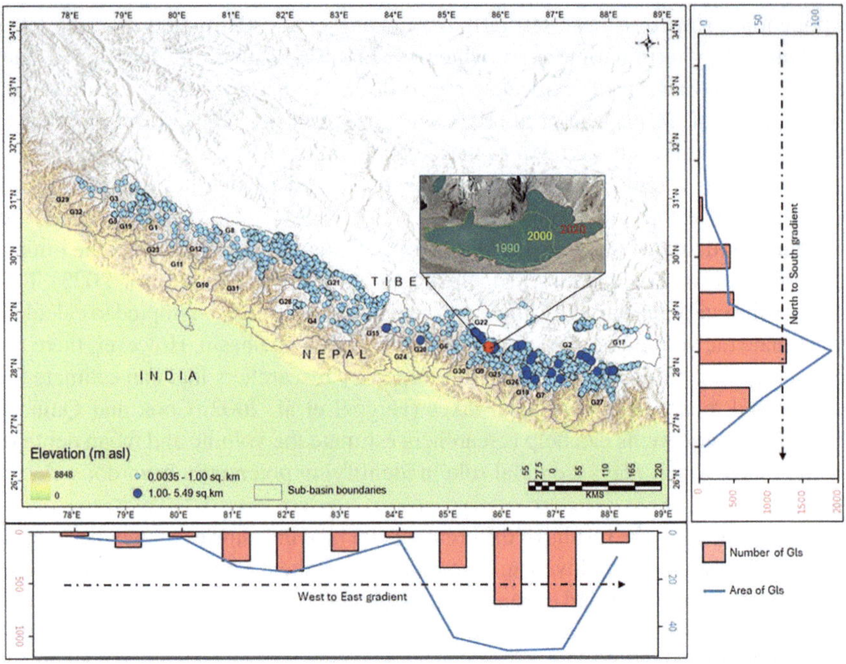

Fig. 4.1 Map of study domain showing the distribution of glacial lakes at the sub-basin level. The left sub-plot show latitudinal distribution of glacial lakes and the lower sub-plot show the longitudinal distribution of glacial lakes. The insert map shows the evolution of the largest glacial lake in the study domain

4.3 Data and Methods

4.3.1 Data

Glacial lakes can be mapped using in situ surveys, toposheets, and remotely sensed aerial/satellite images (Dou et al. 2021). In situ surveys are cost-intensive, time-consuming, and difficult to conduct in unreachable mountainous terrain. In the past decades, due to the recent advancements in satellite-based remote sensing, satellite images have been regarded as the most reliable source to map glacial lakes. Satellite images from several missions (Landsat, IRS, SPOT, Sentinel-1&2) are regularly used to map glacial lakes (Nie et al. 2013; Aggarwal et al. 2017; Mal et al. 2020). In the present study, we used satellite images from the Landsat mission (Landsat 5 & 8) to map the glacial lakes for 1990, 2000, 2010, and 2020 (https://earthexplorer.usgs.gov/). Mainly because it allows us to monitor the continuous growth of glacial lakes since the early 1990s and they are easy to acquire. The majority of the images were collected between September and December, because of the minimum presence of snow and cloud cover in the study domain (Table 4.1). In addition,

Table 4.1 Details of the satellite images used in the study

Year	Satellite mission	Sensor	Spatial resolution (m)	Number of satellite images used for individual inventory
1990	Landsat 5	Thematic mapper (TM)	30	148
2000	Landsat 5	Thematic mapper (TM)	30	138
2010	Landsat 5 & 8	Thematic mapper (TM) and Operational land imager (OLI)	30	88
2020	Landsat 8	Operational land imager (OLI)	30	109

we also used the Shuttle Radar Topography Mission (SRTM) V4 (30 m) Digital Elevation Model (DEM) (https://earthexplorer.usgs.gov/) for better identification of glacial lakes, using hillside, slope, and elevation information.

4.3.2 Methods

4.3.2.1 Glacial Lake Mapping

The methods used to map the glacial lakes using remotely sensed satellite images can be divided into three categories: (i) manual (Rao et al. 2021), (ii) semi-automated (Mal et al. 2020; Zheng et al. 2021), and (iii) automated (Duo et al. 2021). The manual method yields high accuracy, however, it is time-consuming, cost & labor-intensive, and requires a high level of expertise in the identification of glacial lakes. Automated mapping on the other hand is a cost-effective and time-saving technique, however, if left unsupervised, it can lead to misidentification of glacial lakes (Nie et al. 2017; Mal et al. 2020). The semi-automated method is the middle path between the two (manual and automated). It has the cost and time efficiency of automated mapping and the high accuracy of manual mapping (Mal et al. 2020). Therefore, in the present study, we used a semi-automated mapping technique to generate the glacial lake inventories (Fig. 4.2). We used a binary thresholding method to generate the water mask using the Normalized Difference Water Index (NDWI). NDWI is a band ratio between the Green and Near Infrared (NIR) bands (Huggel et al., 2002) (Eq. 4.1). The values range between −1 and 1. An optimal threshold of NDWI ($\geq$0.2) was selected by the hit and trial method to extract the water bodies.

$$\text{NDWI} = \frac{(\text{Band}_{\text{GREEN}} - \text{Band}_{\text{NIR}})}{(\text{Band}_{\text{GREEN}} + \text{Band}_{\text{NIR}})} \tag{4.1}$$

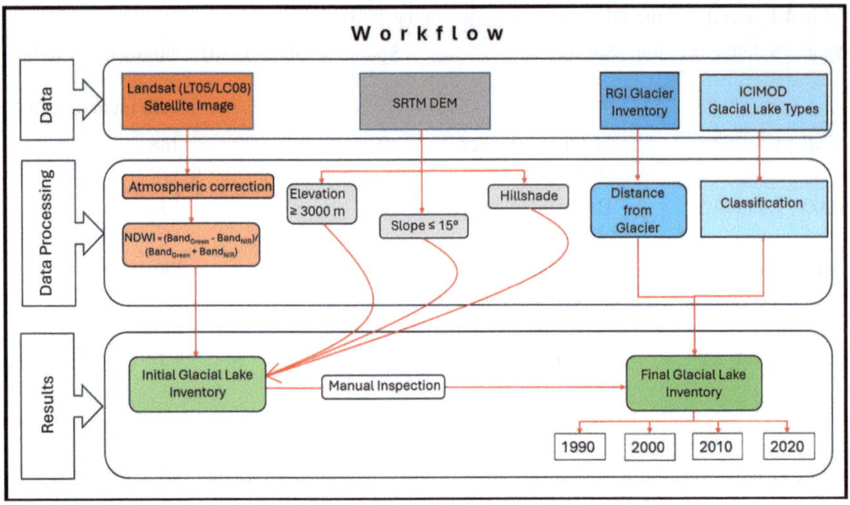

Fig. 4.2 Workflow diagram showing the data sources and methodology used to map the glacial lakes

The next step was to use the elevation threshold of 3000 m asl and slope of less than 15° to identify the potential glacial lakes in the study domain.

After obtaining the initial water layer, next step was to select the minimum mapping area of glacial lakes. We selected the minimum mapping area of 0.0036 km^2. Previous studies showed that, for effective identification of any feature, in our case, glacial lakes, at least 4 pixels of Landsat satellite images (0.0036 km^2) are required for effective identification of glacial lakes (Maharjan et al. 2018; Mal et al. 2020). The last step was to inspect each glacial lake manually using Google Earth Pro for misidentification and final corrections in the glacial lake inventories of 1990, 2000, 2010, and 2020.

4.3.2.2 Glacial Lake Classification

There is no internationally recognized classification available for glacial lakes. However, several studies came up with their classification schemes for glacial lakes (Bhambri et al. 2018; Maharjan et al. 2018). In the present study, we followed the glacial lake classification provided by ICIMOD, because we aim to classify the glacial lakes in the upper Ganga basin based on the dam types. ICIMOD classified glacial lakes based on dam types. Their classification is divided into four major types and eight sub-types (Maharjan et al. 2018).

4.3.2.3 Estimation of Volume and Mean Depth of Glacial Lakes

Information about the volume and mean depth of glacial lakes is of vital importance. The information helps in estimating total water volume, calculating run-off distance, and GLOF modeling (Fujita et al. 2013). However, there is a lack of information about the volume and mean depth of glacial lakes, mainly due to limited bathymetric surveys of glacial lakes in the study domain. Therefore, in the present study, we attempted to estimate the volume and mean depth of glacial lakes based on the empirical equations developed by Huggel et al. (2002) (Eqs. 4.2 and 4.3). The equation was developed by Huggel et al. (2002) using a bathymetric survey of glacial lakes in the Swiss Alps. The equation includes both moraine-dammed and ice-dammed glacial lakes. This is the main reason for selecting the following equations:

$$D = 0.104A^{0.42} \tag{4.2}$$

$$V = 0.104A^{1.42} \tag{4.3}$$

4.4 Results

4.4.1 Spatio-temporal Distribution and Growth of Glacial Lakes at the Sub-basin Level

The total number of glacial lakes in the upper Ganga basin increased was 2554 (a: 170.16 km^2) in 1990, which increased to 2783 (a: 191.04 km^2) in 2000, 2834 (a: 201.14 km^2) in 2010, and 3118 (a: 210.88 km^2) in 2020 (Fig. 4.3). Between 1990 and 2020, total number of glacial lakes increased by 564 (22.08%) and total area increased by 40.72 km^2 (23.93%). In terms of decadal change, 2010–2020 observed the highest increase in total number of glacial lakes of 284 (10.02%), followed by 1990–2000 with an increase of 229 (8.97%), and 2000–2010 observed the lowest increase in the total number of glacial lakes 51 (1.83%). In terms of decadal change in total area of glacial lakes, 1990–2000 observed the maximum expansion of 20.88 km^2 (12.27%), followed by 2000–2010 with an increase of 10.41 km^2 (5.45%), and 2010–2020 observed the minimum expansion of 9.43 km^2 (4.68%) (Table 4.2).

Out of 32 sub-basins, glacial lakes were found in 28 sub-basins in the study domain. In the updated glacial lake inventory of 2020, out of 28 sub-basins with glacial lakes, Arun has the highest number of glacial lakes 740 (23.73%), followed by Humla with 303 (11.45%), Dudh Koshi with 279 (8.95%), Tama Koshi with 251 (8.05%), Mugu with 236 (7.57%), Tamor with 234 (7.50%), Trishuli with175 (5.61%), Sun Koshi with 149 (4.78%), Bheri with 141 (4.52%), Alaknanda with 104 (3.34%), Kali Gandaki with 55 (1.76%), Tila with 51 (1.64%), Bhagirathi with 50

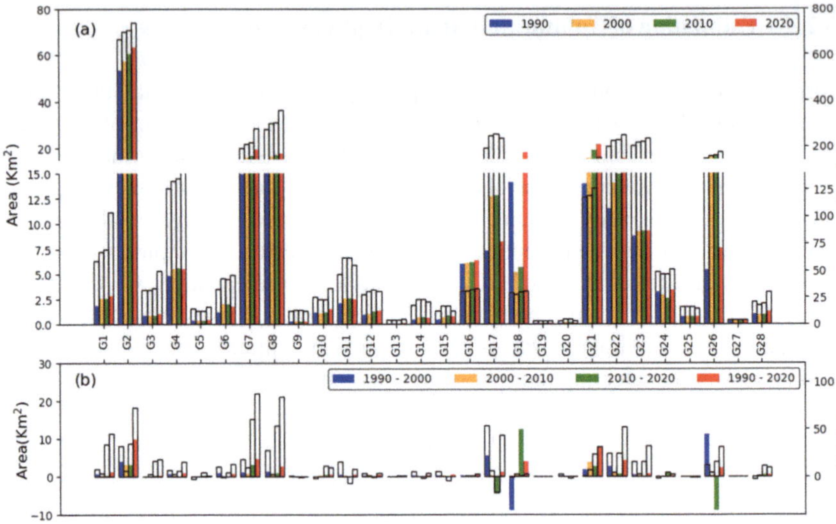

Fig. 4.3 **a** shows total number and area of glacial lakes in individual sub-basins between 1990 and 2020, and **b** shows change in the number and area of glacial lakes in individual sub-basins at the decadal level

Table 4.2 Decadal change in the number and area of glacial lakes

	Number	Area
1990–2000	229 (8.97%)	20.87 (12.27%)
2000–2010	51 (1.83%)	10.41 (10.41%)
2010–2020	284 (10.02%)	9.42 (4.68%)
1990–2020	564 (22.08%)	40.71 (23.93%)

(1.60%), Budi Gandaki with 46 (1.48), Kali with 34 (1.09%), Marshyangdi with 33 (1.06%), Kawari and Pelkhu with 31 (0.99%) each, West Seti with 30 (0.96%), Likhu with 21 (0.67%), Bhilanga with 16 (0.51%), Tons with 15 (0.48%), Indrawati and Mandakini with 12 (0.38%) each, Koshi1 with 5 (0.16%), Trunk Karnali 4 (0.13%), and finally Pindar and Seti have 3 (0.10%) each. However, between 1990 and 2020, the Alaknanda sub-basin (76.27) had the highest growth rate of numbers, and the lowest was in the Tons sub-basin (−6.25). Out of 28 sub-basins with glacial lakes, 22 sub-basins observed a positive growth rate, 5 observed no changes in the number, and 1 sub-basin observed a negative growth rate (Fig. 4.3).

In 2020, Arun have the maximum area of glacial lakes 60.38 km^2 (30%), followed by Sun Koshi with 21.70 km^2 (10.29%), Dudh Koshi with 19.49 km^2 (9.24%), Pelkhu with 17.94 km^2 (8.51%), Humla with 17.60 km^2 (8.35%), Tama Koshi with 15.52 km^2 (7.36%), Tamor with 9.22 km^2 (4.37%), Mugu with 8.20 km^2 (3.89%), Trishuli with 7.53 km^2 (3.57%), Marshyangdi with 6.28 km^2 (2.98%), Bheri with 5.53 km^2 (2.62%), Tila with 3.33 km^2 (1.58%), Alaknanda with 2.81 km^2 (1.33%), Kali Gandaki with 2.45 km^2 (1.16%), Budi Gandaki with 1.76 km^2 (0.83%), Kali with 1.45 km^2 (0.69%), Kawari with 1.29 km^2 (0.69%), West Seti with 1.27 (0.60%), Bhagirathi with 0.99 (0.47%), Tons with 0.75 (0.35%), Mandakini with 0.73 (0.35%), Likhu with (0.29%), Bhilanga with 0.39 (0.19%), Trunk Karnali with 0.36 (0.17%), Indrawati with 0.22 (0.11%), Pindar with 0.08 (0.04%), Koshi1 with 0.06 (0.03%), and Seti with 0.05 (0.03%). Between 1990 and 2020, the Mandakini sub-basins (73.28%) had the highest growth rate of area, and the lowest was in the Tons sub-basins (−1.19). Out of 28 sub-basins, 23 sub-basins observed a positive growth rate in the area of glacial lakes, 4 sub-basins observed no change, and 1 sub-basin observed a negative change in the area of glacial lakes (Fig. 4.3).

4.4.2 Estimated Mean Depth and Volume of Glacial Lakes

For a better understanding of glacial lake dynamics, we estimated the mean depth and volume of glacial lakes using Huggel et al. (2002) empirical equations. The equations are based on the area of glacial lakes. We estimated the average mean depth of glacial lakes was 8.45 m (1990), 8.47 m (2000), 8.49 m (2010), and 8.26 m (2020). Between 1990 and 2020, the average estimated mean depth of glacial lakes decreased by −0.19 m. At the decadal level, estimated mean depth of glacial lakes increased between 1990 and 2000 by 0.02 m. Similarly, 2000 and 2010 show an increase in the average estimated mean depth by 0.02 m. However, the average estimated mean depth reduced by −0.23 between 2010 and 2020 (Fig. 4.4). At the sub-basin level, for the updated glacial lake inventory of 2020, Pelkhu had the highest average estimated mean depth of glacial lakes (18.51 m), and Koshi1 had the lowest of 5.18 m. Between 1990 and 2020, 13 sub-basins showed an increase in the estimated average mean depth, and 15 sub-basins showed a decrease in the estimated average mean depth (Fig. 4.4).

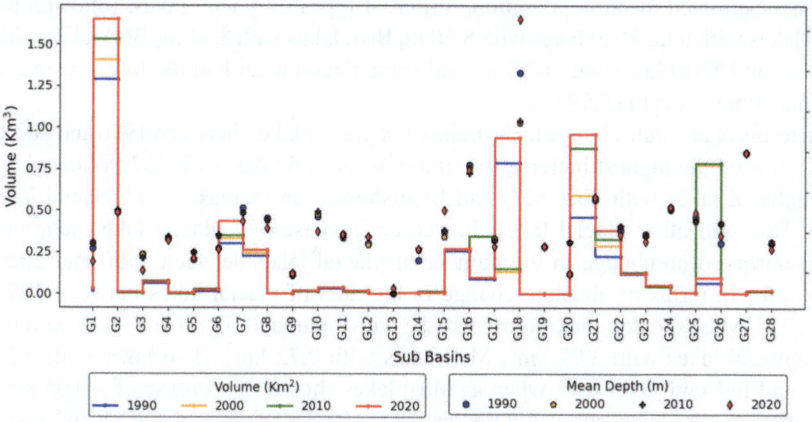

Fig. 4.4 Sub-basin-wise distribution of estimated volume (km^3) and average estimated mean depth (m) of glacial lakes

The total estimated volume of glacial lakes was 4.07 km^3 (1990), followed by 4.8 km^3 (2000), 5.33 km^3 (2010), and 5.54 km^3 (2020). Between 1990 and 2020, the total estimated volume of glacial lakes increased by 1.47 km^3. At the decadal level estimated volume of glacial lakes increased by 0.73 km^3 (1990–2000), 0.53 km^3 (2000–2010), and 0.21 km^3 (2010–2020). At the sub-basin level, in 2020, the Arun sub-basin (1.65 km^3) had the maximum estimated volume and Koshi 1 (0.00035 km^3) had the minimum estimated volume of glacial lakes among the sub-basins (Fig. 4.4).

4.4.3 Typewise Changes in the Glacial Lakes

Based on the ICIMOD's glacial lake classification, in 2020, we were able to find M(e), M(l), M(o), B(c), B(o), supraglacial lakes, and other glacial lakes in the upper Ganga basin. Out of 3118 glacial lakes, 2413 (77.39%) were M(e), followed by supraglacial lakes with 268 (8.6%), B(o) lakes with 227 (7.28%), M(l) with 158 (5.07%), other glacial lakes with 21 (0.67%), M(o) lakes with 19 (0.57%), and B(C) lakes only had 12 (0.35%). In terms of the area, out of 210.88 km^2, M(e) lakes had the highest proportion of 178 km^2 (84.83%), followed by M(l) lakes with 13.42 km^2 (6.36%), B(o) lakes with 11.22 km^2 (5.32%), supraglacial lakes with 4.88 km^2 (2.31%), other glacial lakes with 1.55 km^2 (0.73%), and M(o) lakes with 0.57 km^2 (0.27%), and B(c) lakes only had glacial lake area of 0.35 km^2 (0.16%). In terms of the total estimated volume, out of 5.548 km^3, M(e) had the maximum total estimated volume of 4.986 km^3 (89.86%), followed by M(l) lakes with 0.325 km^3 (5.86%), B(o) lakes with 0.155 km^3 (2.79%), supraglacial lakes with 0.048 km^3 (0.86%), other glacial lakes with 0.024 km^3 (0.43%), and M(o) with 0.006 km^3 (0.10%), and B(c) had the lowest estimated volume of glacial lakes with only 0.003 km^3 (0.05%). In terms of the estimated mean depth, other glacial lakes (9.74 m) had the highest average estimated mean depth among different types of glacial lakes, followed by M(l) lakes with 9 m, M(e) lakes with 8.50 m, B(o) lakes with 8.41 m, B(c) lakes with 7.18 m, and M(o) lakes with 6.96 m, and supraglacial lakes had the lowest average estimated mean depth (5.59 m).

In terms of absolute change in the number of glacial lakes, between 1990 and 2020, M(e) showed the highest increase in number of glacial lakes with 332, followed by supraglacial lakes with 208, M(l) and B(o) showed an increase of 11 glacial lake each, B(c) and other glacial lakes showed an increase of 1 glacial lake each, and M(o) observed no change in the number of glacial lakes between 1990 and 2020 (Fig. 4.5). In terms of absolute change in the area of glacial lakes between 1990 and 2020, M(e) showed maximum increase with an area of 35.80 km^2, followed by supraglacial lakes with 3.97 km^2, M(l) lakes with 0.72 km^2, B(o) lakes with 0.28 km^2, and B© with 0.03 km^2, whereas M(o) lakes showed a decrease of −0.05 km^2 and other glacial lakes showed a decrease of −0.03 km^2 area of the glacial lakes (Fig. 4.5). In terms of absolute change in the volume of glacial lakes between 1990 and 2020, M(e) lakes showed an increase of 1.407 km^3, followed by supraglacial lakes with 0.038 km^3, M(l) lakes with 0.0203 km^3, B(o) lakes with 0.0025 km^3, and B(c) lakes with 0.00019 km^3, whereas M(o) and other glacial lakes observed a decrease of -0.0013 km^3 and -0.0004 km^3, respectively (Fig. 4.5). In terms of changes in the average estimated mean depth between 1990 and 2020, supraglacial lakes showed the highest increase of 0.51 m, followed by M(o) with 0.04 m, and B(c) with 0.01 m, whereas four sub-basins showed a decrease in the average estimated mean depth. Other glacial lakes shown a decline of -0.45 m, followed by M(l) with $-$0.05 m, B(o) with -0.08 m, and M(e) with -0.02 m (Fig. 4.5). To further understand the typewise changes in the distribution of glacial lakes, we analyzed each sub-type at the sub-basin level.

4.4.4 End Moraine-Dammed Lakes (M(e))

M(e) had the largest share among distinct types of glacial lakes. In 1990 there were 2081 M(e) lakes, which increased to 2263 (2000), 2293 (2010), and 2413 (2020). At the sub-basin level in the out of 28 sub-basins, M(e) lakes were present in 27 sub-basins between 1990 and 2020. Arun sub-basin had the highest number of M(e) lakes across the decades. There were 594 M(e) lakes in 1990, which increased to 624 (2000), 632 (2010), and 635 (2020). Seti sub-basin had the lowest number of M(e) lakes across the decades. There were only 2 M(e) lakes in 1990, 3 (2000), 3 (2000), and 2 (2020). Between 1990 and 2020, 22 sub-basins showed an increase in the number of M(e) lakes, 4 sub-basins showed no change, and 1 sub-basin showed a decrease in the number of M(e) lakes (Fig. 4.5).

The total area of M(e) lakes increased from 143.09 km^2 (1990), 162.55 km^2 (2000), 172.38 km^2 (2010), and 178.89 km^2 (2020). Arun sub-basin had the maximum area of M(e) lakes across the decades. The total area of M(e) lakes was 49.92 km^2 (1990), followed by 53.63 km^2 (2000), 56.53 km^2 (2010), and 58.52 km^2 (2020). Seti had the lowest total area among the M(e) lakes across the decades. In 1990 the total area of M(e) lakes in Seti sub-basin was 0.016 km^2 (1990), followed by 0.028 km^2 (2000). Between 1990 and 2020, 25 sub-basins showed an increase in the area of M(e) lakes, 1 sub-basin showed no change, and 1 sub-basin showed a decrease in the area of M(e) lakes (Fig. 4.5).

4.4.5 Supraglacial Lakes

In 2020, supraglacial lakes (268) had the second-highest number of glacial lakes. In 1990, there were 60 supraglacial lakes, followed by 91 (2000), 111 (2010), and 268 (2020). Out of 28 sub-basins with glacial lakes, supraglacial lakes were present in only 8 (1990) sub-basins, which increased to 11 (2000), 13 (2010), and 15 (2020). Among the sub-basins, Dudh Koshi had the highest number of supraglacial lakes across the decades. There were 21 supraglacial lakes in Dudh Koshi sub-basin in 1990, which increased to 35 (2000), 43 (2010), and 86 (2020). Between 1990 and

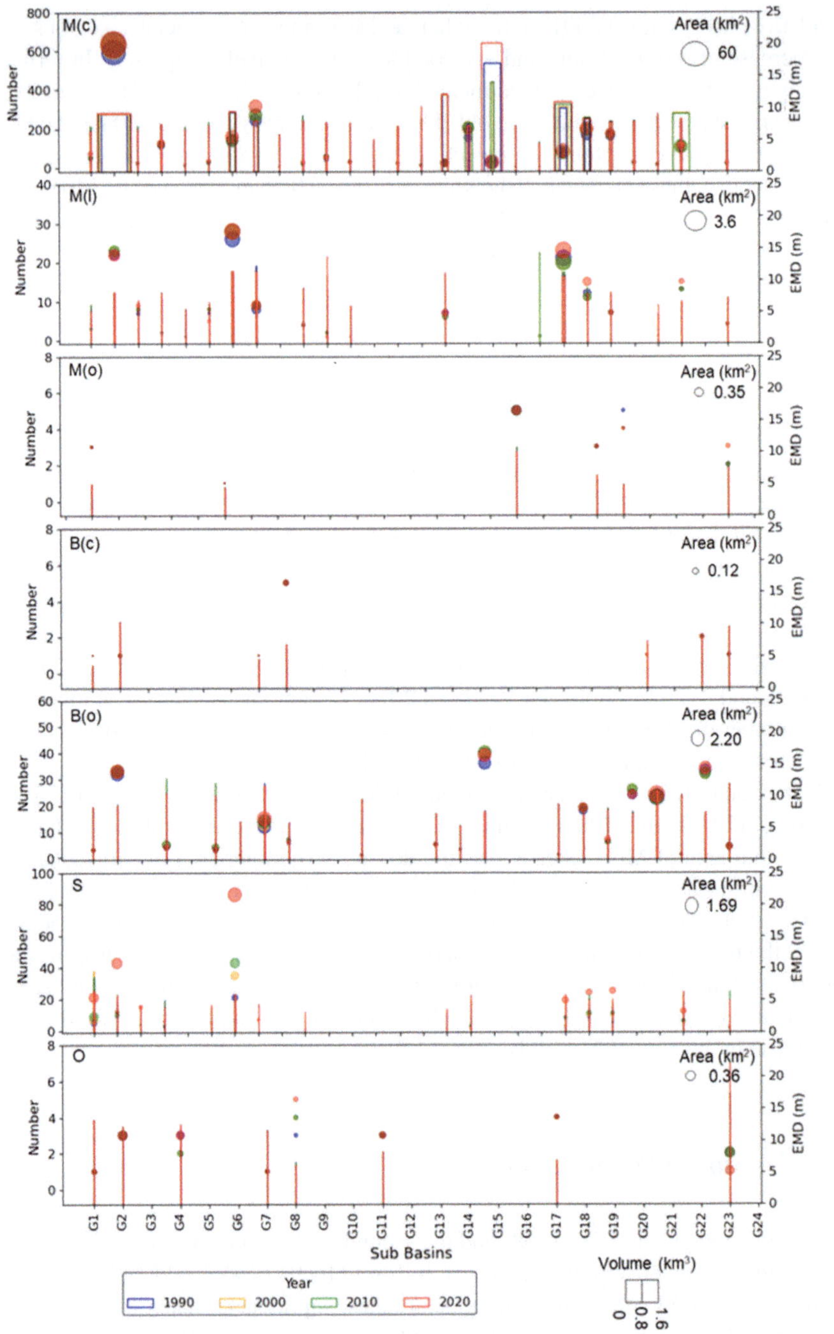

Fig. 4.5 Typewise changes in the glacial lakes in terms of number, area, estimated volume, and average estimated mean depth

2020, 15 sub-basins showed a positive change in the number of glacial lakes, whereas no sub-basins showed a negative change in the number of supraglacial lakes (Fig. 4.5).

In terms of the share in the total area of glacial lakes, supraglacial lakes had 4th largest share. The total area of supraglacial lakes was only 0.91 km^2 in 1990, which increased to 1.81 km^2 (2000), 2.17 km^2 (2010), and 4.88 km^2 (2020). In terms of the sub-basins, Dudh Koshi had the largest share of the total area of supraglacial lakes. In 1990 the area of supraglacial lakes in Dudh Koshi was 0.32 km^2, which increased to 0.62 km^2 (2000), 0.75 km^2 (2010), and 1.63 km^2 (2020). Dudh Koshi sub-basins showed an increase of 1.30 km^2 between 1990 and 2020. Similar to the changes in the number of glacial lakes, all 15 sub-basins showed an increase in the area of supraglacial lakes (Fig. 4.5).

4.4.6 Other Bedrock Dammed Lakes (B(o))

In 2020, there were 227 B(o) lakes, which is the third highest among different types of glacial lakes in the upper Ganga basin. In 1990, there were 216 B(o) lakes, followed by 227 (2000), 228 (2010), and 227 (2020). Between 1990 and 2020, B(o) lakes increased by 11. However, at the decadal time frame, B(o) increased by 11 (1990–2000) and 1 (2000–2010) and decreased by 1 (2010–2020). Out of 28 sub-basins, B(o) lakes were present in 19 sub-basins, across the decades. Among these 19 sub-basins, Mugu had the highest number of B(o) lakes. There were 36 B(o) lakes in Mugu in 1990, which increased to 40 in 2000 and remained the same in 2010, and in 2020 it reduced to 39 (Fig. 4.5). Between 1990 and 2020, out of 19 sub-basins with B(o) lakes, only 7 shown a positive change in the number of glacial lakes, whereas 12 shown no change in the number of B(o) lakes (Fig. 4.5).

In terms of the area, B(o) lakes had the third highest share in the total glacial lake area. In 1990, the total area of B(o) lakes was 10.94 km^2, followed by 11.20 km^2 (2000), and 11.22 km^2 (2010 and 2020). Between 1990 and 2020, the total area of B(o) lakes increased by 0.28 km^2. In terms of the area of B(o) lakes, the Tila sub-basin had the highest area of B(o) lakes. In 1990, the total area of B(o) lakes in the Tila basin was 2.105 km^2, followed by 1.832 km^2 (2000 and 2010), and 2.197 km^2 (2020). The total area of B(o) lakes increased by 0.091 km^2 between 1990 and 2020 in the Tila sub-basin. However, between 1990 and 2000, the Tila sub-basin showed a decrease of the area by −0.27 km^2, whereas there was no change in the area of B(o) lakes in the Tila sub-basin between 2000 and 2010, and the sub-basin showed maximum increase in the area of B(o) lakes between 2010 and 2020 (0.36 km^2). Between 1990 and 2020, out of 19 sub-basins with B(o) lakes, 11 showed an increase in the area of B(o) lakes, whereas 7 sub-basins showed a decrease in the area of B(o) lakes, and 1 sub-basin shown no change in the area of B(o) lakes (Fig. 4.5).

4.4.7 Lateral Moraine-Dammed Lakes (M(l))

In 2020, M(l) had the fourth highest number of glacial lakes among different types of glacial lakes with 158. There were 147 M(l) lakes in 1990, which increased to 152 (2000 and 2010), and 158 (2020). Between 1990 and 2020, M(l) lakes increased by 11 glacial lakes. At the decadal level, M(l) lakes increased by 5 between 1990 and 2000 and increased by 6 between 2010 and 2020, whereas there was no change between 2000 and 2010. At the sub-basin level, out of 28 sub-basins, M(l) lakes were present in 17 sub-basins in 1990, which increased to 19 (2000 and 2010), and 18 in 2020. Among the sub-basins, Dudh Koshi had the highest number of M(l) lakes across the decades. There were 26 M(l) lakes in 1990, which increased to 28 in 2000, 2010, and 2020, respectively. Between 1990 and 2020, out of 18 sub-basins with M(l) lakes, 8 sub-basins showed an increase in the number of glacial lakes, 2 sub-basins showed a decrease, and 8 sub-basin showed no change in number of glacial lakes (Fig. 4.5).

In terms of the area, M(l) lakes had 2nd largest share in the total area of glacial lakes. In 1990, the total area of M(l) lakes was 12.69 km^2, which increased to 13.01 km^2 (2000), 13.11 km^2 (2010), and 13.41 km^2 (2020). Between 1990 and 2020, the total area of M(l) lakes increased by 0.72 km^2. In terms of decadal change, 1990–2000 showed the highest increase in the area of M(l) lakes (0.32 km^2), followed by 2010–2020 (0.29 km^2), and the lowest increase was shown in 2000–2010 (0.09 km^2). In terms of the sub-basins, Dudh Koshi sub-basins had the maximum share in the total area of M(l) lakes across the glacial lake inventories. In 1990, the area of M(l) lakes in the Dudh Koshi sub-basin was 3.582 km^2, which decreased to 3.542 (2000 and 2010), and again increased to 3.584 km^2 (2020). Between 1990 and 2020, at the sub-basin level, out of 18 sub-basins with M(l) lakes, 12 showed an increase in the absolute area of M(l) lakes, 5 showed negative change, and 1 showed no change in the area of M(l) lakes (Fig. 4.5).

4.4.8 Other Glacial Lakes

In terms of the number of glacial lakes, other glacial lakes are 5th largest among the sub-types. There were 20 other glacial lakes in 1990, 2000, and 2010, respectively, which increased to 21 in 2020. Between 1990 and 2020, other glacial lakes increased by 1. At the sub-basin level, out of 28 sub-basins, other glacial lakes were present in 8 sub-basins across the glacial lake inventories of 1990, 2000, 2010, and 2020. Out of 8 sub-basins, Mugu (4) had the highest number of other glacial lakes in 1990. For the years 2000 and 2010, Mugu (4) and Humla (4) shared the highest number of other glacial lakes. In 2020, Humla (5) had the highest number of other glacial lakes. Between 1990 and 2020, out of 8 sub-basins, only 1 sub-basin showed a positive change in the number of another glacial lake, 1 sub-basin showed a negative change in the number of other glacial lakes, and 6 sub-basins showed no change in the number of other glacial lakes (Fig. 4.5).

Other glacial lakes rank 5th in the typewise distribution of glacial lake area. In 1990 the total area of other glacial lakes was 1.577 km^2, which reduced to 1.484 km^2 (2000 and 2010) and 1.546 km^2 (2020). Between 1990 and 2020, the total area of other glacial lakes reduced by −0.03 km^2. In terms of decadal change in the area of other glacial lakes, there was a negative change of −0.03 km^2 (1990–2000), no change in the area (2000–2010), and a positive change of 0.06 km^2 (2010–2020). Out of the 8 sub-basins with other glacial lakes, the Tamor sub-basins had the largest share in the area of other glacial lakes. In 1990, 2000, and 2010, the area of other glacial lakes in the Tamor sub-basin was 0.438 km^2, which reduced to 0.369 km^2 in 2020. Between 1990 and 2020, 4 sub-basins showed an increase, and 3 showed a decrease in the area of other glacial lakes (Fig. 4.5).

4.4.9 Other Moraine-Dammed Lakes (M(o))

According to the updated glacial lake inventory of glacial lake (2020), M(o) lakes rank 6th in terms of the number of different types of glacial lakes. In 1990, there were 19 M(o) lakes, which reduced to 18 (2000 and 2010) and again increased to 19 (2020). Between 1990 and 2020, M(o) lakes showed no change in the number of glacial lakes. However, between 1990 and 2000, M(o) lakes showed a decrease of 1 glacial lake, and between 2010 and 2020, M(o) lakes showed an increase of 1 glacial lake. Out of 28 sub-basins with glacial lakes, M(o) lakes were present in 6 sub-basins across the decades. Out of 6 sub-basins with M(o) lakes, in 1990 Pelkhu (5) and Tama Koshi (5) had the highest number of M(o) lakes, whereas for 2000, 2010, and 2020, Tama Koshi (5) had the highest number of M(o) lakes. Between 1990 and 2020, out of 8 sub-basins, 1 sub-basin showed positive change, 1 sub-basin showed negative change, and 6 sub-basins showed no change in the number of glacial lakes (Fig. 4.5).

In terms of the share in the total area of glacial lakes, M(o) lakes rank 6th. In 1990, the total area of M(o) lakes was 0.622 km^2, which reduced to 0.623 km^2 (2000 and 2010) and 0.574 km^2 in 2020. Between 1990 and 2020, the area of M(o) lakes was reduced by −0.047 km^2. At the decadal level, the area of M(o) lakes increased by 0.00089 km^2 (1990–200), showing no change in the area of M(o) lakes between 2000 and 2010, and the area of M(o) lakes reduced by −0.048 km^2 between 2010 and 2020. In terms of the area of M(o) lakes at the sub-basin level, Pelkhu had the maximum area of 0.404 km^2 (1990, 2000, and 2010) and 0.337 km^2 in 2020. Between 1990 and 2020, out of 8 sub-basins, 3 sub-basins showed an increase in the area of M(o) lakes, and 3 sub-basins showed a decrease in the area of M(o) lakes (Fig. 4.5).

4.4.9.1 Cirque Lakes (B(c))

B(c) had the lowest number of glacial lakes among different types of glacial lakes, across the glacial lake inventories of 1990, 2000, 2010, and 2020. In 1990, there were only 11 B(c) lakes, which increased to 12 (2000, 2010, and 2020). Between 1990 and 2020, total number of B(c) lakes increased by 1. At the decadal level, B(c) lakes increased by 1 between 1990 and 2000, whereas no change was shown in the number of B(c) between (2000 and 2010) and (2010 and 2020). Out of 28 sub-basins with glacial lakes, B(c) lakes were present in 6 sub-basins (1990) and 7 sub-basins (2000, 2010 and 2020). Among the 7 sub-basins with B(c) lakes, the Humla sub-basin had the highest number of B(c) lakes (5) in 1990, 2000, 2010, and 2020, respectively. Between 1990 and 2020, only 1 sub-basin (Sun Koshi) showed a positive change in the number of glacial lakes, whereas the other 6 sub-basins showed no change in the number of B(c) lakes.

B(c) lakes had the lowest area of glacial lakes and ranked 7th among the area of different types of glacial lakes, across the glacial lake inventories of 1990, 2000, 2010, and 2020. In 1990, the total area of B(c) lakes was 0.31 km^2, followed by 0.341 km^2 (2000), 0.341 km^2 (2010), and 0.345 km^2 (2020). Between 1990 and 2020, the total area of B(c) lakes increased by 0.025 km^2. Over the decades, the area of B(c) lakes increased by 0.021 km^2 (1990–2000), with no change in the area of B(o) lakes (2000–2010), and 0.0036 km^2 (2010–2020). At the sub-basin level, out of 7 sub-basins, the Humla sub-basin had the largest share in the total area of B(c) lakes with an area of 0.129 km^2 across the glacial lake inventories of 1990, 2000, 2010, and 2020. Between 1990 and 2020, out of 7 sub-basins with B(c) lakes, 2 showed an increase, 2 showed a decrease, and 3 sub-basins showed no change in the area of B(c) lakes in the upper Ganga basin (Fig. 4.5).

4.4.9.2 Altitudinal Distribution of Glacial Lakes

Altitude plays a crucial role in identifying glacial lakes in the upper Ganga basin. The lowest altitude of glacial lakes was 3331 m asl, and the highest altitude was 6216 m asl. The mean elevation of glacial lakes in the upper Ganga basin was 5045 m asl (1990), which increased to 5053.37 m asl (2000) and reduced to 5053 for 2010 and 2020, respectively. To understand the altitudinal distribution of glacial lakes, we divided glacial lakes into 16 zones based on the elevation of the glacial lakes (Fig. 4.6). The elevation zone of 5200–5400 m had the largest share in the number of glacial lakes across the glacial lake inventories. In 1990 the elevation zone had 461 (18.05%) glacial lakes, which increased to 523 (18.79%) in 2000, 529 (18.66%) in 2010, and 585 (18.76%) (Fig. 4.6). However, between 1990 and 2020, the elevation zone of 5000–5200 m shown the highest increase in the number of glacial lakes (131). At the decadal level, between 1990 and 2000, the elevation zone 5200–5400 m showed the highest increase in the number of glacial lakes (62). Between 2000 and 2010, the elevation zone of 5000–5200 m showed the highest increase in the number of glacial

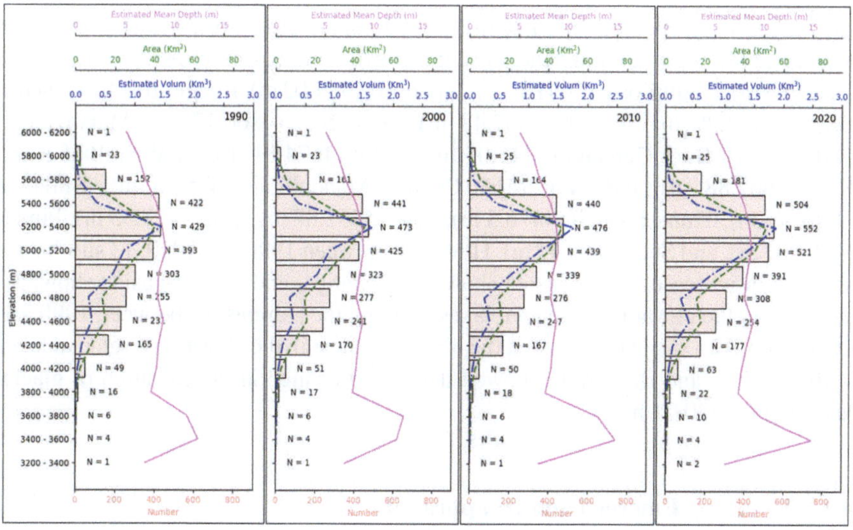

Fig. 4.6 Altitudinal changes in the number, area, estimated volume, and average estimated mean depth of glacial lakes

lakes (19). And between 2010 and 2020, the elevation zone 5000–5200 m showed the highest increase in glacial lakes (66).

In terms of the area of glacial lakes, the elevation zone of 5200–5400 m had the largest share of the total area of glacial lakes across the glacial lake inventories. In 1990, the total area of glacial lakes in the elevation zone of 5200–5400 m was 41.72 km^2 (24.52%), which increased to 47.25 km^2 (24.73%) in 2000, 50.05 km^2 (24.89%) in 2010, and 52.60 km^2 (24.94%) in 2020. However, between 1990 and 2020, the elevational zone of 5000–5200 m asl showed highest absolute change in the total area of glacial lakes with an increase of 12.58 km^2 (Fig. 4.6).

In terms of the estimated volume of glacial lakes, the elevation zone of 5200–5400 m had the largest share of the total area of glacial lakes across the glacial lake inventories. In 1990, the total area of glacial lakes in the elevation zone of 5200–5400 m was 1.47 km^3 (36.26%), which increased to 1.67 km^3 (34.77%) in 2000, 1.82 km^3 (34.17%) in 2010, and 1.89 km^3 (34.11%) in 2020. Between 1990 and 2020, the altitude zone of 5000–5200 m shows the highest absolute change in the total estimated volume of glacial lakes with an increase of 0.576 km^3 (Fig. 4.6).

In terms of the average estimated mean depth of glacial lakes based on the elevation of the glacial lakes, the altitude zone of 3400–3600 m had the highest average estimated mean depth in 1990 (12.48 m), 2010 (14.84 m), and 2020 (14.89 m). However, in the year 2000 the altitude zone of 3600–3800 m had the highest estimated mean depth of 13.11 m. Between 1990 and 2020, the altitude of 3400–3600 m showed the highest increase in the average estimated mean depth of 2.44 m (Fig. 4.6).

In terms of the elevation of the glacial lakes at the sub-basin level, in 2020, KaliGandaki (5580 m) had the highest mean elevation of glacial lakes, followed by

Koshi1 (5558 m), Pelkhu (5406.26 m), Tama Koshi (5219.21 m), Arun (5185.73 m), Humla (5119.94 m), Alaknanda (5105.09 m), Sun Koshi (5085.57 m), Mugu (5053.98 m), Dudh Koshi (5041.61 m), Trishuli (4984.02 m), Bhagirathi (4983.94 m), Bheri (4977.29 m), Tamor (4807.48 m), West Seti (4806.17 m), Marshyangdi (4767.85 m), Budi Gandaki (4745.07 m), Likhu (4674.81 m), Kawari (4665.9 m), Kali (4662.88 m), Indrawati (4630.08 m), Mandakini (4575.25 m), Bhilanga (4509.75 m), Trunk Karnali (4476.5 m), Tons (4452.87 m), Tila (4417.55 m), Pindar (4404.67 m), and Seti (4116.33 m). However, between 1990 and 2020, 13 sub-basins showed an increase, 4 sub-basins showed no change, and 11 sub-basins showed a decrease in the elevation of glacial lakes. Among the sub-basins, between 1990 and 2020, the Likhu sub-basin showed the highest change in the elevation of glacial lakes with 62.97 m, and Bhagirathi showed the lowest change in the elevation of glacial lakes with −220.71 m.

4.4.9.3 Newly Emerged and Disappeared

New emerged and disappeared glacial lakes play a vital role in the growth of glacial lakes in the upper Ganga basin. Between 1990 and 2020, 651 new glacial lakes emerged, which in 2020 had a total glacial lake area of 17.10 km^2, an estimated volume of 0.247 km^3, average estimated mean depth of 6.15 m, and mean elevation of 5095.84 m. At the decadal level, 2010 and 2020 showed the highest increase in the number (443) of glacial lakes, followed by 1990–2000 (238) and 2000–2010 (66) (Fig. 4.7). In terms of the sub-basins, 22 sub-basins shown the appearance of emerged glacial lakes, and 6 sub-basins reported no appearance of emerged glacial lakes between 1990 and 2020. Out of the 22 sub-basins that showed the emergence of new glacial lakes, the Arun sub-basin had the highest number of newly emerged glacial lakes (96) (Fig. 4.7). However, in terms of the area of newly emerged glacial lakes at the sub-basin level, Pelkhu sub-basin had the highest total area of emerged glacial lakes (2.89 km^2) (Fig. 4.7).

In terms of the disappeared glacial lakes, between 1990 and 2020, 72 glacial lakes disappeared, with a total area of 1.02 km^2, total estimated volume of $2.467*10^{-11}$ km^3, average estimated mean depth of 5.16 m, and mean elevation of 5128.23 m. At the decadal level, in 2010–2020, the highest number of glacial lakes disappeared with 71 glacial lakes, followed by 1990–2000 (67) and 2000–2010 (14) (Fig. 4.7). Out of 28 sub-basins, 14 sub-basins observed disappearance and 14 sub-basins observed no disappearance of glacial lakes between 1990 and 2020. Among the 14 sub-basins, that showed the disappearance of glacial lakes, Arun had the highest number of disappeared glacial lakes (17) and disappeared areas of glacial lakes (0.32 km^2) (Fig. 4.7).

Finally, among different types of glacial lakes, M(e) lakes had the highest number of emerged lakes (n: 375, a: 12.38 km^2), followed by supraglacial lakes (n: 239, a: 3.79 km^2), M(l) lakes (n: 19, a: 0.44 km^2), B(o) lakes (n: 12, a: 0.38 km^2), other glacial lakes (n: 3, a: 0.03 km^2), M(o) lakes (n: 2, a: 0.032 km^2), and B(c) lakes had the lowest number of emerged lakes (n: 1, a: 0.032 km^2). In terms of the different

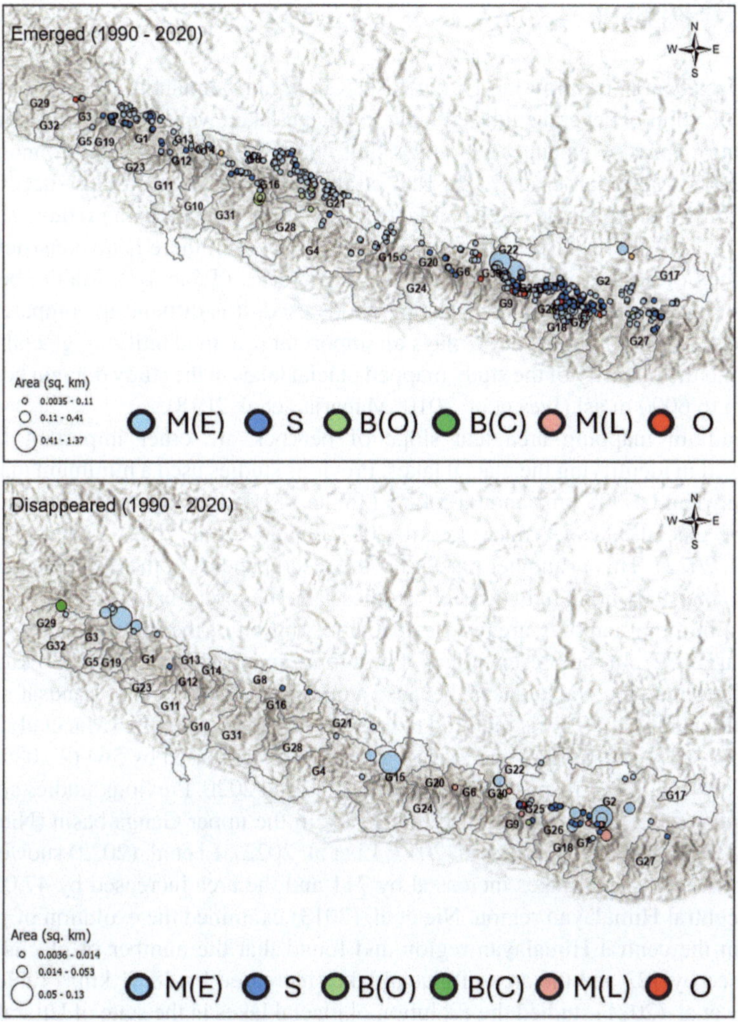

Fig. 4.7 Emerged and disappeared glacial lakes between 1990 and 2020. Different colors show different types of emerged and disappeared glacial lakes. The size of markers indicates the area of emerged and disappeared glacial lakes

types of disappeared glacial lakes, M(e) lakes had the highest number of disappeared lakes (n: 33, a: 0.679 km^2), followed by supraglacial lakes (n: 29, a: 0.221 km^2), M(l) lakes (n: 7, a: 0.107 km^2), M(o) lakes (n: 2, a: 0.013 km^2), and other glacial lakes (n: 1, a: 0.005 km^2), and B(c) and B(o) lakes observed no disappearance between 1990 and 2020 (Fig. 4.7).

4.5 Discussion

The upper Ganga basin in the HinduKush-Karakoram-Himalayas (HKH) has been the focus point of several studies related to glacial lake inventories, highlighting the changing nature of glacial lakes and rising GLOF risk in the study domain (Nie et al. 2013; Maharjan et al. 2018; Rao et al. 2021). Previous studies successfully highlighted the expansion of glacial lakes in the study domain using remote sensing data (Nie et al. 2013, 2017; Zhang et al. 2015). However, there is no consistency in the results of already existing glacial lake inventories (Table 4.3). Mainly because of differences in the data sources and methods used, it is difficult to compare these inventories. Minimum elevation plays an important role in identifying glacial lakes. We found the majority of the study mapped glacial lakes in the study domain between 2400 m to 6000 m asl (Ives et al. 2010; Maharjan et al. 2018).

Minimum mapping area and slope of bedrock are other important factors employed in identifying the glacial lakes. Previous studies used a minimum mapping area between 0.0005 km^2 and 0.05 km^2 (Table 4.3) and a slope of less than 20° to map the glacial lakes (Aggarwal et al. 2017; Pandey et al. 2021; Rao et al. 2021; Li et al. 2022). This is another reason for the discrepancies in the number of glacial lakes in the study domain in different studies. The main reason for the discrepancies in the minimum mapping area is the spatial resolution of the different data sources (Li et al. 2022). In the present study, a minimum mapping area of 0.0036 km^2 was selected to map the glacial lakes, because a minimum of 4 pixels of Landsat images (spatial resolution: 30 m) is required to correctly identify a feature (Mal et al. 2020).

In our study, the number and area of glacial lakes increased by 564 (22.08%) and 40.71 km^2 (23.93%), respectively, between 1990 and 2020. Previous studies confirm the expansion of glacial lakes since the 1990s in the upper Ganga basin (Nie et al. 2013; Zhang et al. 2015; Nie et al. 2017; Li et al. 2022). Li et al. (2022) showed that the number of glacial lakes increased by 711 and the area increased by 47.05 km^2 in the central Himalayan region. Nie et al. (2013) examined the evolution of glacial lakes in the central Himalayan region and found that the number of glacial lakes increased by 123 and the area of glacial lakes increased by 28.81 km^2 (Table 4.3). Pandey et al. (2021) studied the evolution of glacial lakes in the state of Uttarakhand (the western part of the upper Ganga basin) and found that the number of glacial lakes increased by ~ 9% and the area of glacial lakes increased by ~ 57% between 1994 and 2017. Khadka et al. (2018) found that the number of glacial lakes increased by 404 (35.53%) and the area increased by 16.39 km^2 (25%) in the Nepal Himalayas between 1987 and 2017. The present study aligns with the result of previous studies on the expansion of glacial lakes in the central Himalayan region. However, the rate of expansion was different in different studies, mainly because of different methods used to map the glacial lakes.

Furthermore, we estimated the volume and mean depth of glacial lakes in the study domain. To our knowledge, there is no present study that attempted to estimate the volume and mean depth for the central Himalayan region at the sub-basin scale. We used an area-based empirical equation developed by Huggel et al. (2002). There

Table 4.3 Comparison of previous studies

Author(s)	Study area	Data sources	Method	Criteria	Number and area	Year
Nie et al. (2013)	Central Himalayas	Landsat (TM/ETM) (n: 56)	Semi-automated	Area: ≥ 0.0081 km^2	1191 (a: 168.4 km^2) 1290 (a: 185.28 km^2) 1303 (a: 190.84 km^2) 1314 (a: 197.22 km^2)	1990 2000 2005 2010
Worni et al. (2013)	Indian Himalayas	Landsat (ETM+)	Automated	Area: > 0.01 km^2	251	2000 ± 2
Aggarwal et al. (2017)	Sikkim	Resourcesat-2 and Cartosat-2 DEM	Semi-automated		1104 (a: 30.49 km^2)	2013
Khadka et al. (2018)	Nepal Himalaya	Landsat (MSS, TM, and OLI) (n: 45) and SRTM DEM	Semi-automated	Area: ≥ 0.0036 km^2 Slope: $< 10°$ Shaded Relief: > 0.25 Elevation: ≥ 2450 m	1541 (80.95 ± 15.25 km^2)	2017
Zhang et al. (2015)	Ganga basin	Landsat (TM/ETM+)	Manual mapping	Area: > 0.003 km^2 Distance from glacier: 10 km	294 (30 ± 5.2 km^2) 364 (45.8 ± 7.3 km^2)	1990 2010
Maharjan et al. (2018)	Ganga basin	Landsat (TM and ETM+) (n: 151) and SRTM DEM	Semi-automated	Area: > 0.003 km^2	4082 (a: 208.59 km^2)	2005 ± 2
Pandey et al. (2021)	Uttarakhand	Resources-2 (n: 8), Landsat (TM) (n: 9), and SRTM DEM	Automated	Area: > 0.0005 km^2, Distance from a glacier: 2 km	1353 (a: 7.96 km^2)	2015

(continued)

Table 4.3 (continued)

Author(s)	Study area	Data sources	Method	Criteria	Number and area	Year
Rao et al. (2021)	Ganga basin	Resourcesat-2 (n: 105) and Cartosat DEM	Manual mapping	Area: > 0.0025 km^2	4707 (a: 206.85 km^2)	2016–2018
Li et al. (2022)	Central Himalayas	Landsat (TM, ETM +, and OLI) (n: 651) and ASTER DEM	Semi-automated	Distance from Glacier: < 10 km Area:	438 (53.70 ± 6.49 km^2) 677 (70.13 ± 9.12 km^2) 759 (75.88 ± 9.90 km^2) 722 (75.56 ± 9.34 km^2) 833 (80.24 ± 10.73 km^3) 845 (85.01 ± 10.88 km^2) 1149 (100.75 ± 14.33 km^2)	1990 1995 2000 2005 2010 2015 2020

are several other equations developed to estimate the volume and mean depth of glacial lakes (Sakai 2012; Fujita et al. 2013; Cook and Quincey 2015; Qi et al. 2022). However, we used Huggel et al. (2002) equations because they developed two separate equations for estimating volume and mean depth using the same dataset.

4.6 Conclusion

Our results agree with the previous studies and show the expansion of glacial lakes in the upper Ganga basin at a rapid rate. Between 1990 and 2020, the number of glacial lakes increased by 22.08% and the area of glacial lakes increased by 23.93%. We recommend further continued monitoring of glacial lakes at the sub-basin level for developing early warning systems for GLOF predictions and mitigation. We further found that the number of glacial lakes increased at a much faster rate in the western part of the upper Ganga basin as compared to the east part. The reason for the west to east gradient should be further investigated in future studies. Furthermore, the volume and mean depth estimates can help researchers and government agencies in estimating the total water volume of glacial lakes and identification of potentially hazardous glacial lakes in the upper Ganga basin.

Acknowledgements This work is supported by a doctoral research grant provided by the German Academic Exchange Service (DAAD) (funding ID: 57507871). The authors are thankful to the US Geological Survey (USGS) for providing free access to the Landsat satellite imagery.

References

Aggarwal S, Rai SC, Thakur PK, Emmer A (2017) Inventory and recently increasing GLOF susceptibility of glacial lakes in Sikkim, Eastern Himalaya. Geomorphology 295:39–54. https://doi.org/10.1016/j.geomorph.2017.06.014

Ahmed R, Wani GF, Ahmad ST, Sahana M, Singh H, Ahmed P (2021) A review of glacial lake expansion and associated glacial lake outburst floods in the Himalayan region. Earth Systems and Environment 5(3):695–708. https://doi.org/10.1007/s41748-021-00230-9

Bhambri R, Misra A, Kumar A, Gupta AK, Verma A, Tiwari SK (2018) Glacier Lake inventory of Himachal Pradesh. Himalayan Geol 39(1):1–32

Rao BS, Gupta Ankit, Sweta, Maheshwari Ruhi, Raju PV, Rao VV (2021) Glacial Lake atlas of Ganga River Basin. National Remote Sensing Centre, ISRO, Hyderabad, India, pp. 1–194.

Cook SJ, Quincey DJ (2015) Estimating the volume of Alpine glacial lakes. Earth Surf Dyn 3(4):559–575. https://doi.org/10.5194/esurf-3-559-2015

Dou X, Fan X, Yunus AP, Xiong J, Tang R, Wang X, Xu Q (2021) Spatio-temporal evolution of glacial lakes in the Tibetan Plateau over the past 30 years. Earth System Science Data Discussions 1–31:2022. https://doi.org/10.5194/egusphere-2022-586

Fujita K, Sakai A, Takenaka S, Nuimura T, Surazakov AB, Sawagaki T, Yamanokuchi T (2013) Potential flood volume of Himalayan glacial lakes. Nat Hazard 13(7):1827–1839. https://doi.org/10.5194/nhess-13-1827-2013

Huggel et al (2002) Remote sensing based assessment of hazards from glacier lake outbursts: a case study in the Swiss Alps. Can Geotech J 39:316–330. https://doi.org/10.1139/t01-099

Ives JD, Shrestha RB, Mool PK (2010) Formation of glacial lakes in the Hindu Kush-Himalayas and GLOF risk assessment (pp. vi+-56).

Krishnan S, Indu J (2023) Assessing the potential of temperature/vegetation index space to infer soil moisture over Ganga Basin. J Hydrol 621:129611. https://doi.org/10.1016/j.jhydrol.2023.129611

Khadka N, Zhang G, Thakuri S (2018) Glacial lakes in the Nepal Himalaya: Inventory and decadal dynamics (1977–2017). Remote Sensing 10(12):1913. https://doi.org/10.3390/rs10121913

Li W, Wang W, Gao X, Wang X, Wang R (2022) Inventory and spatiotemporal patterns of glacial lakes in the HKH-TMHA Region from 1990 to 2020. Remote Sensing 14(6):1351. https://doi.org/10.3390/rs14061351

Nie Y, Liu Q, Liu S (2013) Glacial lake expansion in the Central Himalayas by Landsat images, 1990–2010. PLoS ONE 8(12):e83973. https://doi.org/10.1371/journal.pone.0092654

Nie Y, Sheng Y, Liu Q, Liu L, Liu S, Zhang Y, Song C (2017) A regional-scale assessment of Himalayan glacial lake changes using satellite observations from 1990 to 2015. Remote Sens Environ 189:1–13. https://doi.org/10.1016/j.rse.2016.11.008

Maharjan SB, Mool PK, Lizong W, Xiao G, Shrestha F, Shrestha RB, ... Baral P (2018) The status of glacial lakes in the Hindu Kush Himalaya-ICIMOD Research Report 2018/1. International Centre for Integrated Mountain Development (ICIMOD).

Mal S, Kumar A, Bhambri R, Schickhoff U, Singh RB (2020) Inventory and spatial distribution of glacial lakes in Arunachal Pradesh, Eastern Himalaya, India. J Geol Soc India 96:609–615. https://doi.org/10.1007/s12594-020-1610-1

Mal S, Allen SK, Frey H, Huggel C (2021) Dimri AP (2021) Sectorwise assessment of glacial lake outburst flood danger in the Indian Himalayan Region. Mt Res Dev 41(1):R1–R12

Pandey P, Ali SN, Champati Ray PK (2021) Glacier-glacial lake interactions and glacial lake development in the central Himalayas, India (1994–2017). Journal of Earth Science 1-12. https://doi.org/10.1007/s12583-020-1056-9

Shugar DH, Burr A, Haritashya UK, Kargel JS, Watson CS, Kennedy MC, Bevington AR, Betts RA, Harrison S, Strattman K (2020) Rapid worldwide growth of glacial lakes since 1990. Nat Clim Chang 10(10):939–945. https://doi.org/10.1038/s41558-020-0855-4

Qi M, Liu S, Wu K, Zhu Y, Xie F, Jin H, Gao Y, Yao X. (2022) Improving the accuracy of glacial lake volume estimation: A case study in the Poiqu basin, central Himalayas. J Hydrol 610:127973. https://doi.org/10.1016/j.jhydrol.2022.127973

Worni R, Huggel C, Stoffel M (2013) Glacial lakes in the Indian Himalayas—from an area-wide glacial lake inventory to on-site and modeling-based risk assessment of critical glacial lakes. Sci Total Environ 468:S71–S84. https://doi.org/10.1016/j.scitotenv.2012.11.043

Zhang G, Yao T, Xie H, Wang W, Yang W (2015) An inventory of glacial lakes in the Third Pole region and their changes in response to global warming. Global Planet Change 131:148–157. https://doi.org/10.1016/j.gloplacha.2015.05.013

Zhang G, Bolch T, Yao T, Rounce DR, Chen W, Veh G, King O, Allen SK (2023) Underestimated mass loss from lake-terminating glaciers in the greater Himalayas. Nat Geosci 16(4):333–338. https://doi.org/10.1038/s41561-023-01150-1

Chapter 5
Water Resources in the Himalayas

Anant Gautam and Suresh Chand Rai

Abstract Hindu-Kush Himalayan region maintains the sustainable livelihood of about one-fourth of the world's population. They are the largest storage house of fresh water in the world. The presence of glaciers all along its arc-shaped areal extension makes it the third pole in the world, after Antarctica and the Arctic. This chapter tries to give a detailed account of the significance of mountains in the hydrological system of the world, highland-lowland interface, sources of water in the Himalayas, major river systems, i.e., Indus, Ganga, and Brahmaputra, and natural and anthropogenic influences on these water resources. Various sources of water in the Himalaya such as glaciers, springs, rivers, dams and reservoirs, and groundwater aquifers not only fulfills the requirements of hilly Himalayan terrain but also renew the soil fertility, boost agricultural yield, provide water for drinking and industrial purposes, facilitate electricity generation through the construction of dams for the people residing in the lower areas. The Himalayan region also facilitates a tourist destination for people all around the world who wish to enjoy snowfall and natural beauty. The hot springs prove medically helpful in curing skin-related diseases and mental calmness. Analyzing the water resources in the Himalayan is crucial as climate change is leading to an increased rate of melting of the glaciers, leading to a decrease in the quantity of water stored in this high region for the perennial supply of water to meet the demands of one of the densely populated regions of the world. It is in this backdrop; that the present chapter attempts to give an account of the hydrological resources present in the Himalayas.

Keywords Hindu-Kush Himalaya · Highland-lowland · Hot springs · Anthropogenic

A. Gautam
Department of Geography, Sikkim University, Gangtok, India

S. C. Rai (✉)
Department of Geography, Delhi School of Economics, University of Delhi, Delhi, India
e-mail: raisc1958du@gmail.com

S. C. Rai (ed.), *Himalaya: Mountains of Destiny*,
https://doi.org/10.1007/978-3-031-94905-0_5

5.1 Introduction

Stretching about 3500 km in length and 3,441,719 km^2 of area, the Hindu-Kush Himalaya (KHK) region comprises the whole or parts of eight countries in an arc shape from Afghanistan to Myanmar and 19 large river systems from Amu Darya in the west to Mekong, Yellow and Tarim in the east. It forms the direct source of livelihood for around 240 million people residing in the mountainous topography of the region and meets the domestic, industrial, and agricultural needs of about 1.9 billion people, i.e., one-fourth population in the world (Scott et al. 2019; Chandel and Ghosh 2021). About 87,340 km^2 of areas covered by glaciers make it the third pole of the Earth. Apart from that, the snow cover areas varying from one season to the other give it a unique aesthetic scenic beauty covering between 951,000 km^2 and 139,000 km^2 areas in winter and between 388,000 km^2 and 481,000 km^2 area in summer. These snow covers along with glacial melt do not only fulfill the water requirements of the people living in basins of the rivers originating in its lap, but they have also created the adjoining plain areas which have sustained a high density of population since time immemorial.

The fertile sediments accumulated by the rivers cutting across these treacherous sediment-loaded mountains have created the fertile agricultural landscape which has not only sustained the most advanced ancient Indus Valley civilization but still sustains the most densely populated segment of the earth (Bandyopadhyay and Gyawali 1994). The three great rivers of India—the Indus, Ganga, and Brahmaputra collectively provide close to 50% of the total of the country's utilizable surface water resources (690 km^3). However, changes in the earth's axis of rotation, orbit of revolution, insolation from the sun, albedo, differential heating of land and water bodies, increase in heat absorbent built-up areas, decrease in forest areas, uncontrolled emission of greenhouse gasses have led to changes in the climatic conditions in HKH region, leading to speedy melting of glaciers at places and making a significant impact on the livelihood of the people. People's desire for development using unsustainable ways has led to an imbalance in the natural ecosystem of the HKH region. This has been illustrated in studies conducted by various scientists and scholars.

5.2 Hydrological Significance of Mountains

Mountains covering only about 27% of the earth's continental surface, are often referred to as natural water towers as they provide essential freshwater to meet the needs of the population living both upstream and downstream (Ives et al. 1997; Viviroli and Weingartner 2004). Experts at an International conference on mountains in 1995 in Lima stated that mountains have acted as the source of about 80% of running freshwater in the world (Viviroli et al. 2003). However, estimates from various studies for the proportion of mountains to the global discharge have varied from 32% (Maybeck et al. 2001), 40–60% (Bandopadhyay et al. 1997) to regionally

95% of the total flow (Liniger et al. 1998). Their role in the hydrological cycle is significant as they represent the blackest of the black box in the hydrological cycle (Klemes 1990). The evaporated waters either from the large ocean surface or land surface mixes into the atmosphere in the form of water vapor. These water vapors condense to form water droplets and precipitate to the ground surface in various forms like rain, snow, ice, hail blizzards, etc. The forms of precipitation depend upon the topographical and climatic conditions in which water vapor has undergone the process of condensation. In general, there are three ways in which this precipitation occurs depending on the processes; Conventional, Frontal, and Orographic. Among these three, orographic is linked to the vertical uplift of air masses when they encounter high lands or mountains as an obstacle in the path of horizontal movement. Whenever this vertical uplift takes place along mountains, the temperature of the airmasses decreases with either a dry adiabatic lapse rate or wet adiabatic lapse rate, resulting in a lowing of the temperature of air masses. With the decrease in temperature, the saturation capacity of the parcel of air is also decreased leading to nearly 100% relative humidity. With the reaching of the due point, condensation occurs and precipitation follows. Most of the precipitation in the ecumene zone of the world where most of the world's population resides is of orographic origin. Cloud processes such as orographic uplift, condensation, precipitation, moisture recycling, and mixing of vapor masses are governed by regional geographical features such as topography, land use, and surface water availability; and in turn, govern the precipitation pattern in the world (Smith 1979).

5.3 Highland-Lowland Interface

The precipitation that occurs in the mountains ultimately meets the ocean flowing through the land surface. In the course of its flow, various landforms are created by its action of erosion and deposition. Most of the landforms of the present world have been created by this action of streams and rivers, known as fluvial landforms. These streams and rivers have been crucial for the formation of large plain areas, especially the Indo-Gangatic Plain, south of the HKH. The sediments eroded and carried by the mighty streams and rivers originating in the Himalayan region have been deposited in the depression between the Indian Peninsular Plateau and Mighty Himalaya, creating a vast alluvial plain. The sediments in these plains are annually renewed after subsequent depositions during monsoons and floods when Himalayan rivers carry enormous sediments along with them. Agricultural productivity in these plain areas of Punjab, North India, and Bangladesh mainly depends upon the sediments deposited annually by the rivers. The Water availability and agricultural productivity decrease in the years when the region is not flooded. Therefore, the HKH region having most of the glaciers in the world and the source of major rivers in the world is closely related to human welfare and ecosystem health because of their role in generating runoff, groundwater recharge, providing flow to the lowlands through river networks and aquifers, transporting nutrients from porous media to the rivers which affect the biological cycle and sustaining both natural terrestrial and aquatic habitats (Immerzeel et al. 2020). However, regional and temporal aspects are also of special importance in considering water availability in a region (Nikkanen et al. 2004).

5.4 Sources of Water in the Himalayas

Rapid uplift of the Himalayas from 0.9 to 0.8 ma was responsible for asynchronous Pleistocene glaciation in these mountains. At least 4 successive glacial advances have been identified in parts of Kashmir, Ladakh, and Tibet. The oldest of the glaciers extended as low as 1675 m asl, whereas the youngest reached an elevation of 2400 m asl. The glaciation and deglaciation occur on Milankovitch timescale. Glaciers occur in the high-altitude region of the mountains and the polar regions of the Earth. They are vital to mankind as they control the global hydrological cycle maintaining the global sea levels and perennial supply of freshwater to the rivers.

The uplift of the Himalayas and the evolution of the South Asian monsoon (SAM) are intrinsically linked. The proto-SAM was established during the early Miocene. A major intensification of SAM took place c13-15 million years ago, by which time the Himalaya had obtained a height of c > 5000 m asl. The Himalayan mountains act as an effective barrier to the summer monsoon winds and western disturbances in the winter to facilitate orographic rainfall (Bandyopadhyay and Gyawali 1994). These higher mountains of the Himalayas generate higher run-off from orographic precipitation and store it in the form of snow, glaciers, and lake reservoirs (Immerzeal et al. 2020). Due to this storing capacity, these snow and glaciers after melting in hot seasons provide constant water supply to the downstream areas.

Despite a clear understanding of the major rain-bearing system in the Himalayas, the detailed hydrometeorological processing varying across the length and breadth of the Himalayas has not been clearly understood yet (Jeelani and Deshpande 2017; Jeelani et al. 2018). In the central Himalayas, western disturbances cause severe snowstorms in high-altitude regions due to their interaction with unique topographic features (Pang et al. 2014). However, the actual contribution varies from the general perception that Western disturbances derive their water from the Mediterranean Sea. Instead, as per a study conducted by Oza et al., about 98% of moisture contribution during western disturbances in the Jammu region is from convention-driven continental recycling, and the negligible contribution of about 2% from the Arabian Sea. This is because about 4300 km traveling of air parcel during western disturbances reduce the specific humidity to less than 5 g/kg which is too low; as a result, the actual moisture reaching northern India from the Mediterranean region is less than 1%. Although the moisture content does not reach the Himalayan region, the depressions do travel. This causes conventional uplift and continental recycling of the waters from Himalayan rivers, irrigation surplus, and canals. A similar process prevails during the southwest monsoon too in western India even though about 80% of the precipitation in the western Himalayan region is marked with south-west monsoon. The examination of monthly climatological (1981–2010) means outgoing longwave radiation, mean vertical wind velocity, and isotopic signatures of precipitation in northern India indicated thick clouds and increased convention over the Indo-Gangatic plains during July-Aug, making this process responsible for maximum precipitation in monsoon months. The contribution of the Arabian Sea in monsoon precipitation in western Himalaya ranged between 1 to 21%, and that of the Bay of Bengal was between 1 and 4% only.

The major source of water resources in the HKH is precipitation (snow and rain) which varies in time, location, and extent. At lower elevations, it occurs as rainfall, whereas in higher locations it occurs as snowfall. As a result, some of the rivers originating from the Himalayas are rainfed and some are snowfed. The heaviest rainfall occurs in the eastern parts feeding the rivers like Mekong, Brahmaputra, Irrawaddy, Salween, Yangtse, etc. Whereas in the western parts of the Himalayas, most of the precipitation occurs in the form of snowfall; thereby feeding the rivers like Indus and its tributaries. Hence it can be said that the rainfall pattern along the Himalayas decreases from east to Northwest. Similarly, the rainfall decreases horizontally from Siwalik to the Greater Himalayas, the latter receiving precipitation mostly in the form of snow. Therefore, the rivers originating from the central Himalayas are fed by both, initially the snow and glaciers, and later in downstream by the heavy rainfall. Receiving the first wind aloft during monsoon, the windward side of the Siwalik receives more rainfall than the leeward side.

However, no precise quantification of the total volume in these rivers is possible, Bandyopadhyay and Gyawali (1994) have compiled a preliminary assessment of the annual runoff in the major rivers (Table 5.1).

- The figures do not include the contribution of tributaries that do not originate in the Himalayan region.
- The flow approximates the total Himalayan output of the main tributaries of the Indus.

Table 5.1 Annual runoff in the major Himalayan rivers (Bandyopadhyay 1992; Dasgupta 1984; Shahjahan 1983; Bandyopadhyay and Gyawali 1994)

River system	Average annual rainfall (BCM)
Jinsha (Yangtse)	143.7
Lan Chang (Mekong)	58.3
Nu Jiang (Salween)	52.3
Irrawaddy	–
Meghna	141.9
Yarlungtsangpo	139.5
Brahmaputra	606.8
Ganga	371.6
Indus	143.6

5.5 Water Resources in the Himalayas

Water resources in the Himalayas are present in the forms of glaciers, springs, rivers, dams and reservoirs, and groundwater aquifers.

5.5.1 Glaciers

Glaciers, frozen rivers of ice flowing from high mountain ranges have a significant role in sustaining the perennial flow of rivers in the Himalayan region. Almost 30–50% of the annual flow of all the rivers originating from the higher Himalayas comes from snow melts (Agarwal et al. 1983; Jain et al. 2010; Rathore et al. 2018). Numbering about 15,000, the glaciers lie between the two syntaxial bends of the Hindu Kush-Karakoram Ranges (The Pamir) and the Eastern Himalaya-Patkai Bum Ranges. Covering about 17% of the Himalayas, the ice-clad region is much more than the proportional ice-covered area in any of the mountains of the world, i.e., only about 2.2% of the Alps is snow-covered (Ahmad et al. 2004; Lemke et al. 2007). However, glacier size differs from as giant as more than 60 km to puny ones only 3–5 km long. Its density varies across the Himalayas, being maximum in the Saltuj and Teesta basins, moderate in the Jhelum and Bhagirthi basins, and low in the river basins of Arunachal Pradesh (Ahmad et al. 2004). The Himalayan glaciers cover at a much greater altitude than mountain glaciers elsewhere in the world because of their location in lower latitudes. However, even within the Himalayan extent, the snowline varies considerably in different sections depending upon latitude, amount of precipitation received, and local topography. In the north-western Himalayas, the southwest-to-northeast gradient of decreasing annual precipitation controls the lower altitude of the slow line and the extent of glaciation (Brozovic et al. 1997). Although no regular pattern has been observed concerning snowline from the east to west in the Himalayan range; in most general way, the average snowline is about 4200 m in Eastern Himalaya to 4000 m in the Karakoram range. A parallel pattern was also seen during the Pleistocene, but then the snow line was lower by some 200 to 800 m from the east to the west (Mathur 1991). Monitoring of 28 sub-basins of the Ganga and Indus revealed the snowline at 2480 m (Kulkarni et al. 2010). In Himachal Himalaya, the summer snow line fluctuates between 4000 to 5000 m (Dobhal and Kumar 1996). The seasonal analysis of snowline from 2003 to 2021 illustrated snowline fluctuation from 2349 ± 117 m to 4802 ± 51.7 m in the Upper Indus River basin, 3303 ± 286 m to 4871 ± 64.4 m in the Ganga River basin, and 3238 ± 127 m to 5047 ± 73 m in Brahmaputra River basin between winter and summer seasons (Dixit et al. 2024). The snowline is also not the same vertically. The snowline is lower on the southern slope than the northern slope due to relatively much greater precipitation and much steeper topography of the southern slope than the northern which is gentler and much drier (Jain et al. 2009).

The glaciers of the Himalayas can be found under the following grouping within Himalayas;

1. Eastern Himalaya: There are not many glaciers in the eastern part of the Himalayas and information about them is also very scanty due to extremely inhospitable terrain and difficult access. A few small transverse glaciers descend from the Namcha Barwa peak (7756 m) in the easternmost terminal of the Himalayas. However, between Namcha Barwa and Kangta Peak (7089 m), the region is devoid of glaciers. Then again after a gap, glaciers are present in northern Bhutan on Kurla Kangri peak (7593 m) with westward continuation.
2. The Kanchenjunga-Everest Region: This region has several large glaciers above 4000 m elevation with steep gradients. River Arun (Kosi) divides the Kanchenjunga and Everest group of glaciers in the region. The important glaciers in the Kanchenjunga group are Zemu, Kanchenjunga, Rambang, Alukthang, Talung, and Yalund; whereas, in the Everest group, important glaciers are Khumbu, Rongbuk, Kangshung, Barun, Kyebrak, Ngojumba, Lunak and Tolam.
3. Central Nepal: In the central Nepal Himalaya (between the Cho Oyu and Dhaulagiri), most of the glaciers are confined to the surroundings of Gosainthan, Manaslu, Annapurna, and the Dhaulagiri mountains. Among them, prominent are the Yepokangara, Lidanda, Chhuling, South Annapurna, and Mayondi glaciers.
4. The Kumaon-Garhwal Region: The region extending from western Nepal to the western parts of Uttarakhand has numerous small and large glaciers with base elevations as low as 3600 m. Significant among them are Pindari, Poting, Milam, Sankalp, Gangotri, Mana, Bhagirthi-Kharak, Satopanth, and Raikane glaciers.
5. The Pir Panjal Range: Although the glaciers of this region are not as extensive as the Kumaon-Garhwal region; prominent among them are Sonapani, Bara Shigri, Gangstung, Gangri, Buzhi, Tashan, Chungphar, and Rakhiot glaciers.
6. The Karakoram Range: Home to the largest glaciers outside the polar and subpolar regions, two groups of glaciers are found in this region, i.e., the Baltistan-Ladakh group and Hunza-Nagar group. The significant glaciers in the Baltistan-Ladakh region are the Baifo, Baltoro, Siachen, and Rimo glaciers. Some of the prominent glaciers in the Hunza-Nagar group are the Hispar, Bahura, Kurdopin, and Virjerab glaciers.
7. The Hindu-Kush range: The snowline in the Hindu-Kush region varies in different parts of the mountains. The snowline is at 4000–4500 m in the western Hindu-Kush, 4300 m in the eastern side, 4900–5100 m in the northern, and 4500 m at the southern side. Some of the prominent glaciers are Dara-i-Quala-i-Panji, Anoshah, Kotagaz, Hurusko Kuh, Atrak, Tirich, Barun, Gazikistan, Rakhug, Chiantar, and Chatiboi glaciers.

Perennial ice-covers in the catchment areas of various Indian Himalayan rivers have also been shown in Table 5.2. However, in the broadest sense, the glaciers of the Himalayas have been in a general state of retreat since 1850.

Table 5.2 Perennial ice cover on the Himalayas (Ahmad et al. 2004)

River	Glacier area (m^2)	% Glaciation
Indus	7890	3.3
Jhelum	170	5
Chenab	2944	10
Ravi	206	2.5
Satluj	1295	2.7
Beas	638	4.4
Jamuna	125	1.1
Ganga	2312	10
Ramganga	2	0.04
Kali	997	6.01
Karnali	1543	2.9
Gandak	1845	4.9
Kosi	1281	2.1
Teesta	495	4
Raikad	195	0.7
Manas	528	1.7
Subansiri	725	4
Brahmaputra	108	0.4
Dibang	90	0.7
Luhit	425	2

5.5.2 *Springs*

Most of the human population residing in the hilly and mountainous terrain, especially in the Himalayas rely on springs to meet their daily water requirements. Springs are a part of the natural landscape with a heterogeneous presence of rising water from lithospheric surfaces. They emerge in most of the ecosystems on earth, including a wide range of subaerial terrestrial settings and as subaqueous discharge from the floors of freshwater and marine bodies of water (Springer et al. 2008). They are located in the specific physiographic division of the earth, i.e., Hilly and mountainous areas (Gautam et al. 2019). They are present mainly along the fault lines which are linked to some kind of water reserves. In general, the source of water in them is either stored groundwater or surface ponds and lakes fed by precipitation water. Water in the springs comes out either following the fault line, joints, interspaces from sedimentary rocks (Young et al. 2007; Gupta et al. 2017), along faults through hot igneous and metamorphic rocks (Pradhan and Jena 2016), caverns or points of intersection between piezometric level and ground surface (Gautam et al. 2018). They are homogenously spread along the whole range of the Himalayas from east to west and north to south and form a major source for many streams flowing in the Himalayas.

These springs are divided into mainly two types based on the temperature of water coming out at these spots, i.e., hot springs and Cold springs. In general, this nomenclature depends upon the temperature of the water coming out at these places in comparison to the temperature of the immediate surroundings. A hot water spring or a hydrothermal spring is a place where warm or hot groundwater comes out from the earth regularly for at least a periodical part of the year and is significantly above the ambient ground temperature (Bisht et al. 2011). Spring temperature is largely controlled by Aquifer lithology (Shi et al. 2017). Water oozing out of the hot springs gets its thermal character from the heat trapped in the rocks of the Himalayas during its formation, and the presence of Sulfur to a great extent. The springs are seasonal and perennial depending upon the continuity of the flow of water through them. Some of the springs dry out in summer due to a lack of water source that mainly depends upon rainfall. Some of the perennial springs have also dried out with time due to emptying their water source in the rocks. However, government programs have focused on the rejuvenation of these dried springs at various places in the Himalayan region.

5.5.3 Rivers

There are three major river systems in the Indian Sub-continent; Indus, Ganga, and Brahmaputra. The drainage area and water discharges of some of the major rivers originating in the Himalayas and flowing toward the Indian Sub-continent have been illustrated in Table 5.3.

5.5.3.1 Indus River System

The origin of the River Indus can date back to the early Tibetan uplift following the India-Asia collision. The river is said to have remained at the stationary position in the suture since early Eocene time. It continued cutting down through its earlier depositions which uplifted in subsequent orogenic processes and deformed while northward folding and thrusting associated with the Zanskar backthrust at c 20 Ma. The Indus appears to have been located close to its present position within the foreland basin since at least mid-Meocene time (c 18 Ma), and to have migrated only c 100 km east since early Eocene time (Cliff 2002). Unlike the rivers of East Asia, which strongly interacted as a result of eastward propagating deformation, the Indus has remained uninterrupted and represents the oldest known river in the Himalayan region. The total basin area of River Indus is 1,165,500 km^2, of which 321,289 km^2 is in India. Of the total water flow in the Indus River basin, about 69% is from the catchments of India, 19% from Pakistan, and 12% from Tibet and Afghanistan. Among various rivers and streams draining its water in the Indus River, the basin

Table 5.3 Drainage area and water discharge of major rivers of India (Modified after Rao 1975; Singh 2005)

River	Station	Area (10^5 km^2)	Discharge (km^3 $year^{-1}$)
Indus			
Jhelum	Mangla	1.42	28
Chenab	Merala	0.26	29
Beas	Mandi	0.18	16
Satluj	Rupar	–	17
Indus	Thatta	11.65	197
Ganga			
Ganga	Devprayag	0.20	22
Yamuna	Batamandi	0.10	11
Gomti	Near confuence with Ganga	0.30	7
Yamuna	Pragayraj	3.66	93
Ghaghra	Near confuence with Ganga	1.28	94
Gandak	Near confuence with Ganga	0.46	52
Kosi	Near confuence with Ganga	0.75	62
Ganga	Farakka	9.35	380
Brahmaputra			
Tsangpo	Yangcun	6.36	29
Siang	Pasighat	2.46	200
Dibang	Sadeya	0.13	63
Lohit	Sadeya	0.24	60
Brahmaputra	Dibrugarh	3.23	323
Subansiri	Near confluence with Brahmaputra	0.33	54
Burhi Dihang	Near confluence with Brahmaputra	0.08	14
Jia Bhareli	Near confluence with the Brahmaputra	0.12	26
Dhansiri	Near confluence with Brahmaputra	0.12	20
Kopili	Near confluence with Brahmaputra	0.16	28
Manas	Near confluence with the Brahmaputra	0.38	32
Brahmaputra	Bahadurabad	6.30	670

comprises the main Indus River and its six major tributaries, i.e., Kabul, Jhelum, Chenab, Rabi, Beas, and Satluj. Except River Kabul which originates in Afghanistan, the other major tributaries have their basins mainly in India.

The principal source of water supply in these rivers is the precipitation (snow and rain) in the lower Himalayan Mountains at an elevation ranging from 3000 m to 8000 feet above sea level. Snowfall at higher altitudes above 2500 m accounts for most of the river runoff. There is very little contribution to the river flow after the river enters the plains. Originating from Bokhar-chu glacier near Mansarovar lake at about 22,000 feet, river Indus travels west-ward for about 200 miles in Tibet; enters into India, flows for about 500 miles in North Kashmir and Gilgit; enters Pakistan and takes a south-west turn and travels about 1200 miles to empty its water in Arabian sea at Karachi. It enters into the plains from the mountains at Kalabagh. River Kabul, flowing for about 300 miles, originates about 40 miles west of Kabul city, enters Pakistan near Warsak, and meets River Indus in Peshawar valley at Attock. River Jhelum originates from Verinag Spring at the bottom of the high scarp of a mountain spur. Originating in the vale of Kashmir, Jhelum enters Wular Lake, flows in a narrow George across Pir Panjal, and sharply turns south after reaching Muzaffarabad. The river is navigable for about 160 km near Srinagar. It enters into Plains at Mangla after flowing for about 400 km in mountains from its source. Further, after traveling for about 300 km in plains, it joins River Chenab at Trimmu. River Chenab gets its name after the river Chandra and Bhaga confluences in Lahul and Spiti regions in Himachal Pradesh. It is also known as Chandrabhaga (mentioned in the ancient Indian texts). The headstreams of Chanda and Bhaga have their source on the opposite sides of Bara Lacha at an elevation of about 4900 m. It flows through the Pir Panjal Range and enters Jammu Hills and the foothill plains. It enters the plains from the mountains at Merala. After the confluence with River Jhelum, it flows for about 50 km joined by River Ravi and Satluj. Flowing for a total length of about 1232 km, it meets the River Indus at Panchnad. River Ravi originates from the joint streams formed by the glacial-fed Bhadal and Tant-Gari. Forming various Georges, the river reaches Rohtang Pass in the Kangra district and further drains the southern slopes of Dhauladhar. Flowing for about 95 km in Himachal Pradesh, the river enters Jammu province and then forms the boundary between India and Pakistan. It enters the plains at Madhopur and meets River Chenab after flowing for a total length of about 880 km. River Beas rises in the Pir Panjal Range near Rohtang Pass, travels through the city of Manali, Himachal Pradesh, and enters the Mandi plains from the hills. After flowing for a total length of about 464 km, the river meets Satluj at Harike in Punjab (India). River Satluj originates from Rakas taal, west of Mansarovar, flows for about 676 km parallel to the Indus, and then cuts right through the Great Himalaya. It enters India from Tibet through Shipkila Pass where it is joined by Spiti River. It enters into the plains from the Mountainous regions of Himachal Pradesh at Rupar in Punjab (India). It enters Pakistan near Suleimanke and joins the River Indus at Punchnad after flowing a total length of about 1536 km. The average water resources potential of the whole Indus River basin is 73,310 million m^3, of which 46,000 million m^3 is utilizable. The water in the rivers of the basin increases in the months from Feb to

April due to the start of snow melting after its accumulation in winter. By mid-May, the water in these rivers begins to fall and its minimum in September.

5.5.3.2 Ganga River System

Ganga river basin covers a total area of about 1,086,000 km^2 across India, Nepal, and Bangladesh. Among its major rivers originating in Himalaya include, River Ganga, Yamuna, Ghaghara, Kali, Gandak, and Kosi. The flow of the River Ganga is in the form of surface and subsurface flow in the high reaches of snow-dominated areas (Krishan et al. 2019), the link of which may sometimes be said near the Kailash Mansarovar. However, studies are still confirming the link through water isotopes. River Ganga gets its name at Devprayag after the confluence of Rivers Alaknanda and Bhagirthi. In general public sentiment, River Bhagirthi originating from the snouts of Gaumukh at 3892 m amsl in Gangotri Glacier is considered the origin point of River Ganga. Forming deep Gorges, it travels to Bhaironghati, Harshil, and enters a wide valley. Traversing through Uttarkashi, the Bhagirthi River meets the Bhilangna River (originating at the foothill of the Khatling glacier) at Old Tehri. With a basin area of about 6921 km^2 and flowing for about 205 km, River Bhagirthi meets River Alaknanda at Devprayag. River Alaknanda originates from the confluence of snow melt from Satopanth and Bhagirth Kharak glaciers at about 3880 m amsl. After the confluence with the Saraswati River at Mana village, the river traverses the Badrinath valley. Covering a total basin area of about 10,882 km^2, the river flows down and meets various other rivers originating in the glaciers at the left limb of its watershed. River Dhauliganga originating near the Niti pass between Nanda Devi and Kamet glaciers empties its water into River Alaknanda at Vishnuprayag. After crossing Joshimath, River Alaknanda crosses the Main Central Thrust at Helang and enters into Lesser Himalaya. At Nandprayag, it is joined by River Nandakini originating at the western edge of Nanda Ghunti peak. Further, River Pindar originating from Pindari Glacier joins River Alaknanda at Karnprayag. Flowing downward, it is again met by River Mandakini from its right at Rudraprayag. River Mandakini originates from the glacial melt in the Kedarnath valley. Flowing ahead, river Alaknanda enters a wide valley near Srinagar (Uttarakhand) and finally meets River Bhagirthi at Devprayag. All the above-mentioned Prayag along the way of River Alaknanada is called Panch Prayag. The river then enters a deep V-shaped valley before entering into plains via wide valleys at Rishikesh and Haridwar. Quantifying the contribution of glacial meltwater and rainwater surface run-off at Rishikesh, Maurya et al. (2011) found that about 70–90% contribution is from glacial meltwater in the winter season while glacial met to the streams discharge peaks during the winter season.

Originating from a few kilometers west of Yamunotri glacier at about 4500 m amsl, River Yamuna cuts across Nag Tibba, Mussoorie, and Shivalik Ranges before entering into the plains in Himachal Pradesh. Flowing for about 1376 km and along the cities of Delhi, Agra, and Mathura, the Yamuna River meets the River Ganga at Prayagraj. During the late Pleistocene and early Holocene, the River Yamuna was a part of the Indus River system. It was joined with Saraswati and flowed as Ghaghar emptied its water into the Indus or Rann of Kutch. However, the tectonic disturbance along the Aravalli Axis changes its course toward the east making it a tributary of River Ganga (Mathur 1991).

Ghaghara River, known as River Karnali in Western Nepal originates near Gurla Mandhata peak, South of Mansarovar in Tibet at about 3962 m amsl. From its origin, it flows southeast and then southwest cutting across the higher Himalayas. Flowing for about 1080 km, it empties its water into the River Ganga near Chhapra in Bihar. Kali River later known as Sharda River originates from Nanda Devi massif at an elevation of about 3600 m in the state of Uttarakhand. Forming a boundary between Nepal and India, it reaches plains near Tanakpur. The river joins River Ghaghara southwest of Bahraich in Uttar Pradesh after flowing for about 480 km. River Gandak originating in the Nepal Himalaya empties its water in Ganga near Sonpur (Bihar). Fed by various streams from the glaciers of the Gauri Shankar-Everest-Makalu-Kanchenjunga group of peaks, River Koshi enters the plains in Bihar. It is the river known for continuously shifting its bed east and west across 400 km in the North Bihar plains. It empties wits water in the River Ganga near Katihar (Bihar). At Farakka, through a canal, the water of River Ganga is linked with River Hoogly to empty its water in the Bay of Bengal. River Ganga becomes Padma with a host of small streams in Lower Bengal (Bangladesh), where it joins the Jamuna River and Meghna River finally meeting with the Bay of Bengal.

5.5.3.3 Brahmaputra River System

Brahmaputra river basin is spread over about 580,000 km^2 in China, India, Bangladesh, and Bhutan with major tributaries such as Gya Chhu, raga Zangbo, Gyamada Chu, Nyang, Lohit, Dibang, Subansiri, Kameng, Manas, Raidak, Dharla, and Teesta Rivers. About 50% of its basin lies in China, followed by India (33.6%), Bangladesh (8.1%) and Bhutan (7.8%). About 96% of Bhutan comes under the Brahmaputra Basin (Sharma and Paithankar 2014). Called Tsangpo in Tibet, the river originates in the glaciers of the Kailash Range about 150 km eastward to the source of the Indus River and 35 km from the origin of River Satluj on the other side of the water divide. It flows for about 1100 km eastward within the depression formed by the Indus-Tsangpo suture zone. Despite the great elevation, River Tsangpo flows through Tibet as a navigable river through a wide valley. In the upper reaches, River Tsangpo receives waters from Gya Chu flowing past Lhasa, joining at Chushul (Uuxu). Another tributary Raga Zangbo originating from western Nyainqengtang first runs parallel to Tsangpo but meets it at Pindzoling (Puncogling). At Tselha Dzong, Tsangpo is met with another tributary Gyamada Chu which is about 3 km wide. At about 2450 m north of Namcha Barwa the river loops southward through Dihand or Siang Gorge and enters Assam Valley to only 135 m amsl at Sadiya. From Sadiya, River Brahmaputra flows westward through the narrow Assam valley and turns southward entering Bangladesh. In the Upper Assam valley, River Lohit and Dibang join with several streams draining from the Eastern Himalayas. Originating in Tibet, River Subansiri flows through Miri hills in Arunachal Pradesh and joins River

Brahmaputra at Jamurighat in Assam. Another tributary River Kameng originates in the Tawang district of Arunachal Pradesh from the glacial lake at about 6300 m amsl. After flowing in the treacherous hilly terrain of eastern Himalaya, the river joins the Brahmaputra at Tezpur in Assam. Further westwards, River Brahmaputra is joined at Jogighopa by River Manas originating in Tibet, flowing through Bhutan and India. Rising in the upper reaches of the Himalayas and crossing through Thimpu and other parts of Bhutan, River Raidak joins River Brahmaputra at Chainage in Bangladesh., where it is also referred to as Dudhkumar River. Dharla River, another tributary of Brahmaputra originates in East Sikkim and is known as Jaldhaka River. It then flows to Bhutan and re-enters India in Kalimpong district, West Bengal. Flowing through Jalpaiguri, the river enters Bangladesh through the Lalmonirhat district and flows as the Dharla River. It empties its water in the Kurigram district in Bangladesh. Further ahead, River Brahmaputra is joined by River Teesta, which is a 414 km long river originating from Teesta-Khangtse Glaciers in Sikkim Himalaya. Fed by the streams of Lachung Chu, Lachen Chu, Ranikhola, Rangpo Chu, and Ranjit, it enters into plains at Sevoke, north of Siliguri in West Bengal. It enters Bangladesh in Dahagram and joins the River Brahmaputra at Haripur port near Gaibandha. River Brahmaputra after meeting several streams and rivers from Himalaya and Meghalaya hills, joins River Ganga at Goalundo in Bangladesh after traveling a distance of about 2900 km. The Brahmaputra is a very young river and its present configuration into Assam valley took place only during the Quaternary period (Mathur 1991).

5.5.4 *Dams and Reservoirs*

Dams and reservoirs have been used as a water storage structure for human consumption since time immemorial (Table 5.4). They have been constructed mainly to provide water to meet agricultural needs, domestic needs, flood control, and generation of hydroelectricity. Major dams having at least 15 m in height number about 4300 in India and 1855 in China. The constitution of major dams has increased in the KHK region of China and India in the last six decades (ICIMOD 2009). Some of the dams with water storage capacity have been listed in Table 4. Site suitability is a must while constructing dams in Himalayan terrains due to their fragile nature. These dams have a great potential to submerge a large area in the plains during failure or brake. However, they have proved as a boon to meet the water requirements of a large section of people living in cities of North Indian Plains. The water deviated from these dams through canals is linked to the agricultural fields of the states of Punjab, Haryana, and Uttar Pradesh in India, leading to high agricultural yield. The water requirements of the Capital city of India (Delhi) are also met with one such canal, i.e., the Satluj Yamuna canal bringing waters from Bhakra dam to the city of Delhi. Apart from that, these dams are largely used to generate hydroelectricity to meet the power demands of the whole of South Asia. In this way, the potential energy of water resources present in the highlands of the Himalayas stored in the dams meet

Table 5.4 Water storage capacity of some of the large Dams in the Himalayan region of India

Dams (River)	Water storage capacity (million m^3)
Tehri (Bhagirathi)	3540
Lakhwar	588
Bhakra (Satluj)	9867
Ranjit Sagar (Ravi)	3280
Baglihar (Chenab)	475
Pong (Beas)	8570
Subansiri Lower (Subansiri)	1643
Ramganga (Ramganga)	2448
Salal (Chenab)	285
Mangla (Jhelam)	10,150
Tarbela (Indus)	11,960

Source India-WRIS WebGIS

most of the human needs when converted into kinetic energy while sloping down the valley.

5.5.5 Groundwater Aquifers

A body of rocks and sediments that hold water in its pores at saturation level are called Aquifers. The Himalayan Mountain is largely composed of sediments accumulated from the previously uplifted structures as a result of a collision between Indian and Eurasian lithospheric plates. Considered as a young mountain, it is still in the process of building. In the form of rock mountains. This leads to the suitable conditions for groundwater aquifers in the Himalayan zones. The water in these aquifers is the result of water infiltration, percolation, and groundwater recharge occurring in suitable geological conditions that vary from region to region. Infiltration occurs during snowfall, rainfall, or surface run-off. The infiltrated water percolates between various sedimentary layers in these mountains saturating the pores and creating and recharging the groundwater aquifers in the Himalayan geological environment. Majorly three types of aquifers could be found in this setting; unconfined, confined, and perched. Unconfined aquifers are mostly common in mountainous terrain because of some opening to the aquifers in the included lithological setup in the Himalayan terrain. Confined aquifers could be found but either deep in the aquifers or in between impermeable clay layers. Perched aquifers are also common in this region; where a less extent impermeable clay layer with depression stores an ample amount of water in its environment. The groundwater stored mainly in these perched aquifers are source of water for most of the springs in the region. The pore pressure in the Himalayan lithology provides stability to the above-lying structures. Sometimes, there is a rupture in the aquifers due to earthquakes, landslides, or tunneling; leading to the free flow of water from these aquifers. This further destabilizes the structures above, and the chances of subsidence of the above structure increase. This situation has been experienced in the Himalayan states of Uttarakhand, Sikkim, Himachal Pradesh, etc. However, the storage of water in the groundwater aquifers of the Himalayas also provides water to the rivers flowing from these mountains during seasons of low rainfall or minimum snow melting. In the long term, they also contribute to the groundwater of the plain regions, due to internal percolation along slope and sub-surface run-off.

5.6 Natural and Anthropogenic Impact on Water Resources

In the Himalayan region, the average temperature is rising faster than the global temperature at a rate of 0.6 °C per year (Wester et al. 2019). Present projections predict that a global temperature rise of 1.5 °C will lead to a warming of 2.1 ± 0.1 °C in the region of Himalayan glaciers; and that only $64 \pm 7\%$ of the present-day ice mass stored in the high mountains of Asian Glaciers will remain by the end of the century (Kraaijenbrick et al. 2017). This scenario if not checked by nature itself would lead to the extinction of glaciers from the Himalayas. The clean and pure waters would meet the saline water uninterruptedly. This would dry out various headstreams that are feeding water to the mighty rivers. The lessening of the water in the rivers would affect the sustainability of human civilizations in the densely populated zone of the world. Apart from this natural process, the increase in human population also affects the water resources in the region. Domestic, Industrial, and agricultural demands of people are putting extra pressure on the existing water resources in both plain and hilly regions. Construction of large dams is obstructing the natural flow of the rivers, leading to a reduction in base flow in rivers downstream and putting extra pressure on dam sites in hilly areas, which might attract catastrophic events in future devastating the whole region. Various studies have been carried out to study this pressure in various drainage basins. Intense use of discharge in the Indus River catchment by dams, barrages, river links, and canals, illustrates the pressure of water resources in arid, semi-arid, and densely populated regions (Viviroli and Weingartner 2004). The decrease in river run-off in the Indus and Amu Darya rivers may be due to high consumption to meet irrigation needs in lower catchment densely populated areas than sparsely populated mountainous areas (Siebert et al. 2002). Growing upstream use and intensive downstream irrigation use have led to a decrease in run-off in the Euphrates and Tigris rivers (Viviroli and Weingartner 2004). The agricultural dependency of downstream countries on mountain waters can be seen in the proportion of annual water withdrawals which ranges from 92% in India and Iraq, 97% in Pakistan, 95% in Nepal to 79% in Egypt (World Bank 2001; Yang and Zehnder 2002; Merz et al. 2003).

5.7 Conclusion

The increasing water demands of densely populated plains of southeast Asia are met with the waters from the Himalayan rivers. Be it for domestic, agricultural, or industrial purposes, the human population is entirely dependent upon either the rivers having snow-melt water from Himalayan glaciers or precipitation occurring in the Himalayan region. The imagination of the high density of human population in this region was impossible without the presence of the Himalayas, as it acts as an obstacle in the way of air masses crossing the equator and moving toward the northern hemisphere between May and August. Orographic precipitation either from monsoon winds or western disturbances facilitates snowfall and rainfall in the Himalayan region. It also plays a significant role in sustaining the hydrological system in the Indian Sub-continent. Major Hydrological resources in the HKH region include a large number of glaciers all along the 3500 km Himalayan Arc feeding a large number of rivers draining the plain areas. Enormous springs sprouting out at the foothills or through cracks in the rocks provide a means of sustainability to the human population thriving in the macro and micro watersheds of the streams and rivers in the HKH region. The dense vegetation cover acts as a CO_2 absorber for purifying the air and maintaining the required level of O_2 in the atmosphere. They also facilitate water infiltration for groundwater recharge in Himalayan aquifers that act as a source of recharge for groundwater aquifers in the adjoining plains. The surface and sub-surface run-off not only completes the hydrological cycle but also renews the polluted water in the plains to meet good quality water demands for domestic consumption. Site suitability for dam construction and generation of electricity fulfills the water supply requirements at the household level in towns and cities suffering from water scarcity, irrigation requirements in the agricultural fields, high water demands in all the industries, and power demands in the current pace of human development.

References

Agarwal KG, Kumar V, Das T (1983) Melt runoff for a sub-catchment of Beas basin. In Proceedings of the First National Symposium on seasonal snow cover, 28–30 April, New Delhi, India. Manali, Snow, and Avalanche Study Establishment, 43–63.

Ahmad S, Hasnain SI, Selvan MT (2004) Morpho-metric characteristics of glaciers in the Indian Himalayas. Asian J Water Environ Pollut 1(1–2):109–118

Bandopadhyay J, Kraemer D, Kattelmann R, Kundzewicz ZW (1997) Highland waters—a resource of global significance. In: Messerli B, Ives JD (eds) Mountains of the world: a global priority. Parthenon, New York, pp 131–155

Bandyopadhyay J (1992) The Himalaya: prospects for and constraints on sustainable development. In Stone P (eds) The state of world's mountains. Zed Books, London

Bandyopadhyay J, Gyawali D (1994) Himalayan water resources: ecological and political aspects of management. Mt Res Dev 14(1):1–24

Bisht SS, Das NN, Tripathi NK (2011) Indian hot-water springs: a birds eye view. J Energy, Environ Carbon Credits 1(1):1–15

Brozavic N, Douglas WB, Meigs AJ (1997) Climatic limits on landscape development in the northwest Himalaya. Science 276(5312):571–574

Chandel VS, Ghosh S (2021) Components of Himalayan rivers flows in changing climate. Water Resources Research 57

Clift PD (2002) A brief history of the Indus River. Geol Soc London, Spec Publ 195:237–258

Dasgupta SP (1984) The Ganga basin: parts I and II (eds) center for the study of man and environment. Calcutta

Dixit A, Goswami A, Jain S, Das P (2024) Assessing snow cover in the Indus-Ganga-Brahmaputra River basins of the Hindu Kush Himalayas using snow persistence and snow line as metrics. Environ Challenges 14:100834

Dobhal DP, Kumar S (1996) Inventory of glacier basins in Himachal Himalaya. J Geol Soc Ind 48:671–681

FAO (2011) AQUASTAT Transboundary River basin—Indus River basin. Food and Agricultural Organization of the United Nations (FAO). Rome, Italy

Gautam A, Rai SC, Shekhar R (2018) Hot-springs of Rajgir: spatial re-distribution and conservation. Sree J Spat Perspect 2(1):1–9

Gautam A, Rai SC, Shekhar R (2019) Spatial distribution of hot-springs in India: an overview. Int J Soc Sci Econ Res 4(2):848–861

Gupta BK, Prakash P, Ahmad MF (2017) Water quality of Rajgir Hot Springs, Nalanda Bihar. Int J Emerg Trends Sci Technol 04(8):5447–5464

ICIMOD (2009) Water storage: a strategy for climate change adaptation in the Himalayas. Sustainable Mountain Development No. 56, ICIMOD, Winter 2009

Immerzeel WW, Lutz AF, Andrade M, Bahi A, Biemans H, Bolch T, Hyde S, Brundy S, Davies BJ, Elmore AC, Emmer A, Feng M, Fernandez A, Haritashya U, Kargel JS, Koppes M, Kraaijenbrink PDA, Kulkarni AV, Mayewski PA, Nepal S, Pachheco P, Painter TH, Pellicciotti F, Rajaram H, Rupper S, Sinisalo A, Shrestha AB, Viviroli D, Wada Y, Xiao C, Yao T, Baillie JEM (2020) Importance and vulnerability of world's water towers. Nature 577(7790):364–369

Ives JD, Messerli V, Spiess E (1997) Introduction. In: Messerli B, Ives JD (eds) Mountains of the world: a global priority. Parthenon, New York, pp 1–15

Jain SK, Goswami A, Saraf AK (2009) Role of elevation and aspect in snow distribution in western Himalaya. Water Resour Manage 23:71–83

Jain SK, Goswami A, Saraf AK (2010) Snowmelt runoff modeling in a Himalayan basin with aid of satellite data. Int J Remote Sens 31(24):6603–6618

Jeelani G, Deshpande RD (2017) Isotope fingerprinting of precipitation associated with western disturbances and Indian summer monsoon across the Himalaya. J Earth Syst Sci 126(8):1–13

Jeelani G, Deshpande RD, Galkowski M (2018) Isotope composition of daily precipitation along the southern foothills of the Himalaya: impact of marine and continental sources of atmosphere. Atmos Chem Phys 18(12):8789–8805

Klemes V (1990) Foreward. In: Molnar L (ed) Hydrology of mountainous areas. In: Proceedings of a workshop; June 1988; Strbske Pleso, Slovakia. IAHS Publication No. 190. IAHS Press, Wallingford, CT

Krishan G, Singh S, Thayyen RJ, Ghosh NC, Rai SP, Arora M (2019) Understanding river-subsurface water interaction in upper Ganga basin, India. Int J River Basin Manage https://doi.org/10.1080/15715124.2019.1683853

Kulkarni AV, Mathur P, Rathore BP, Alex S, Thakur N, Kumar M (2002) Effect of Global warming on snow ablation pattern in the Himalaya. Curr Sci 83(2):120–123

Kulkarni AV, Rathore BP, Singh SK, Ajai (2010) Distribution of seasonal snow cover in central and western Himalaya. Ann Glaciation 51(54):123–128

Kumar A, Srivastava P (2018) Landscape of the Indus River. In Singh DS (eds) The Indian rivers, Springer Hydrogeology, Springer Nature Singapore

Laghari AN, Vanham D, Rauch W (2012) The Indus basin in the framework of current and future water resources management. Hydrol Earth Syst Sci 16:1063–1083

Lemke P, Ren J, Alley RB, Allison I, Carrasco J, Flato G, Fujji Y, Kaser G, Mote P, Thomas RH, Zhang T (2007) Observations: changes in snow, ice, and frozen ground. In: Solomon SD, Manning Q, Chen Z, Marquis M, Averyt KB, Tignor M, Miller HL (eds) Climate Change 2007: the physical science basis. Contribution of working group I to the fourth assessment report of the intergovernmental panel on climate change. Cambridge University Press, Cambridge, United Kingdom and New York, NY, USA

Liniger HP, Weingartner R, Grosjean M (1998) Mountains of the world: water towers for the 21st century—a contribution to global freshwater management. Department of Geography, University of Berne, Switzerland, Mountain Agenda

Lozan JL, Grassl H, Hupfer P (2001) Climate of the 21st Century: changes and risks (eds) Hamburg, Wissenschaftliche Auswertungen

Maurya AS, Shah M, Deshpande RD, Bhardwaj RM, Prasad A (2011) Hydrograph separation and precipitation source identification using stable water isotopes and conductivity: River Ganga at Himalayan foothills. Hydrol Proc 25(10):1521–1530

Merz J, Nakami G, Weingartner R (2003) Potential solution to water scarcity in the watershed of Nepal's middle mountains. Mt Res Dev 23:14–18

Meybeck M, Green P, Vorosmarty CJ (2001) A new typology for mountains and other relief classes: an application to global continental water resources and population distribution. Mt Res Dev 21:34–45

Nikkanen R, Wisser D, Schumann A (2004) Application of water management indicators on different spatial and temporal scales. Hydrolie und Wasserbewirtschaftung 48:22–27

Pang H, Hou S, Kaspari S, Mayewski P (2014) Influence of regional precipitation patterns on stable isotopes in ice cores from the Central Himalaya. Cryosphere 8(1):289–301

Pradhan B, Jena R (2016) Spatial Relationship between Earthquake, Hot-Springs, and Faults in Odisha, India. Earth Environ Sci 37:1–13

Prince RJ (1973) Glacial and fluvioglacial landforms. Edinburgh, Oliver, and Boyd

Rao KL (1975) India's water wealth. Orient Longman, London

Rathore BP, Bahuguna IM, Singh SK, Brahmbhatt RM, Randhawa SS, Jani P, Yadav SKS, Rajawat AS (2018) Trend of snow cover in western and west-central Himalayas during 2004–2014. Curr Sci 111(4):800–807

Ruddiman WF (2005) How did humans first alter global climate? Scientific America 292(3):34–41

Raup BH, Keiffer HH, Hare TM, Kargel JS (2000) Generation of data acquisition requests for the ASTER satellite instrument for monitoring a globally distributed target: glaciers. IEEE Trans Geosc Remote Sens 38:1105–1112

Scott CA, Zhang F, Mukherji A, Immerzeel W, Mustafa D, Bharti I (2019) Water in the Hindu Kush Himalaya. Springer International Publishing, The Hindu Kush Himalaya Assessment

Shahjahan M (1983) Regional cooperation in the utilization of water resources of the Himalayan Region. In: Zaman M (ed) River basin development. Tycooly, Dublin.

Sharma JR, Paithankar Y (2014) Brahmaputra Basin. Version 2. Ministry of water resources, Government of India

Siebert S, Doll P, Hoegeveen J (2002) Global map of irrigates areas. FAO, Rome, Italy

Singh SK (2005) Chemical weathering in the river basins of the Himalaya India. Current Sci 89(5):841–849

Shi Z, Liao F, Wang G, Xu Q, Mu W, Sun X (2017) Hydrogeochemical characteristics and evolution of hot springs in eastern tibetan plateau geothermal belt, Western China: Insight from Multivariate Analysis. Geofluids, Hindawi

Smith RB (1979) The influence of mountains on the atmosphere. Adv Geophys 21:87–230

Springer AE, Stevens LE, Anderson DE, Partnell RA, Kreamers DK, Levin LA, Flora S (2008) A comprehensive springs classification system. integrating geomorphic, hydrogeochemical and ecological criteria. In: Stevens LE, Meretsky VJ (eds) Aridland Springs in North America. University of Arizona Press and the Arizona-Sonora Desert Museum, Tucson, pp 49–75

Viviroli D, Weingartner R (2004) The hydrological significance of mountains: from regional to global scale. Hydrology and Earth Sciences 8(6):1016–1029

Viviroli D, Weingartner R, Bruno M (2003) Assessing the hydrological significance of the World's mountains. Mt Res Dev 23(1):32–40

World Bank (2001) World development indicators. International Bank of Reconstruction and Development. Washington DC, USA 142–145

Yang H, Zahnder AJB (2002) Water scarcity and food impact: a case study for southern Mediterranean countries. World Dev 30:1413–1430

Young MC (2007) Hot Springs and Hot Pools of the Southwest Publisher: Aqua Thermal Access 4:8.

Part II
Ecosystem, Resource and Development

Chapter 6
Assessing the Channel Planform Dynamics of River Jhelum from Sangam to Asham, Kashmir Valley, Jammu and Kashmir, India

Hakim Farooq Ahmad and Bashiqa Farooq

Abstract The alteration of the River Jhelum's course in recent years, particularly in areas characterized by lithological susceptibility, underscores the intricate geomorphic dynamics influenced by geological conditions. The river's tendency to shift its path can be attributed to varying lithological strengths, where areas with geological vulnerabilities experience heightened erosion and channel adjustments. The channel planform dynamics of the River Jhelum during 1992, 2001, and 2015 carry substantial significance. The temporal analysis across these periods provides insights into the river's evolution and allows for trend identification. The spatial deviation assessment, measuring changes, quantifies how the river has shifted, offering valuable information about geomorphic characteristics. The present study amply indicates that the Jhelum Basin has undergone changes over the 30 years between 1972 and 2015 and that its equilibrium is being disturbed continuously. The comprehensive temporal analysis of River Jhelum's planform and the sinuosity coefficients across different segments provide valuable insights into the dynamic nature of the river system. The observed deviations from the base map, analyzed from 1972 to 2015, reveal both stable and fluctuating patterns in various locations, indicating the river's adaptability to changing conditions. The sinuosity analysis further unveils spatial variations in meandering behavior along different segments, showcasing the intricate geomorphic dynamics. The analysis of floodplain distances along the river between 2001 and 2015 reveals a consistent trend of decreasing values for both the right and left sides. This suggests a notable shift in the river's course or floodplain dynamics during this period. The results from the study will be a step forward in framing policy and plans to reduce the flood risk in the valley.

Keywords River Jhelum · Planform · Lithology · Sinuosity · Flood risk

H. F. Ahmad (✉) · B. Farooq
Department of Geography and Disaster Management, University of Kashmir, Srinagar, India
e-mail: hakimfarooq.geo@gmail.com

S. C. Rai (ed.), *Himalaya: Mountains of Destiny*,
https://doi.org/10.1007/978-3-031-94905-0_6

6.1 Introduction

A river basin is a distinct hydrological unit and has its own 'geo-bio-cultural' interactions (Alabyan and Chalov 1998; Alvarez 2005). Rivers play a fundamental role as vital lifelines, providing essential freshwater resources for ecosystems, agriculture, and human populations. These dynamic and complex natural systems continuously undergo geomorphic changes in their planform, which refers to the spatial configuration and pattern of the river channel. Understanding these planform changes is crucial for comprehending broader riverine ecosystem dynamics, hydraulic processes, and the impact of human activities on fluvial landscapes. The intricate interplay of natural forces, climatic variations, and human interventions contributes to the evolving nature of river planforms. River dynamics are closely linked to hydrology, involving the movement of water through the system. The hydrological cycle, driven by processes like precipitation, evaporation, runoff, and infiltration, directly shapes river flow and water levels. Morphology, representing the physical configuration of river channels, varies from meandering to braided, influenced by factors like slope and sediment load. Sediment transport is vital, balancing erosion and deposition to shape riverbed composition. Fluvial processes, including erosion and deposition, mold the landscape over time. Natural events such as floods reshape channels, while human interventions like dam construction alter flow patterns. River dynamics profoundly impact ecosystems, influencing biodiversity and health (Bandyopadhyay et al. 2013; Benda et al. 2004a, b; Nelson et al. 2013). Climate change adds complexity, affecting precipitation patterns and flow regimes. Monitoring and modeling, aided by technology like remote sensing, enable prediction, impact assessment, and effective management strategies. The orderly presentation of these interconnected aspects contributes to a comprehensive understanding of the intricate dynamics shaping river systems.

A river changes its course continuously to adjust itself to the topography. A river course, in general, is also sensible to major natural as well as anthropogenic calamities. Any major or continued hindrance originating from such calamities leads to changes in the river course. These changes not only affect the main river course but also the surrounding landforms (Bertrand et al. 1998). The formation of meanders is a common process in rivers and varies between different rivers based on the lithology of the river valley floors. Meanders, characterized by sinuous curves in a river or watercourse, occur as the watercourse erodes sediments on the outer concave bank, creating a cut bank or river cliff, while simultaneously depositing sediments on the inner convex bank, known as a point bar. This dynamic process results in the development of a sinuous course, prompting the channel to shift across the floodplain. The region where a meandering stream periodically alters its channel is termed a meander belt, typically spanning 15–18 times the width of the channel. Over time, meanders migrate downstream, presenting challenges for infrastructure maintenance. Sinuosity, measured as the ratio of the channel length to the straight-line down-valley distance, serves as an indicator of the degree of meandering, with watercourses having a single channel and sensuosities of 1.5 or more classified as meandering streams or rivers. Old maps and qualitative records over the past eight centuries have been used by various authors to describe meandering (Alabyan and Chalov 1998; Bandyopadhyay et al. 2013; Benda et al. 2004a, b; Bertrand et al. 1998; Magliulo and Valente 2015).

The Himalayan rivers in India stand out for their unique characteristics shaped by substantial seasonal fluctuations in both flow and sediment load. These rivers have adapted to a diverse range of discharges, showcasing morphologies influenced by substantial flood events. While the flow patterns are somewhat influenced by snowmelt from the Himalayas, the rivers also experience notable discharge fluctuations, characterized by peak flows during the summer monsoon and reduced discharges in other months. One of the striking characteristics is that the rivers originating from the Himalayan orogenic belt are highly sediment-laden and contribute 20% of the global sediment input into the oceans (Stolum 1998; Vorosmarty et al. 2003). Among the major Himalayan river systems, the Jhelum is one of the most important rivers of Kashmir. The river rises from a spring at Verinag in the south-eastern part of the Valley of Kashmir. It flows north-westward for about 110 km where it enters the Wular Lake. Further downstream from the Baramulla, it enters a gorge 2130 m deep and moves toward Muzaffarabad Pakistan (Williams 1978, 1986).

6.2 Significance of the Study

The channel planform dynamics of the River Jhelum during 1970, 1992, 2001, and 2015 carry substantial significance. The temporal analysis across these periods provides insights into the river's evolution and allows for trend identification. The spatial deviation assessment, measuring changes from a base map, quantifies how the river has shifted, offering valuable information about geomorphic characteristics. Calculating the meandering coefficient contributes to understanding the river's sinuosity and response to various factors. Beyond scientific understanding, the research has environmental implications, aiding in assessing the impact on habitats, ecosystems, and communities. Additionally, it establishes a baseline for future studies, supporting ongoing monitoring and river management strategies. Overall, the dissertation enhances our knowledge of hydrological processes, fostering sustainable development, and conservation efforts in the River Jhelum region.

6.3 Study Area

The Jhelum River, a major tributary of River Jhelum, originates from the spring of Verinag, on the north-western side of Pir Panjal and flows in a direction parallel to the Indus at an average elevation of 5500 feet (1677 m). Flowing in a northwest direction, it is joined by many streams before reaching the Wular Lake (Fig. 6.1).

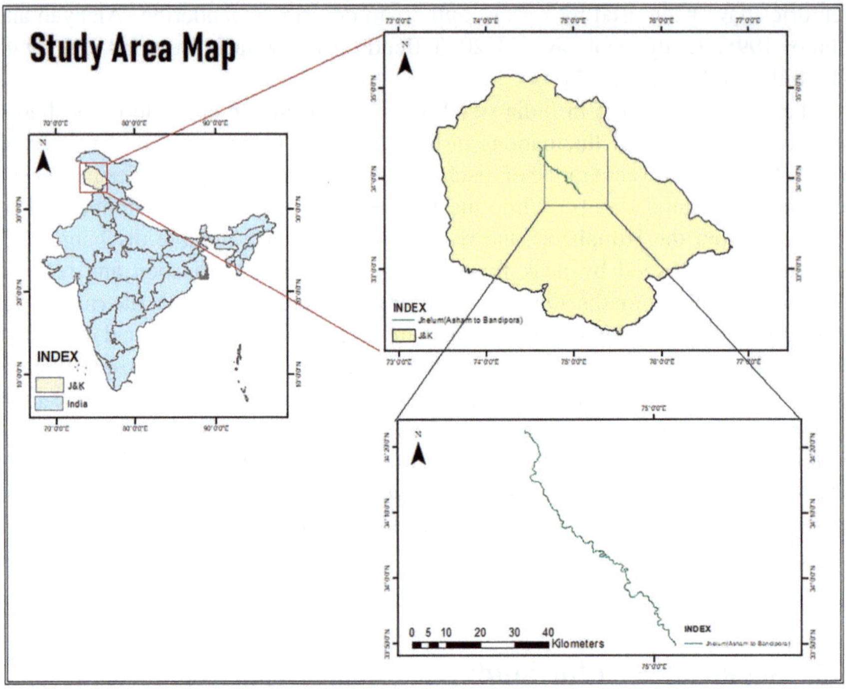

Fig. 6.1 Map showing study area

It covers approximately 165 km in India, with a catchment area of about 17,622 km^2. The river meanders through the picturesque Kashmir Valley, flanked by mountain ranges rising to 5487 m in the northeast. Peaks in this range are snow-covered from October to May. The valley, perched at an average elevation of 1829 m, is 135 km in length and 32–40 km in breadth, with notable peaks like Kuti and Brahmasakal at 4675 m in elevation.

The physiography of the alluvial valley is characterized by higher riverbanks, a common feature in alluvial plains. The Jhelum River is fed by major tributaries like the Lidder, Sandran, Bringi, Arapath, Vishow, and Rambiara. Its course is marked by high banks, except for some swampy stretches prone to floods. The river navigates a zig-zag path from Khanabal to Srinagar, abutting the hills on its right side. Before entering the city of Srinagar, it is supplemented by a channel to protect against floods, and downstream, Dudh-ganga and Nallah Sindh join. At Baramulla, the river enters a gorge, taking a sharp bend below Khadanyar, where additional streams like Wingle, Pohru, and Vij meet. Noteworthy tributaries include Lidder, Sandran, Bringi, Arapath, Vishow, and Rambiara (Dar et al. 2017).

The study area covers the Jhelum River from its entry into Wular Lake to the confluence with the Veshaw River, with significant towns like Sopore, Baramulla, and areas within the Kashmir Valley. The physiography encompasses the high mountain ranges and valleys, while the river system includes the intricate network of tributaries contributing to the Jhelum's flow dynamics.

6.4 Database and Methodology

The assessment of River Jhelum's channel planform dynamics was conducted through a systematic approach. Satellite images from three different years, 1992, 2001, and 2015, were used, and a meticulous digitization process at a scale of 1:30,000 was undertaken using ArcGIS, aligning them with a reference base map (SOI Toposheet 1970). The deviation distances were quantified employing the measurement tool to precisely determine alterations from the base map. For a more in-depth analysis, River Jhelum was divided into segments. The meandering coefficient, a crucial parameter for understanding the sinuosity of a river, was calculated using established formulas.

$$\text{Sinuosity} = \text{Segment length/Total straight line distance.}$$

The sinuosity, reflecting the winding course of the river, was also computed. This methodology aligns with established practices in geomorphological studies and builds upon previous knowledge, ensuring a standardized and rigorous approach to the assessment of channel planform dynamics.

6.5 Results and Discussion

The alteration of the River Jhelum's course in recent years, particularly in areas characterized by lithological susceptibility, underscores the intricate geomorphic dynamics influenced by geological conditions (Fig. 6.2). The river's tendency to shift its path can be attributed to varying lithological strengths, where areas with geological vulnerabilities experience heightened erosion and channel adjustments. This geomorphic response may be influenced by a combination of factors, including lithological variations, climate fluctuations, and human activities (Table 6.1). Understanding these intricacies is pivotal for regional planning, infrastructure development, and environmental management. It also contributes to scientific discourse on fluvial systems' responses to geological variables, shedding light on the adaptability and resilience of riverine landscapes to evolving conditions.

The graph displays (Fig. 6.3) the temporal changes in the planform of the Jhelum River at various locations, measured as the deviation distance from the base map (1792) for the years 1992, 2001, and 2015. Locations h and i show a decrease in deviation distance from the base map over the years, with location h reaching 0 in 2015. This could indicate stabilization or return to the original path at these locations. Locations f, g, and j also exhibit decreasing deviation distances, suggesting a trend

Base map reference points	Map of 1992 showing deviation of distance from the base map	Map of 2001 showing deviation of distance from the base map	Map of 2015 showing deviation of distance from the base map
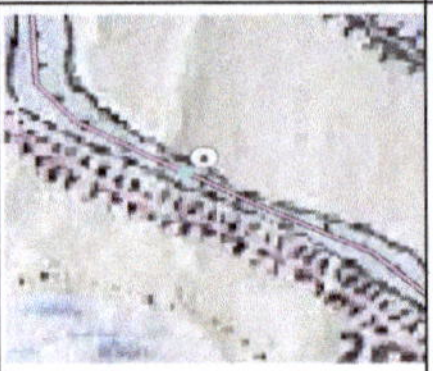	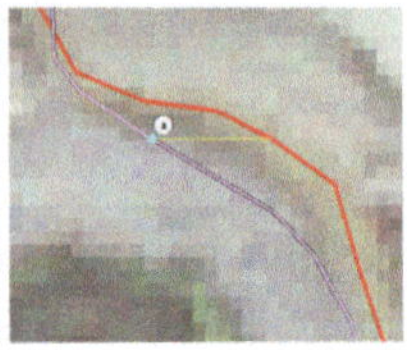	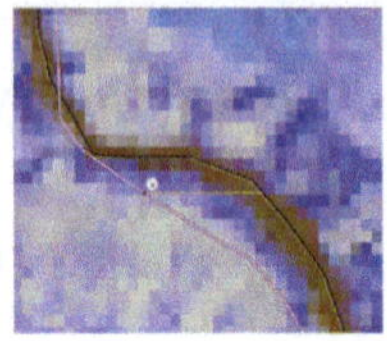	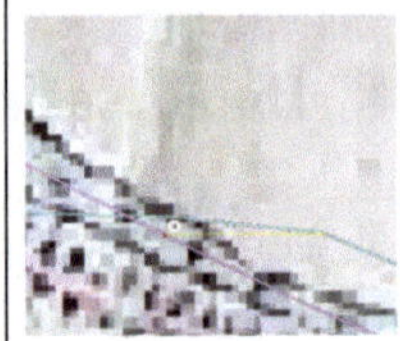
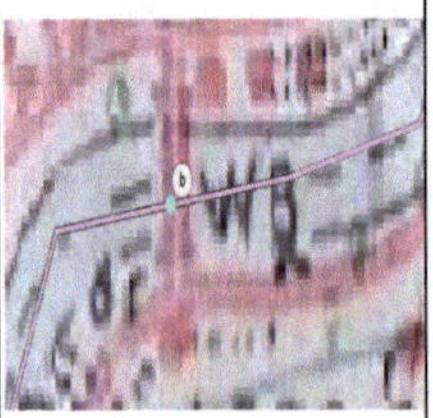		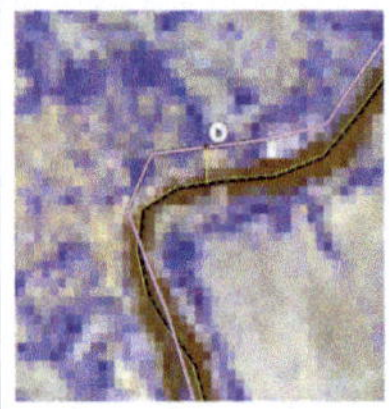	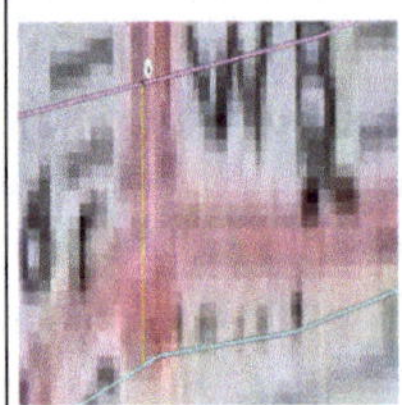
	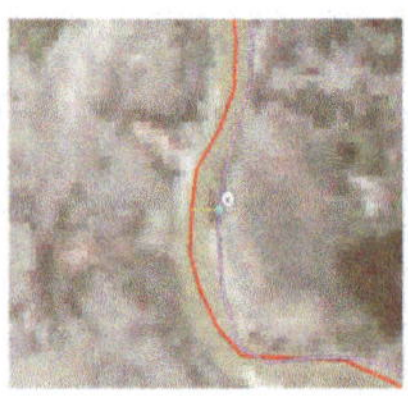	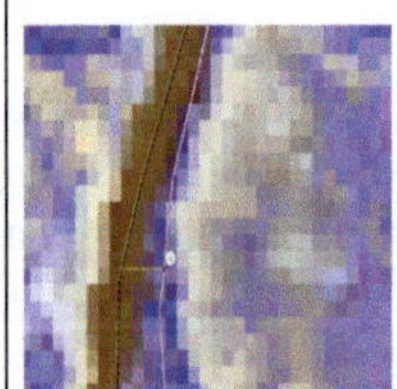	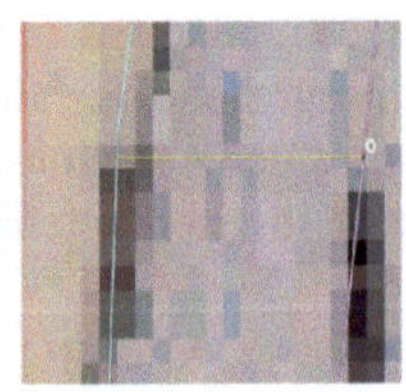
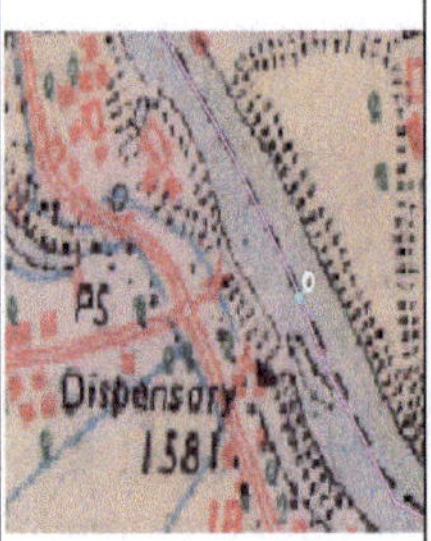	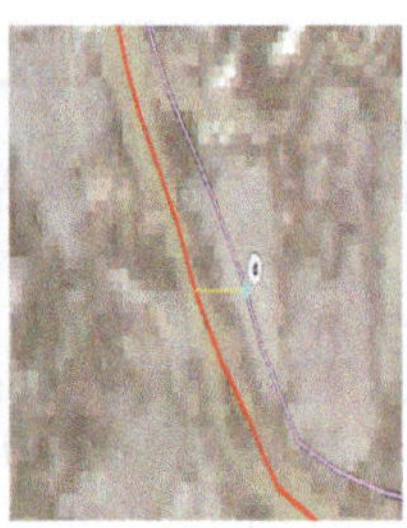	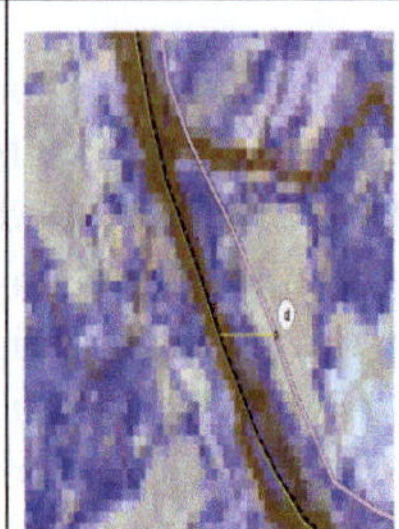	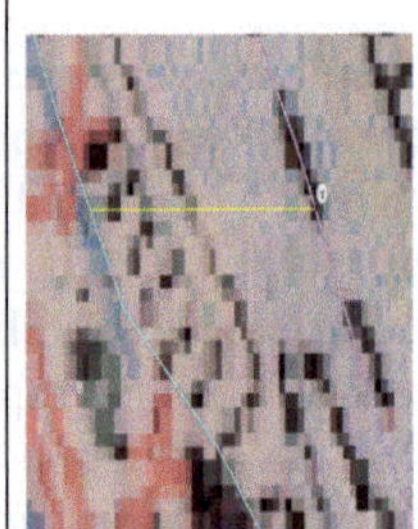
Base map reference	Map of 1992 showing	Map of 2001 showing	Map of 2015 showing

Fig. 6.2 Meandering segments of River Jhelum at different locations showing distance deviation from base map across a temporal scale

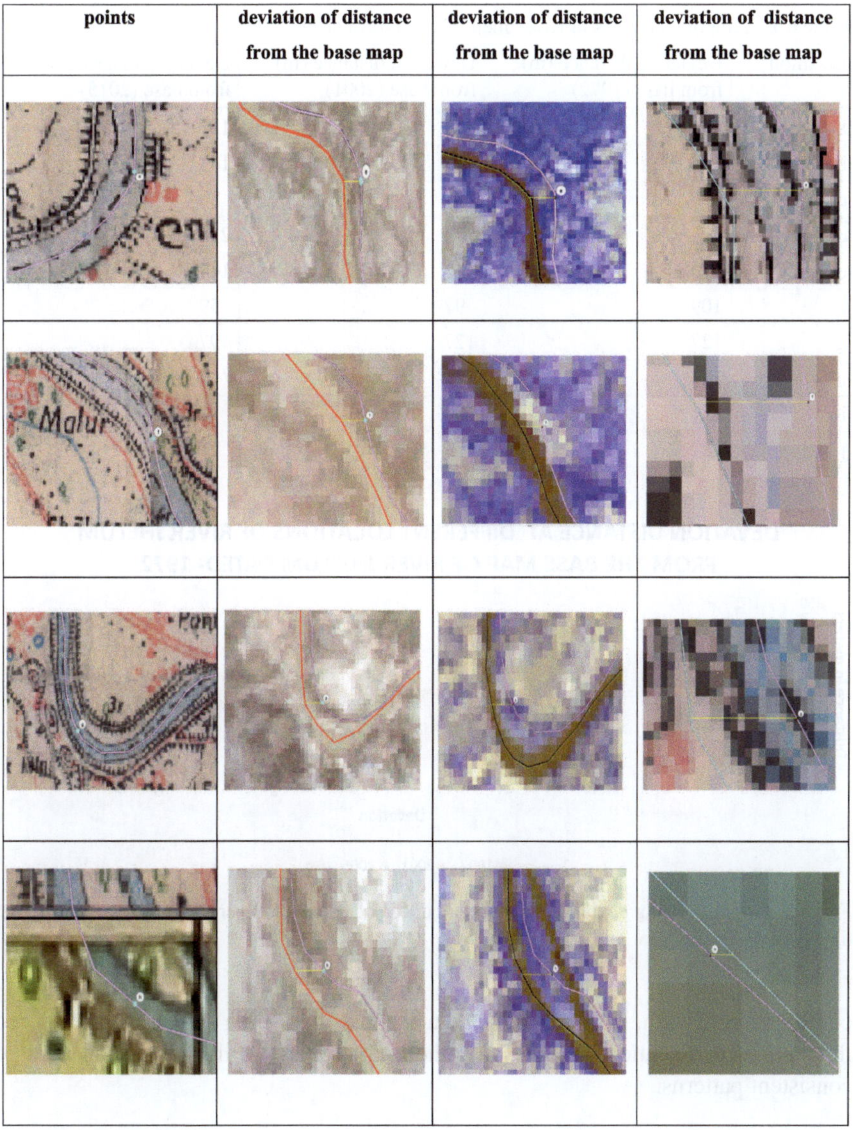

Fig. 6.2 (continued)

toward the original course. Locations b and g have relatively consistent deviation distances across the three-time points, suggesting a stable planform in these areas. Location h shows a consistent decrease, potentially indicating efforts to restore the river to its original path. Locations a, d, e, and j have varying degrees of deviation, showcasing the dynamic nature of the river planform. The variations may be influenced by factors like natural geomorphic processes, human interventions, or a

Table 6.1 Temporal changes in River Jhelum's planform (1992, 2001, 2005)

Location	Deviation distance (m) from base (1972)	Deviation distance (m) from base (2001)	Deviation distance (m) from base (2015)
a	281	315	139
b	113	113	141
c	140	108	88
d	272	272	226
e	179	166	144
f	109	97	59
g	127	127	97
h	164	179	0
i	151	154	12
j	182	153	17

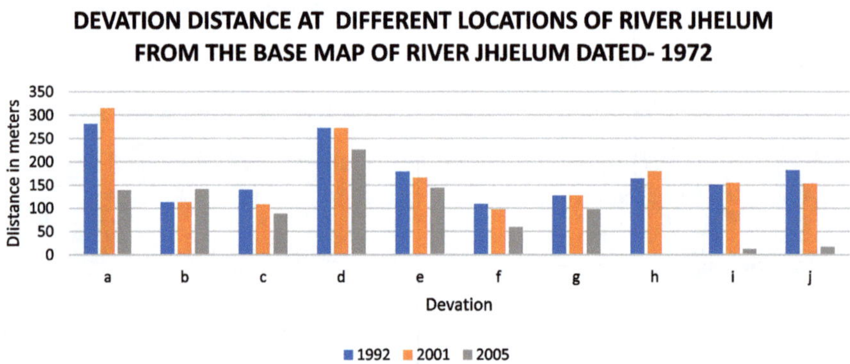

Fig. 6.3 Graph showing deviation distances at different points of River Jhelum of three different time periods

combination of both. In general, the data suggests that some locations experienced fluctuations in deviation distances over the studied years, while others showed more consistent patterns.

6.5.1 Calculation of Sinuosity of River Jhelum

The study instigates the temporal shifts in the planform of the Jhelum River across various locales, with deviation distances measured from a reference point established in 1972. Covering the years 1972, 2001, and 2015, the data provides a nuanced exploration of alterations in the river's course (Fig. 6.4).

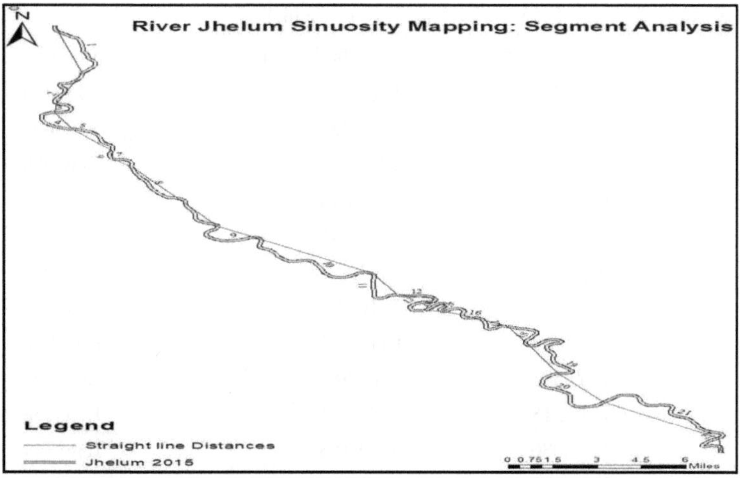

Fig. 6.4 Sinuosity profile of River Jhelum

The subsequent analysis delves into location-specific trends, the constancy of deviation distances, and variable changes observed, unveiling a mosaic of patterns that invite consideration of both natural processes and anthropogenic influences (Table 6.2). This study thus serves as a valuable lens through which to discern the complex dynamics governing the Jhelum River's planform evolution over time.

Segmented Analysis: The table breaks down the river into distinct segments, allowing for a detailed analysis of sinuosity variations along the course.

Meander Length: Indicates the cumulative length of the river's winding path within each segment, providing insight into the overall sinuous nature of that particular part of the river.

Straight-Line Distance: This represents the direct distance between the start and end points of each segment, serving as a reference for comparison with the meandered length.

Sinuosity: Quantifies the degree of meandering in each segment. Sinuosity values above 1 suggest a winding or meandering river path, while values below 1 indicate a more linear or straight path.

Variations in sinuosity values across segments highlight spatial patterns of meandering behavior along the river. Higher sinuosity values (above 1) in segments 3, 13, 14, and 15 suggest significant meandering, indicating that the river's path is substantially longer and more sinuous compared to a straight line. Segments with sinuosity values around 1.5 (e.g., Segment 11) indicate a moderately meandering river path. Segments with sinuosity values below 1.2 (e.g., Segments 5, 6, 12, and 16) suggest less meandering, closer to a straight-line path (Fig. 6.5).

Precisely the table serves as a valuable tool for understanding how sinuosity varies along different segments of the River Jhelum, providing insights into the river's meandering behavior and geomorphological characteristics.

Table 6.2 Sinuosity analysis of river segments of River Jhelum

Segment	Meander length	Straight_Line_Distance	Sinuosity
1	6589.313	5683.654	1.159344
2	4527.73	3825.226	1.18365
3	2234.978	1083.842	2.062089
4	3424.408	2142.894	1.59803
5	3803.342	3402.746	1.117727
6	1008.29	941.4274	1.071023
7	1768.825	1416.898	1.248379
8	9104.853	8415.241	1.081948
9	4169.573	2283.885	1.825649
10	9746.518	7780.171	1.252738
11	4237.086	2845.188	1.489211
12	2100.399	2035.162	1.032055
13	3004.496	1504.695	1.996747
14	4435.307	890.7075	4.979533
15	4435.307	1212.268	3.658686
16	1445.535	1253.124	1.153546
17	2286.975	1505.535	1.519045
18	3673.359	2616.091	1.40414
19	7564.157	3610.177	2.095232
20	6917.541	4200.157	1.646972
21	10,182.09	7529.955	1.352211
22	3217.911	1811.467	1.776411

Table 6.3 Meandering geometrics of River Jhelum at random locations

Location	Right side distance (m) 2015	Left side distance (m) 2015	Right side distance (m) 2001	Left side distance (m) 2001
Point 1	Data not available	10,698.04	Data not available	10,811.851
Point 2	Data not available	8566.99	Data not available	8645.84
Point 3	8836.73	11,000	8724.38	11,119
Point 4	8056.47	1038.34	7941.09	1155.60
Point 5	7795.01	823.82	7627.71	991.40
Point 6	4850	548	747	4567

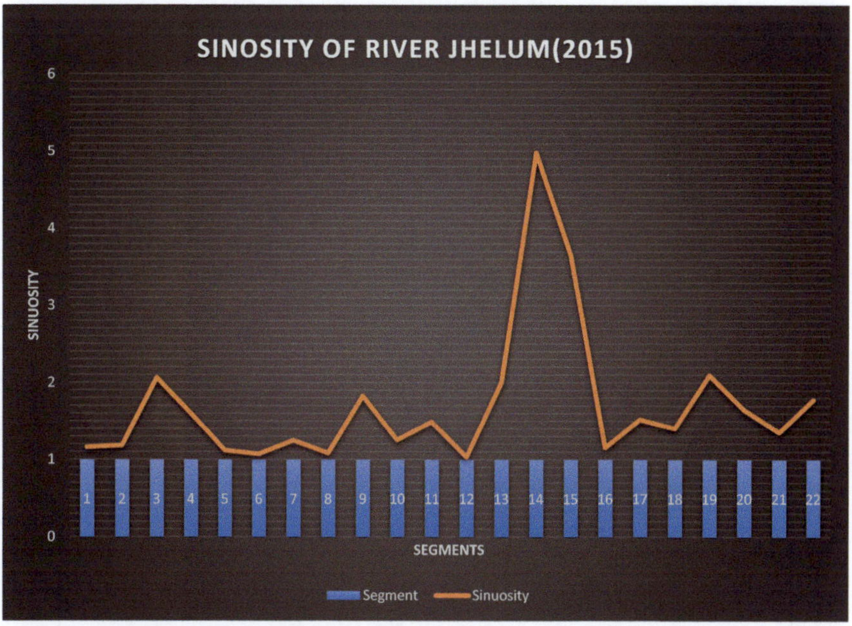

Fig. 6.5 Sinuosity of River Jhelum at different segments

6.5.2 Spatial Dynamics of the Jhelum River: (1992, 2001, 2015)

The Jhelum River floodplain is a significant geographical feature associated with the Jhelum River. The floodplain is the flat, low-lying area (Fig. 6.6) adjacent to the river that is subject to periodic inundation during periods of increased water flow, such as heavy rainfall or snowmelt.

Fluctuations in deviation distances over the studied years imply a dynamic river planform, influenced by natural geomorphic processes, climate changes, and human interventions. Segments with higher sinuosity values (e.g., 3, 13, 14, and 15) indicate significant meandering, while others with lower values (e.g., 5, 6, 12, and 16) suggest less meandering, contributing to a nuanced understanding of river behavior.

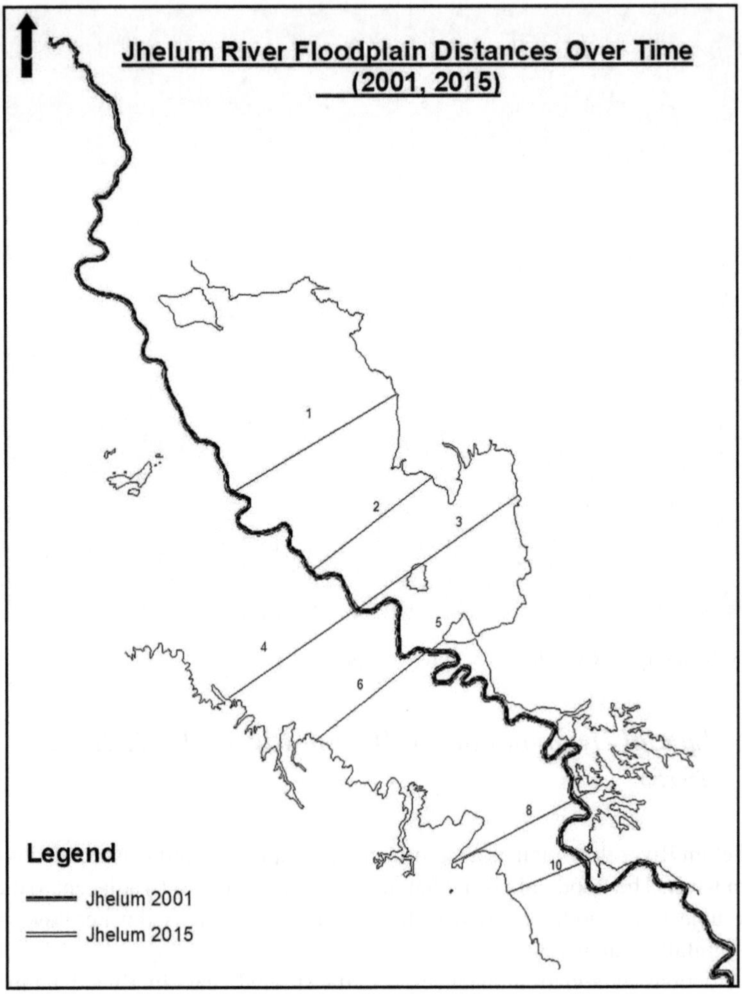

Fig. 6.6 Jhelum flood left/right bank stretches

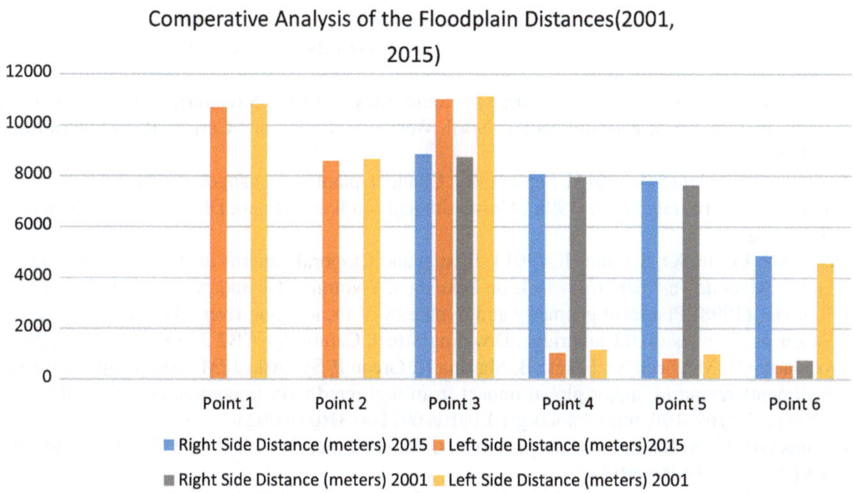

Fig. 6.7 Graphical representation of Jhelum River meandering

6.6 Conclusion

The present study amply indicates that the Jhelum Basin has undergone changes over the 30 years between 1972 and 2015 and that its equilibrium is being disturbed continuously. The comprehensive temporal analysis of River Jhelum's planform and the sinuosity coefficients across different segments provide valuable insights into the dynamic nature of the river system. The observed deviations from the base map, analyzed from 1972 to 2015, reveal both stable and fluctuating patterns in various locations, indicating the river's adaptability to changing conditions. The sinuosity analysis further unveils spatial variations in meandering behavior along different segments, showcasing the intricate geomorphic dynamics.

References

Alabyan A, Chalov R (1998) Types of river channel patterns and their natural controls. Earth Surface Proc Landforms 23:467–474. https://doi.org/10.1002/(SICI)1096-9837(199805)23:53.0.CO;2-T

Alvarez A (2005) Channel planform dynamics of an alluvial river. PhD Thesis, Texas A&M University, USA

Bandyopadhyay S, Saha S, Ghosh K, De SK (2013) Channel planform change and detachment of tributary: a study on the Haora and Katakhal Rivers, Tripura, India. Elsevier B.V

Benda L, Andras K, Miller D, Bigelow P (2004a) Confluence effects in rivers: interactions of basin scale, network geometry, and disturbance regimes. Water Resour Res 40(5):1–15

Benda L, Poff NL, Miller D, Dunne T, Reeves GH, Pess GH, Polluck M (2004b) The network dynamics hypothesis: how channel networks structure riverine habitats. Bioscience 54:413–427

Bertrand G, Rangin C, Maury RC, Htun HM, Bellon H, Guillaud JP (1998) The Singu basalt (Myanmar): new constraints for the amount of recent offset on the Sagaing Fault. Earth and Planetary Sci 327:479–484

Dar AA, Bali BS, Tahir TA (2017) Morphotectonic study of Bringi Stream right bank tributary of River Jhelum, SE of Kashmir Valley, North West Himalayas. Int J Current Res Modern Educ 2(2):318–329

Magliulo OP, Valente A (2015) September) Channel planform changes of the Lower Calore River, Southern Italy (1870–1998). UniversitàdegliStudi del Sannio, Dipartimento di Scienze e Tecnologie

Nelson NC, Erwin SO, Schmidt JC (2013) Spatial and temporal patterns in channel change on the Snake River downstream from Jackson Lake dam, Wyoming. Geomorphology 200:132–142

Stølum HH (1998) Planform geometry and dynamics of meandering rivers. Department of Earth Sciences, University of Cambridge, Downing Street, Cambridge CB2 3 EQ, United Kingdom.

Vorosmarty CJ, Meybeck M, Fekete B, Sharma K, Green P, Syvytski JPM (2003) Anthropogenic sediments retention: major global impact from registered river impoundments. Global Planet Change 39:169–190. https://doi.org/10.1016/s0921-8181(03)00023-7

Williams GP (1978) Bankful discharge of rivers. Water Resour Res 14:1141–1154. https://doi.org/10.1029/WR014i006p01141

Williams GP (1986) River meanders and channel size. J Hydrol 88:147–164. https://doi.org/10.1016/0022-1694(86)90202-7

Chapter 7
Ecosystem, Resources, and Development Synergy in the Northwestern Himalayas

Saiema Ali and Manzoor A. Wani

Abstract Set against the awe-inspiring backdrop of the Himalayan peaks, this study explores the multifaceted dynamics of North Kashmir, a region rich in natural beauty and cultural heritage. Using vibrant imagery that reflects its agricultural diversity, forestry, and burgeoning tourism, the research delves into the complex interactions between physical and cultural elements that influence population distribution, socio-economic structures, and environmental conditions. The study presents a detailed analysis of topographical features, climate conditions, land use, and cultural influences, revealing critical insights into population dispersal, literacy challenges, and gender imbalances. With a predominantly rural population of 2,171,508 people and a literacy rate of just 52%, the research highlights the struggle for educational attainment and the skewed gender ratio of 864 females per 1000 males. The paper advocates for strategic enhancements in productivity and sustainability across agriculture, horticulture, and sericulture. It emphasizes the need for robust infrastructure development, particularly in tourism hubs, to maximize the region's natural and cultural potential. Through addressing these challenges, North Kashmir can pursue a trajectory of inclusive development and sustainable growth. The paper advocates for strategic enhancements in productivity and sustainability across agriculture, horticulture, and sericulture. It emphasizes the need for robust infrastructure development, particularly in tourism hubs, to maximize the region's natural and cultural potential. Through addressing these challenges, North Kashmir can pursue a trajectory of inclusive development and sustainable growth.

Keywords Himalayan peaks · North Kashmir · Socio-economic dynamics · Environmental sustainability

S. Ali · M. A. Wani (✉)
Department of Geography and Disaster Management, University of Kashmir, Srinagar, India
e-mail: manzoorwani8559@gmail.com

S. C. Rai (ed.), *Himalaya: Mountains of Destiny*,
https://doi.org/10.1007/978-3-031-94905-0_7

7.1 Introduction

The resource is that component of man's physical environment on which he relies for assistance to gain the means to meet specific needs (Zimmermann 1933). Mostly, the resources it controls and the ability of its people to use and protect these resources for future generations determine the power of a region. Resources are the vital for the functions of human survival and acts as a corner stone for the economic development and abundant resources and its proper utilization are regarded as a great advantage for country's economic development (Zhao 2019). Utilization of resources and enhancing their growth in terms of productivity and efficiently are pivotal for moving toward the sustainable economic growth in the major global economies (Shah et al. 2024). Development plans for every region must therefore include the locally available resources and the possibility of maximizing their exploitation to promote a better level of rural development.

From the economic perspective, resources are considered as factors of production that can be extracted, mined or harvested as raw materials that after efficiently processed and boosting our economies, the resource can be classed as high, medium, or poor quality depending on its inherent capacity and technical ease (cost) of satisfying its demands and aspirations. The majority of the resources we consume, such as land, minerals, fish, crops, and people, are of high quality. We consume more medium or low-quality resources when technology advances, demand rises, or depletion occurs (Neo 2009). The assessment of North Kashmir's resource base is critical for proposing a region-specific resource development plan for the best usage of the area's existing resources. Human population, agriculture, woods, horticulture, tourism, and other resources are vital to the region. The purpose of this study is to analyze the resource base and its spatial distribution in the area.

7.2 Study Area

Kashmir valley a half-closed ecosystem, opening up slowly in space (Moonis et al. 1978). Bewitching Kashmir being the cynosure of all eyes. This landscape is thought to be the paradise on Earth, adorned with tall and towering mountains having snow-clad peaks, which surrounded it from all sides. The Kashmir valley lies between the Pir Panjal range from the south-west and Zanskar range in the northeast. Geographical coordinates of north Kashmir is 34°26′N and 74°39′E. The total area of the region is estimated approximately of 6967 km^2 of which district Kupwara shares 2379 km^2 (34.10%), Baramulla with 4243 km^2 shares 60.90% of the total area and district Bandipora shares the least area of 345 km^2 which accounts for only 5% of the total area (Census 2011). Baramulla, the oldest and the most important town of the north princely state of Jammu and Kashmir. The district shares its border with Srinagar and Budgam in the south and Bandipora in the east. District Baramulla and Kupwara share a long boundary with Pakistan along northwestern side, which up to 27 October

1947, was the *Gateway of Kashmir Valley'* by the Rawalpindi-Murree-Muzaffarabad-Baramulla Road. The district is also considered to be the gateway of all the major weather phenomenon such as moisture-laden winds from western side known as western disturbances.

7.3 Human Resources

The population is as much a part of the habitat as other elements of the natural environment in any resource analysis in any place. The economic growth of any place is the outcome of the natural environment and resources on the one hand, and the territory's existing human resource structure on the other. As a result, the pattern and process of population and development are inextricably intertwined. A population pattern is determined by the degrees and sizes of development and reacts to the requirements of developmental processes. The primitive man is generally on the receiving end, adapting himself to fit into the extremely sophisticated system of secondary and tertiary activities in a well-developed superstructure. The characteristics of the population highlight the process of such adaptation. Humans as both producers and a consumer of resources, are an essential component of the resource system. The utility of human resources is well evident in the development process with all of its positive aspects, such as participation rate, physical fitness, literacy, and so on, which mostly focus on the region's human resources.

7.3.1 Spatial Distribution of Population

The physio-cultural features of the area have a considerable influence on population dispersal. Topography, climate conditions, land fertility, and water availability are the physical elements influencing population distribution in the research area. Agriculture, irrigation facilities, industries, accessibility, and modes of communication are the most important cultural influences. According to the 2019 statistical digest, the overall population of the study area is 2,171,508 people, with a rural population of 1,983,595 people and an urban population of roughly 187,913 people. As per Table 7.1, approximately 91% of the population in the Study area was rural, whereas approximately 8.6% was urban. Out of 50 blocks in the study area, 42 blocks have 100% rural population while 8. The two blocks such as Baramulla and Sopore had more than the National Average Urban population. Figure 7.1 shows the population distribution of both rural–urban blocks in the study area.

Sex Ratio

Figure 7.2 shows the distribution of sex ratio in the study region. North Kashmir has a female population of 46% compared to a male population of 54%. North Kashmir's female population is lower than the national average (48.04%). According to the

Table 7.1 Block-wise rural–urban population of North Kashmir (2019–2020)

Blocks	Population	Rural pop	Percentage	Urban pop	Percentage
Baramulla	140,913	67,998	48	72,915	52
Narvah	44,416	44,416	100	0	0
Sopore	92,958	48,773	52	44,185	48
Zaingeer	58,507	58,507	100	0	0
Tujersharif	40,476	40,476	100	0	0
Wagoora	79,939	79,939	100	0	0
Sangrama	43,884	43,884	100	0	0
Rafiabad	61,997	61,997	100	0	0
Kandibelt	10,171	10,171	100	0	0
Nadihal	54,510	54,510	100	0	0
Rohama	65,325	65,325	100	0	0
Tangmarg	45,478	45,478	100	0	0
Kunzer	30,323	26,581	88	3724	12
Khaipora	13,648	13,648	100	0	0
Wanigam	14,933	14,933	100	0	0
Pattan	149,606	134,037	90	15,569	10
Lalpora	22,245	22,245	100	0	0
Wailoo	19,963	19,963	100	0	0
Hardbora	27,684	27,684	100	0	0
Uri	49,718	42,918	86	6800	14
Parenpila	36,128	36,128	100	0	0
Boniyar	25,013	25,013	100	0	0
Bijhama	23,177	23,177	100	0	0
Noorkhah	30,771	30,771	100	0	0
Singhpora	67,371	67,371	100	0	0
Sherabad Khore	52,277	52,277	100	0	0
Drugmulla	30,879	30,879	100	0	0
Handwara	40,552	26,952	66	13,600	34
Hayhama	37,015	37,015	100	0	0
Kalarooch	44,614	44,614	100	0	0
Keran	12,026	12,026	100	0	0
Kralpora	46,116	46,116	100	0	00
Kupwara	72,849	51,078	70	21,771	30
Langet	26,467	17,118	65	9349	35
Machil	17,030	17,030	100	0	0
Magam	21,134	21,134	100	0	0
Mawar	43,354	43,354	100	0	0

(continued)

Table 7.1 (continued)

Blocks	Population	Rural pop	Percentage	Urban pop	Percentage
Meliyal	25,294	25,294	100	0	0
Nutnoosa	25,663	25,663	100	0	0
Qadirabad	44,686	44,686	100	0	0
Qaziabad	42,755	42,755	100	0	0
Rajwar	29,077	29,077	100	0	0
Ramhal	22,564	22,564	100	0	0
Chowkibal	29,551	29,551	100	0	0
Sogam	78,659	78,659	100	0	0
Tangdar	34,745	34,745	100	0	0
Trathpora	22,777	22,777	100	0	0
Teetwal	25,384	25,384	100	0	0
Trehgam	33,506	33,506	100	0	0
Wavoora	63,398	63,398	100	0	0
Total	2,171,508	1,983,595	91	187,913	8.6

Source District Statistical Handbook, 2019–2020

2019 census, the study area's sex ratio is 864 females per 1000 males, which is significantly lower than the national sex ratio of 947.

Literacy Rate

According to the 2019 Census, the literacy rate in North Kashmir was 52.00%, which is much lower than the state average of 67.2%. Tujersharif block has the highest overall literacy rate of 74.6% among the 50 blocks studied, while block Noorkhah has the lowest total literacy rate of 17.7%. All of the blocks had literacy rates that were lower than the national and state averages. The maximum male literacy rate reported in the study region was 77% in Tujersharif Block, and the lowest male literacy rate was 12.5% in Bijhama. Figure 7.3 shows the literacy rate block-wise.

7.3.2 Land-Use/Cover

The land use analysis of 2020 has been generated from satellite data and the statistics of the same are presented in Fig. 7.4. The statistics reveal that dense forest is the dominant land use with a total area of 2404.3 km^2, which constitutes 45.9% followed by cropland, which occupies 2196.8 km^2, which is 42% of the total area of North Kashmir. Barren land is the next major land cover category spread over an area of 283.547 km^2 occupying 5.4% of the study area as indicated in (Fig. 7.4). Shrub Land constitutes 183.9809 km^2 constituting about 3.5% of the total area. The next major land cover class in the study area is the built-up which covers an area of 147.27 km^2

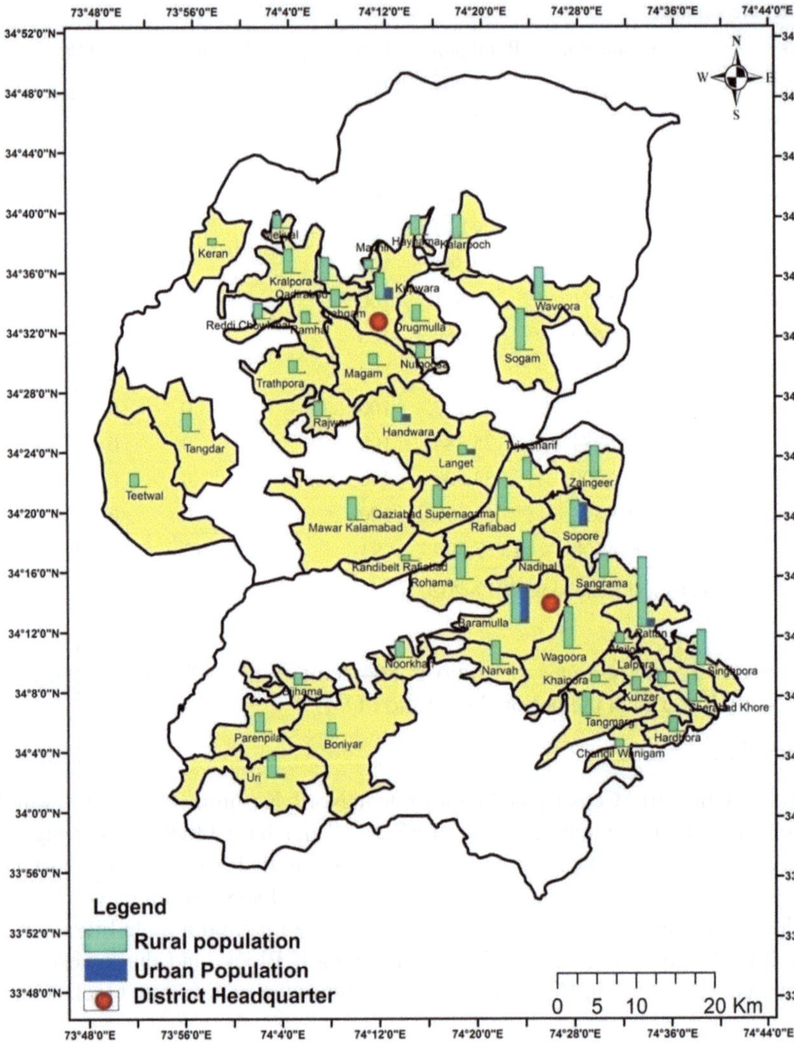

Fig. 7.1 Block-wise rural–urban population (2019–2020)

which is 2.7%. A total of 16.1021 km^2 is occupied by the water bodies which is only 0.31%, whereas Snow/ Glacier cover an area of 3.1 km^2 which constitutes 0.05% of the study area. Figure 7.5 shows the spatial distribution of Land use and land cover (LULC) in the study region (2019–2020).

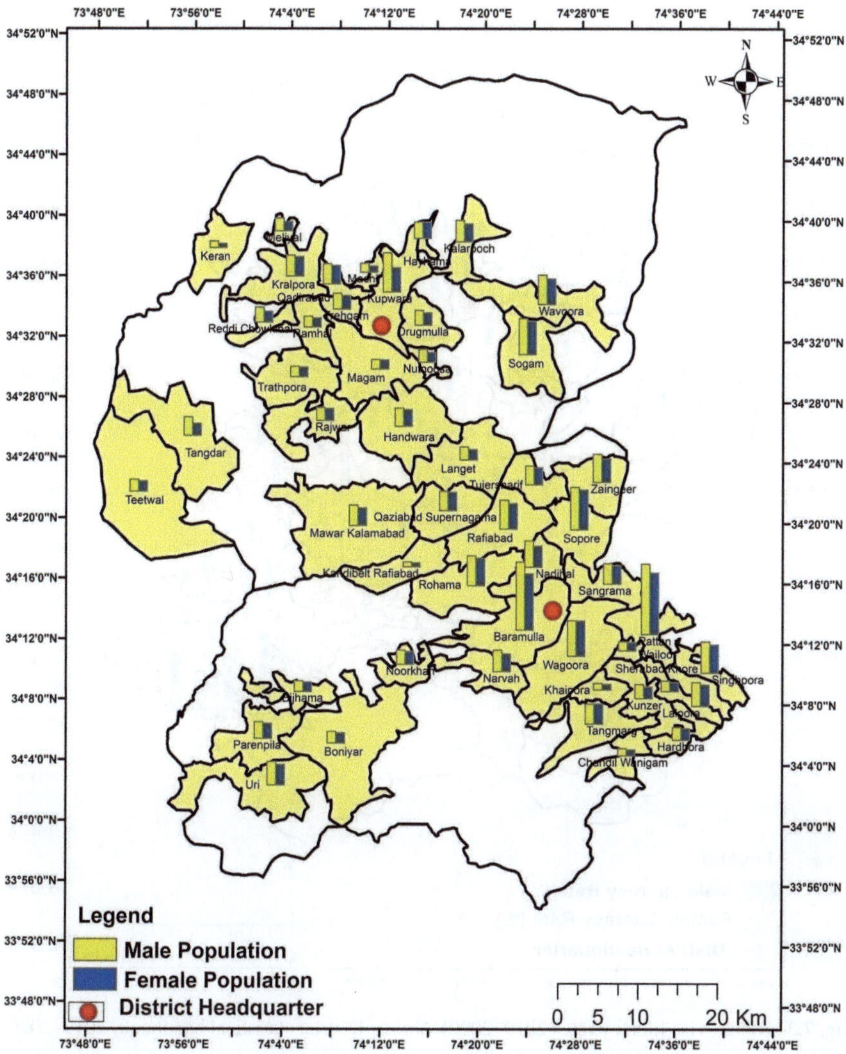

Fig. 7.2 Sex ratio block-wise (2019–2020). *Source* District Statistical Handbook, 2019–2020

7.3.3 Agriculture

Cattle and Sheep Population

It has been observed that the livestock sector provides different and diversified benefits to the rural poor to alleviate the food insecurity challenges of the country's ever-growing population. More precisely, cattle can be the most stable and long-term source of cash income for small and landless farmers, who are believed to be more vulnerable to poverty. Livestock is an important sub-sector of agriculture in

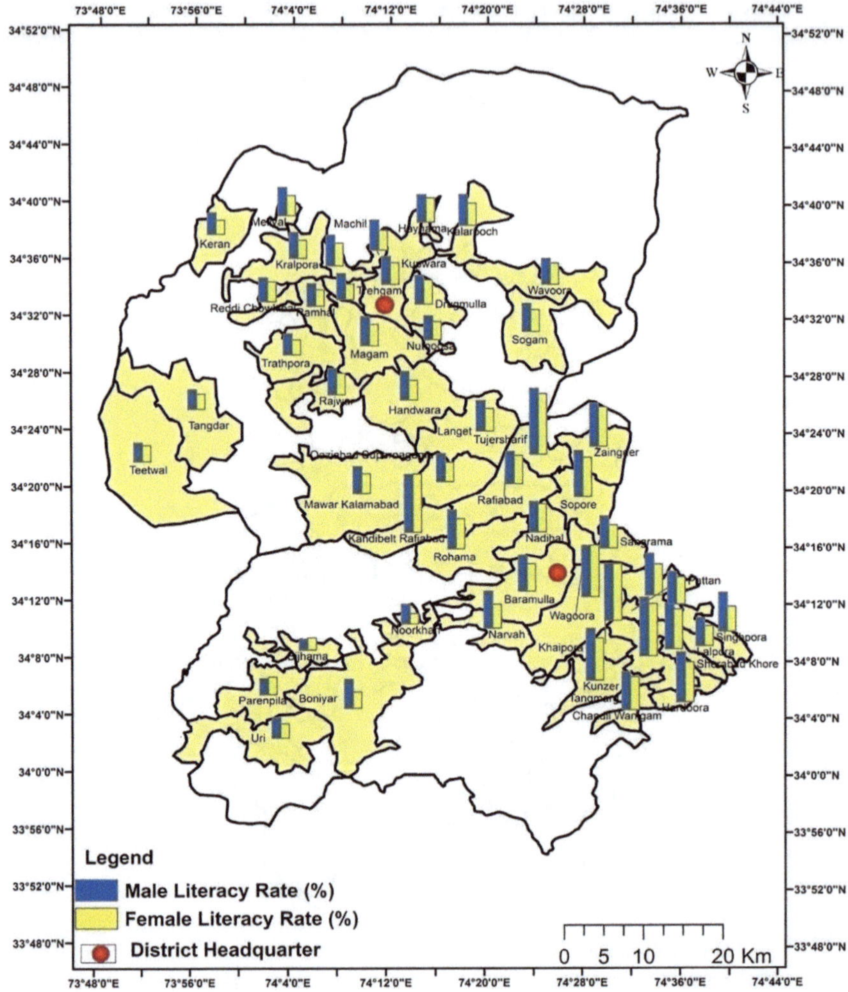

Fig. 7.3 Block-wise literacy rate (2019–2020). *Source* District statistical handbook, 2019–2020

India. Agriculture employed more than half of the Indian labor force in 2018 and generated 17–18% of the country's GDP. (Government of India 2019–20). Despite its relevance at the individual/household and national levels, the country's overall performance in this sector is subpar. Concerning the concerns of the livestock sector in India, many other scholars noted that this sector suffers several issues that impair its actual potential and advancement. These issues impede the country's efforts to achieve the goals of sustainable agricultural and rural development.

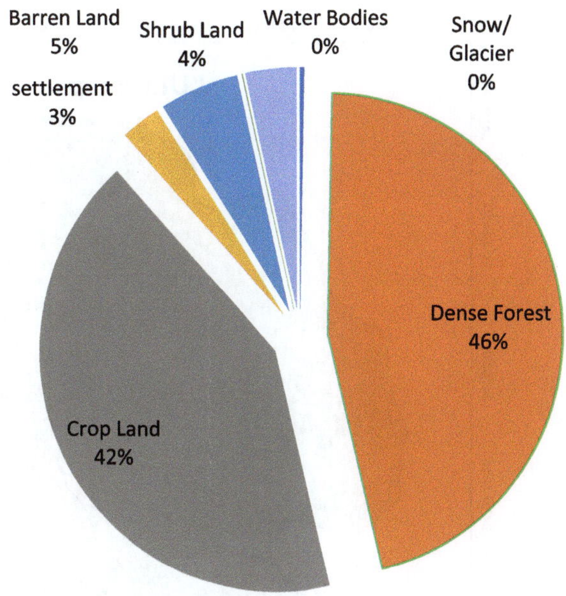

Fig. 7.4 LULC of North Kashmir (2019–2020). *Source* Generated from IRS P6 LISS III Image, 2020

The development of the livestock sector is critical to the country's overall economic success. Crop production alone may not be sufficient to address the problems of poverty and food insecurity at the household level. In the current circumstances, the need of the hour may be to develop the livestock sector to worldwide standards to meet the meat and milk demands of the country's quickly growing population. With this goal in mind, the first step must be to identify the fundamental issues/ problems of livestock producers so that they may contribute more effectively to the livestock sector by implementing the most recent and up-to-date production and management practices. There are 14 veterinarian blocks and 23 sheep development blocks in the research region. Figures 7.6 and 7.7 show the total cattle population and population of sheep and goat population block, respectively (2019–2020).

Horticulture

Horticulture plays an important role in the global food value chain for the world's growing population. Its prospects are favorable for a long-term and environmentally friendly ecosystem. Ornamental plants are developed for their aesthetic appeal, which plays an important role in people's lives. In the study area, the majority of the population is dependent on the agriculture & Horticulture sectors and considered the backbone of the region. The region with a total area under horticulture of 69,964 hectares with a total production of 885,841 tons. Area-wise Singapore, Handwara, Sopore, and Pattan Rafi Abad blocks have having largest area under horticulture while hilly areas such as Parenpil, and Machil blocks contribute a low percentage of area under horticulture. Production-wise blocks such as Sopore, Langet, Rafi Abad, and Kunzer contribute the highest (Chief Horticulture Office Kashmir). Table 7.2

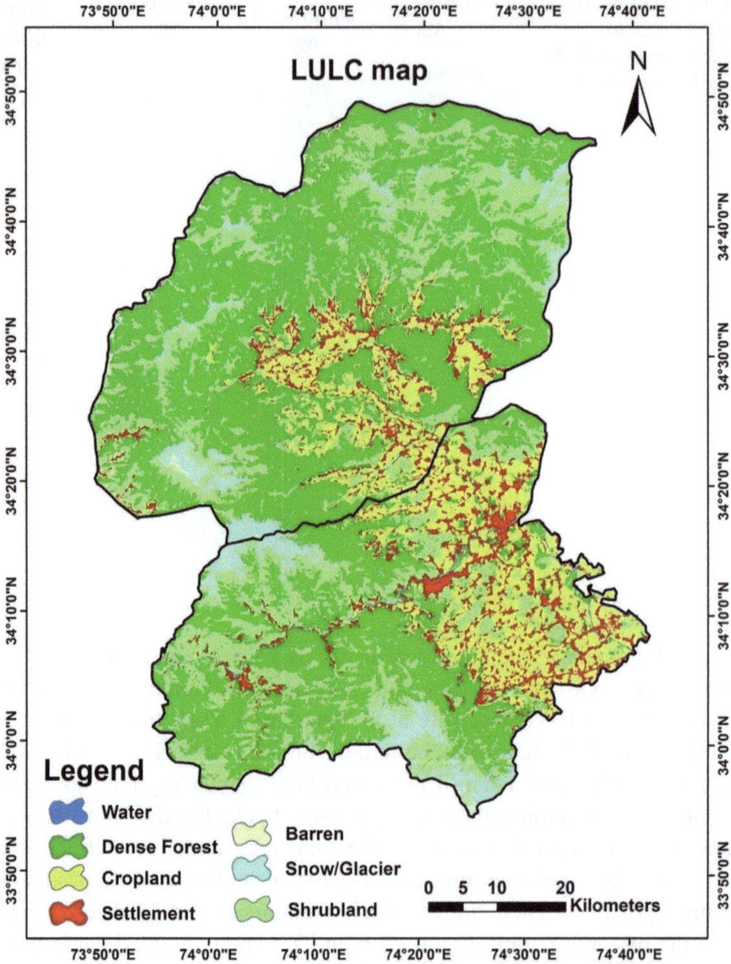

Fig. 7.5 Spatial distribution of LULC (2019). *Source* Generated from IRS P6 LISS III Image, 2019

shows the area and production of horticulture crops in the study region. Figure 7.8 shows the spatial distribution of the area under horticulture (hectares), production of horticulture (tons), and area under non-agriculture use block-wise. Figure 7.9 shows the total irrigated and non-irrigated land block-wise.

Cropping Pattern

The area's cropping pattern is essentially determined by four natural elements: physiography, climate, soil, and water resources (Rai 1988). It will reflect the character of the said natural components in combination with the two main agricultural seasons, Rabi and Kharif. Kharif crops such as rice, maize, vegetables, pulses, and feed require adequate moisture and are therefore sown in June and July. Rabi crops, on the other

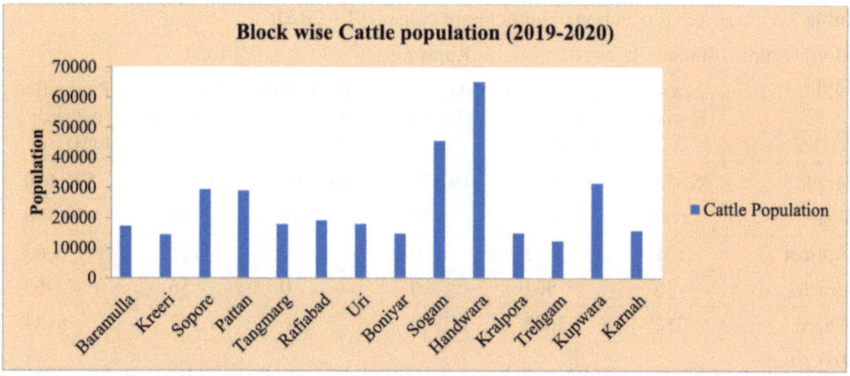

Fig. 7.6 Blockwise cattle population in North Kashmir (2019–2020). *Source* District Animal Husbandry Offices (2019–2020)

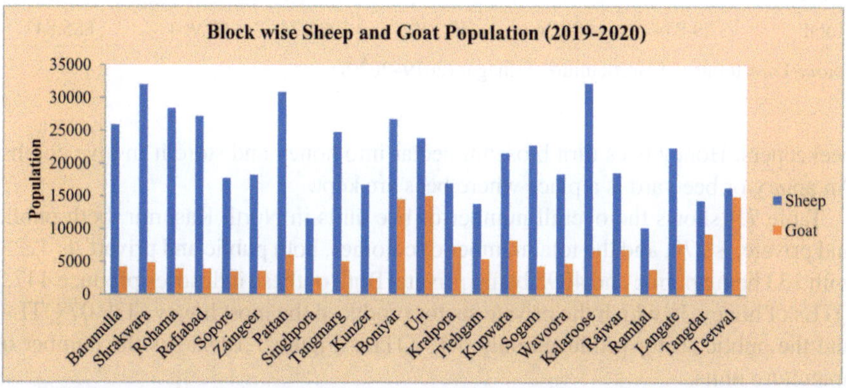

Fig. 7.7 Block wise sheep and goat population in North Kashmir (2019–2020). *Source* District Sheep Husbandry Offices (2019–2020)

hand, such as oilseeds, wheat, pulses, vegetables, and fodder, which can withstand a lack of moisture, are planted in October and November when the rains are receding. Interspersed between these two crop seasons is another crop season, Zaid, which has little impact in the area. Kharif and Rabi crops are used to examine the cropping pattern of the research area. Crop area production and productivity are examined. Table 7.3 shows the area under food crops and production of each crop (2019–2020).

Apiculture in North Kashmir

Apiculture, also known as beekeeping, is the practice of keeping honeybee colonies, most commonly in hives, by beekeepers. A hive is only a rectangular box. A beekeeper or apiarist raises bees to harvest honey and other hive products (such as beeswax, pollen, and royal jelly) for pollination of crops or to breed bees for sale to other

Table 7.2 Area and production of horticulture crops (2019–2020)

Horticulture crops	Baramulla		Kupwara		Total	
	Area (Hectares)	Production (metric tons)	Area (Hectares)	Production (metric tons)	Area (Hectares)	Production (metric tons)
Apple	25,353	359,677	19,525	166,963	44,878	526,640
Pear	660	11,286	478	3494	1138	14,780
Apricot	121	2063	0	0	121	2063
Peach	56	963	0	0	56	963
Cherri	212	3625	11,300	566	11,512	5154
Dry fruits						
Walnut	3144	10,199	8812	26,406	11,956	36,605
Almond	258	553	13	39	271	592
Other dry fruits	30	1	2	6	32	7
Total	29,834	688,367	40,130	197,474	69,964	885,841

Source Directorates of Horticulture, Srinagar (2019–2020)

beekeepers. Honey bees turn blossom nectar into honey and store it in hive combs. An apiary or bee yard is a place where bees are kept.

Table 7.4 shows the overall number of bee units in North Kashmir, both public and private, is 270, and the total number of colonies, both public and private, is 4228, with 133 being public and 4095 being private. Both of these industries produce 447.2 QTLs of honey. In which the private sector produced the most honey (439.07 QTLs) and the public sector produced only 8.13 QTLs. Figure 7.10 shows the number of apiculture units.

Sericulture in North Kashmir

Sericulture is an agricultural industry. It entails the cultivation of host plants as well as the rearing of silkworms for the manufacture of the cocoon, which is used to make raw silk. Sericulture's main tasks include food-plant cultivation to feed the silkworms that spin silk cocoons and reeling the cocoons to unwind the silk thread for processing and weaving to generate the product. Figure 7.11 shows the area and production of sericulture block-wise in the study area. Sericulture is not cultivated in every block of north Kashmir, but it is grown in some regions. Uri, Baramulla, Sopore, Pattan, Tangmarg, Lolab, Handwara, Kupwara, Langet, and Karnah are the ten sericulture divisions in North Kashmir. The overall sericulture area is 328 hectares, with a total production of 115,802. Lolab Division has the most sericulture land (86 hectares), followed by Uri (40 hectares), Pattan (38 hectares), Sopore (36 hectares), Tangmarg (35), Baramulla (31), Handwara (20), Kupwara (19), Langet (16), while Karnah Division has the least (7).

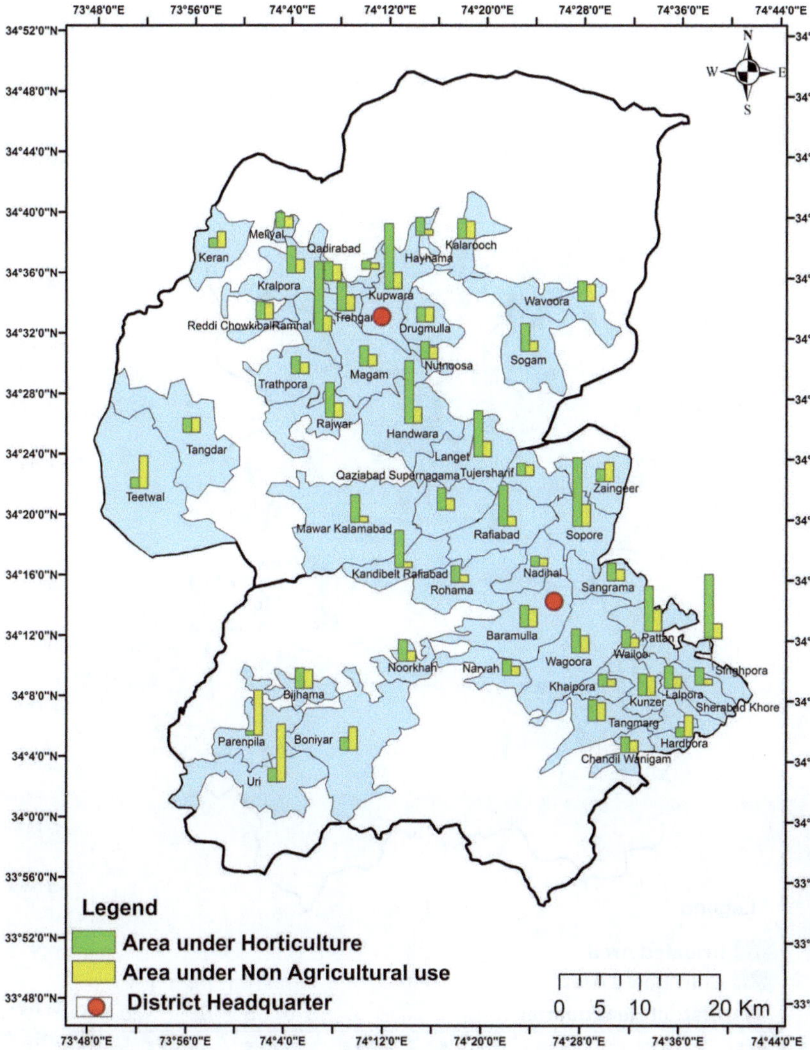

Fig. 7.8 Area under horticulture and non-agriculture use. *Source* Directorates of Horticulture, Srinagar (2019–2020)

7.3.4 *Tourism Sector*

Kashmir is India's most northern state. Jammu and Kashmir is included in this state. Agriculture, horticulture, animal husbandry, forestry, and tourism are the key economic sectors in J&K. Tourism are a significant industry. Because industrial development is limited to steep terrains, J&K sees tourism as a competitive advantage. J&K is a popular Asian tourist destination. Tourism in J&K has resulted in

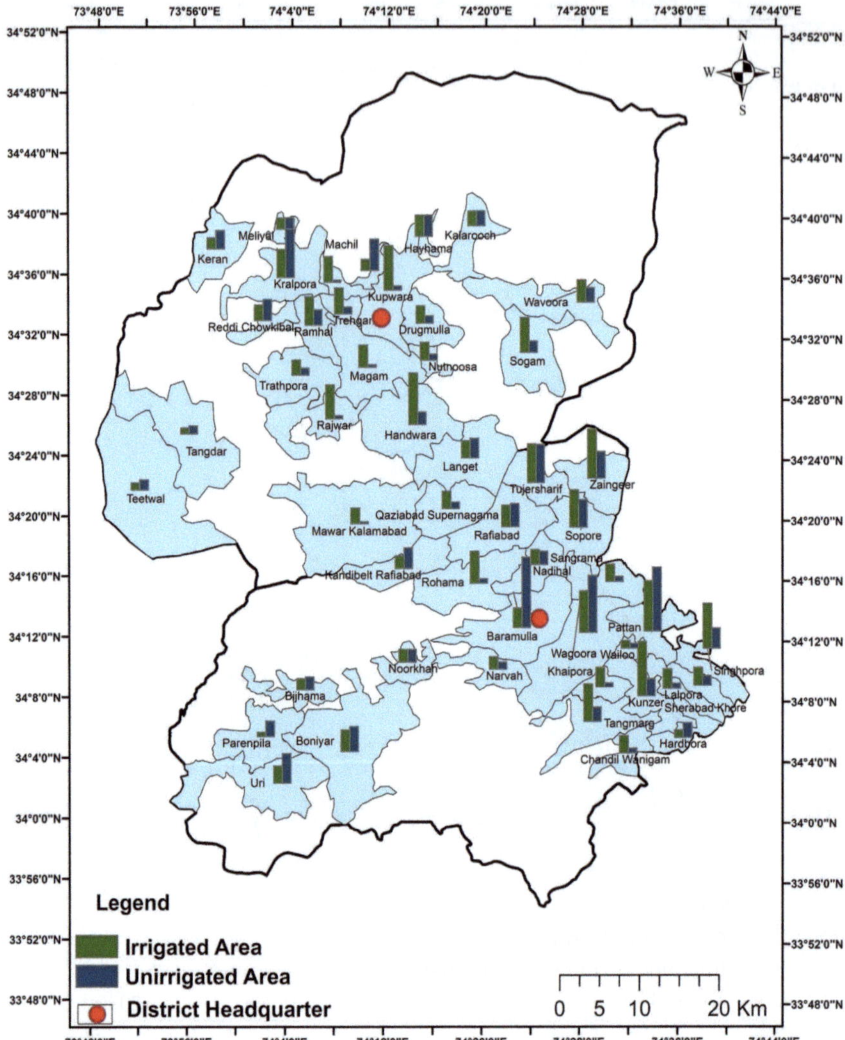

Fig. 7.9 Area under irrigated and non-irrigated land. *Source* District Chief Agriculture Offices (2019–2020)

Table 7.3 Area under food crops in the study area (2019–2020)

	Rice	Maize	Vegetables	Fodder Qtls/ha	Oil seeds kg/ha	Pulses kg/ha	Other millets Qtls/ha
Area (ha)	38,011	38,943	32,856	78	442	1324	120
Production (tons)	262,276	81,780.3	726,117.6	31,200	284,206	673,916	1800

Source Financial commissioner office Srinagar (2019–2020)

Table 7.4 Distribution and Production of Bee units in North Kashmir (2019–2020)

North Kashmir	No.of units/Beekeepers		No. of Colonies		Production of Honey in QTLs	
	Public	Private	Public	Private	Public	Private
	7	263	133	4095	8.13	439.07

Source Directorate of apiculture Srinagar (2019–2020)

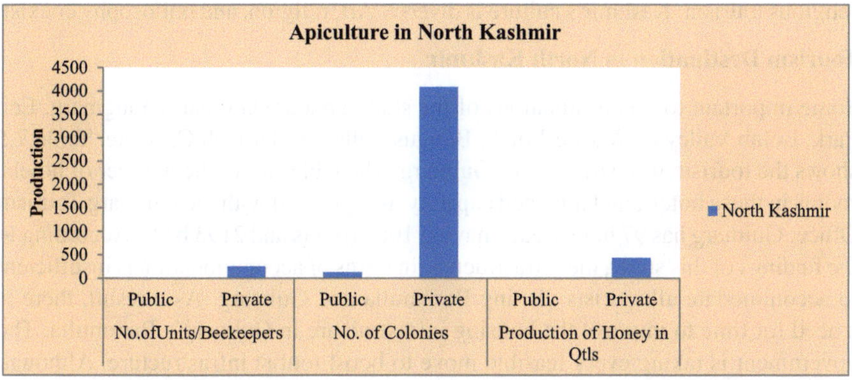

Fig. 7.10 Apiculture units in the study area. *Source* Directorate of apiculture Srinagar (2019–2020)

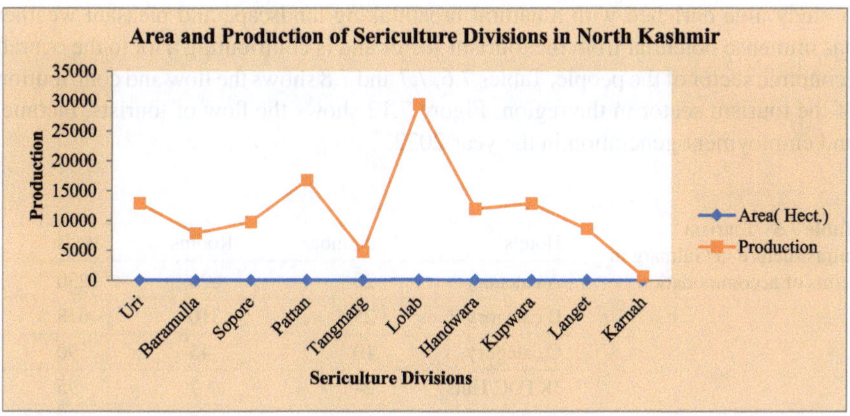

Fig. 7.11 Area and production of sericulture. *Source* Directorate of Sericulture Srinagar (2019–2020)

numerous attractions. Jammu and Kashmir is known for its beautiful backwaters, hill towns, and landscape. Historical landmarks, forts, religious sites, and hill resorts enhance the state's attractiveness. They attract international visitors. Tourism is a significant economic contributor and employer in Jammu and Kashmir. This labor-intensive industry provides many jobs. Hotel managers, receptionists, tour guides, tour operators, travel companies, and photographers are among those employed in the fragmented tourist sector. If government officials focus on this potential, J&K might become an important medical tourism destination. The state has numerous forests, lakes, and rivers. Jammu and Kashmir is noted for its breathtaking valleys, lakes, snow-capped peaks, energizing weather, valleys, chilly climate, trekking, fishing, and skiing, as well as its diverse archaeological, historical, cultural, and religious sites, calm atmosphere, and friendly people. The Kashmir valley in J&K offers a variety of tourist attractions. Examples include skiing, rafting, adventure travel, and religious tourism. Kashmir's culture is diverse. Art, religion, and philosophy coexist.

Tourism Destination in North Kashmir

Some important tourist destinations of the study area are Gulmarg, Tangmarg, Eco Park, Lolab Valley or Wadi-e-Lolab, Bungus Valley, Kalarooch Cave etc. Table 7.5 shows the tourism infrastructure of Gulmarg. The table shows the number of hotels, rooms in these hotels, and total bed capacity as registered with the Gulmarg Tourism Office. Gulmarg has 97 hotels, each having 1062 rooms and 2193 beds. According to the findings of this study, the infrastructure in terms of accommodation is insufficient to accommodate all tourists visiting Baramulla and Gulmarg. As a result, there is a need for time to improve the housing infrastructure in Gulmarg's Baramulla. The government is taking every feasible move to boost tourist infrastructure. Although significant progress has been made, much more work needs to be done.

Tourism Potential

A study area enriched with a natural breathtaking landscape and pleasant weather has immense potential from the tourism sector and is contributing a lot to the overall economic sector of the people. Tables 7.6, 7.7 and 7.8 shows the flow and contribution of the tourism sector in the region. Figure 7.12 shows the flow of tourists, income, and employment generation in the year 2022.

Table 7.5 Tourism infrastructure in Gulmarg in terms of accommodation

Hotels	Number	Rooms	Beds
A category	20	620	1230
B category	25	310	618
C category	10	45	90
JKTDC Huts	24	7	95
Guest house	18	80	160
Total	97	1062	2193

Source Directorate of Tourism, Govt. of J & K

Table 7.6 Tourist flow in the study area

Year	Tourist Inflow		
	Total	Foreign	Domestics
2022	26.73(in lakhs)	20,000	2,653,000

Source Directorate of tourism, Govt. of J & K

Table 7.7 Total benefit from tourist arrivals in the study area

Year	Total direct benefit (Rs. In crore)	Total indirect benefit (Rs. In crore)	Total benefit
2022	5202.64	6150.70	11,353.34

Source Directorate of Tourism, Govt. of J & K

Table 7.8 Generation of employment in the study area

Year	Tourist flow	Direct employment (in lakhs)	In-direct employment (In lakhs)	Total employment
2022	26.53 (in Lakhs)	2.67	10.38	13.05

Source Directorate of tourism, Govt. of J & K

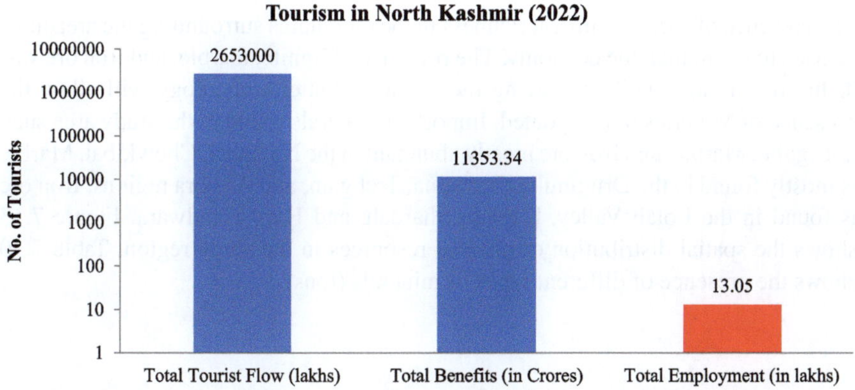

Fig. 7.12 Tourism flow, income, and employment generation. *Source* Directorate of Tourism, Govt. of J & K

7.3.5 Industries

At the moment, the district of Baramulla has only two industrial estates: Baramulla (58 Kanals) and Sopore (82 Kanals). Approximately 60% of the Industrial Estates are occupied by security troops, which they want to be removed. The Department of Industries and Commerce is working hard to get land for the creation of new industries in the district of Baramulla, and the identification process is in full swing. Table 7.9 shows the existing micro and small industries in the study area.

Table 7.9 Existing micro and small enterprises and artisan units in North Kashmir (2019–2020)

Type of industry	No. of units	Investment (lakh)	Employment
Agro based	201	685	625
Cotton textile	10	5.12	21
Woolen, silk and artificial thread-based clothes	23	3.20	30
Ready-made garments embroidery	260	228	780
Wood/wooden based furniture	142	501.5	721
Paper and Paper products	48	339.87	215
Leather-based	01	4.50	02
Chemical/chemical based	09	210	70
Mineral based	07	170	150
Metal-based	379	2543.87	1492
Repairing and servicing	622	1972	1613

Source Ministry of MSME, Govt. of India (2019–2020)

7.3.6 Mineral Resources

The presence of various mineral richness in the mountains surrounding the area must be used to strengthen the economy. The presence of lignite, marble, and iron ore was deduced long ago while surveying the territory; better technology will allow the presence of Minerals to be updated. Important minerals found in the study area such as Lignite, Marble, and Iron ore mostly abundant in the Nichama, Chowkibal. Marble is mostly found in the Drugmulla, Zurhama, Trehgam, and Awoora regions. Iron ore is found in the Lolab Valley, Karnah, Shardah, and Haril Handwara. Figure 7.13 shows the spatial distribution of mineral resources in the study region. Table 7.10 shows the presence of different types of minerals (tons).

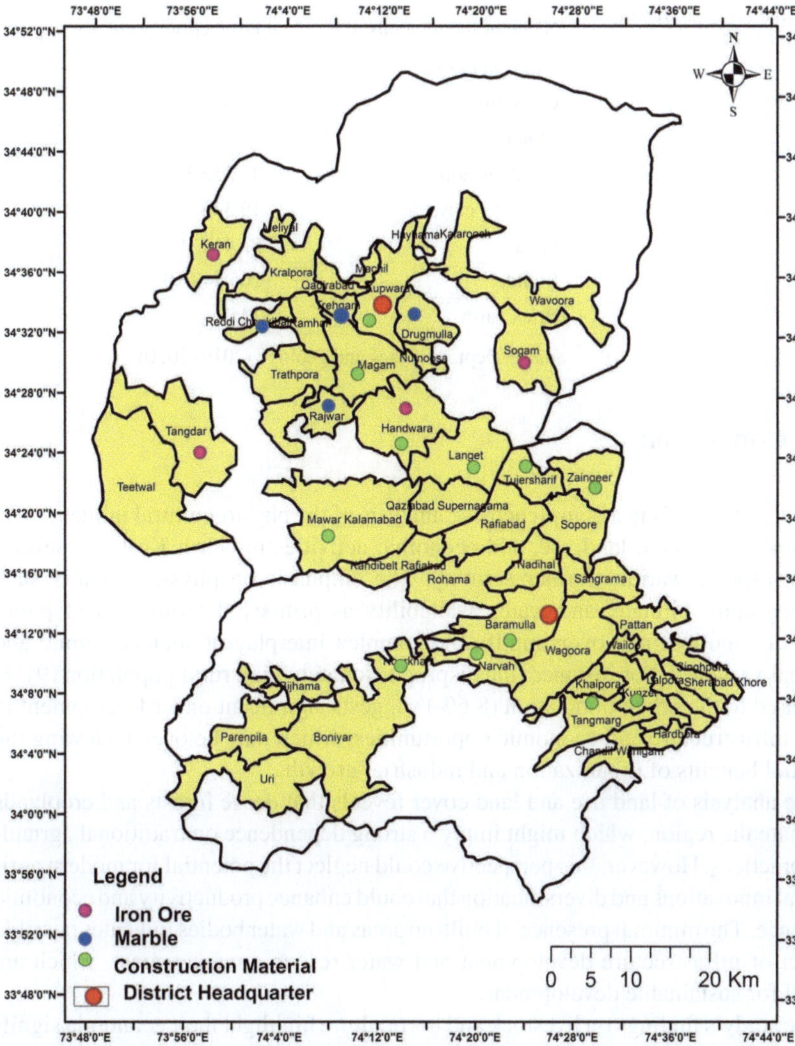

Fig. 7.13 Spatial distribution of minerals. *Source* Dept. of Mines and Geology

Table 7.10 Availability of minerals (2019)

Name of the minerals	Production in tones
Major minerals	
Gypsum	13,666.05
Minor	
Building stones	47,335.33
Ordinary clay	19,346
Sand	82,416.194
Boulder	205,443.10
Brick earth	14.00

Source Dept. of mines and geology (2019–2020)

7.4 Conclusion

While the study offers a comprehensive analysis of the physio-cultural influences on population dispersal, land use, and economic activities in North Kashmir, several critical aspects warrant further scrutiny. The emphasis on physical factors such as topography, climate, and water availability as primary determinants of population distribution may oversimplify the complex interplay of socioeconomic and political variables. For instance, the disproportionately high rural population (91%) compared to the urban population (8.6%) suggests significant underdevelopment in urban infrastructure and economic opportunities, which may be overshadowing the potential benefits of urbanization and industrial growth.

The analysis of land use and land cover reveals that dense forests and croplands dominate the region, which might imply a strong dependence on traditional agricultural practices. However, this perspective could neglect the potential for modern agricultural innovations and diversification that could enhance productivity and economic resilience. The minimal presence of built-up areas and water bodies indicates possible neglect of infrastructure development and water resource management, which are crucial for sustainable development.

The study's findings on livestock and horticulture highlight their economic significance, yet the broader context of the livestock sector's challenges and the need for modernization might be underestimated. The focus on local benefits of livestock for small farmers should also consider the potential for integrating advanced practices to boost productivity and economic impact. Similarly, the horticultural sector's performance, although promising, might be constrained by limited market access and technological constraints that warrant deeper analysis.

Tourism, while identified as a major economic contributor, faces significant infrastructural limitations. The insufficiency of accommodation facilities in key tourist destinations such as Gulmarg underscores a critical gap in the region's ability to capitalize on its tourism potential fully. Despite the government's efforts, the pace of infrastructural development may need to accelerate to meet the growing demand and fully exploit tourism's economic benefits.

Lastly, the study's findings on mineral resources present a promising but underdeveloped aspect of the region's economy. The potential for mineral extraction could be a significant economic driver if coupled with advanced technology and sustainable practices. However, the focus on current mineral wealth without addressing the broader environmental and social implications could lead to skewed development priorities.

In summary, while the research provides valuable insights into the physio-cultural and economic dimensions of North Kashmir, it is crucial to consider the broader implications and potential areas for improvement in infrastructure, modern agricultural practices, and sustainable economic development strategies.

References

Census India (2011) Census of India 2011—data on population distribution. www.Censusindia.gov.in

Chief Horticulture Office Kashmir (2020) Horticulture statistics and land use. https://hortikashmir.gov.in

Department of Geology and Mines. (2020) Mineral Resources of Jammu and Kashmir. https://geologymining.jk.gov.in

Director of Tourism JK (2020) Tourism statistics and infrastructure in Jammu and Kashmir. https://www.tourismjk.in

JK Statistical Digest (2019) Jammu and Kashmir statistical digest 2019. Link to JK. https://ecostatjk.nic.in

Ministry of MSME GOI (2019) Annual report 2019–20. https://msme.gov.in

Moonis R, Aijazuddin A, Ali M (1978) The valley of Kashmir. Vikas publishing House PVT LTD, ISBN: 0706905253

Neo H (2009) Resources and environmental economies. Int Encyclopedia of Human Geography. https://doi.org/10.1016/B978-008044910-4.00225-X

Rai SC (1988) Spatial organization and rural development. Seema Publications, New Delhi, p 183

Shah WH, Hao G, Yan H, Yasmeen R (2024) Natural resources utilization efficiency evaluation, determinant of productivity change, and production technology heterogeneity across developed and developing G20 economies. Technol Soc 77. https://doi.org/10.1016/j.techsoc.2024.102507

Zhao H (2019) Energy: "The Achilles' Heel" of China. The economics and politics of China's energy security transition. Academic Press. https://doi.org/10.1016/B978-0-12-815152-5.00007-5 .

Zimmermann EW (1933) World resources and industries: a functional appraisal of the availability of agricultural and industrial materials. Revised edition. Harper and Brothers

Chapter 8
Nature-Based Livelihood Improvement Strategies in the Mountainous Regions of Ladakh

Ajay Kumar Gupta, Lalit Giri, and Sandipan Mukherjee

Abstract Livelihoods in Trans-Himalaya, particularly in Ladakh, generally depend on ecosystem-based natural resources. The present chapter explores the nature-based livelihood strategies in Ladakh, a high-altitude region in the Indian Himalayas. It examines how the communities in this challenging environment use their natural resources, traditional knowledge, and innovative approaches to sustain livelihoods historically. However, in the past few decades, the livelihoods of the communities have been affected due to the changing climate and development activities, especially tourism. The chapter documents and analyses the data related to promoting nature-based livelihood strategies adopted in the past two decades. Further, the chapter discusses the effectiveness of these strategies in the face of climate change and socio-economic challenges by implementing region-specific nature-based solutions, and how these nature-based solutions provide an alternative livelihood. Further, it is elaborated that these solutions can be replicated in other Trans Himalayan regions, subsequently; recommendations are provided for achieving different sustainable development goals in similar mountainous regions.

Keywords Trans-Himalaya · Natural resources · Traditional knowledge · Climate change

8.1 Introduction

The mighty Himalayas, known as the "water tower of Asia" are the world's youngest and loftiest mountain chain. It stretches across five countries: India, Nepal, Bhutan, China, and Pakistan, covering approximately 2400 km. Around 70% of the total geographical region of Himalaya is a part of the Indian Himalayan Region (IHR) which is around 18% of India's landmass (Wani et al. 2023). IHR includes 13 states

---

A. K. Gupta (✉) · L. Giri · S. Mukherjee
Ladakh Regional Centre, G.B. Pant National Institute of Himalayan Environment, Leh, Ladakh UT, India
e-mail: ajay.gupta23@gbpihed.nic.in

S. C. Rai (ed.), *Himalaya: Mountains of Destiny*,
https://doi.org/10.1007/978-3-031-94905-0_8

and union territory, stretches from Arunachal Pradesh in the east to Jammu and Kashmir in the west, characterized by a diverse demographic, versatile economic, environmental, social, and political systems. It is the home of nearly 52 million people (GoI 2019). IHR is classified as Greater Himalaya (above 6000 m), Lesser Himalayas (1500–4500 m), Outer Himalayas (below 1500 Meters), and Trans-Himalayas which includes Ladakh and some parts of Himachal Pradesh experiencing harsh climate. The Trans Himalayas are also known as cold deserts.

The Trans-Himalayan Mountain region is located north of the Great Himalayas, consisting of the Karakoram, Ladakh, Zanskar, and Kailash Mountain ranges. Ladakh, under Trans-Himalayas, is the northernmost part of India and is characterized by its high-altitude desert landscape, unique cultural heritage, and ecologically susceptible and fragile ecosystem. Ladakh Union territory is comprised of Leh (45,110 km^2) and Kargil (14,036 km^2) districts. The region's extreme climatic conditions, including long, harsh winters and short summers, pose significant challenges to the livelihoods of its inhabitants. It is one of the most sparsely populated areas in India, characterized by its remote location. The region lies at an altitude of over 3000 m asl, with a landscape dominated by extreme climatic conditions, including long, harsh winters and short summers, which pose significant challenges to the livelihoods of its inhabitants. However, the local communities have developed distinct ways of adapting to this environment. In this remote and ecologically sensitive area, the sustainability of traditional livelihoods such as agriculture, pastoralism, and handicrafts are crucial.

However, these livelihood activities are increasingly threatened by climate change, environmental degradation, and socio-economic shifts. This chapter explores how Ladakh adapts to these challenges through nature-based strategies. The present chapter aims to describe the traditional livelihood practices and threats to these livelihood options from climate change. Subsequently, a way forward is provided for nature-based solutions to adapt to the changing climate.

8.2 Traditional Livelihoods in Ladakh

Traditional livelihoods of Ladakh primarily revolve around agriculture, pastoralism, and handicrafts. The self-subsistence agricultural production of Ladakh is limited to one crop per year and is mainly focused on barley and wheat with limited vegetable production. Ladakh has a vast geographical area of around 59,000 km^2 but the lowest population density in the country with 59.6% and 82.5% of the households having less than one hectare of cultivable land, respectively, in Leh and Kargil districts (Stobdan 2023). The short growing season limited arable land, and dependency on glacial melt water for irrigation make agriculture practices particularly challenging. Despite these challenges, the region produces several essential crops that contribute to the livelihood of its population. A detailed overview of the major crops, their cultivation, and production in Ladakh is mentioned in Table 8.1. Barley is one of the most extensively grown crops in Ladakh, thriving in the region's high-altitude,

Table 8.1 Major crops and their production in Ladakh during the year 2022

S. No.	Major crops	Cultivation (ha)	Production (tons)
1	Barley	10,842	17,056
2	Wheat	4512	7020
3	Vegetables	1163	17,572
4	Apricot	2612.6	15,864
5	Apples	946.6	5191

Source District Statistical Handbook, Leh, 2022–23

cold desert environment. It is a staple food for the local population, primarily used to make Tsampa (roasted barley flour), a traditional Ladakhi dish. In 2022, around 10,842 hectares of barley cultivation yielded approx. 17,056 tons of product. Its short growing season makes barley ideal for Ladakh's harsh climate, as it requires less water than other crops.

Additionally, barley straw is used as fodder for livestock, further enhancing its importance to the rural economy. Wheat is another key staple in Ladakh, though it is less widely cultivated than barley. It is mainly used to make Khambir (a local fermented bread) and other types of bread, as well as for trade. Wheat farming in Ladakh is labor-intensive, largely due to the challenging terrain and limited access to mechanized equipment. In 2022, approx. 4512 hectares of wheat cultivation produced 7020 tons. The region's cold, arid climate necessitates specialized irrigation methods, often relying on glacial melt water to support the crops. Therefore, in the late 1960s, only a few types of vegetables were grown in the region which includes potatoes, carrots, radishes, peas, spinach, and leafy greens. Today, over 100 varieties are cultivated, due to government interventions and the introduction of greenhouses that extend the region's short growing season. In 2022, approx. 1163 hectares of vegetable cultivation yielded 17,572 tons, reflecting the significant progress in agricultural productivity. Apricots are among Ladakh's most renowned and economically significant crops, particularly in lower-altitude areas like the Sham region. They are consumed fresh, dried, or processed into products like jams and oils, with dried apricots being a valuable item for trade. In 2022, 2,612.6 hectares of land produced 15,864 tons of apricots. Moreover, the *Raktsey Karpo Apricot* variety of Ladakh received its first-ever Geographical Indication (GI) for a period of twenty years in 2022. Additionally, apples are also being cultivated in certain regions of Ladakh. In 2022, a total of 5191 tons of apples were harvested from 946.6 hectares of land.

Pastoralism remains a significant livelihood option in Ladakh, particularly in higher-altitude regions where agriculture is less viable. The pastoral communities rear livestock such as yaks, sheep, goats, and Dzos (yak-cattle hybrids). These animals provide milk, meat, and wool, which are essential for the communities for sustenance and economic activities. The pastoral lifestyle is closely linked to the environment of the region, with traditional practices such as rotational grazing and seasonal migration ensuring the sustainability of pasturelands. At present the total number of sheep and goats in the Leh districts of Ladakh is 2,59,921 which includes 89,158 sheep,

1,62,134 Pashmina goats, and 8599 non-pashmina goats (LAHDC 2023). The current pashmina production in the Leh districts of Ladakh is estimated to be between 37,000 to 50,000 kg annually (Ladakh Pashmina Vision 2021). The harsh winters when the temperature can plummet to − 30 °C require pastoralists to be well-prepared, storing fodder and ensuring that their herds are in good health to survive the cold. Traditional knowledge plays a key role in managing livestock health, with the use of medicinal plants and natural remedies being widespread. In the past few decades, decreasing trends in pastoralism have been observed in the trans-Himalayan region mainly due to the changing climate leading to habitat fragmentation and fodder shortage along with the transformation of the pastoral communities to settlers due to increased tourism activities leading to easier economic benefit. Moreover, the enhanced tourism in the region is reported to exacerbate additional pressure on the fragile ecosystems of the regions (Clemens and Nüsser 2000; Wu et al. 2014).

Ladakh's rich cultural heritage is reflected in its traditional handicrafts. These crafts are not only a source of income but also play a vital role in preserving the region's cultural identity. These include wood carving/carpentry, carpet weaving, Ladakhi garment making, embroidery, Namdamaking, silver filigree/blacksmith, pottery, fresco painting (Thangka painting), clay molding (sculpture), Papu (shoe) making, pottery making, basket making, stone (Mani carving), dragon painting, hosiery/hand knitting, etc. Among all, the production of pashmina wool is particularly significant. Pashmina, derived from the undercoat of the Changthangi goat, is one of the finest and warmest wools in the world. The craft of spinning and weaving pashmina has been passed down through generations and is a vital source of income for many families. In recent years, there has been resurgence in the interest in Ladakhi handicrafts, driven by both local initiatives and external demand. However, the industry faces challenges such as competition from machine-made products, access to markets, and the need for quality improvement.

8.3 Challenges Facing Traditional Livelihoods

The present Ladakhi societies are experiencing two broad categories of challenges through climate change and socio-economic shifts. A brief description of these challenges is provided below.

8.3.1 Climate Change

Climate change poses one of Ladakh's most significant threats to traditional livelihoods. In this context, this is to note that approx. 50% of all the glacier wealth of the Indian Himalayan region is situated within the Ladakh region. The region is witnessing a reduction in snowfall and the accelerated melting of glaciers with increasing temperature (Dimri and Chevuturi 2014; Mandal et al. 2024; Abdullah et al. 2020), affecting critical water sources for agriculture and pastoralism (Fig. 8.1). Moreover, the ensemble climate model projections have indicated that under enhanced warming of more than 2 °C, the entire north-western Himalaya is expected to receive fewer winter-season wet days (Ballav et al. 2021). Similarly, the exacerbated warming is expected to marginally increase the summer monsoon rainfall of June over the Ladakh region, but a gradual decline is expected during August and September (Mukherjee et al. 2019). However, the average number of heavy rainy days during the monsoon period, which is around 0.27 ± 0.09 days over Jammu, Kashmir, and Ladakh region is expected to increase to 0.82 ± 0.09 days under a much warmer climate (Mukherjee 2021). Changing precipitation patterns with more erratic rainfall is disrupting the traditional agricultural calendar and leading to water shortages during the growing season. Moreover, the warming climate is affecting the health of livestock, with warmer winters leading to the proliferation of pests and diseases that were previously controlled by the cold (Sapkota et al. 2022; Tuladhar 2023). This led to a decrease in the population of the pashmina goat in the region (Table 8.2), which alone is responsible for the production of 90% of the pashmina wool in the country (LAHDC 2023). The changing climate also poses a risk to the traditional knowledge systems that have been developed over centuries, as these systems are based on predictable seasonal patterns that are now being disrupted (Ladon et al. 2023).

Fig. 8.1 Different challenges facing traditional livelihood in Ladakh

Table 8.2 Year-wise livestock (sheep and goats) in the Leh, Ladakh

Year/Block	Sheep	Goat		Others	Total
		Pashmina	Non-pashmina		
2015–16	84,196	222,575	2787	0	340,158
2016–17	103,375	202,561	19,099	0	325,035
2017–18	110,943	219,198	16,199	1520	347,860
2018–19	120,637	204,495	17,489	2086	344,707
2019–20	98,983	186,431	11,516	821	297,751
2020–21	98,983	186,431	11,516	821	297,751
2021–22	94,734	180,690	12,133	759	288,316
2022–23	89,158	162,164	8599	0.00	259,921

(*Source* District Statistical Handbook, Leh, 2022–23)

8.3.2 Socio-Economic Shifts

Ladakh is experiencing significant socio-economic changes, driven by factors such as the growth of tourism, increasing connectivity, and migration. Tourism, which opened for Ladakh in the late 1970s, while providing new income opportunities, also brings challenges, including environmental degradation, cultural commodification, and increased pressure on local resources. In the past two decades, Ladakh has witnessed rapid land use changes especially in terms of productive natural resources which as decreased drastically. For example, the total area of settlement has increased by 123 km^2 and wetlands have decreased by 28 km^2 between 1999 and 2019 similarly agriculture area in 1999 was estimated to be 374.579 km^2 which reduced to 270.63 km^2 in 2019 (Bhasin et al. 2022). The decrease in the range of land and agricultural land is mainly due to an increase in tourism activities (Bhasin et al. 2022) which is directly linked to the livelihood of pastoralists and small landholders. In this context, this is to highlight that while in 1974 only 527 numbers of tourists arrived in Leh, the number significantly changed to 523,921 in 2022 (Data source: Department of Tourism, Ladakh UT). A recent study by Geneletti and Dawa (2009) indicates that major tourist destinations of Ladakh such as Nubra, Changthang are experiencing a substantial impact on their wetland biodiversity that hosts some of the unique and IUCN red-listed flora and faunas.

Along with increased tourism, out-migration, particularly among the younger generation, is another challenge. Many young Ladakhi are leaving their villages to seek education and employment opportunities in urban areas, leading to a decline in traditional practices and a loss of cultural knowledge. The shift toward a cash economy is also changing consumption patterns, with a growing reliance on imported goods and food, which can undermine local food security.

8.4 Nature-Based Livelihood Improvement Adaptation Strategies

To address the challenges posed by climate change and environmental degradation, Ladakhi farmers are adopting a range of nature-based agricultural innovations. One such innovation is the introduction of high-altitude resilient crop varieties that are better suited to the changing climate. For example, new barley and wheat strains are being cultivated that are more resistant to drought and pests (Hussain et al. 2024). Organic farming practices are also gaining traction, with an emphasis on maintaining soil fertility and reducing dependency on chemical inputs. These practices include the use of natural fertilizers such as compost and green manure, as well as traditional crop rotation techniques.

8.4.1 Promotion of Solar-Based Greenhouse for Vegetable Production

Additionally, the construction of greenhouses, known locally as 'solar poly-houses,' has allowed farmers to extend the growing season, enabling the production of vegetables throughout the year. These solar-based polyhouses enhanced vegetable production in Ladakh from 250 tonnes during 1971–72 to nearly 2500 tonnes in 1981 and 17,572.2 tonnes in 2022 (Stobdan 2023). Although solar-based greenhouse cultivation in Ladakh was started in the early 1980s by using glasses as the glazing material, due to limited knowledge and initial high-cost factors, it remained limited in its extent. In the last three decades, due to the advancement in technology and innovations, different types of greenhouses have been tested and tried in the region. Some of them are constructed as glass houses, gable type, gothic type, walk-in tunnels, trench, Ladakhi poly-house, fiber reinforced polyester (FRP) greenhouses, double layer polycarbonate, triple layer polycarbonate and polyench greenhouses (Angmo 2019). As a result, the area under vegetable and fruit production in Leh, Ladakh has considerably increased from around 338 hectares in 2001–02 to around 550 hectares in 2018–19, whereas, the area under wheat production has decreased from 2604 hectares (2001–02) to 2245 hectare (2019–20) as shown in Fig. 8.2.

Recently, the Ladakh Regional Centre of G.B. Pant National Institute of Himalayan Environment (NIHE), Leh, has successfully demonstrated and up-scaled the soil-less hydroponic-based agriculture system. The system was further improved to use treated wastewater from the Municipal Faecal Sludge Treatment Plant of Leh for vegetable cultivation (Giri et al. 2024). Although in its primary stage, hydroponic-based agriculture practices are gaining attention across the country. In addition, traditional water management systems are being revitalized and adapted to current needs. The concept of 'Ice Reservoir' was pioneered by Shri Chewang Norphel (Norphel and Tashi 2015). and local engineer Shri Sonam Wangchuk is involved in creating artificial glaciers that store winter snow and slowly release water during the spring

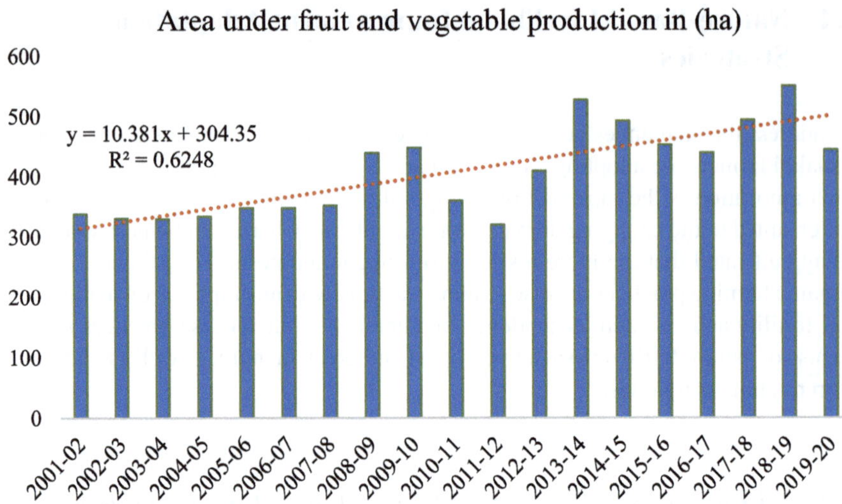

Fig. 8.2 The year-wise area under fruit and vegetable production in Leh, Ladakh. *Source* District Statistical Handbook, Leh, 2020–21

and summer (Nüsser et al. 2019). This innovative approach has helped mitigate water shortages and ensured a more reliable water supply for agriculture. In this context, this is also to highlight that artificial ice reservoirs constructed by NIHE, Leh along with people participation and site-specific modifications were able to provide additional water for irrigation approx. 14 days in advance which resulted in the plantation to meet futuristic wood and fodder requirements (Sharma et al. 2022).

8.4.2 Sustainable Pastoralism

Agro-pastoralism has been practiced in Ladakh since time immemorial. However, pastoralism is primarily noted for the nomadic communities of the region. The Changthang region of eastern Ladakh is renowned for pastoralism and is associated with the Changpa community. Approximately, 90–95% households of in the community rely on pastoralism for their annual income (Pandit et al. 2024). Sustainable pastoralism practices are being promoted within the nomadic communities to ensure the long-term viability of livestock rearing in Ladakh. This includes the revival of traditional grazing practices, such as rotational grazing, which helps prevent overgrazing and maintains the health of pasturelands. The establishment of community-managed grazing reserves has also been successful in some areas, allowing for better management of common resources. The use of indigenous knowledge to maintain livestock health is being reinforced, with local herbal remedies and medicinal plants playing a key role in preventing and treating livestock diseases. These practices reduce the reliance on expensive veterinary medicines and support the conservation of local biodiversity. There is also a growing emphasis on improving the quality of livestock products, such as wool and dairy, to increase their market value. Initiatives to enhance the processing and marketing of pashmina wool have helped local pastoralists secure better prices for their products, providing a more stable income.

8.4.3 Revitalizing Handicrafts

Efforts to revitalize Ladakh's traditional handicrafts have focused on improving product quality, expanding market access, and promoting eco-friendly production methods. Local cooperatives, local governments, and NGOs have been instrumental in providing training and resources to artisans, helping them improve their skills and adapt to changing market demands. The revival of pashmina wool production has been particularly successful, with increased demand for high-quality, handwoven products both locally and internationally. Initiatives such as the *Pashmina Project*, supported by the Indian government, have provided technical and financial assistance to Pashmina producers, helping them improve the quality of their wool and access new markets. In the recent few years, many training centers related to handicrafts and handlooms have been set up by the government and training has been provided to the locals (Table 8.3).

Eco-labeling and fair-trade certifications are also being explored as ways to enhance the marketability of Ladakhi handicrafts, ensuring that artisans receive fair compensation for their work while promoting environmentally sustainable practices. To promote nature-based handicraft manufacturing at the village level, the G B Pant National Institute of Himalayan Environment has imparted several training programs over the last two years. These training programs are related to (Basket making training, Seabuckthorn Products, Mushroom Cultivation). These trainings are being imparted to the women's self-help groups to enhance their skills and livelihood options.

Table 8.3 Year-wise handicraft and handloom training centers and no. of individuals trained

Year	Number of training centers			Number of persons trained in		
	Handicraft	Handloom	Total	Handicraft	Handloom	Total
2014–15	32	31	63	32	31	63
2015–16	32	29	61	32	29	61
2016.17	31	17	47	31	17	48
2017–18	27	17	44	284	375	659
2018–19	23	0	23	129	0	129
2019–20	52	47	99	412	588	1000
2020–21	52	49	101	485	580	1065

Source District Statistical Handbook Leh, 2021–22

8.4.4 Eco-Tourism and Cultural Preservation

Eco-tourism has emerged as a significant livelihood strategy in Ladakh, providing an alternative source of income while promoting the conservation of the region's natural and cultural heritage. Ladakh's unique landscape, combined with its rich cultural traditions, makes it an attractive destination for tourists seeking authentic experiences. Local communities are increasingly involved in eco-tourism initiatives, offering homestays, guided treks, and cultural tours that provide tourists with an immersive experience while ensuring that the benefits of tourism are distributed within the community. These initiatives are designed to minimize the environmental impact of tourism, promote cultural exchange, and generate income for local families. In addition to economic benefits, eco-tourism plays a role in cultural preservation by creating incentives for the maintenance of traditional practices and knowledge. For example, traditional festivals, crafts, and food are being promoted as part of the tourist experience, helping to keep these traditions. In response to climatic and environmental challenges, Ladakhis have adopted several nature-based agricultural innovations. These include the introduction of high-altitude resilient crop varieties, organic farming practices, and the revival of traditional water management systems. The use of greenhouses to extend the growing season has also been a significant development.

8.5 Discussion

The chapter discusses the importance of integrating traditional knowledge with modern innovations to enhance livelihood resilience. It highlights how the success of nature-based strategies in Ladakh depends on the active participation of local communities, access to resources, and supportive policies. The discussion also addresses the challenges faced by these strategies, such as the need for better infrastructure, market access, and policy support. It identifies opportunities for scaling up successful initiatives and replicating them in other mountainous regions facing similar challenges. In the end, it is recommended that the policymakers focus on supporting nature-based livelihoods through targeted investments in infrastructure, capacity building, and market development as depicted in (Fig. 8.3).

We emphasize the importance of preserving traditional knowledge and practices while encouraging innovation. Recommendations are made to promote community-led development initiatives that empower local populations to take ownership of their resources and livelihoods. This includes strengthening local institutions, enhancing access to financial and technical resources, and fostering partnerships with NGOs and the private sector. Further, it is also imperative to assess the carrying capacity based on the productivity of the grassland of Ladakh for sustainable pastoralism. It is suggested that strategies for promoting sustainable tourism that balance economic development with environmental conservation are needed. Further, it is advocated

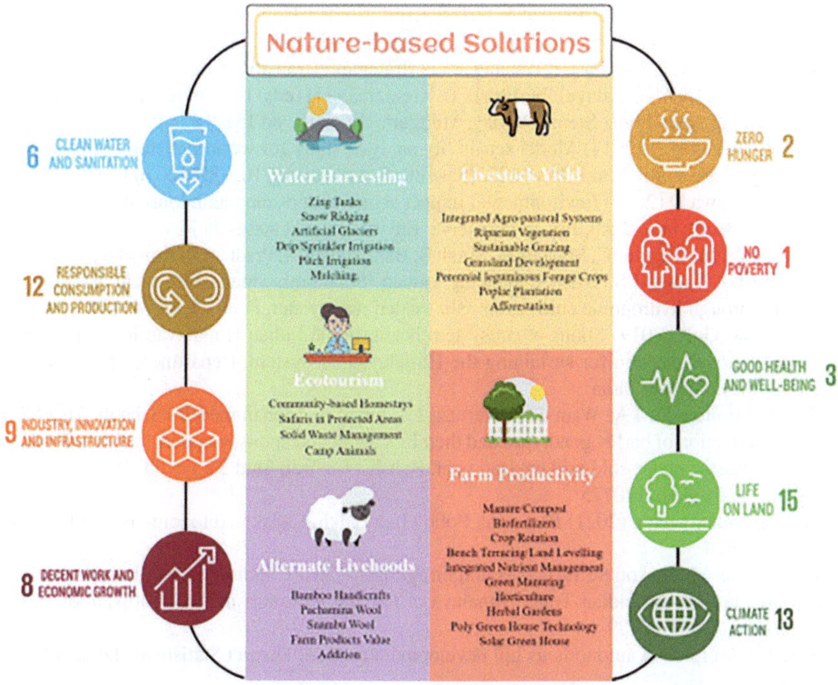

Fig. 8.3 Nature-based livelihood supporting different SDGs as options in Ladakh

that there is a need for the development of policies for regulating tourism growth, protecting natural and cultural assets, and ensuring that tourism revenues benefit local communities. We conclude by summarizing the key insights from the study of nature-based livelihood strategies in Ladakh. It emphasizes the potential of these strategies to enhance the resilience and sustainability of livelihoods in mountainous regions. The chapter calls for continued research, policy support, and community engagement to build on the successes and address the challenges identified.

References

Abdullah T, Romshoo SA, Rashid I (2020) The satellite observed glacier mass changes over the Upper Indus Basin during 2000–2012. Sci Rep 10:14285. https://doi.org/10.1038/s41598-020-71281-7

Angmo P, Dolma T, Namgail D, Tamchos T, Norbu T, Chaurasia OP, Stobdan T (2019) Passive solar greenhouse for round-the-year vegetable cultivation in trans-Himalayan Ladakh Region, India. Defence Life Sci J 4(2):103–116

Ballav S, Mukherjee S, Gosavi V et al (2021) Projected changes in winter-season wet days over the Himalayan region during 2020–2099. Theor Appl Climatol 146:883–895. https://doi.org/10.1007/s00704-021-03765-z

Bhasin A, Dolker P, Raina P, Ghosal S (2022) Land use and land cover change detection using remote sensing in the Trans Himalayan Region of Ladakh, India. ECS Trans 107(1):2985

Clemens J, Nüsser M (2000) Pastoral management strategies in transition: indicators from the Nanga Parbat region (NW-Himalaya). In: Ehlers E, Kreutzmann H (eds) High mountain pastoralism in Northern Pakistan. Franz Steiner Verlag, Stuttgart, Germany, pp 151–187

Dimri AP, Chevuturi A (2014) Model sensitivity analysis study for western disturbances over the Himalayas. Meteorol Atmos Phys 123:155–180. https://doi.org/10.1007/s00703-013-0302-4

Geneletti D, Dawa D (2009) Environmental impact assessment of mountain tourism in developing regions: a study in Ladakh, Indian Himalaya. Environ Impact Assessm Rev 29(4):229–242

Giri L, Hussain M, Angmo JC, Mustafa G, Singh B, Bahukhnadi A, Pradhan R, Kumar R, Mukherjee S, Bhatt ID, Nautiyal S (2024) Enhancing tomato (Solanumlycopersicum) yield and nutrition quality through hydroponic cultivation with treated wastewater. Food Chem 141079

Govt of India (GoI) (2019) Status of ecosystem health in the Indian Himalayan Region: a report under National mission for sustaining the Himalayan ecosystem. Department of Science and Technology, Govt of India

Hussain B, Mohuiddin FA, Wani SH, Murtaza I, Ahmad S, Mohammed R, Sabagh AE (2024) Characterization of barley genotypes and their biochemical responses against leaf rust (Puccinia hordei) disease under cold arid environment. Polish J Environ Stud 33(1):185–195. https://doi.org/10.15244/pjoes/170775

Ladakh Pashmina Vision (2021) Pashmina Policy LehLadkh, Ladakh Autonomous Hill Development Council

Ladon P, Nüsser M, Garkoti SC (2023) Mountain agropastoralism: traditional practices, institutions, and pressures in the Indian Trans-Himalaya of Ladakh. Pastoralism 13:30. https://doi.org/10.1186/s13570-023-00289-1

LAHDC (2023) Ladakh autonomous hill development council, District Statistical Handbook year 2022–23

Mandal A, Vishwakarma BD, Angchuk T, Azam MF, Garg PK, Soheb M (2024) Glacier mass balance and its climatic and non-climatic drivers in the Ladakh region during 2000-2021 from remote sensing data. J Glaciol 1–23. https://doi.org/10.1017/jog.2024.19

Mukherjee S (2021) Nonlinear recurrence quantification of the monsoon-season heavy rainy days over northwest Himalaya for the baseline and future periods. Sci Total Environ 789:147754

Mukherjee S, Hazra A, Kumar K et al (2019) Simulated projection of ISMR over Indian Himalayan region: assessment from CSIRO-CORDEX South Asia experiments. Meteorol Atmos Phys 131:63–79. https://doi.org/10.1007/s00703-017-0547-4

Norphel C, Tashi P (2015) Snow water harvesting in the cold desert in Ladakh: an introduction to artificial glacier. In: Nibanupudi H, Shaw R (eds) Mountain hzards and disaster risk reduction. Disaster Risk Reduction. Springer, Tokyo. https://doi.org/10.1007/978-4-431-55242-0_11

Nüsser M, Schmidt S, Dame J (2012) Irrigation and development in the Upper Indus Basin: characteristics and recent changes of a socio-hydrological system in Central Ladakh, India. Mountain Res Developm 32(1):51–61

Nüsser M, Dame J, Kraus B et al (2019) Socio-hydrology of “artificial glaciers” in Ladakh, India: assessing adaptive strategies in a changing cryosphere. Reg Environ Change 19:1327–1337. https://doi.org/10.1007/s10113-018-1372-0

Pandit A, Mir M, Mir M, Wani Y, Bisati I, Nisa SU, Khan HM, Shah RA (2024) Pastoralism in Changthang, Ladakh: adaptations, challenges, and pathways for sustainability. Mt Res Dev 44(1):A1–A7

Sapkota S, Acharya KP, Laven R, Acharya N (2022) Possible consequences of climate change on survival, productivity and reproductive performance, and welfare of Himalayan Yak (Bosgrunniens). Vet Sci 22;9(8):449. https://doi.org/10.3390/vetsci9080449

Sharma S, Tamchos S, Kumar R et al (2022) Low-cost mitigation strategies for community resilience in the Trans-Himalaya to address food and water scarcity in a changing environment. Environ Monit Assess 194:879. https://doi.org/10.1007/s10661-022-10546-2

Stobdan T (2023) Agriculture in Ladakh: a steptowards sustainable mountain development. Beeja House, New Delhi, India

Tuladhar S, Hussain A, Baig S, Ali A, Soheb M, Angchuk T, Dimri AP, Shrestha AB (2023) Climate change, water and agriculture linkages in the upper Indus basin: a field study from Gilgit-Baltistan and Leh-Ladakh. Front Sustain Food Syst 6:1012363

Wani SA, Mugal MA, Dar FA, Reddy CS, Rashid I, Khuroo AA (2023) Biodiversity data synthesis on trees of Indian Himalayan Region: policy and management implications. Global Ecol Conservat 48:e02698

Wu N, Ismail M, Joshi S, Shaoliang Y, Shrestha RM, Jasra AW (2014) Livelihood diversification as an adaptation approach to change in the pastoral Hindu-Kush Himalayan region. J Mt Sci 11:1342–1355. https://doi.org/10.1007/s11629-014-3038-9

Chapter 9
Patterns and Projection of Public Healthcare Facilities in the Kashmir Division of North-West Himalayas Using Geospatial Technology

Rajender Singh, Gursheen Kour, and Nazma Koser

Abstract Equitable accessibility to public healthcare facilities (HCF) is critical for healthy communities and societies. However, factors like physical barriers and uneven spatial distribution of the HCFs devoid people from availing these services. This has become many-fold with the stress of the growing population, making it more difficult to ensure affordable and quality healthcare services to all sections of society. The present study analyzes the spatial accessibility and pattern of HCFs in a district-wise manner for the Kashmir Division using geospatial techniques and also the population–HCF projection for the years 2021, 2031, and 2041, taking 2011 as the base year. Service of HCFs based on travel time (minutes) and travel distance (km) using the road network was prepared for each district. The Average Neighbor Ratio (0.643) depicts a highly clustered pattern of HCFs with a high statistical significance of such a spatial clustering (P-value < 0; Z-score = – 25.795). The central HCF, which is the most accessible from all directions, came out to be the Sub-center (SC) Bather located in the Budgam District. None of the districts, except Bandipore, have the District Hospital serving as the central HCF. Overall kernel density of the HCFs ranged from 0 to 0.647 per km^2 with Srinagar District having the highest density of all. In addition, the future requirement of new HCFs in proportion to the growing population from the years 2021 to 2041 was also computed using two methods i.e geometric and exponential projection. Results using both projection methods reveal the highest increase in the new HCFs for the Anantnag and Kupwara districts both having more than 420 HCFs by the end of 2041.

Keywords Accessibility · Geo-statistics · Healthcare facilities · Population projection · Service area

R. Singh (✉) · G. Kour · N. Koser
Department of Geography, University of Jammu, J&K, India
e-mail: rajsinghju@gmail.com

S. C. Rai (ed.), *Himalaya: Mountains of Destiny*,
https://doi.org/10.1007/978-3-031-94905-0_9

9.1 Introduction

Adequate access to public healthcare is one of the most important factors contributing to a healthy society. Equitable accessibility to healthcare, in terms of geographic, social, and economic dimensions, is pertinent for the overall well-being and development of a society. It is estimated that at least 400 million people around the globe lack access to basic and essential healthcare services (McGrail et al. 2009; Wooley et al. 2018). With the growing population, it has become more daunting to guarantee equitable access and quality of healthcare services to different groups and sections of society. Access to the HCFs can be spatial (geographic) and non-spatial depending upon several factors (Cressie 2015). Factors such as availability and distance are some of the spatial accessibility attributes (Khan 1992) and can be used to analyze the pattern and distribution of the features over space. Longer distances to the public HCFs could adversely affect the physically challenged and elderly people and may also increase the risk of fatalities (Hare and Barcus 2007).

The Geographic Information System (GIS) has great potential in effectively mapping and assessing the pattern and distribution of HCFs and also for studying spatial epidemiology (McLafferty 2003; Luo et al. 2003; Pui-Jen et al. 2009). In the field of health geography, numerous studies all over the world have utilized geospatial technologies for computing the pattern and distribution of both public and private HCFs over space and time. Masoodi et al. (2015) utilized GIS in creating buffers and catchment areas of the healthcare centers for the coastal city of Bander Abbas in Iran by also incorporating the traffic data and average driving speed of the vehicles in the city. Using geospatial techniques such as GIS, remote sensing, and GPS, Shaba et al. (2023) did buffer and cluster analysis of both public and private HCFs in the plateau state of Nigeria. The accessibility measurements for the healthcare centers in India are dominated by urban regions rather than rural ones (Dutta et al. 2021; Minutha 2021; Das et al. 2023). Khandelwal et al. (2020) did a spatial accessibility analysis of the most populous district of Ahmedabad in the Gujarat state using GIS. In the Murshidabad District of West Bengal, Parvin et al. (2020) did the physical accessibility and location-suitability assessment of the HCFs using both GIS and a hybrid decision-making method.

In India, the framework of healthcare service delivery consists of three tiers. At the primary level, there are Sub-centers (SC) and Primary Health Centers (PHC). At the secondary and tertiary levels, the Community Health Centers (CHC) the Sub-District Hospitals (SDH) and District Hospitals (DH) provide healthcare services, respectively. The Union Territory (UT) of Jammu and Kashmir (J&K) consists of a total of 2882 HCFs, with 1560 HCFs in the Jammu division and 1322 HCFs in the Kashmir Division (https://www.data.gov.in/catalog/all-india-health-centers-directory). Being the most populated division, Kashmir has a total number of 15 DHs; 38 CHCs; 327 PHCs; and 942 SCs spread across its ten districts (https://www.data.gov.in/catalog/all-india-health-centers-directory). Despite having a large number of HCFs, there is uncertainty in asserting whether or not the people residing in Kashmir have equitable geographical access to these HCFs because of complex terrain, physical barriers, and uneven distribution. Some studies in the past have attempted to analyze the pattern and distribution of HCFs while other studies have assessed the infrastructural and manpower deficiencies among various HCFs. Tali et al. (2017a, b) also studied the spatial pattern and distribution of HCFs in the Pulwama District of Kashmir by applying two methods, namely location quotient and Lorenz curve. At the medical-block level, the study found uneven distribution of HCFs, particularly for PHCs and SCs, with the Pulwama block having more than the required number of PHCs and the Tral block having a lesser number of PHCs in proportion to their respective populations. Using GIS and statistical methods, Tali et al. (2017a, b) estimated the travel time required to reach the CHCs and DHs in the Pulwama District of Kashmir Division. The results depicted huge disparities in travel time among various medical blocks wherein 10.64% of the total population required at least 48 min or 20 km to reach a CHC. In the case of DHs, 21.04% of the total population of the district had to travel for at least one hour or 36 km to reach a DH. Bhat et al. (2019) assessed the district-wise disparities in the availability of healthcare infrastructure for the whole of Jammu & Kashmir. The results revealed that 40–50% of the total population of the UT faced inaccessibility toward the HCFs and also the dearth of required qualified doctors, para-medical staff, and nurses in rural Kashmir. Another recent study by Rather et al. (2023) regarding the primary healthcare centers in the UT of Jammu & Kashmir revealed a shortage of 65.3% of para-medical staff in the PHCs while only 0.98% of the total PHCs had a 'Kayakalp' score above 70% for the base year 2018–19, thus rendering the PHCs unable to ensure affordable and equitable healthcare to all the people of the UT. It is evident from the above review of literature that most of the studies have focused on estimating the spatial accessibility of HCFs in individual districts of the Kashmir Division. In the present study, an attempt has been made to utilize GIS in depicting the pattern of public HCFs while also estimating the requirement of new HCFs in proportion to the growing population in Kashmir at the tehsil level for the coming decades.

The medical and healthcare history of Kashmir dates back to the pre-independence era. A significant milestone came in the 1860s with the establishment of the Kashmir Medical Mission, which marked the introduction of Western medicine to the valley. The first mission hospital played a pioneering role in modern healthcare and was instrumental in offering organized medical care (https://www.dhskashmir.org/history.php). Till 1895, the region had one hospital at Srinagar and six different dispensaries located in its various districts (https://www.dhskashmir.org/history.php). During the reign of Maharaja Hari Singh, the Shri Maharaja Hari Singh (SMHS) hospital was inaugurated at Srinagar in 1945. At present, the Directorate of Health Services (DHS)-Kashmir is the apex institution responsible for the management and planning of all the public HCFs in the region to ensure quality and affordable services to the people (https://www.dhskashmir.org/index.php). To promote equitable accessibility in the region, it is important to analyze the present distribution and accessibility of the HCFs that will leverage to identify suitable locations for establishing new HCFs in proportion to the increasing population in the region. Keeping this in mind, the present study comprises the following objectives: (i) to analyze the spatial pattern of HCFs in the Kashmir Division at the Tehsil level, (ii) to understand the population–HCF ratio in various Tehsils of Kashmir Division, (iii) to estimate the requirement of new HCFs in proportion to the growing population over three decades (i.e., from 2011 to 2041).

9.2 Study Area

Kashmir Division is one of the administrative divisions of the Union Territory of Jammu & Kashmir. It is bounded by the Jammu division to its south, Ladakh to its east and north, and Pakistan to its west (Fig. 9.1). Geographically, the division is bordered by the Karakoram Range in the north, the Zanskar Range in the south and southeast, and the Pir Panjal Range in the south. The division comprises the famous valley of Kashmir apart from other surrounding regions and has a total of ten districts with one district, namely Muzzafrabad in the Pakistan Occupied Kashmir (POK) at its north-west end. With a total population of more than 6.8 million, the region has a population density of about 431 persons/km^2 and dominated by four ethnic groups Kashmiri, Pahari, Gujjar-Bakkerwal, and Shina. Bounded by Himalayan ranges in its north and south, Kashmir has an average elevation of 1620 m. The Middle and Greater Himalayan Mountain ranges pass through the region with Machoi (5458 m) serving as the highest peak. Jhelum is the most important river flowing through the region along with its numerous tributaries, namely Sind, Lidder, Kishenganga, Pohru, Erin, Arpath, Sukhnag, etc. The climate is generally temperate, and the region receives a mean annual precipitation of 1200mm with most of the annual rainfall during the

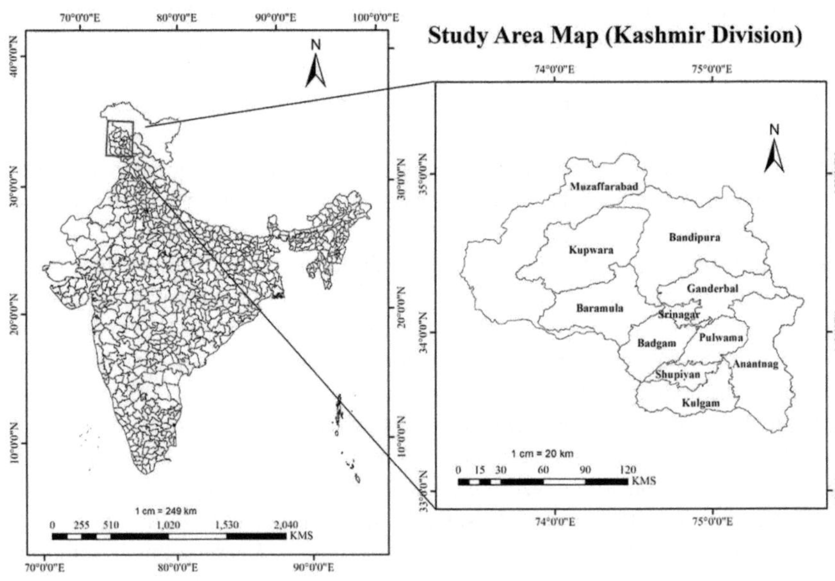

Fig. 9.1 Location map of Kashmir Division

spring season (March–May). Demographically, the region is dominated by Muslim population (97.06%), followed by Hindus (2.11%), Sikhs (0.58%), and Christians (0.11%).

9.3 Methodology

Table 9.1 illustrates the data set used in the study. To achieve the stated objectives, the methodology employed an integrated approach that incorporated multiple spatial analysis techniques, namely overlay analysis; network analysis; average nearest neighbor analysis; kernel density; geo-statistical techniques, and also statistical methods of population projection. The results were effectively illustrated through maps prepared in GIS environment. A brief description of some techniques is given below:

Table 9.1 Datasets used in the study

Data	Source	Uses
Tehsil and district boundary data	Survey of India[1]	In identifying the administrative boundary of tehsils and districts as well as for overlay analysis and understanding the pattern of HCFs within these boundaries
Location of tehsil headquarters and coordinates	Google Earth Pro	To determine the central points of tehsils for spatial analysis
Road network information	Open Street Map[2]	For network analysis consider road lengths and attributes, to calculate travel distances and times to healthcare facilities
Healthcare facilities data	All India Health Centers Directory[3]	To identify the coordinates and locations of healthcare facilities for overlay analysis, service area network analysis, and nearest neighbor analysis
Population data	Census of India[4]	To assess the population-healthcare facility ratio and estimate future healthcare needs based on population growth

[1] https://onlinemaps.surveyofindia.gov.in.

[2] https://www.geofabrik.de.

[3] https://www.data.gov.in/catalog/all-india-health-centers-directory.

[4] https://censusindia.gov.in.

9.3.1 Network Analysis

The network analysis included service area maps based on road networks using the network analyst add-in tools. To determine the journey time, a standard vehicle speed of 30 km/hr was assumed, and the resulting time in hours was converted into minutes.

9.3.2 Average Nearest Neighborhood Analysis

The average nearest neighborhood tool gives the nearest neighbor ratio, z-value, and p-value[5]. Z-scores are standard deviations. A z-score indicates the significance of the observed nearest neighbor distance compared to random expectations, with negative values suggesting clustering and positive values indicating dispersion. The p-value is a probability. The p-value represents the probability of obtaining results as extreme as the observed data under the null hypothesis of complete spatial randomness.[6]

9.3.3 Geo-statistics and Spatial Analysis

Geo-statistics and spatial analysis involved the calculation of key spatial metrics, including the central feature, the mean center, the standard distance circle, and the directional distribution ellipse. Central features were identified to pinpoint significant locations or clusters within the region. The mean center is the average x and y coordinate of all the features in the study area, providing a central reference point for spatial analysis.[7] The standard distance circle measures the compactness of a distribution, providing a single value representing the dispersion of features around the center.[8] A directional distribution ellipse was utilized to visualize the orientation and dispersion of features, offering insights into spatial patterns.[9]

[5] https://desktop.arcgis.com/en/arcmap/latest/tools/spatial-statistics-toolbox/h-how-average-nearest-neighbor-distance-spatial-st.htm..

[6] https://pro.arcgis.com/en/pro-app/latest/tool-reference/spatial-statistics/what-is-A-Z-score-what-is-A-P-value.htm.

[7] https://desktop.arcgis.com/en/arcmap/latest/tools/spatial-statistics-toolbox/mean-center.htm.

[8] https://desktop.arcgis.com/en/arcmap/latest/tools/spatial-statistics-toolbox/standard-distance.htm.

[9] https://pro.arcgis.com/en/pro-app/latest/tool-reference/spatial-statistics/h-how-directional-distribution-standard-deviationa.htm.

9.3.4 Kernel Density Analysis

This analysis used a kernel function to distribute influence from each point, resulting in a continuous surface that emphasized areas with high and low density.[10] By aggregating the influence of nearby locations, the kernel density tool produced a visual representation of the spatial distribution of healthcare facilities, emphasizing areas with concentrated or sparse facility presence. Apart from these spatial techniques, the following statistical techniques were used for population and HCF projection.

9.3.5 Projected Healthcare Facility

The methodology for projecting healthcare facilities involved employing population projection methods in which geometric and exponential population projection methods were used. The geometric projection method assumes a constant rate of growth over time, while the exponential projection method considers a changing growth rate. The following formulae were used to find the population projection from 2011 to 2021, 2031, and 2041 using geometric and exponential methods.

$$P_n = P_0\left[1 + \frac{r}{100}\right]^n \text{ – Geometric Method}$$

$$P_n = P_0\left(e^{Rn}\right) \text{ – Exponential Method}$$

where P_n is the population after 'n' number of decades; P_0 is the last known population; 'r' is the growth rate; and 'R' is the growth rate expressed as decimal. The number of HCFs was determined using the ratio method which involves calculating the ratio of HCFs to the projected population figures for each district.

$$\text{New HCFs} = \text{Past number of HCFs} \times (1 + P)$$

$$\text{where } P = \frac{\text{Present Population} - \text{Past Population}}{\text{Past Population}} \times 100$$

10 https://desktop.arcgis.com/en/arcmap/latest/tools/spatial-analyst-toolbox/kernel-density.htm.

9.4 Results and Discussion

9.4.1 Service Area of HCFs Using Network Analysis

To understand the accessibility of HCFs, tehsil-wise service area maps for each district were created based on travel time (4, 8, and 12 min) and travel distance (1, 2, and 3 km). The northern part of Anantnag District (Fig. 9.2a), i.e., Pahalgam tehsil has limited HCFs, which is due to poor road connectivity in the region. Tehsils of Anantnag, Shangus, and Bijbehara have better accessibility to the HCF. This is due to the high density of well-connected roads in this region. Most of the service area in Anantnag District falls within the 3 km category having a total area of 666.86 km^2 (Table 9.2). For the Budgam District (Fig. 9.3e), a total service area of 580.17 km^2 requires a 4-min travel time by road to reach the nearest HCF. Also, the south-western part of the Budgam (Fig. 9.3f) has limited HCFs and negligible service areas which are due to poor road connectivity. Tehsils of Chadoora, Beerwah, and Budgam have better accessibility to the nearest HCF which is due to well-connected roads in this region. Further, the tehsils of Sonawari and Bandipora (Fig. 9.2c) have better accessibility to the nearest HCF since the road networks in these tehsils are better compared to other tehsils of the district. Limited numbers of HCFs are present in tehsil Boniyar, Tangmarg, and north-west Uri (Fig. 9.3g) due to limited accessibility in these areas.

Ganderbal tehsil (Fig. 9.3i) has well-connected road network; hence, the service areas for most HCFs are within 3 km of travel distance and 12 min of travel time. The western part of the tehsil Damhal and Hanjipora (Fig. 9.3i) shows the absence of HCFs and road networks. In the Kulgam District (Fig. 9.3k), the average distance and time to reach the nearest HCF for most of the region district are 3 km and 4 min, respectively (Table 9.2). For the Shopian District (Fig. 9.4q), the western and southern parts have larger service areas requiring an average distance of 2 km to access the nearest HCF while the time taken to reach the nearest HCF is 4 min having a total area of 392.57 km^2 (Table 9.2) in mostly the entire district. For the Srinagar District (Fig. 9.4s), its eastern region, i.e., tehsil Srinagar (north), has limited HCFs and poor road connectivity while tehsil Srinagar (south) has a larger service area.

9.4.2 Spatial Pattern of HCFs Using Nearest Neighbor Ratio

Table 9.3 represents the 'Average Nearest Neighbor Analysis' for various districts of the Kashmir Division. If the ratio is less than 1, the pattern exhibits clustering, while a ratio greater than 1 indicates dispersion. The districts of Budgam and Shopian have a nearest neighbor ratio of 1.002 and 1.040, thereby showing a random clustered pattern. The districts of Kulgam and Pulwama have a nearest neighbor ratio of 0.897 and 0.817, respectively, which shows a moderately clustered pattern. All other districts have a nearest neighbor ratio between 0.542 and 0.789; thus, these districts

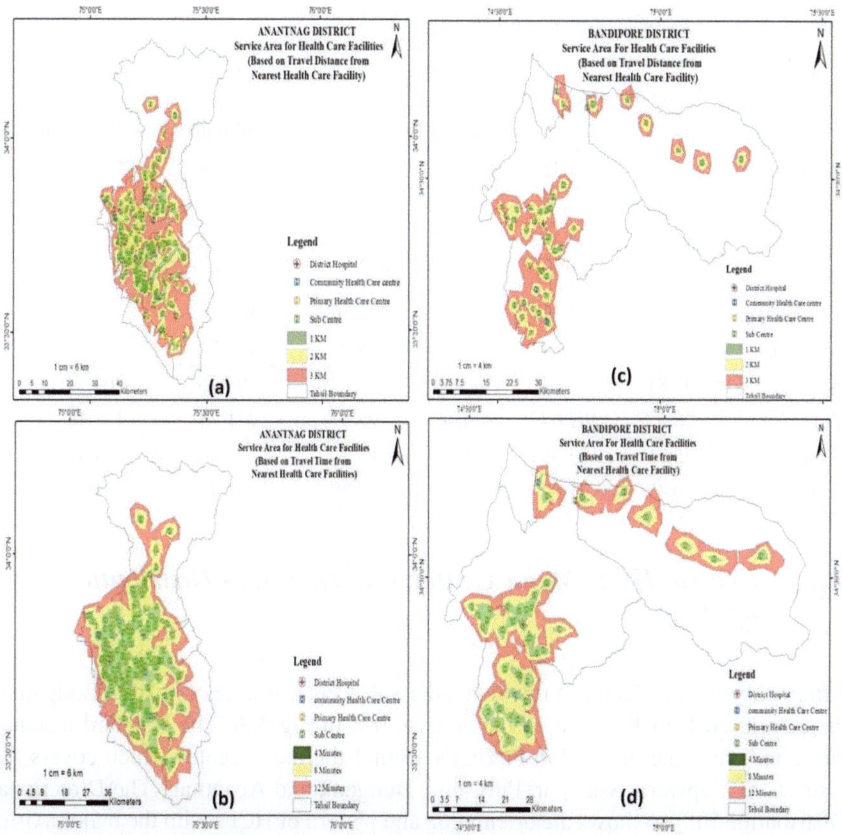

Fig. 9.2 Service area based on travel time (in minutes) and travel distance (in km)—**a** and **b** Anantnag **c** and **d** Bandipore

have a highly clustered pattern. Overall, the division as a whole has a nearest neighbor ratio of 0.643, which shows a highly clustered pattern. The z-scores provided in Table 9.3 are associated with the nearest neighbor ratio. The z-scores of all the districts are negative showing clustering in all districts except for districts Budgam and Shopian showing random patterns. A z-score value of − 25.795 is observed for the entire division which means that there is a high negative deviation from the expected ratio under randomness.

The p-value ranges from 0.000 to 0.953 (Table 9.3) and several districts have p-values less than 0.05, indicating statistically significant clustering or dispersion. The Anantnag District has a p-value of 0.00, which is less than the common 0.05 significance level. This strongly suggests that the observed clustering is unlikely to have occurred by chance and is a result of an ongoing spatial process in the region. Overall, Kashmir Division has a nearest neighbor ratio of 0.643798 which is less than 1, exhibiting a high clustering pattern for all the HCFs (Fig. 9.5).

Table 9.2 District-wise service area for HCFs

District	Service Area for HCFs (in km^2)					
	Travel distance			Travel time		
	1 km	2 km	3 km	4 Minutes	8 Minutes	12 Minutes
Anantnag	261.14	464.71	666.16	580.17	545.28	535.41
Budgam	171.34	315.74	284.93	444.31	306.95	205.04
Bandipore	45.21	167.3	350.15	148.98	278.68	365.29
Baramula	221.15	562.3	590.42	596.43	453.09	275.14
Ganderbal	59.39	142.61	187.05	170.16	163.88	172.57
Kulgam	45.69	110.83	209.69	285.92	152.5	114.7
Kupwara	297.81	512.34	480.94	503.02	296.76	264.74
Pulwama	128.37	307.67	260.46	392.57	241.73	131.18
Shopian	75.76	155.49	143.61	216.7	132.5	79.46
Srinagar	50.37	71.62	95.61	105.58	82.97	81.28

9.4.3 Central HCF, Mean Center, and Direction Distribution of HCFs

After analyzing the data, it is found that the Sub-Center Bather of District Budgam is the central HCF of Kashmir Division as shown in Fig. 9.6. The standard distance circle indicates the dispersion of HCFs around the mean center which covers the districts of Kupwara, Srinagar, Pulwama, Budgam, and Anantnag. The Directional Distribution Ellipse shows the orientation and pattern of HCFs with the major axis in the north-west and southeast direction from Kupwara to Anantnag District. Table 9.4 depicts the central HCF and mean center for all the districts. Overall, the sub-centers acted as the central HCF for most of the districts signifying their centrality and accessibility. For the Anantnag District (Fig. 9.7a), the Central HCF is Sub-Center Wantrag with the directional ellipse in a north–south direction. In the Budgam District (Fig. 9.7b), the central HCF is again a Sub-Center, namely Kokerbag. The standard distance circle covers all the tehsils of the district, and more than half of the facilities in Budgam are found within this circle. However, for the Bandipore District (Fig. 9.7c), the central feature is the District Hospital (DH) Bandipora, meaning that the DH is the most accessible out of all the HCFs from all directions. The mean center is also located in Bandipora tehsil. In the Baramulla District (Fig. 9.7d), the Sub-Center Uplina came out to be the central HCF; most accessible from all directions. The standard distance circle covers all the tehsils of the district Baramulla except for Uri, the border tehsil, which is far away from the mean center. For the Ganderbal District (Fig. 9.7e), the central HCF is Sub-Center Nunnar with the mean/geographic center in Kangan tehsil. For the Kulgam District (Fig. 9.7f), both the central HCF and mean center are located within the Kulgam tehsil. Both the standard distance and directional distribution ellipse cover somewhat similar areas of the HCFs around the mean center. In the Kupwara District (Fig. 9.7g), the standard distance circle covers

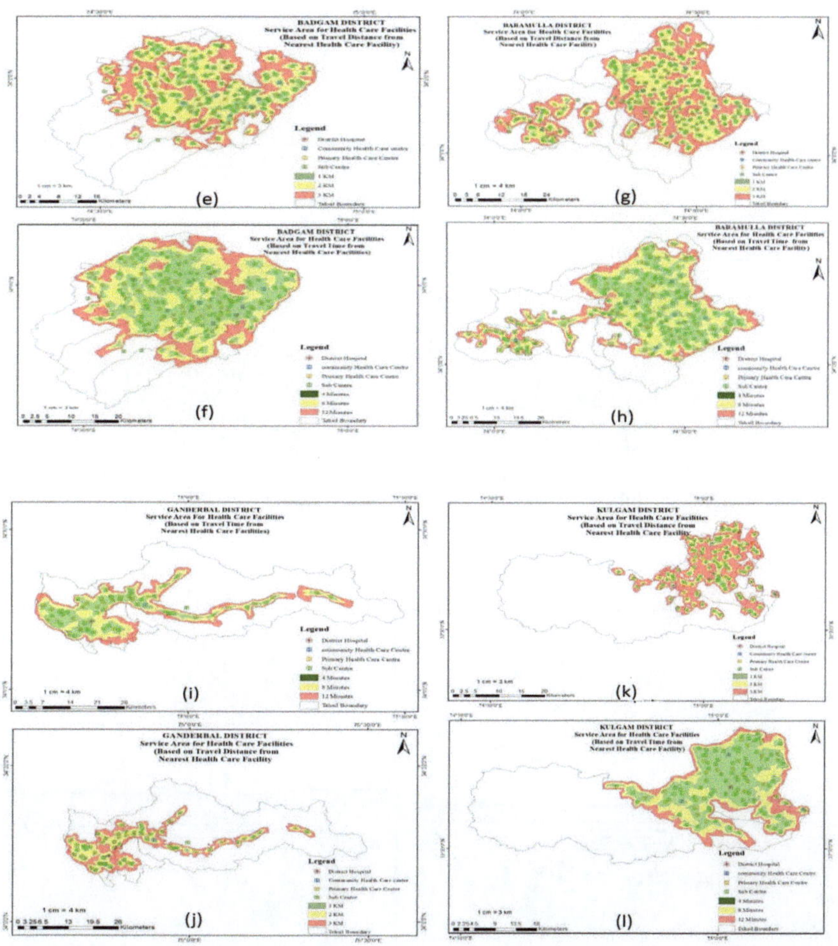

Fig. 9.3 Service area based on travel time (in minutes) and travel distance (in km) for **e** and **f** Budgam, **g** and **h** Baramulla, **i** and **j** Ganderbal, **k** and **l** Kulgam

all the tehsils except Karnah tehsil which is far away from the mean center. NTPHC Puchhal is the central HCF of Pulwama District and the mean center is found near this facility, both the central and mean centers being within Pulwama tehsil (Fig. 9.7h). The PHC Nadigam in Shopian tehsil is the central HCF for the Shopian District (Fig. 9.8a). In the Srinagar District (Fig. 9.9b), the central HCF is Nowhata of tehsil Srinagar (South) which is a sub-center. The standard distance circle of the district covers the south-western parts of the district and excludes all facilities that are in Srinagar (North) tehsil. The reason for the exclusion of HCFs might be due to the clusters of HCFs found in the south-western parts as observed in the overlay analysis.

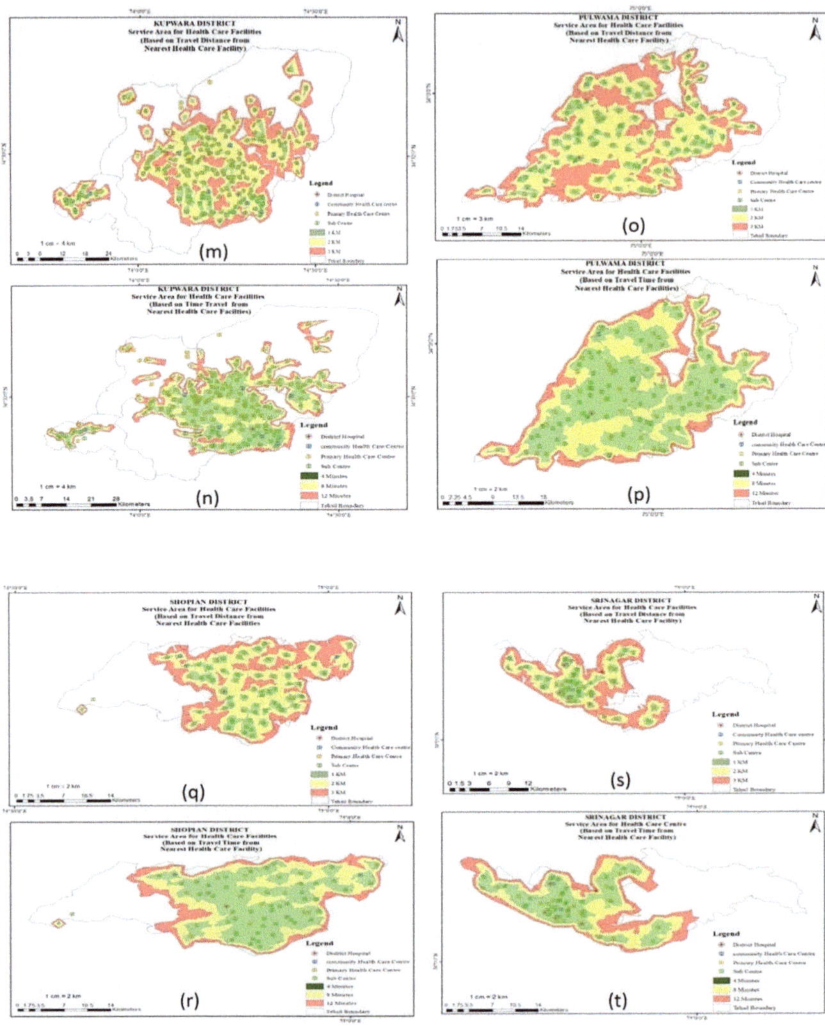

Fig. 9.4 Service area based on travel time (in minutes) and travel distance (in km) for **m** and **n** Kupwara, **o** and **p** Pulwama, **q** and **r** Shopian, **s** and **t** Srinagar

9.4.4 Density Per Unit Area of HCFs

The kernel density map of the Kashmir Division (Fig. 9.9) depicts the distribution and concentration of HCFs per unit area within the region. The darker and lighter shades in the map indicate higher and lower concentrations of HCFs, respectively. Overall, the kernel density for the entire division ranged from 0 to 0.647 HCFs per unit area. The district of Srinagar shows the highest concentration of HCFs while the regions of the complex terrains in districts Budgam, Pulwama, Shopian, and

Table 9.3 Average nearest neighbor distribution pattern of districts and division

Average nearest neighbor analysis				
District	Nearest neighbor ratio	Z-score	P-value	Distribution pattern
Anantnag	0.747	– 6.254	0.000	Highly clustered
Budgam	1.002	0.057	0.953	Random
Bandipore	0.542	– 6.127	0.000	Highly clustered
Baramulla	0.789	– 5.811	0.000	Highly clustered
Ganderbal	0.775	– 3.485	0.000	Highly clustered
Kulgam	0.897	– 2.028	0.042	Moderately clustered
Kupwara	0.710	– 8.261	0.269	Random clustered
Pulwama	0.817	– 4.075	0.000	Highly clustered
Shopian	1.040	0.627	0.530	Random
Srinagar	0.570	– 6.925	0.042	Moderately clustered
Kashmir Division	0.643	– 25.795	0.000	Highly clustered

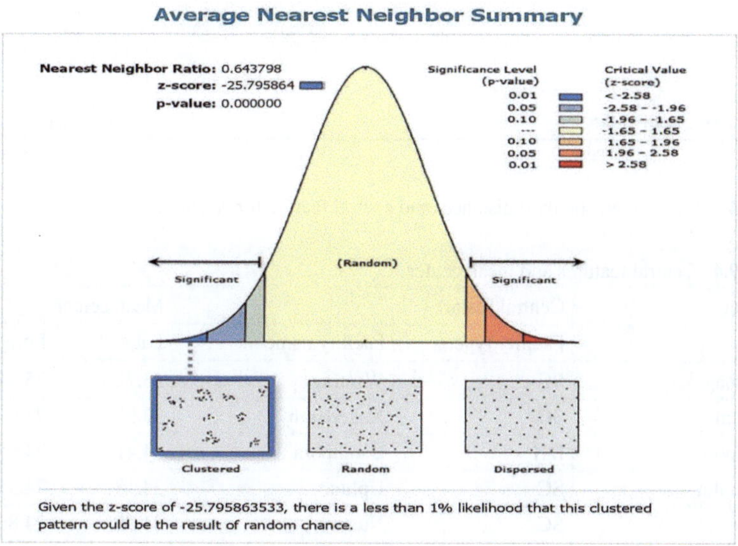

Fig. 9.5 Results of average nearest neighbor ratio for the study area

Anantnag where the Greater Himalayan ranges (in the north) and the Pir Panjal Range (in the south) are located show lower concentration of HCFs. Due to geopolitical constraints, data on HCFs for Muzaffarabad District was not available. The northeastern and southern parts of the division have almost negligible concentrations of HCFs due to the presence of mountain ranges and fewer inhabitants.

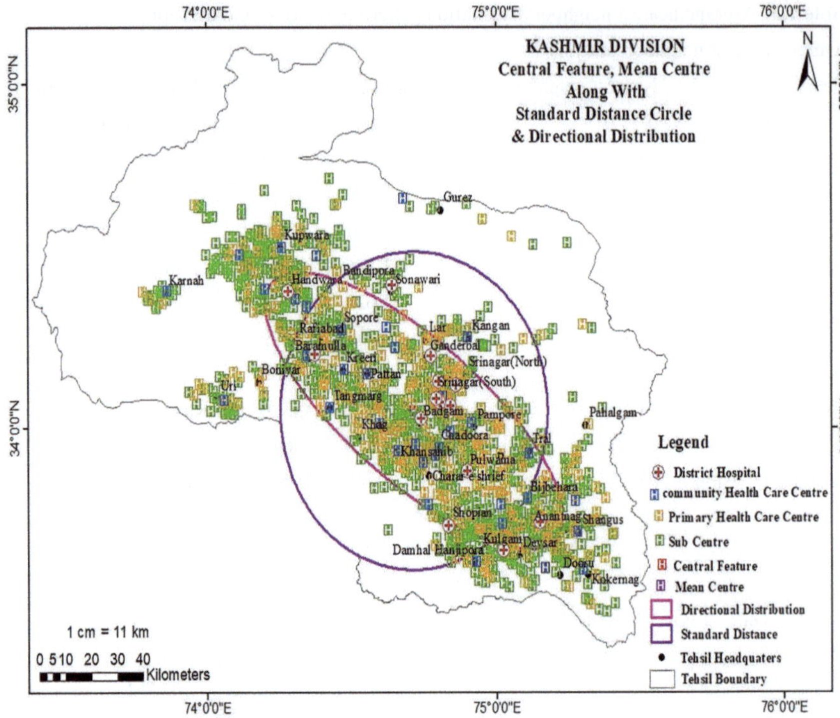

Fig. 9.6 Mean center, standard distance, and central feature for Kashmir Division

Table 9.4 Central features and mean center

District	Central feature		Mean center	
	Facility type	Facility name	Lat	Long
Anantnag	SC	Wantrag	33.72	75.24
Budgam	SC	Kokerbagh	33.97	74.68
Bandipore	DH	Bandipora	34.41	74.68
Baramulla	SC	Uplina	34.18	74.35
Ganderbal	SC	Nunnar	34.25	74.86
Kulgam	SC	Sangas	33.65	75.01
Kupwara	SC	Pazalpora	34.48	74.19
Pulwama	NTPHC	Puchhal	33.91	74.97
Shopian	PHC	Nadigam	33.72	74.87
Srinagar	SC	Nowhata	34.10	74.81
Kashmir Division	Sub-Center	Bathar (Budgam)	34.05	74.71

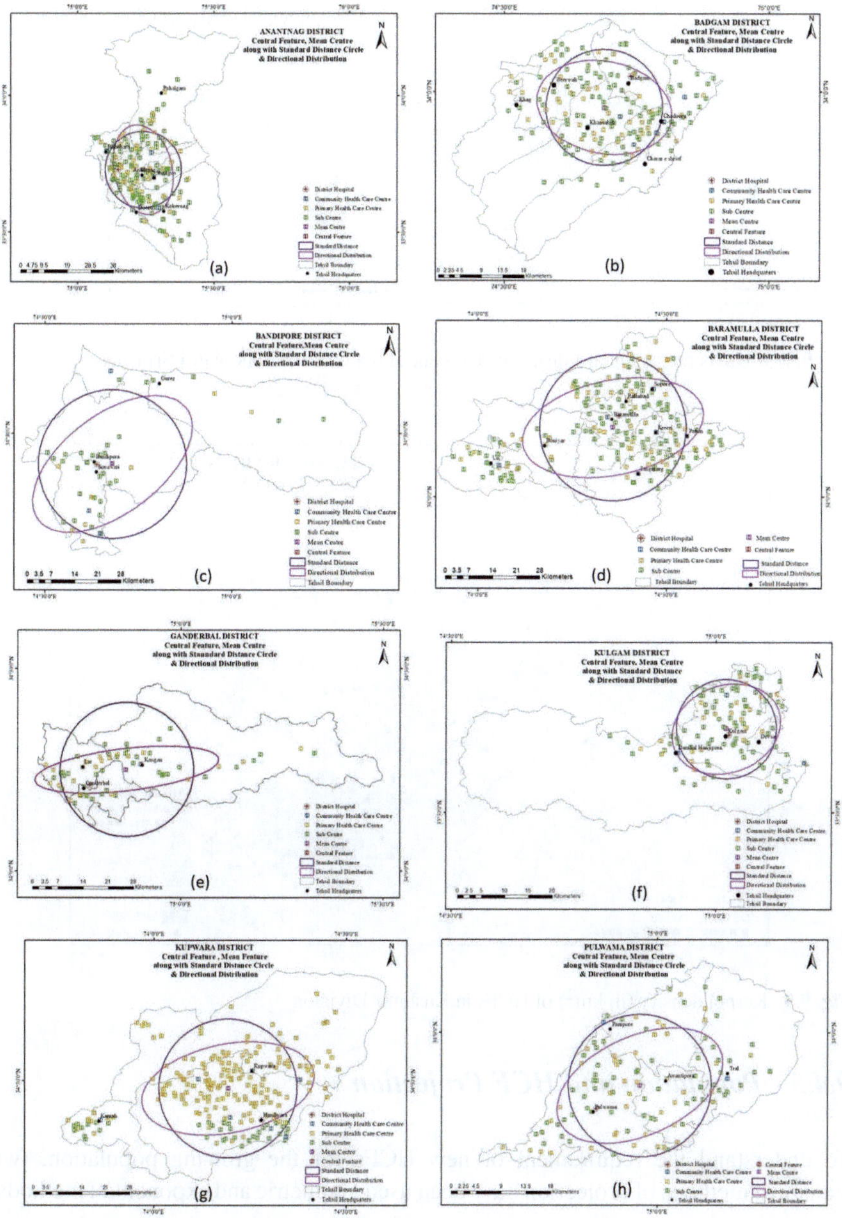

Fig. 9.7 Mean center, standard distance, and central features for **a** Anantnag **b** Budgam **c** Bandipore **d** Baramulla **e** Ganderbal **f** Kulgam **g** Kupwara **h** Pulwama

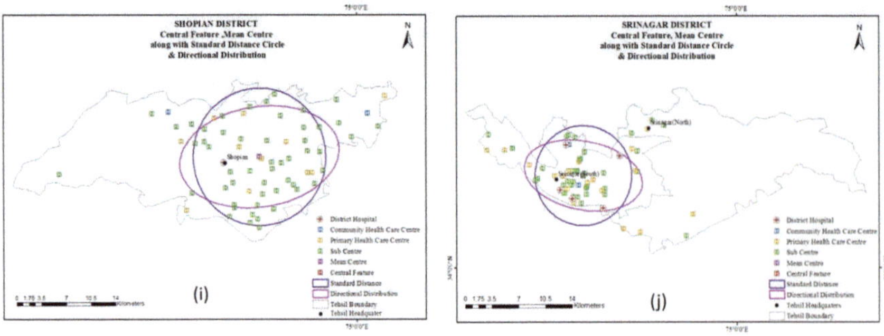

Fig. 9.8 Mean center, standard distance, and central features for **i** Shopian **j** Srinagar

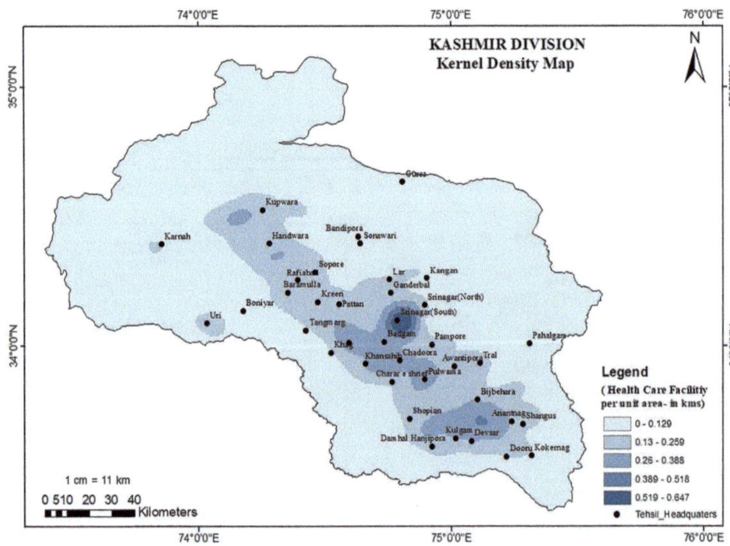

Fig. 9.9 Kernel density (in km^2) of HCFs in Kashmir Division

9.4.5 *Population and HCF Projection*

To understand the requirement of new HCFs for the growing population, two statistical methods of projection have been used geometric and exponential methods.

Geometric Method

The district-wise projected HCFs (Fig. 9.10) concerning the Census-2011 over 30 years, from 2011 to 2041, using the geometric method of population projection are shown in Table 9.5. In 2011, all districts had HCFs less than 250. Overall, the increase in the projected number of HCFs in each decade is the highest for the Kupwara and Anantnag districts. Both these districts show an increase in projected

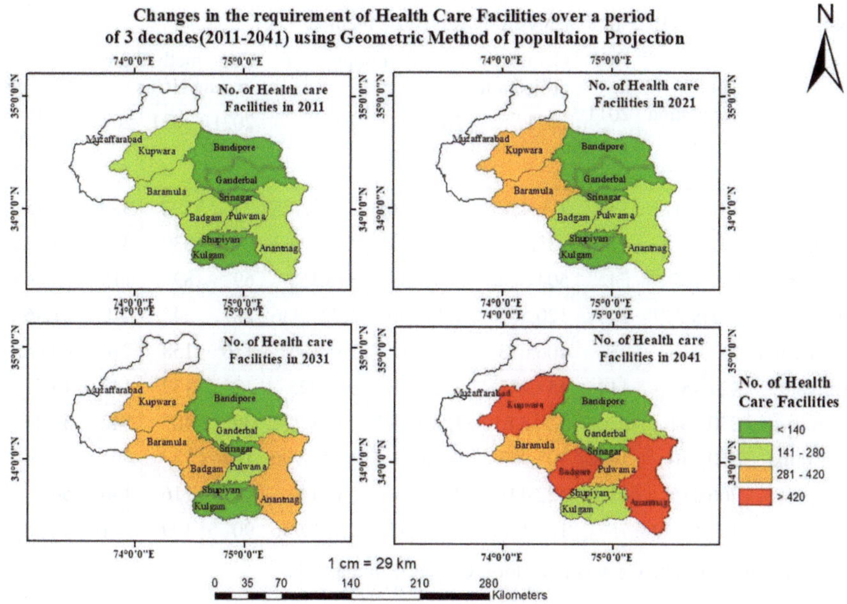

Fig. 9.10 Change in the number of HCFs in proportion to population growth from 2011 to 2041 using the geometric method

HCFs of more than 300 over three decades. The projected HCF is lowest in Bandipore wherein the projected number of HCFs increased by several 15 in 2021 and 61 in 2041. The highest and lowest percentage change in the number of HCFs was observed for the Anantnag (62.45%) and Kulgam (19.71%) districts, respectively (Table 9.5). Srinagar District also witnessed a moderate increase of 41.98% HCFs from 2011 to 2041.

Exponential Method

It can be observed from Fig. 9.11 that all the districts had less than 300 HCFs in 2011. By 2021, the projected map shows an overall increase in the number of HCFs. In Kupwara, the HCFs increased from 226 in 2011 to 317 in 2021. On looking forward to 2031, the projected distribution of HCFs continues to evolve, with more areas transitioning to higher facility counts, as Anantnag, and Kupwara reaching a number above 300 HCFs. Finally, the 2041 projection shows a significant expansion in the Anantnag District, representing the highest facility counts of over 633. Bandipore District in all four years is projected to have lower HCFs, i.e., below 130. The highest and lowest percentage change from 2011 to 2014 in the number of HCFs was observed for Anantnag (68.56%) and Kulgam (20.83%) districts, respectively (Table 9.6). Srinagar District also witnessed a moderate increase of 45.65% HCFs from 2011 to 2041.

Table 9.5 Projected HCF ratio analyses by geometric method

District	Census 2011 population	HCF in 2011	Geometric method of population projection			Ratio method for finding HCFs			% Change from 2011 to 2041
			2021	2031	2041	2021	2031	2041	
Anantnag	1,078,692	199	1,494,851	2,071,565	2,870,775	276	382	530	62.45
Budgam	753,745	220	935,699	1,161,577	1,441,981	273	339	421	47.74
Bandipore	392,232	54	504,606	649,176	835,165	69	89	115	53.04
Baramulla	1,008,039	240	120,501	1,440,469	1,721,936	287	343	410	41.46
Ganderbal	297,446	85	406,013	554,208	756,495	116	158	216	60.64
Kulgam	424,483	114	457,295	492,644	530,725	123	132	142	19.71
Kupwara	870,354	226	1,558,612	1,558,612	2,085,734	302	405	542	58.30
Pulwama	560,440	147	903,933	903,933	1,147,996	187	237	301	51.16
Shopian	266,215	73	422,441	422,441	532,149	92	116	146	50
Srinagar	1,236,829	75	1,791,438	1,791,438	2,155,996	90	109	131	41.98

Source All India Health Centers Directory; Census of India-2011

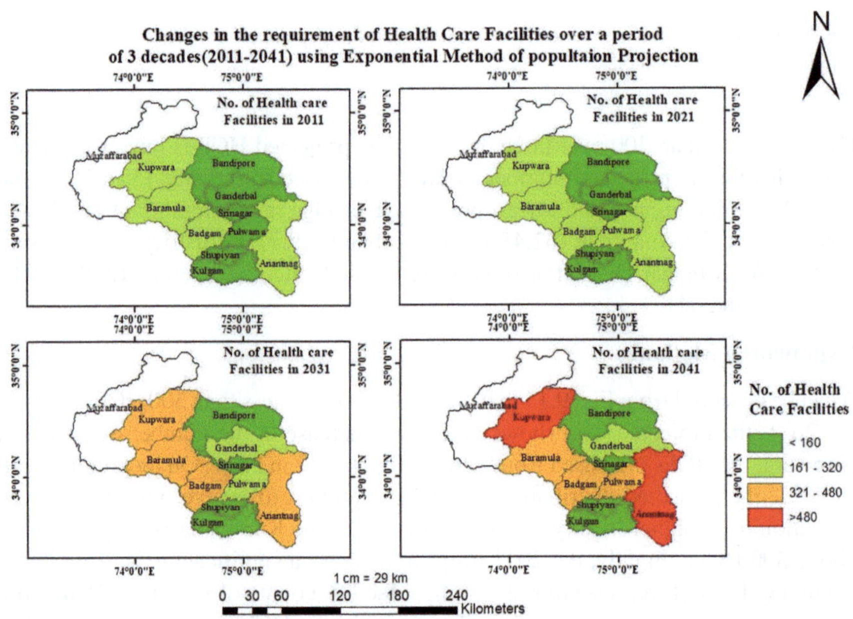

Fig. 9.11 Change in the number of HCFs in proportion to population growth from 2011 to 2041 using exponential method

Table 9.6 Projected HCF ratio analyses by the exponential method

District	Census 2011 population	HCF in 2011	Exponential method of population projection			Ratio method for finding HCFs			% Change from 2011 to 2041
			2021	2031	2041	2021	2031	2041	
Anantnag	1,078,692	199	1,586,529	2,333,453	3,432,020	293	430	633	68.56
Budgam	753,745	220	959,540	1,221,523	1,555,036	280	357	454	51.54
Bandipore	392,232	54	522,358	695,654	926,443	72	96	128	57.81
Baramulla	1,008,039	240	1,225,571	1,490,045	1,811,593	292	355	431	44.32
Ganderbal	297,446	85	428,475	617,224	889,120	122	176	254	66.54
Kulgam	424,483	114	458,597	495,452	535,270	123	133	144	20.83
Kupwara	870,354	226	1,220,602	1,711,798	240,062	317	444	623	63.72
Pulwama	560,440	147	734,156	961,718	1,259,817	193	252	330	55.45
Shopian	266,215	73	345,158	447,512	580,218	95	123	159	54.09
Srinagar	1,236,829	75	1,515,962	1,858,093	2,277,437	92	113	138	45.65

Source All India Health Centers Directory; Census of India-2011

9.5 Conclusion

The distribution pattern of HCFs across various districts of the Kashmir Division reveals significant spatial disparities. Areas such as Phalgam, Dooru, and Gurez exhibit limited overall HCFs, possibly attributed to sparse populations and challenging terrains. It has been observed that most of the HCFs are located around administrative capitals, whereas the far-flung areas show limited presence of HCFs. Moreover, cluster analysis using the nearest neighbor ratio and z-scores indicates significant clustering across all districts, except for Budgam and Shopian, suggesting a non-random distribution pattern. Geo-statistical techniques reveals that SC Bather in Budgam District is the central HCF for the whole division. Except for Bandipore, Pulwama, and Shopian districts, all the other districts have a sub-center as their central HCF. The whole division has a highly clustered pattern of HCFs having a nearest neighbor ratio of 0.643 (z-score of − 25.795 and p-value of 0). The overall density per unit area of the HCFs, calculated using Kernel density, ranged from 0 to 0.64, with Srinagar District having the highest density of HCFs. Moreover, the district-wise requirement of HCFs till year 2041 was calculated using the geometric and exponential methods. Using the geometric method, the requirements of HCFs in proportion to population growth were estimated at 90, 109, and 131 for the years 2021, 2031, and 2041, respectively. Similarly, using the exponential method, the estimated requirement of HCFs stood at 92, 113, and 131 for the years 2021, 2031, and 2041 respectively aligning with the population growth projection. While both methods project an overall increase in healthcare facilities, discrepancies in growth rates among the districts underscore the importance of targeted infrastructure development initiatives. Overall, the findings in the present study provide a comprehensive understanding of the current healthcare landscape and future trends in the Kashmir Division. By leveraging spatial analysis and projection techniques, policymakers can develop evidence-based strategies to address disparities, improve access, and ensure equitable healthcare delivery across the region.

References

Cressie N (2015) Statistics for spatial data. Wiley, New York

Das M, Dutta B, Roy U, Das S, Rath S (2023) Spatial accessibility modeling to healthcare facilities in the case of health shocks of Midnapore municipality, India. GeoJournal 88:3749–3772. https://doi.org/10.1007/s10708-023-10838-1

Dutta B, Das M, Roy U, Das S, Rath S (2022) Spatial analysis and modeling for primary healthcare site selection in Midnapore town, West Bengal. GeoJournal 87:4807–4836. https://doi.org/10.1007/s10708-021-10528-w

Hare TS, Barcus HR (2007) Geographical accessibility and Kentucky's heart-related hospital services. Appl Geogr 27(3–4):181–205

Khan AA (1992) An integrated approach to measuring potential spatial access to health care services. Socioecon Plann Sci 26(4):275–287. https://doi.org/10.1016/0038-0121(92)90004-O

Khandelwal S, Kanchan R (2020) An investigation into spatial accessibility of primary health centers in Ahmedabad district, Gujarat, India. Annals of the National Assoc Geographers India 40(2):220–233. https://doi.org/10.32381/ATNAGI.2020.40.02.2

Luo W, Wang F (2003) Measures of spatial accessibility to healthcare in a GIS environment: synthesis and a case study in Chicago region. Environ Plann B 3:865–884

Masoodi M, Rahimzadeh M (2015) Measuring access to urban health services using geographical information system (GIS): a case study of health service management in Bander Abbas, Iran. Int J Health Policy Managem 4(7):439–445. https://doi.org/10.15171/ijhpm.2015.23

McGrail MR, Humphreys JS (2009) The index of rural access: an innovative integrated approach for measuring primary care access. BMC Health Serv Res 9(124). https://doi.org/10.1186/1472-6963-9-124

McLafferty SL (2003) GIS and health care. Annual Rev Public Health 24:25–42

Minutha V (2021) Accessibility of public healthcare services in Mysore City-A GIS approach. Int J Health Sci Res 11(5):290–298

Parvin F, Ali SA, Hashmi SNI, Khatoon A (2020) Accessibility and site-suitability for healthcare services using GIS-based hybrid decision-making approach: a study in Murshidabad, India. Spat Inform Res 29(1):1–18. https://doi.org/10.1007/s41324-020-00330-0

Pui-Jen T, Men-Lung L, Chein-Min C, Cheng-Hwang P (2006) Spatial autocorrelation analysis of healthcare hotspots in Taiwan in 2006. BMC Public Health 9:464

Rather WA, Sharma D, Khan MA (2023) Primary health care in Jammu and Kashmir: a critical analysis. European Chemical Bulletin 12(5):553–559. https://doi.org/10.31838/ecb/2023.12.si5.066

Shaba HA, Omomoh E, Nannim S, Danlami GRR, Omirinde MO, Davou GY, Mairiga B, Gwamzhi EP (2023) Cluster and buffer analysis of the distribution of healthcare facilities in Plateau State. Int J Med Sci Clinical Res Stud 3(10):2420–2431. https://doi.org/10.47191/ijmscrs/v3-i10-58

Tali JA, Nusrath A, Nengroo A (2017a) Accessibility to higher order public health centers a study of District Pulwama (J&K). Int J Scient Res Sci Technol 3(8):55–61

Tali JA, Divya S, Nushrat S (2017b) Spatial distribution of health centers: a study of district Pulwama (Jammu & Kashmir). Int J Health Sci Res 7(11):185–190

Woolley T, Cristobal F, Siega-Sur J, Ross S, Neusy AJ, Halili, S, Reeve C (2018) Positive implications from socially accountable, community-engaged medical education across two Philippines regions. Rural Remote Health 18: 1–10. https://doi.org/10.22605/RRH4264

Chapter 10
Implications of Climate Change on Forest Resources: Insights from Kinnaur District, Northwestern Himalaya

Nidhi Kanwar and Jagdish Chandra Kuniyal

Abstract The forest ecosystem plays a pivotal role in providing essential ecosystem services and supporting local livelihood options. However, these forests are increasingly vulnerable to the impacts of climate change, which are already altering their structure and function. This study focuses on the Kinnaur District, Himachal Pradesh within the Indian Himalayan Region (IHR) to assess the decadal change in forest cover, change in climatic variables, and experience sharing of local communities regarding climate change impacts on forests. Using a combination of analysis of satellite data and perception surveys conducted from 2017 to 2019, this research examines trends in forest cover and climatic variables. A report from the Forest Survey of India reveals a fluctuating trend in forest cover dynamics, with noticeable changes in dense and open forest areas over the past two decades. Climatic trend analysis highlights significant increases in winter, pre-monsoon, and post-monsoon temperatures, alongside a decrease in winter precipitation. The perception survey among 250 households across 18 villages in Kinnaur District underscores a strong awareness among local communities of climate change impacts on forests. Respondents observed changes such as declining forest area, reduced water availability from streams, and increased frequency of extreme weather events. These perceptions align with scientific findings indicating shifts in seasonal patterns and ecological disruptions affecting flora and fauna. Integrating local perceptions with scientific data is essential for developing context-specific adaptation policies that enhance the resilience of forest communities in the fragile Himalayan landscapes.

Keywords Forest cover · Climate change · Perception survey · Kinnaur District · Indian Himalayan region

---

N. Kanwar
Govind Ballabh Pant National Institute of Himalayan Environment, Kosi-Katarmal, Almora, Uttarakhand, India

J. C. Kuniyal (✉)
Department of Forest Resource Management, College of Forestry, Veer Chandra Singh Garhwali Uttarakhand University of Horticulture and Forestry (VCSG UUHF), Ranichauri, Tehri Garhwal PIN 249 199, Uttarakhand, India
e-mail: jckuniyal@gmail.com

S. C. Rai (ed.), *Himalaya: Mountains of Destiny*,
https://doi.org/10.1007/978-3-031-94905-0_10

10.1 Introduction

The Indian Himalayan Region (IHR) is globally recognized among 36 significant biodiversity hotspots (Myers et al. 2000). The forests of the IHR play a vital role in regulating the region's climate, sequestering carbon, and supporting the livelihoods and cultural practices of indigenous and local communities (Rawat and Rawat 2013; Chao 2012). These forests are home to a rich diversity of plant and animal species, many of which are endemic to the region. The ecosystem services provided by the forests in the IHR are essential resources (i.e., fuel, fodder, and non-timber forest products) for the well-being of the inhabitants of the region. However, these forests are increasingly under threat from the impacts of climate change, which have been observed and experienced by the people living in the IHR (Shrestha and Bawa 2014; Kanade et al. 2018; Duraisami et al. 2022).

Climate change has a significant impact on the forests of the IHR, with rising temperatures, changing precipitation patterns, and an increase in the frequency and intensity of extreme weather events, reduced tree productivity, and altered growth patterns (Shrestha and Bawa 2014; Negi et al. 2021; Seidl et al. 2017). Studies have shown that the recent warming trends in the Himalaya have resulted in decreased forest productivity, particularly affecting tree growth (Srivastava 2022). The forest cover of the Indian Central Himalayan Region is estimated to be 63% with a marginal loss of up to 4.9% from 1985 to 2005 but a considerably higher loss from 10.6 to 52% from 2005 to 2015) (Sati 2021; Parida et al. 2022). Perception-based climate change impact study in the different parts of the IHR conducted by Kuniyal et al. (2021) reported a shift in vegetation species to higher altitudes. The mentioned tree species among these include *Betula utilis* and *Pinus wallichiana.* Also, 67% of the total respondents reported a change in the Satluj Basin in terms of plant species composition in genus Pinus and Betula which were rarely seen at higher altitudes. Extreme climate events, such as heavy snowfall, cloudbursts, flash floods, forest fires, etc., have damaged a significant portion of the region's trees, creating large treefall gaps and possibly defining the upper altitudinal boundaries of certain species (Kuniyal et al. 2021; Thadani 2023). Additionally, the sensitivity of different age-class trees to climate change varies with younger trees experiencing a decline in productivity due to rising temperatures and lower levels of precipitation (Dhyani et al. 2022). These changes in forest dynamics have wider implications for biodiversity, ecosystem stability, and the livelihoods of local communities in the diversified Himalayan landscapes.

The Himalayan forests are more vulnerable to forest fires, evapotranspiration, and decreased productivity as a result of climate change (Chitale et al. 2014a, b). Strategies for adaptation are essential to the region's sustainability and resilience. Forests are impacted by the disruption of natural ecosystems and species distribution caused by climate change in the Indian Himalayan region. These changes are leading to shifts in the distribution and composition of forest communities, increased risk of wildfires, and reduced forest productivity (Shrestha and Bawa 2014; Negi et al.

2021; Seidl et al. 2017). The degradation and loss of forests in the IHR may have far-reaching consequences for the region's ecosystems and the communities that depend on them.

Understanding the perceptions and experiences of the local communities is crucial for developing effective climate change adaptation strategies for the forests in the IHR (Shrestha and Bawa 2014; Negi et al. 2021). The studies have shown that indigenous and local communities in the region have observed the impacts of climate change on their forests, including changes in the timing and intensity of rainfall, drying up of water sources, and shifts in the distribution of plant and animal species (Kuniyal et al. 2021; Shrestha and Bawa 2014; Negi et al. 2021). These observations and experiences can provide valuable insights into the local-level impacts of climate change and indicate the development of context-specific adaptation strategies.

Because of the impacts of climate change on forests in the IHR, it is essential to conduct comprehensive research that integrates the perceptions and experiences of the local communities with scientific data and analysis (Shrestha and Bawa 2014; Negi et al. 2021; Seidl et al. 2017). This study focuses on the Kinnaur District, Himachal Pradesh within the IHR to assess the decadal change in forest cover, change in climatic variables, and perception of the local communities regarding climate change impacts on forests. Engaging the local stakeholders in the research and decision-making process is crucial for ensuring the long-term sustainability of the forest resources.

10.2 Study Area

The Kinnaur District is located between 31º05′50″ N and 32º05′15″ N latitudes and 77º45′0″ E and 79º00′35″ E longitudes (Fig. 10.1). Whole of the district is mainly mountainous. The largest portion of the area was classified as moderately steep (27.32%), and steep (25.42%) slope. Relative relief varied from 1192 to 6724 metre (m) above sea level. The study region encompassing Nichar, Pooh, and Kalpa blocks with shaded slopes featuring the highest forest density, while sunny slopes were used for horticultural purposes. The soil texture of the study region is coarse, medium, fine, and rocky. The lithology composition of the area includes limestone, siltstone, shale, quartzite, and calcareous shale. The Kinnaur District is rich in ecological and cultural significance and has a diverse range of vegetation. Due to its delicate topography when compared with the lower Himalayan regions, the region faces geological vulnerability (Kuniyal et al. 2019; Lata et al. 2017).

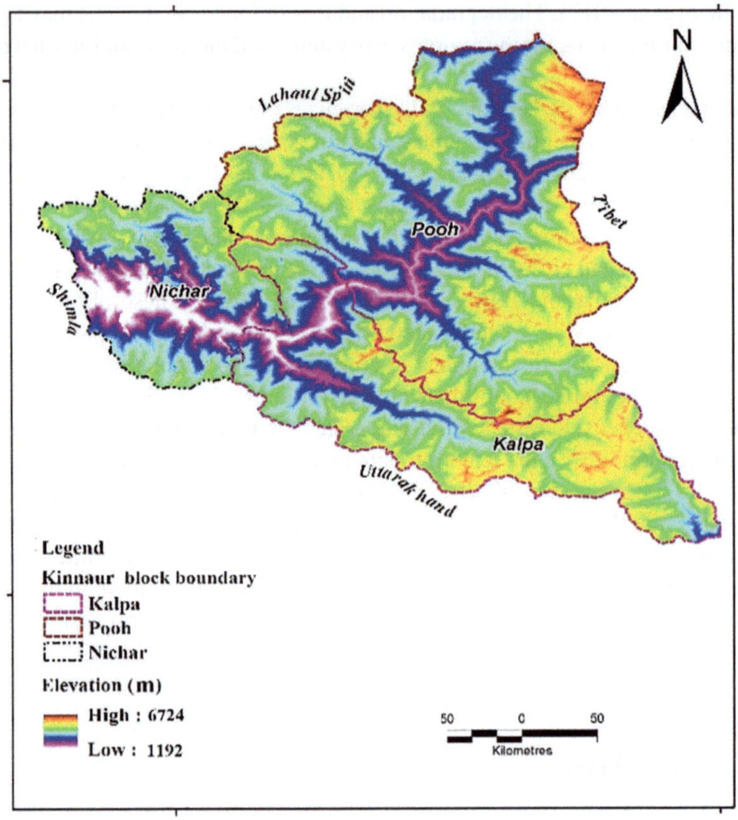

Fig. 10.1 Study area map of the Kinnaur District

10.3 Methodology

10.3.1 Forests Cover and Change

The evaluation of alterations in forest cover was conducted by analyzing all Forest Survey of India (FSI) reports published from 2001 to 2021. Further LISS-III satellite dataset was used for generating the forest cover layer of 2021. The digital image processing of satellite data for forest cover mapping is based on considering the reflectance behavior of the canopy cover to ensure the accuracy of the analysis.

10.3.2 Trend Analysis

Gridded datasets Climate Research Unit (CRU), version TS3.21 (Harris et al. 2013) to obtain monthly mean temperature and precipitation data analyzed the data to acquire extreme climatic indices for 122 years (from 1901 to 2023). The obtained gridded datasets have a spatial resolution of 0.5° × 0.5°. The yearly and seasonal temperature and precipitation were then calculated using an arithmetic mean technique. However, the whole country experiences almost all types of climate due to its physiographic location (Martinez-Austria et al. 2016). The district experiences four distinct seasons, i.e., (i) winter which spans from December to February, (ii) pre-monsoon from March to May, (iii) monsoon prevails from June to September, and (iv) the post-monsoon season occurs during October and November (Kanwar et al. 2024). To analyze the trends of the chosen variables, the continuous daily time series data were arranged into monthly and annual time series after the indices were chosen. Trend analysis was carried out using the Mann–Kendall test (non parametric test) and Sen's slope estimator and its level of significance was calculated. The Mann–Kendall Test (Mann 1945; Kendall 1975), a generally accepted statistical test for trend analysis in climatologic and hydrologic time series, was used to analyze seasonal patterns in climatic variables such as minimum, maximum, and diurnal temperatures, as well as rainfall (quantity and days) (Pohlert 2018). The Mann–Kendall Test works on the basic null hypothesis 'Ho' of no trend, *i.e.*, data is independent with a random order that is tested against the alternative hypothesis 'Ha'. Sen's Slope, which computes the linear rate of change and intercept, measures the magnitude of the pattern. After evaluating a range of linear slopes, the Sen's Slope is determined as the median of all linear slopes, resulting in the magnitude of the observed seasonal pattern. The *P*-value* is another statistic correlated with the Mann–Kendall T-test (Bharti et al. 2024).

10.3.3 People's Perception of Climate Change and Its Impact on Forest Resources

The study was conducted from December 2017 to August 2019. Households live close to forest areas or adjacent to the boundaries of the villages that adjoin the forests. A total of 250 households were randomly selected from 18 villages across 3 blocks in the Kinnaur District (Annexure I). Because of understanding how the perception of climatic characteristics shifts over a considerable period, we consider age.

Thirty percent of the households were surveyed. Sixty percent of the respondents were older than sixty. The majority of the respondents (69%) were literate, meaning they had received education up to the primary level. However, 41% of the respondents were not educated or had not attended primary schools at all. The impact of climate change was perceived based on variations in the severity (degree of strength

Table 10.1 Criteria of evaluation of the perception study

Category level of change	Average mean score range	Level of change
No change	0	None
Changed somewhat	0.01–0.50	Positive or negative
Changed	0.51–1.00	Positive or negative

and duration) and regularity (consistency and uniformity in distribution) of climatic factors and occurrences (such as temperature, precipitation, floods, droughts, etc.). The perception of climate change was recorded using both group discussions and personal interviews. The interviews were carried out with the selected participants using a structured questionnaire consisting of two sections containing statements on general opinions on the impact of climate change and the impact of climate change on forests. The views were assessed using a three-point Likert scale (Dey et al. 2016b, a), with response options of 'agree' (+ 1), 'undecided' (0), and 'disagree' (– 1). The average score for each statement was obtained using the following formula (Eq. 10.1).

$$\text{Mean score (MS)} = \frac{\text{S1} - \text{S2}}{\text{S}} \tag{10.1}$$

where S1 number of respondents agreed with the statement, S2 number of respondents disagreed with the statement, and S is the total number of respondents.

To evaluate the perceived change in a given statement, the average mean score was calculated by considering all statements and grouped according to the following categories as shown in Table 10.1.

10.4 Results

10.4.1 Forest Cover Status in the Indian Himalayan Region

In 2001, the forest cover of the Indian Himalayan Region (IHR) was 2,16,407 km^2 which increased to 0.02% (2,20,294 km^2) in 2021 (FSI 2021). Much of this increase in forest cover was recorded in the Western Himalaya. For example, this increase was 9.47% from 2001 to 2021 in Jammu & Kashmir (including Ladakh). While, it increased by 5.77% in Himachal Pradesh, but decreased by 0.12% for Uttarakhand's Forest cover during the same period. Talking about the North Eastern Region (NER) of the IHR, there was found increasing trend in terms of forest cover. Among the different states of the NER, Tripura recorded the highest increase in forest cover (8.15%) followed by Meghalaya (7.72%). On the contrary, Nagaland lost 9.56% of its forest cover, followed by Arunachal Pradesh (4.01%) (Fig. 10.2a, b).

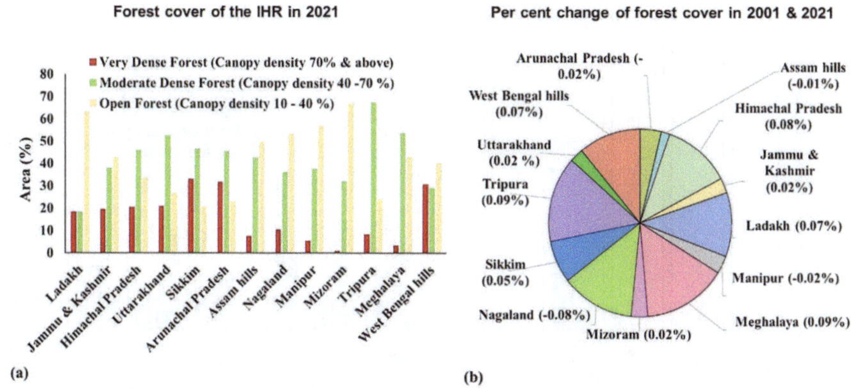

Fig. 10.2 **a** Forest cover of the IHR in 2021, and **b** percent change of forest cover in 2001 and 2021 2021 (*Source* ISFR 2001,)

10.4.2 Fire-Prone Forest Area in the Indian Himalayan Region

Significant differences are found when forest fire-prone areas from the Forest Survey of India, 2021 reports are analyzed in various IHR states and union territories. Jammu & Kashmir has a very small percentage (0.08%) of its forest area classified as very highly fire-prone, while the majority of the forest area (94.4%) is less fire-prone (Fig. 10.3). Ladakh has almost negligible extremely or very highly fire-prone areas having its forest cover falling under the less fire-prone category. Himachal Pradesh has a small portion of its forests classified as very highly (0.04%) and highly (1.98%) fire-prone. The majority of forests (91.51%) in Himachal Pradesh are less fire-prone. When compared to other Himalayan states, Uttarakhand has a higher percentage of its forests 3.12 and 16.75% classified as extremely highly and very fire-prone, respectively. However, more than half (55.71%) of the forest of Uttarakhand is less fire-prone (see Fig. 10.3). The Northeastern states, particularly Mizoram, Nagaland, Manipur, and Tripura, have a significant portion of their forests classified as extremely (26.28%) or very highly (49.73%) fire-prone. Large portions of Meghalaya and Arunachal Pradesh are between 4.13 and 21.8% in the category of extremely to highly fire-prone (ISRF 2021).

10.4.3 Forest Cover Status in Himachal Pradesh

According to the latest report by the Forest Survey of India 2021, the forest cover in Himachal Pradesh has increased up to 9.43% since the last assessment in 2019 (ISFR 2019). The total forest cover in the state is now 15,442 km^2, which is 27.73% of the total geographical area of the state (ISFR 2021). The forest cover change matrix over

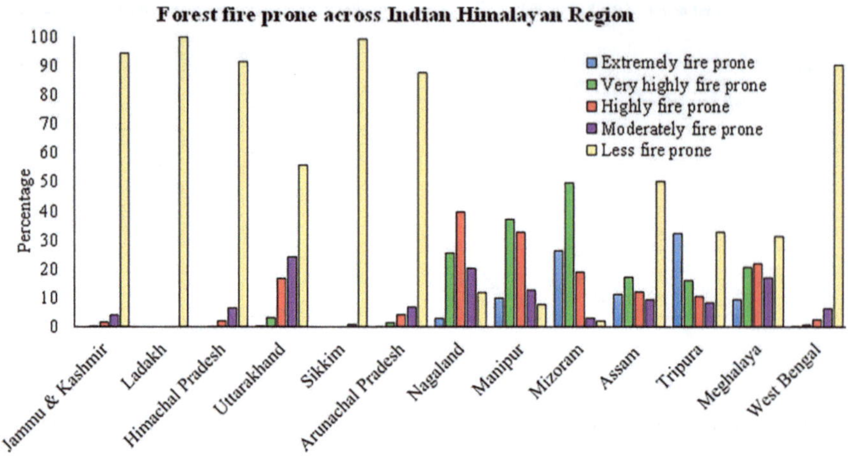

Fig. 10.3 Forest fire-prone region across the Indian Himalayan States

two assessment years (i.e., 2019 and 2021) shows that the very dense forest cover has a net increase of 0.016% (50 km^2), while the moderately dense forest cover has decreased by 0.004% (26 km^2) (ISFR 2019, 2021). Similarly, the open forest cover has decreased by 0.003% (15 km^2), while the scrub forest cover has a net increase of 0.02% (7 km^2). The total carbon stock of the state is 258.07 million tons which is 3.58% of the total forest carbon of the country (ISFR 2021).

10.4.4 Forest Cover of Kinnaur District

The temporal analysis of the forest has been done for Kinnaur District from 2001 to 2021. The various parameters analyzed were total forest cover, dense forest, open forest cover, scrub area, and total percent change in the area under different classes. The total forest cover of Kinnaur District showed a decrease of 50 km^2 (2.011%) in 2005 (597 km^2) when compared with 2001 (647 km^2). Ever since 2005, there has been a constant increase in forest area. According to the assessment year 2021, the forest cover of the district covers 10.08% of the total geographical area of the state (Table 10.2). The change assessment between 2019 and 2021 was a decreasing trend of 0.68% (ISFR 2021). It has been observed that moderate and open forests have demonstrated a declining trend from the year 2017 (see Table 10.2).

(i) **Dense Forest cover**: The dense forest area showed a significant increase by 2.16% (14 km^2) from 2001 to 13.58% (82 km^2) in 2015. In 2017, a decrease of 0.90% was found, which again in 2021 increased by 0.02% (Fig. 10.4a).

(ii) **Moderate Forest Cover**: The moderate dense forest area showed a significant decrease of 64.61% (418 km^2) from 2001 to 43.38% (262 km^2) in 2015. However, since then, a consistent increase has been observed (see Fig. 10.4a).

Table 10.2 Temporal enumeration (2001–2021) of the total forest cover in Kinnaur District

Total forest cover (km^2)									
Forest cover type	2001	2005	2009	2011	2013	2015	2017	2019	2021
Dense	14 (2.16)	16 (2.68)	82 (13.62)	82 (13.67)	82 (13.53)	82 (13.58)	79 (12.68)	80 (12.38)	80 (12.40)
Moderate	418 (64.61)	324 (54.27)	263 (43.69)	262 (43.67)	262 (43.23)	262 (43.38)	266 (42.70)	329 (50.93)	329 (51.01)
Open	215 (33.23)	257 (43.05)	257 (42.69)	256 (42.67)	262 (43.23)	260 (43.05)	278 (44.62)	237 (36.69)	236 (36.59)
Total	647	597	602	600	606	604	623	646	645
Scrub	113	67	70	70	69	72	71	60	62

Source FSI Reports (2001–2021)
Values in parentheses are in percent

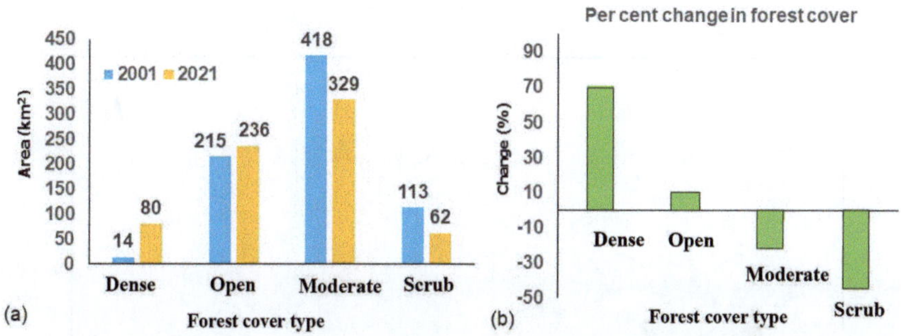

Fig. 10.4 **a** Temporal trend of forest cover in Kinnaur District during 2001 and 2021; **b** Percent change in forest cover type

The percentage decline in moderate dense forests was 21% for the whole period from 2001 to 2021 (Fig. 10.4b).

(iii) **Open Forest**: The open forest area showed a significant increase of 33.23% in 2001 to 44.62% in 2017. But since then, the region covered by its forest cover has kept decreasing (see Fig. 10.4a). Between 2001 and 2020, there was a 10% rise in open forests (see Fig. 10.4b).

(iv) **Scrubs**: According to the FSI report, the area under scrubs has decreased substantially from 2001 (113 km^2) to 2021 (62 km^2) (see Fig. 10.4a, b).

Altitudinal Distribution of Forest Cover in Kinnaur District

The distribution of forest cover in Kinnaur District was evaluated based on forest cover raster data acquired from the Forest Survey of India (2021) and Alos Palsar (10 m) Digital Elevation Model (DEM) data. The analysis shows that the altitude zone of 2500–3500 m has the highest forest cover (70.4%), followed by the altitude zone of 1500–2500 m with 16.5% forest cover. The altitude zone between 3500 and 4500 m has the lowest percentage of forest cover (4.6%) (Table 10.3; Fig. 10.5).

Table 10.3 Distribution of forest cover (km^2) along different altitude gradients in Kinnaur District

Altitude zone (metre)	Area (km^2)					Percent	Area (km^2) scrub	Percent
	Geographical area	Very dense forest	Moderate dense forest	Open forest	Total			
0–1500	8	5	32	19	55	8.6	2	3.23
1500–2500	248	18	49	39	106	16.5	1	1.61
2500–3500	1012	55	233	166	454	70.3	21	33.87
3500–4500	2030	1	16	13	30	4.6	13	20.97
> 4500	3102	0	0	0	0	0.0	25	40.32
Total	6401	80	330	236	646	100	62	

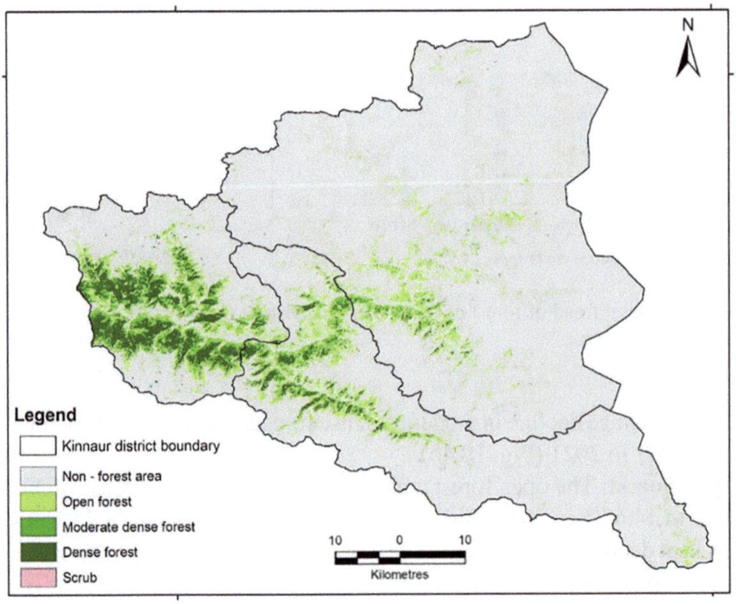

Fig. 10.5 Forest cover of Kinnaur District in 2021 (*Source* LISS-III satellite image, https://bhuvan.NRSC.gov.in)

10.4.5 Trend Analysis of Climatic Variables

The analysis of climatic variables in Kinnaur District reveals several notable trends. Based on the statistical analysis of the Mann–Kendall Trend test, the variations in climatic variables were observed in the winter, pre-monsoon, monsoon, and post-monsoon seasons (Table 10.4).

Precipitation Trends

According to the analysis, the winter precipitation exhibits a statistically significant decreasing trend, as indicated by the *Z*-value of 2.209 and the *P*-value of 0.027. The magnitude of this trend is reflected in the Sen's slope of − 0.413. In contrast, the pre-monsoon precipitation shows an increasing trend, with a *Z*-value of 1.38, but the trend is not statistically significant at a 95% confidence level, as the *P*-value of 0.167 exceeds the threshold (see Table 10.4; Fig. 10.6). The Sen's slope of 0.284 represents the magnitude of this trend. The monsoon and post-monsoon precipitation do not exhibit statistically significant trends. The *Z*-values of − 0.191 and − 0.278, along with the *P*-values of 0.848 and 0.78, respectively, suggest no significant changes in these seasons. The Sen's slopes of − 0.11 and − 0.067 indicate slight decreasing trends, but they are not statistically significant.

Temperature Trends

The analysis reveals statistically significant increasing trends in winter, pre-monsoon, and post-monsoon temperatures. The *Z*-values of 5.513, 3.151, and 5.454, along with the corresponding *P*-values of 0.0000035, 0.0016, and 0.0000049, respectively, confirm the significance of these trends. The Sen's slopes of 0.012, 0.009, and 0.0087 quantify the magnitudes of the increasing trends in these seasons. In contrast, the monsoon temperature does not exhibit a statistically significant trend, as indicated

Table 10.4 Seasonal trend analysis of precipitation and temperature from 1901 to 2023

Climatic variables	*Z*-Value	*P*-value	Sen's slope
Precipitation			
Winter (December–February)	− 2.209*	0.027	− 0.413
Pre-monsoon (March–May)	1.38	0.167	0.284
Monsoon (June–September)	− 0.191	0.848	− 0.11
Post-monsoon (October–November)	− 0.278	0.78	− 0.067
Temperature			
Winter (December–February)	5.513*	0.0000035	0.012
Pre-monsoon (March–May)	3.151*	0.0016	0.009
Monsoon (June–September)	− 0.352	0.725	− 0.00037
Post-monsoon (October–November)	5.454*	0.0000049	0.0087

* Indicates trend statistical significance at 95% confidence level (+ for increasing and − for decreasing)

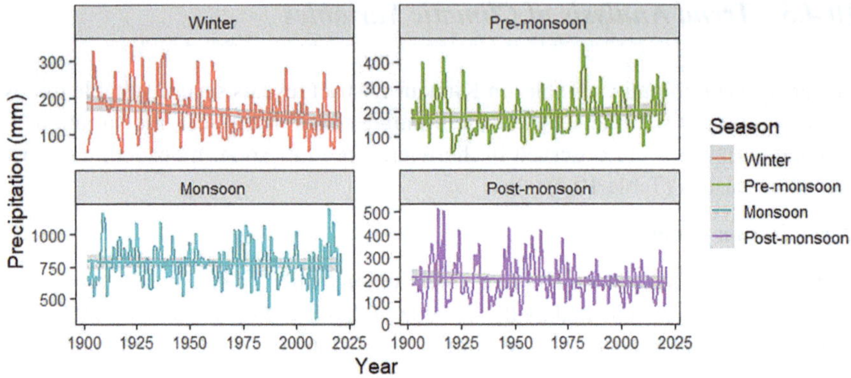

Fig. 10.6 Seasonal variation in precipitation of Kinnaur District from 1901 to 2023

by the *Z*-value of – 0.352 and the *P*-value of 0.725 (see Table 10.4; Fig. 10.7). The Sen's slope of – 0.00037 suggests a slightly decreasing trend, but it is not significant.

These findings provide valuable insights into the changing climatic patterns in the region, highlighting the need for further investigation and potential adaptation strategies to address the observed trends.

In summary, the analysis reveals statistically significant decreasing trends in winter precipitation and increasing trends in winter, pre-monsoon, and post-monsoon temperatures. The trends in pre-monsoon precipitation, monsoon precipitation, and monsoon temperature are not statistically significant at the 95% confidence level.

Perception Survey Regarding Climate Change Impact on Forest Resources

The local community in Kinnaur District has a strong awareness of the enormous impact of climate change on the forest and other vegetation. More than 60% of

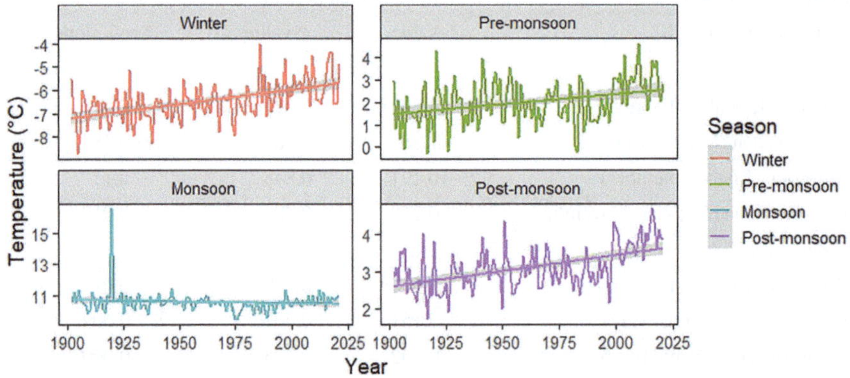

Fig. 10.7 Seasonal variation in temperature of the Kinnaur District from 1901 to 2023

respondents reported a decline in the extent of forested areas. Their firsthand experience of depending on these sources for their residential and irrigation water needs has given them an understanding of decreased stream and river flow as well as the quick drying of seasonal streams and water bodies (Fig. 10.8a). About 93% of respondents perceived that the intensity and frequency of climatic extreme events (e.g., flash floods, landslides, drought, cloud bursts, etc.) have increased over the years. Perception of climate change's impact on forest fauna and flora also seems to be positive, as they have experienced such situations based on their past and present situations. Of the total respondents, 50% respondents said that one of the indirect causes of animal mortality could be low crop output, crop failure, and declining feed (Fig. 10.8b, c).

Also, respondents think that throughout time, there have been fewer incidences of noticing wild animals in the forest of the district.

Temporal analysis of forest cover in Kinnaur District reveals an overall positive trend, with fluctuations in dense, and open forest covers, while moderate forest and scrub areas have shown a consistent decrease over the years. The Kinnaur District

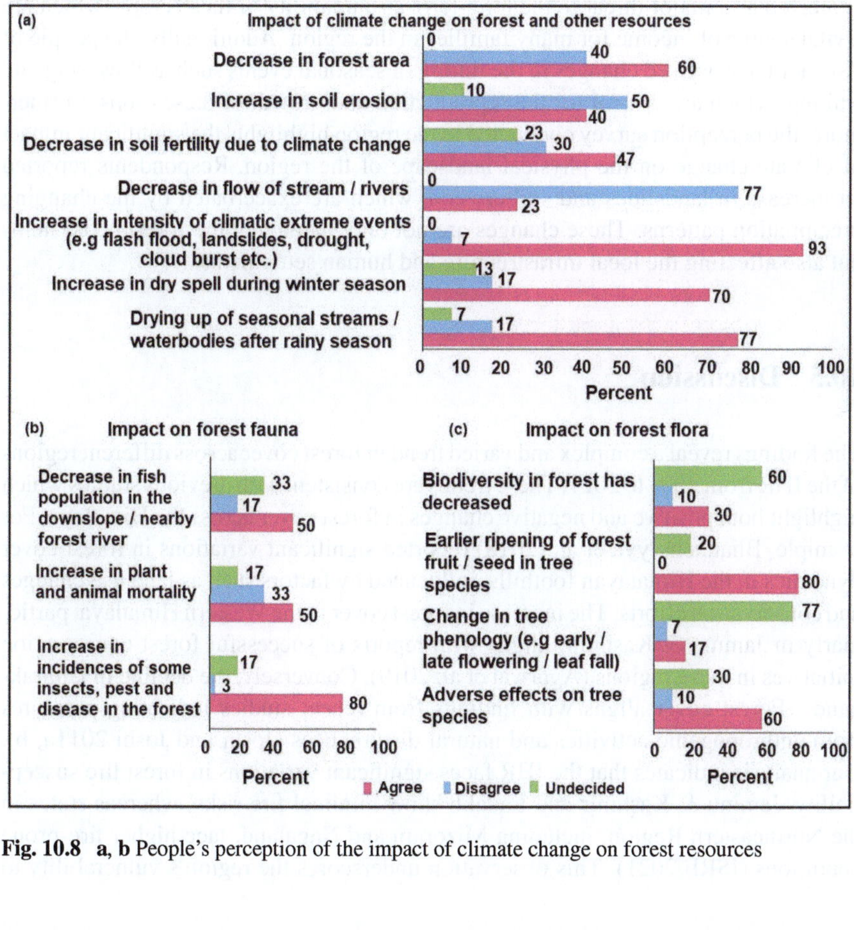

Fig. 10.8 **a, b** People's perception of the impact of climate change on forest resources

exhibits a diverse range of forest types, including subtropical pine, Himalayan moist temperate, dry temperate, sub-alpine, and alpine meadows, influenced by altitude, climate, and anthropogenic factors. Dominant species vary across ecosystems, from *Pinus roxburghii* in subtropical pine forests to *Cedrus deodara* in moist temperate forests, showcasing the intricate interplay of biophysical and human elements in shaping vegetation distribution. Forests in Kinnaur District provide crucial ecosystem services, including fresh air, carbon sequestration, and cultural significance, with specific emphasis on the importance of Chilgoza pine and pastures for local livelihoods. Depletion of forest resources is attributed to a combination of natural and human-induced factors such as deforestation, over-exploitation, and climate change. Local perception in Kinnaur reflects a strong awareness of climate change impacts on forests, with reported declines in forested areas, reduced water sources, increased climatic extremes, and observed impacts on flora, fauna, and livelihoods, emphasizing the urgent need for conservation efforts.

Residents in the Kinnaur District have reported a range of impacts on their forests due to climate change. For instance, they have observed a decline in chilling hours, which is essential for fruit formation in crops like apples and apricots. This decline is perceived as a major threat to the long-term sustainability of these crops, which are a vital source of income for many families in the region. Additionally, the people of Kinnaur have noticed changes in the timing of seasonal events such as flowering and fruiting, which are critical for the reproduction and growth of these crops. Furthermore, the perception survey conducted in the region highlights the significant impact of climate change on the physical landscape of the region. Respondents reported an increase in landslides and soil erosion, which are exacerbated by the changing precipitation patterns. These changes are not only damaging the forest ecosystems but also affecting the local infrastructure and human settlements.

10.5 Discussion

The findings reveal a complex and varied trend in forest cover across different regions of the IHR from 2001 to 2021. These trends are consistent with previous studies which highlight both positive and negative changes in forest cover across the Himalaya. For example, Bhattacharyya et al. (2018) reported significant variations in forest cover dynamics in the Himalayan foothills, influenced by factors such as land use changes and conservation efforts. The increase in forest cover in the Western Himalaya, particularly in Jammu & Kashmir, aligns with reports of successful forest conservation initiatives in these regions (Agarwal et al. 2019). Conversely, the decline in Uttarakhand's Forest cover aligns with findings from recent studies indicating pressures from anthropogenic activities and natural disturbances (Joshi and Joshi 2011a, b). Our analysis indicates that the IHR faces significant variations in forest fire susceptibility. Jammu & Kashmir and Ladakh show minimal fire risks, whereas states in the Northeastern Region, including Mizoram and Nagaland, face higher fire-prone conditions (ISRF 2021). This observation underscores the region's vulnerability to forest fires, which are exacerbated by climate change-induced factors like increased temperatures and prolonged dry spells (Chitale et al. 2014a, b). This finding is consistent with studies that have identified increasing fire risks in the Himalayan region due to changing climatic conditions (Tiwari et al. 2018). Additionally, the increased frequency of forest fires in the northeastern states could be linked to both natural factors and human activities, such as shifting agriculture and forest management practices (Kumar et al. 2020). In Himachal Pradesh, the overall increase in forest cover by 9.43% from 2019 to 2021 reflects successful forest conservation efforts and improved forest management practices (ISFR 2021). This increase contrasts with the observed fluctuations in forest cover in the Kinnaur District, where dense forest cover increased significantly from 2001 to 2015 but declined thereafter (ISFR 2019). These trends in Kinnaur are indicative of the dynamic nature of forest ecosystems, which are influenced by both natural processes and anthropogenic activities (Bharti et al. 2024; Gairola et al. 2011). The increase in dense forests and the decrease in moderate and open forests could be attributed to a combination of conservation efforts and changes in land use patterns over the years (Rai et al. 2021). However, the recent decline in dense forest cover and the increasing trend in scrub areas reflect ongoing challenges related to climate change and human interventions (Sati 2021). The climatic trend analysis reveals significant increases in winter, pre-monsoon, and post-monsoon temperatures, alongside a significant decrease in winter precipitation. These findings are consistent with broader studies on the Himalayan climate change, which have documented warming trends and changes in precipitation patterns (Kanwar et al. 2024; Shrestha and Bawa 2014; Seidl et al. 2017). The increase in temperatures during winter and pre-monsoon periods could contribute to reduced snow cover and altered hydrological cycles, affecting water resources and forest ecosystems (Khan et al. 2018). The decrease in winter precipitation could lead to reduced snow accumulation, which is crucial for maintaining seasonal water flows (Lutz et al. 2014). These changes can impact forest health, lead to increased incidences of pests and diseases, and alter species distributions (Bajgain et al. 2016). The perception survey reveals that local communities in Kinnaur are highly aware of climate change impacts on their forests and livelihoods. Respondents reported changes in forest cover, increased climatic extremes, and impacts on flora and fauna. These observations reflect findings from similar studies that have highlighted local awareness of climate change effects in the Himalayan regions (Shrestha and Bawa 2014; Negi et al. 2021). The vegetation and tree species appear to be impacted by early ripening and flowering, as well as a rise in pests and illnesses (Supit et al. 2010; Joshi and Joshi 2011a, b; Logan et al. 2003). For instance, Shrestha and Bawa (2014) noted that the communities in the Himalaya observed changes in seasonal patterns and the health of forest ecosystems. The scientific studies from the Himalayan region support the assertion that the weather has become uncertain and variable, with the melting of snow and drying up of water sources (Kuniyal et al. 2019; Chaudhary and Bawa 2011; Rawat 2013; Shukla et al. 2015). The new study adds more evidence to the widely accepted belief that populations that depend on forests for their livelihoods and forest ecosystems are seriously threatened by decreased access to water supplies (Dey et al. 2016b, a). Our study extends these findings by documenting specific impacts such as reduced water availability, changes in crop yields, and increased frequency of extreme weather events. The strong awareness among local communities about these issues emphasizes the importance of incorporating local knowledge into climate change adaptation strategies (Dey et al. 2016b, a).

10.6 Conclusion

The study of the forest cover in Kinnaur District over 20 years, from 2001 to 2021, reveals a complex pattern of declining total forest cover followed by a rebound after 2005. However, the number of moderate forests and scrub regions has continuously reduced. Although there were variations, dense forests typically increased, whereas open forests showed an initial rise followed by a drop. Climatic variables, particularly temperature, have shown significant increasing trends in winter, pre-monsoon, and post-monsoon seasons, coupled with a significant decrease in winter precipitation. Local perceptions corroborate these findings, highlighting observed declines in forested areas, water sources, and increased climatic extremes, alongside impacts on flora, fauna, and livelihoods. These results highlight the urgent need for focused conservation initiatives to address issues brought on by both humans and climate change, ensuring the long-term management of crucial and diversified forest ecosystems in Kinnaur District in the northwestern Himalaya.

Acknowledgements The authors are thankful to the Director, G.B. Pant National Institute of Himalayan Environment (NIHE), Kosi-Katarmal, PIN 263643, Almora, Uttarakhand for providing facilities in the institute which could make the present study possible. The corresponding author (JCK) also gratefully acknowledges the Hon'ble Vice Chancellor, Veer Chandra Singh Garhwali Uttarakhand University of Horticulture and Forestry (VCSG UUHF), Bharsar, PIN 246 123, Pauri Garhwal, Uttarakhand, India and Dean, College of Forestry, VCSG UUHF, Ranichauri, PIN 249 199, Tehri Garhwal, Uttarakhand, India for providing facilities which could make the present study possible to be finalized.

References

Agarwal A, Narain S, Kothari A (2019) State of forests in India: a critical review. J Environ Manage 227:52–65

Bajgain R, Bawa KS, Bhatta RP (2016) Impact of climate change on the Himalayan Forests. Environ Res Lett 11(5):055007

Bharti H, Panatu A, Sharma P, Randhawa SS, Dhiman S, Rana RS (2024) Impact of climatic variations in horticulture sector, Kinnaur, Himachal Pradesh, India. Mausam 75(2):313–326

Bhattacharyya S, Sharma SK, Singh KK (2018) Forest cover dynamics in the Himalayan foothills. Ecol Ind 90:621–634

Chao S (2012) Forest peoples: numbers across the world. Forest Peoples' Programme

Chaudhary P, Bawa KS (2011) Local perceptions of climate change validated by scientific evidence in the Himalayas. Biolog Lett 7:767–770

Chitale VS, Raut S, Katti S (2014a) Forest fire vulnerability in the Himalayas: a review. Int J Wild Fire 23:642–654

Chitale VS, Shrestha HL, Agarwal NK, Choudhury D, Gilani DHK, Murthy MSR (2014b) Forest climate change vulnerability and adaptation assessment in Himalayas. ISPRS—Int Arch Photogram Rem Sens Spat Inform Sci 11(8):1291–1294

Dey SK, Singh AK, Bhattacharya P (2016a) Local perceptions of climate change in the Indian Himalayan Region. Clim Change 136(1):191–203

Dey T, Pala NA, Shukla G, Pal PK, Das G, Chakarvarty S (2016b) Climate change perceptions and response strategies of forest fringe communities in Indian Eastern Himalaya. Environ Dev Sustain 1–17

Dhyani R, Joshi R, Ranhotra PS, Shekhar M, Bhattacharyya A (2022) Age dependent growth response of Cedrus deodara to climate change in temperate zone of Western Himalaya. Trees Forest People 8:100221

Duraisami M, Das P, Segarin S, Singh R, Jayahari KM, Balasubramanian K (2022) Opportunity for forest protection and landscape restoration in the Indian Himalayan Region, 1st edn. Handbook of Himalayan ecosystems and sustainability, p 156

Forest Survey of India (FSI) (2021) India state of forest report 2021. Ministry of Environment, Forest & Climate Change, Government of India; Dehradun. Accessed on 19 March 2020

Gairola S, Muniappan KS, Singh SK (2011) Temporal changes in forest cover in the Indian Himalayan Region. Environ Monit Assess 179(1):221–234

Harris IC, Jones PD, Osborn TJ, Lister DH (2013) Updated high-resolution grids of monthly climatic observations—the CRU TS3.10 dataset. Int J Climatol 34:623–642

Harris I, Osborn TJ, Jones P, Lister D (2020) Version 4 of the CRU TS monthly high-resolution gridded multi-variate climate dataset. Sci Data 7(1):109

ISFR (2001) India state of forest report. Forest Survey of India, Ministry of Environment. Accessed on 20 Mar 2020

ISFR (2005) India State of Forest Report. Forest Survey of India, Ministry of Environment, Forest and Climate Change, Government of India, Dehradun. Accessed on 20 Mar 2020

ISFR (2009) India State of Forest Report. Forest Survey of India, Ministry of Environment, Forest and Climate Change, Government of India, Dehradun. Accessed on 20 Mar 2020

ISFR (2011) India State of Forest Report. Forest Survey of India, Ministry of Environment, Forest and Climate Change, Government of India, Dehradun. Accessed on 20 Mar 2020

ISFR (2013) India State of Forest Report. Forest Survey of India, Ministry of Environment, Forest and Climate Change, Government of India, Dehradun. Accessed on 20 Mar 2020

ISFR (2015) India State of Forest Report. Forest Survey of India, Ministry of Environment, Forest and Climate Change, Government of India, Dehradun. Accessed on 20 Mar 2020

ISFR (2017) India State of Forest Report. Forest Survey of India, Ministry of Environment, Forest and Climate Change, Government of India, Dehradun. Accessed on 20 Mar 2020

ISFR (2019) India State of Forest Report. Forest Survey of India, Ministry of Environment, Forest and Climate Change, Government of India, Dehradun. Accessed on 20 Mar 2020

ISFR (2021) India State of Forest Report. Forest Survey of India, Ministry of Environment, Forest and Climate Change, Government of India, Dehradun. Accessed on 20 Mar 2020

Joshi AK, Joshi PK (2011a) A rapid inventory of indicators of climate change in the middle Himalaya. Curr Sci 100:831–832

Joshi PK, Joshi DP (2011b) Climate change impacts on forest ecosystems in the Himalayas. For Ecol Manage 262(3):548–559

Kanade R, Badola R, Reddy S (2018) Climate change impacts on forest ecosystems in the Indian Himalayan Region. Climate Policy 18(5):619–632

Kanwar N, Kuniyal JC, Rautela KS, Singh L, Pandey DC (2024) Longitudinal assessment of extreme climate events in Kinnaur district, Himachal Pradesh, north-western Himalaya, India. Environ Monit Assessm 196:557

Kendall MG (1975) Rank correlation methods. Oxford University Press, New York, NY

Khan AA, Ali S, Hasan M (2018) Snow cover dynamics and climate change in the Indian Himalayas. J Hydrol 558:249–261
Kumar S, Singh K, Mahanand N (2020) Forest fires in the Indian Himalayan region: trends and management. Forest Policy Econ 118:102218
Kuniyal JC, Pandey SK (2019) Climate change impacts on the Himalayan Forests. Clim Dyn 52(9):5977–5992
Kuniyal JC, Jamwal A, Kanwar N (2019) Vulnerability assessment of the Satluj catchment for sustainable development of hydroelectric projects in the northwestern Himalaya. J Mt Sci 16:2714–2738
Kuniyal JC, Kanwar N, Bhoj AS, Rautela KS, Joshi P, Kumar K, Sofi MS, Bhat SU, Rashid I, Lodhi MS, Devi CA, Singh HB (2021) Climate change impacts on glacier-fed and non-glacier-fed ecosystems of the Indian Himalayan region: people's perception and adaptive strategies. Curr Sci 120(5):888–899
Lata J, Singh R, Pandey S (2017) Geological vulnerability in the Indian Himalayan Region. J Mt Sci 14(3):565–576
Logan JA, Régnière J, Powell JA (2003) Assessing the impacts of global warming on forest pest dynamics. Front Ecol Environ 1:130–137
Lutz AS, Immerzeel WW, Dery S (2014) Snow cover and climate change in the Himalayas. Cryosphere 8(3):983–999
Mann HB (1945) Nonparametric tests against trend. Econometrica 13:245–259
Martinez-Austria P, Schmid C, Hess T (2016) Climate variability and trends in the Indian Himalayan Region. Int J Climatol 36(7):2821–2833
Myers N, Mittermeier RA, Mittermeier CG, Da Fonseca GA, Kent J (2000) Biodiversity hotspots for conservation priorities. Nature 403:853
Negi VS, Thakur S, Dhyani R, Bhatt ID, Rawal RS (2021) Climate change observations of indigenous communities in the western Himalaya of India. Weather Clim Soc 13(2):305–320
Parida BR, Kumar P, Chakraborty JS, Bag N, Kumar N (2022) Dynamics of land use/land cover over Indian Himalayan Region (IHR) using satellite data from 1985 to 2015 coupled with local perceptions, 1st edn, vol 1. Handbook of Himalayan Ecosystem and Sustainability, pp 147–152
Pohlert T (2018) Non-parametric trend tests and change-point detection. Technical Report
Rai A, Sharma R, Kumar A (2021) Investigating forest fragmentation through earth observation datasets and metric analysis in the tropical rainforest area. SN Appl Sci 3(1):1–14
Rawat VS (2013) People's perception on climate change and their influence on various aspects around Tones Valley of Garhwal Himalaya. Environ Ecol Res 1:150–154
Rawat GS, Rawat YS (2013) Forests of the Indian Himalayan region: A review of their ecological and socio-economic significance. J Environ Sci Health B 48(1):1–14
Sati VP (2021) Forests' management for sustainable livelihoods in the Himalaya: a review of existing literature. Forest Resour Resil Confl 321–328
Seidl R, Thom D, Kautz M, Martin-Benito D, Peltoniemi M, Vacchiano G, Reye CP (2017) Forest disturbances under climate change. Nat Clim Chang 7(6):395–402
Shrestha AB, Bawa KS (2014) Forests and forest-dependent communities in the Indian Himalayan region: a review of their ecological and socio-economic significance. J Environ Sci Health B 49(1):1–14
Shukla G, Kumar A, Pala NA, Chakravarty S (2015) Farmer's perception and awareness of climate change: a case study from Kanchandzonga biosphere reserve. Environ Dev Sustain 17:4
Srivastava N (2022) Influence of global warming on Himalayan region of India: a review. Plant Archives 22(1):360–363
Supit I, Van Diepen CA, de Wit AJW, Kabat P, Baruth B, Ludwig F (2010) Recent changes in the climatic yield potential of various crops in Europe. Agric Syst 103:683–694
Thadani R (2023) Snow damage in a Himalayan Forest maintains the dominance of evergreen oaks. Biotropica 55(4):737–741
Tiwari K, Kumar M, Kumar S (2018) Hotspot and trend analysis of forest fires and its relation to climatic factors in the western Himalayas. Forest Ecosyst 5(1):1–13

Chapter 11
Habitat Ecology and Harvesting of Caterpillar Fungus (*Ophiocordyceps sinensis*): A Medicinal Plant in the Higher Himalayan Region of Mandakini River Valley, Kumaun Himalaya

Kavindra Singh Kathayat and R. C. Joshi

Abstract Himalayan mountain livelihood systems generally rely on natural resources in the higher Himalayan regions. Keedajadi (*Ophiocordyceps sinensis*) is a natural insect fungus mainly found in the *Bugyals* (meadows) of high Himalayan regions at an altitude of 3000–5000 m above mean sea level. The medicinal properties of Keedajadi have made it one of the most expensive natural resources in the area. The Keedajadi availability and its commercial value have been explored in the recent past. The high demand and value of Keedajadi in the markets of Tibet and China have played an important role in strengthening the economy of the rural areas of the study area. There was no information about Keedajadi availability in the Mandakini River valley until two decades ago, which has emerged as an important resource. This resource has become very influential and successful in bringing economic and social change to the rural economy of seventeen (17) revenue villages in the area. This study shows that the number of families, which was 1064 in 2011, increased to 1287 in 2022. Total production was 59.5 kg (value Rs. 417 lakhs (629,719$) and 43.3 kg (value Rs. 433 lakhs (581,677$) in 2016 and 2022 respectively. There has been a change of -16.2 kg (27.2%) in keedajadi production from 2016 to 2022. There are a total of eight (8) *Bugyals* (*Dhunkhan, Kokan dhunga, Northern Blati, Southern Blati, Selapani Lambaladi, Khathapu (A, B), Janthri*) from where keedajadi collection is made. In general, the income is used by most families for building construction, educating children, starting their employment-oriented businesses, and meeting their day-to-day needs.

Keywords Keedajadi (*Ophiocordyceps sinensis*) · Land use · Bugyal (alpine meadows) · Soil · Income · Himalaya · Mandakini valley

K. S. Kathayat · R. C. Joshi (✉)
Department of Geography, DSB Campus, Kumaun University, Nainital, Uttarakhand, India
e-mail: prof.rcjoshi@kunainital.ac.in

S. C. Rai (ed.), *Himalaya: Mountains of Destiny*,
https://doi.org/10.1007/978-3-031-94905-0_11

11.1 Introduction

The collection and marketing of medicinal plants in the higher Himalayan region has played an important role in supplementing family livelihoods (Olsen and Larsen 2003). In tropical and subtropical countries, natural products account for 28% of household income (He et al. 2022). This study is focused on the Mandakini, a tributary of the Goriganga River. A British mycologist, Berkely first named it *Sphaeria sinensis Berk*in in 1843. Later by, Saccardo, in 1878 named it *Cordyceps sinensis*. Finally, it is named *Cordyceps sinensis Berk Sacc.* The molecular analysis was used to revise the *Cordycipita-ceaeand* and *Clacicipitaceae classification resulting in the creation of the Ophiocordycipitaceae* (Sen et al. 2023). As a result, *Ophiocordyceps sinensis* is another name for cordyceps sinensis until 2007.

It is locally known as *Keedajadi*. Caterpillar fungus grows on larvae of host insects, an entire stage that remains three to 4 years underground (Wei et al. 2021; Negi et al. 2020). Generally, Cordyceps die in winter after being infected by the fungus, and the fungal stoma (Fig. 11.1) comes in earlier, from April last to May first (Winkler 2008; Li et al. 2011; Wei et al. 2021). *Keedajadi* is a dormant stage in the life cycle that can resist unfavourable snow-cold conditions. *Ophicordyceps sinensis* can grow at about two °C with the optimum temperature between 15 and 18 °C (Negi et al. 2020; Wei et al. 2021). *Keedajadi* is generally located in the subtropical high Himalayan regions with an altitude of 3000–5000 m. *Keedajadi* grows in meadows and scrubland along Himalayan hill slopes, with an annual rainfall of 350 mm between tree and snow lines (Winkler 2009). In China, the maximum availability of *Keedajadi* is found between an altitude of 2660–5000 m. Yunan, Xinjiang, Jiling, Shanxi, Shaanxi, Hubei, Zhejiang, Jiangxi, Guizhou, Taiwan, Guangdong, Guangxi, Sichuan, and Hainan Province, and Lhasa Shannan in Tibet are the areas to collect it (Winkler 2009; Li et al. 2011; Negi et al. 2020; Dai et al. 2020). In Nepal, the areas with an altitude of 3540–5050 m include Darchula, Jumla, Dolpa, Baijura, Kalikot, Mugu, humla, Rukum, Bajhang, Mustang, Gorkha, Lanjung, Rasuwa, Dolakh, Dhanding, Sindhuchowk, Solukhumbu, Sankhuwasabha, and Taplejung localities for the collection of *Keedajadi* (Shrestha and Sung 2005; Adhikar 2008; Devkota 2008, 2010; Negi et al. 2020). In Bhutan, the *Keedajadi* availability area is found between 4200 and 5200 m height above mean sea level. It includes Namna, Bumthang Valley, and Bumdeling wildlife sanctuary areas (Cannon et al. 2009; Negi et al. 2020). In India, *Keedajadi* availability areas lie in the height of 3200–4800 m including Uttarakhand (Goriganga Valley, Darma Valley, Choudas Valley, Pindari Valley, Niti Valley, Nanda Devi Biosphere reserve), Sikkim, and Arunachal Pradesh (Sharma 2004; Garbyal et al. 2004; Negi et al. 2006, 2014, 2015, 2016; Kuniyal and Sundriyal 2013; Caplins and Hlvorson 2017; Laha et al. 2018; Pradhan et al. 2021).

Ophiocordyceps sinensis is known as *Yartsagunbu* in Tibet (Winkler 2008; Shrestha et al. 2010), *Yarsogumbub* in Bhutan, *Dongchong xia cao* in China, *Yarsagumba* in Nepal, and *Keedajadi* in India (Yadav et al. 2016). The main aim of this work is to identify the families involved in collecting Keedajadi and using the amount for the well-being of their families.

Fig. 11.1 **a** Origin of Keedajadi as well as melting snow; **b** Larva and fungus part of *keedajadi*; **c** *keedajadi* ready to sell in the local market after cleaning the soil; **d** *keedajadi* collection at Selapani Bugyal. *Source* Primary data collected during a field visit

11.2 Study Area

The Mandakini River valley area is spread over an area of 212 km^2 between 30°2′28″ N to 30°15′14″ N, latitude, 80°16′15″ E to 80°28′20″ E, longitude in the Kumaun Himalaya of Uttarakhand state in India (Fig. 11.2). The study area is composed of rugged and undulating terrain with an altitude of 6778–1153 m. Mandakani is one of the tributaries of river Goriganga which originates from Northern Blati (3262–5519 m) and southern Blati Glaciers (3422–6345 m) located at the foothills of the Panchachuli mountain ranges. From its origin, the Mandakini River flows 20 km and forms a confluence with the Gori Ganga River at a height of 1200 m at a place called Madkot. The eastern boundary of the Mandakini River valley is formed by five peaks of Panchachuli (1–6904 m, 2–6312 m, 3–6334 m, 4–6437 m, and 5–6072 m above

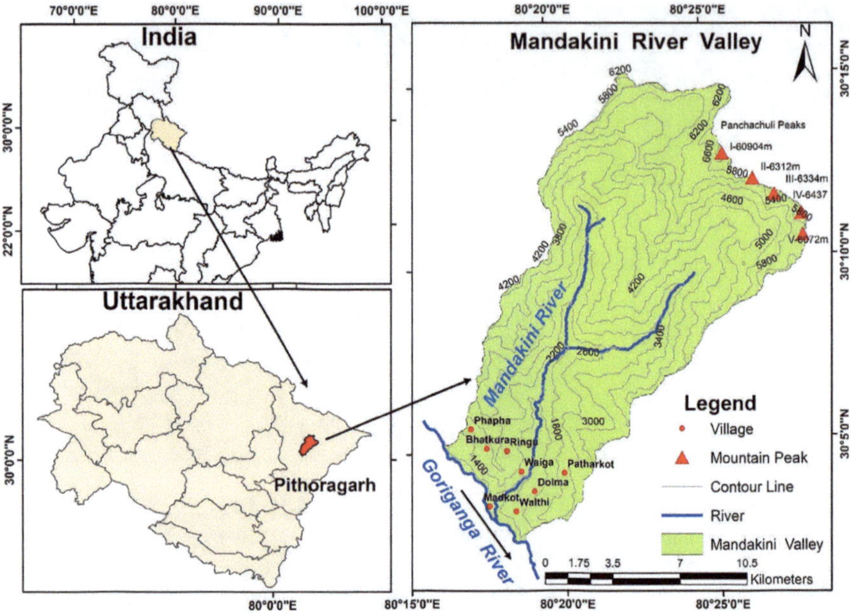

Fig. 11.2 Location of study area

sea level mountain ranges). Northern, western, and southern boundaries are formed by Dhunkhandhar Ridge (4312 m), Chulkot Dhar Ridge (2230 m), and Walthi Top (2700 m), respectively.

People from seventeen (17) villages collect Keedajadi from eight (8) *Bugyals (Dhunkhan, Kokan dhunga, Northern Blati, Southern Blati, Selapani Lambaladi, Khathapu (A, B), Janthr*i) from where *Keedajadi* collection is made. Out of seventeen villages, four, i.e. Phapa, Badni Dhar, and Bhatkura are located in the Gori Ganga River basin, but the people of these villages are also collecting from the *Bugyals* lying in the Mandakini river valley.

11.3 Materials and Methods

Data regarding families of seventeen (17) revenue villages of the year 2011 is collected from the secondary data whereas for the year 2022 is based on the primary data. The amount and income of the families with Keedajadi collection are compared with the year 2016–2022. In this study, seventeen (17) villages are studied. 2011 census data are used for the household study of villages. In the year 2022, the number of families involved in Keedajadi collection, total production, total income, and Bugyals being used for Keedajadi collection is based on secondary data. Interviews were conducted with the village head and Van Panchayat Sarpanch (community forest

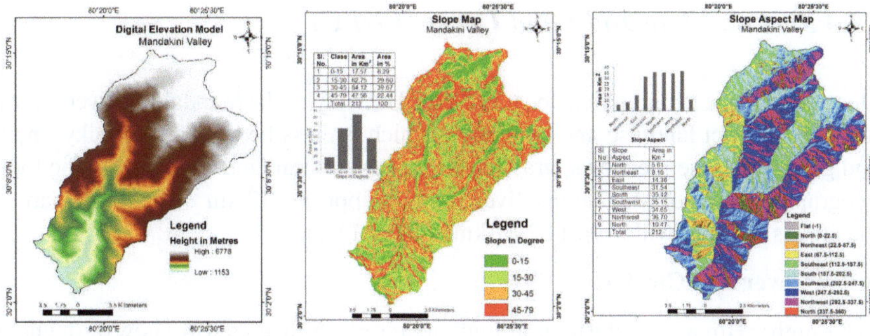

Fig. 11.3 DEM, slope, and aspect map

head) through a questionnaire for each village. Boundaries of eight (8) *Bugyalas* are delineated with the help of Google Maps and satellite data.

Slope (in degree and percentage) and aspect map are prepared using Alos Palsar DEM of 12.5 m using ArcGIS 10.8.2. Further fieldwork was conducted by visiting all the *Bugyals* to identify the area of *Keedajadi* collection. Soil sampling, average slope, altitude, and slope aspects are assessed using GPS, Google Earth map, and topographical map. ArcGIS is used for the final map preparation.

11.4 Results

11.4.1 Slope

The slope in degree is ranging in between 0 and 79° which is further sliced into (four) 4 categories based on the slope angle. The area with slope < 15°, 15–30, 30–45 and 45–79 is covering an area of 17.57 km^2 (8.29%), 62.75 km^2 (29.60%) 30–45°, 84.12 km^2 (39.67%), and 47.56 km^2 (22.44%), respectively (Fig. 11.3).

11.4.2 Slope Aspect

Aspect is the direction of the slope face of the land. The aspect of the slope is an important aspect of understanding the amount of radiation received by which direction of land. The Mandakini River valley has a high-altitude semi-arid environment and rugged terrain where aspect plays an important role in the growth of vegetation, moisture content, etc. In the study area aspect, 135–315° covers the major geographical area around 81% (Fig. 11.3).

11.4.3 Land Use and Land Cover (LULC)

Due to its unique geographical location in the Himalayas, the Mandakini River Valley has very distinct land use and land cover which is classified as barren rocky, snow and glacier, Alpine meadows, grassland, and agricultural land (Fig. 11.4). The total geographical area of Mandakini River valley is approx. 212 km^2. The major part of the area is covered with forest 82.47 km^2 (Table 11.1).

Snow Cover and Glacier

In the higher Himalayas, the high-altitude areas remain covered with snow throughout the year. Along with this, all the glaciers also part of them and this region experienced extreme cold. The snow cover and glacier in Mandakini valley is covered with an area of 38.55 km^2 which is 18.19% of the total area. Three small glaciers are Nothern Blati, Southern Blati, and Kathapu Glacier in Mandakini Valley.

Barren Rocky

This land is an arid and semi-arid region also called barren rocky, where soil is absent and seasonal grass is scattered. It is a rocky and residual rock of different lithology.

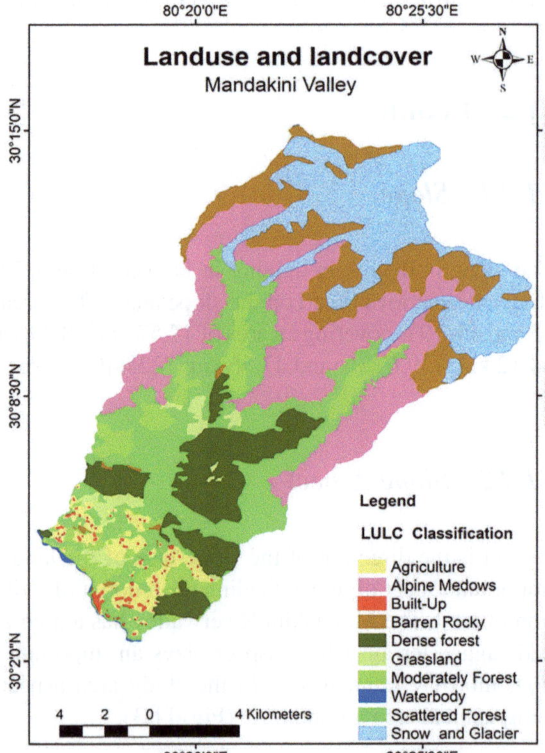

Fig. 11.4 Land use/cover map, 2022

Table 11.1 Land use classification Mandakini River Valley 2022

S. No.	Classes	Area (km^2)	Area (%)
1	Snow and glacier	38.55	18.18
2	Barren rocky	26.57	12.54
3	Agriculture	7.44	3.51
4	Dense forest	23.35	11.01
5	Moderately forest	14.69	6.93
6	Scattered forest	44.43	20.96
7	Built-up	0.84	0.39
8	Waterbody	0.48	0.23
9	Alpine meadows	51.66	24.37
10	Grassland	3.99	1.88
Total		212	100

This class covers a total area of 26.57 which is 12.53% of the total land area. This area is found below snow and glaciers.

Agriculture Land

This is the land that is mainly used for farming, food, and horticultural crops. Most of this land is found in areas near rivers. The total land under this class is 7.44 km^2 which is 3.51% of the total area.

Forest

This class has been divided into three classes based on the density of the tree. This class is told in three classes based on the number of trees. The forest land cover in Mandakini Valley is spread over 82.47 km^2 of area which is 38.90% of the total area.

Dense

This class includes vegetation with maximum density in the forest. In this class, higher-density plants are included in the forest. This is mainly found in mixed forests. In this forest, the density of the forest is very high. The spread of this class is 23.35 km^2 which is 11.01% of the total land area.

Moderately Dense Forest

Generally, the density of vegetation cover in this class is between 40 and 70%. This class includes dense forests and low-density forests, which are spread over 14.69 km^2, which is 6.93% of the total land.

Scattered Forest

Very low-density forests are found in this type of class and along with trees, small thorny bushes are also found. The area of this class in Mandakini Valley is spread over 44.43 km^2 which is 20.96% of the total land.

Alpine Meadows (Bugyal)

This class mainly includes the high Himalayan Bugyal region which is situated in the land above the tree line. In this, the land is spread over 51.66 km^2 which is 24.37% of the total land area.

Built-up Land

It covers every area of human habitation developed due to non-agricultural use, from settlements or commercial areas to areas with various buildings, home industries, or construction areas. The spread of this class is 0.84 km^2, which is 0.39% of the total area. This class is spread in the lowest strata of the land in the Mandakini Valley.

Water Bodies

In this class river systems and lakes are included. The area of water bodies in the Mandakini watershed is spread over 0.48 km^2 which is 0.23% of the total area.

Grassland (Grazing)

In this category, mainly pasture land has been ruled, its total area is 3.99 km^2 and it is spread in total 1.88%.

11.4.4 Keedajadi Collection Bugyal

In the Mandakini River valley, there are eight (8) *Bugyals* with an area of 30.36 km^2 from where *Keedajadi* is collected (Fig. 11.5). Local people informed that the first time *Keedajadi* collection was started by the Nepali labourers as seen by the Indian sheep herders in *Dhunkhan Bugyal* of this valley. These Nepalese labourers used to come to make wooden utensils in the forests, bordering Bugyals.

Due to the pre-existing skill of collecting Keedajadi with these Nepali labourers, they were collecting Keedajadi in the Bugyals here as well. Some sheep herders of village Ringu learned the skill of identification and collection technique of Keedajadi from Nepalese labourers in Dhunkhan Bugyal. The commercial collection was started in the year 2001 by the villagers of Mandakini Valley. A total of eight (8) Bugyals are situated in the study area from where Keedajadi collection is made. Some of the characteristics of each Bugyal are given in Table 11.2. Four samples are collected out of a total of eight Bugyals to have a cursory idea about the characteristics of soil (Fig. 11.5). Soil depth of the area ranges from 30 to 50 cm. The pH value ranges from 5.1 to 5.3 which reflects the acidic nature of the soil. The highest organic carbon amount, i.e. 2.92%, is found in *Lamba Ladi Bugyal* where the average slope is lower and the slope direction is South-East. Probably less degree of slope and more mount of sunshine might have been responsible. The highest amount of nitrogen, i.e. 388.9 kg/h, is found in *Dhunkhan Bugyal* (Table 11.3).

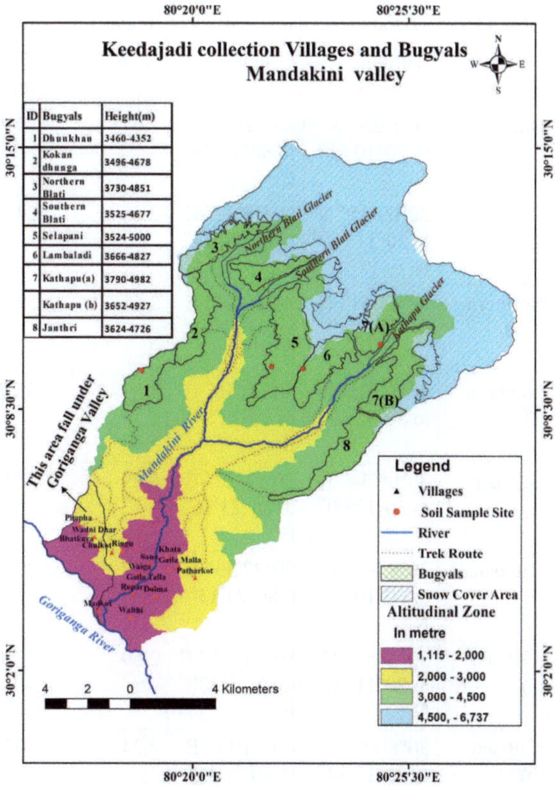

Fig. 11.5 Bugyals and villages of Mandakini Valley

The cause of the high amount of nitrogen is because of the grazing of thousands of sheep and horses causing a high amount of manuring the area. The texture of the soil indicates the sandy nature where the clay amount is very low.

11.4.5 Keedajadi Collection

The period of Keedajadi collection is from the middle of April to the middle of July as the snow starts melting in the Bugyals, with the arrival of the summer season. The collection first starts at Lambaladi and Kokan Dhunga Bugyals in the valley. These Bugyals are south and south-east facing as a result snow melting takes place early. Keedajadi collectors for a duration of 40–60 days, from the last week of April to the last week of June live in tents in the velvety delightful.

Bugyals appear as the city of tents. For this subtle period, one has to go through very difficult routes on foot to reach the Bugyals passing through treacherous mountains. Food items and other logistics materials are transported to the Bugyals on the

Table 11.2 Geographical characteristics of Bugyals

Name of Bugyal	Latitude	Longitude	Area (km^2)	Elevation (m)	Slope (°)	Slope aspect
Dhunkhan	30°8′25″ N to 30°10′01″ N	80°18′13″ E to 80°19′33″ E	2.64	3460–4352	34.1°	SE
Kokan dhunga	30°09′35″ N to 30°11′ 55″ N	80°19′32″ E to 80°20′44″ E	2.55	3496–4678	27.1°	E
Northern Blati	30°11′51″ N to 30°13′8″	80°19′40″ E to 80°21′41″ E	2.65	3730–4851	23.4°	SE
Southern Blati	30°11′25″ N to 30°12′07″	80°20′56″ E to 80°23′8″ E	1.89	3525–4677	31.9°	SE
Selapani	30°08′34″ N to 30°12′00″ N	80°21′22″ E t0 80°23′36″ E	6.29	3524–5000	18°	W
Lambaladi	30°8′32″ N to 30°10′23″ N	80°22′33″ E to80°24′13″ E	3.37	3666–4827	18.4°	SE
Kathabu (a)	30° 9′51″ N to 30°11′26″ N	80°24 ′55″ E to80°25′17″ E	2.69	3790–4982	24.9°	SE
Kathabu (b)	30°8′11″ N to 30°10′37"	80°24′11″ E to 80°26′08″ E	4.24	3652–4927	25.3°	W
Janthri	30°06′13″ N to 30°8′27″ N	80°21′55″ E to 80°24′53″ E	4.03	3624–4726	19.8°	NW

Table 11.3 Physicochemical characteristic

S. No.	Bugyal's Name	pH	Organic Carbon (%)	Nitrogen (Kg/h)	Phosphorus (%)	Potassium (%)	Texture (%)		
							Sand	Silt	Clay
1	Selapani	5.2	2.77	276	0.0033	0.0031	80	14	6
2	Lamba Ladi	5.3	2.92	250.9	0.0028	0.0066	70	26	4
3	Kathapu	5.2	2.73	225.8	0.0028	0.0088	72	24	4
4	Dhunkhan	5.1	2.38	388.9	0.0054	0.0139	82	14	4

Source Primary data collected during a field visit

Fig. 11.6 **a** Logistics materials are delivered in Bugyals by local sheep goat; **b** logistics stuff is taken in Karbachh; **c** Keedajadi collectors worship in Selapani Bugyal; **d** Keedajadi collectors temporary tents in Kathapu Bugyal. *Source* Primary data collected during a field visit

backs of sheep and goats (Fig. 11.6). A goat or sheep carries a load of about 10 kg at a time in a Karbachh (a special bag made of sheep skins Jute used to keep items).

Nature and man's relationship has been very strong from ancient times in this area and they worship nature. People used to come to Bugyals to worship God and formed a shrine in the natural cave, locally known as Gupha. The deities worshiped are Panchchuli Mountain, Bugyals, Pari Devi, and Aidi Devta. They worship all God and Goddesses for their safe and healthy stay during the collection of Keedajadi. In June they start to return to their home with their sheep and goats.

In (Table 11.4) the total number of families in the Mandakini River basin in the year 2011 was 1064 which increased to 1287 in the year 2022 an increase of 20.96%. Nowadays there is a tendency in the Central Himalayas that people are migrating towards cities and foothill areas in search of jobs, good education, and health services. After the introduction of the *Keedajadi* collection and marketing, people are earning a good amount of money which is making them stay in their villages. Shows that there are some villages where the percentage of families engaged in the collection is very low.

During the field survey, it was found that these families have government, private, or self-employed occupations. In addition to this, some families do not have young people to go to *Bugyal*. The number of families engaged in *Keedajadi* collection

Table 11.4 Village-wise details of total families in Mandakini River Valley 2011–2022

S. No.	Village Name	No. of families[a] (2011)	No. of the family[b] (2022)	Change in the no. of families (%)	Families engaged in the Keedajadi collection (2022)	Families engaged in Keedajadi collection (2022) (%)
1	Phapa	68	85	25.00	75	88.24
2	Badni Dhar	43	80	86.05	70	87.50
3	Chulkot	53	60	13.21	42	70.00
4	Bhatkura	29	32	10.34	16	50.00
5	Ringu	135	160	18.52	95	59.38
6	Sana	17	18	5.88	15	83.33
7	Khata	4	6	50.00	6	100.00
8	Gaila malla	19	28	47.37	20	71.43
9	Patharkot	28	35	25.00	33	94.29
10	Gaila talla	20	28	40.00	21	75.00
11	Madkot	353	360	1.98	8	2.22
12	Dobri Narki	11	13	18.18	8	61.54
13	Waiga	16	18	12.50	7	38.89
14	Rapti	42	40	– 4.76	20	50.00
15	Ropar	13	16	23.08	7	43.75
16	Walthi	192	288	50.00	12	4.17
17	Dolma	21	20	– 4.76	8	40.00
	Total	1064	1287	20.96	463	35.98

[a] Census of India (2011)
[b] Field Survey (2022)

is less than 50% in villages Madkot, Walthi, Waiga, Dolma, and Ropar. The main reason for this is that the location of these villages is very far away from the Bugyals.

In Table 11.5, there is a change of – 67 (– 12.6%) in the total number of families involved in Keedajadi collection in the year between 2016 and 2022. The total number of villages with positive change is 06. There is no change in several families engaged in Keedajadi collection in the village of Waiga. In the remaining 10 villages there is a negative change in the number of families involved in Keedajadi collection (Fig. 11.7).

Table 11.5 Year-wise number of families involved in *Keedajadi* collection (2016–2022)

S. No.	Villages	2016	2017	2018	2019	2020	2021	2022	Change (2016–22)	Change in % (2016–2022)
1	Phapa	70	70	71	74	74	75	75	5	7.1
2	Badni dhar	50	55	58	68	69	70	70	20	40.0
3	Chulkot	45	45	45	43	43	42	42	– 3	– 6.7
4	Bhatkura	23	20	18	17	17	16	16	– 7	– 30.4
5	Ringu	120	110	105	100	98	98	95	– 25	– 20.8
6	Sana	16	15	15	15	15	15	15	– 1	– 6.3
7	Khata	5	6	6	7	6	6	6	1	20.0
8	Gaila malla	18	19	19	19	19	20	20	2	11.1
9	Patharkot	30	30	31	30	31	32	33	3	10.0
10	Gaila talla	20	20	21	21	21	21	21	1	5.0
11	Madkot	10	7	10	10	8	8	8	– 2	– 20.0
12	Dobri Narki	10	10	9	9	9	8	8	– 2	– 20.0
13	Waiga	7	8	8	4	5	7	7	0	0.0
14	Rapti	25	22	21	22	20	20	20	– 5	– 20.0
15	Ropar	9	8	8	7	7	7	7	– 2	– 22.2
16	Walthi	60	53	45	30	20	12	12	– 48	– 80.0
17	Dolma	12	12	10	10	10	8	8	– 4	– 33.3
	Total	530	510	500	486	472	465	463	– 67	– 12.6

Source Field Survey (2022)

The details of the change in village-wise Keedajadi collection and income between the year 2016 and the year 2022 are given in Table 11.6. The total Keedajadi collection in this valley has changed by – 16.2 (– 27.2%) kg and income by Rs.16 (4%) lakhs. The total number of villages with positive change in Keedajadi collection and income is only two.

In Table 11.7, the utilization area of income obtained from Keedajadi from the year 2001 to 2022 has been studied in 17 villages of Mandakini River valley. The maximum income is used by 308 (66.52%) families for building construction in the village. 262 (56.8%) families in the field of higher education, 34 (7.34%) families purchased personal vehicles. 32(6.9%) in house construction outside the village. 31(6.69%) in animal husbandry. 27 (5.88%) in the shop, 14 (3.02%) in the hotel business, and 9 (1.94) families have used their income to buy taxis (Table 11.7).

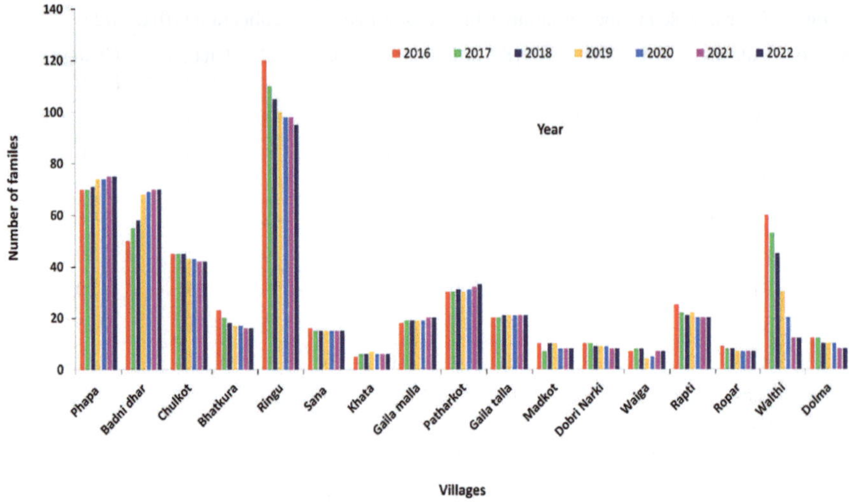

Fig. 11.7 Year-wise engaged family in Keedajadi collection

11.5 Discussion

The supply of Keedajadi in the international market has now brought a decrease in the Keedajadi production and collection in the Himalayas and the Tibetan Plateau. A study carried out by different scholars highlighted that the Keedajadi supply has decreased from 70 to 90% in China's Yunnan Province (Stone 2008), 37% in the Dhorpatan Hunting Reserve, 28% in Dolpa District in Nepal (Thapa 2014), and 25% in Pithoragarh District of India during 2008–2012 (Negi et al. 2015).

This study of seventeen villages shows that there is a 27.2% decrease in the production of Keedajadi for the duration of 2016 to 2022 in the Mandakini River basin (Table 11.8). Due to the excessive exploitation of Keedajadi its existence is being threatened creating a crisis in the international market. The Keedajadi is listed as a vulnerable species in China's Biodiversity Red List 2018 (Dai et al. 2020). *Ophiocordyceps sinensis* was recently listed as vulnerable on the IUCN Red List due to "its strict host–specialty on moth insects, and confined geographical distribution, and over-exploitation by humans in recent decades" (Yang 2020; He et al. 2022).

According to Gurung (2019) in Sephu, there were five chewogs with only 265 households in 2004 but the number of households increased to 314 in 2005. In 2018 the number of households increased to 319 as reported by Gup of Sephu Gewog. The finding says that there is a positive impact of Cordyceps collection on the socio-economic status of the respondents. In the Garhwal region of Central Himalaya, the income is made out of Keedajadi collection on education, food, wedding ceremonies, and house construction (caplins and Halvorson 2017). In Tibet income was used to start entrepreneurship, buying cattle and gold ornaments (Winklar 2009).

Table 11.6 Village-wise *Keedajai* collection and income (2016–2022)

Villages	Collection in Kg (2016)	Total Income in Rs (Lakhs)	Collection in kg (2017)	Total Income in Rs (Lakhs)	Collection in kg (2018)	Total Income in Rs (Lakhs)	Collection in kg (2019)	Total Income in Rs (Lakhs)	Collection in kg (2020)	Total Income in Rs (Lakhs)	Collection in kg (2021)	Total Income in Rs (Lakhs)	Collection in kg (2022)	Total Income in Rs (Lakhs)	Change in collection (in kg)	Change in the collection (%)	Change in Income in RS (Lakhs)	Change income %
Phapa	12	84	10	80	8	64	4.5	40.5	4.5	36	4.1	32.8	5	50	– 7	– 58.3	– 34	– 40.5
Badni dhar	6	42	6	48	5	40	3	27	2.8	22.4	2.8	22.4	3	30	– 3	– 50.0	– 12	– 28.6
Chulkot	5	35	5	40	5	40	4	36	4	32	4	32	5	50	0	0.0	15	42.9
Bhatkura	1.5	11	1	8	1	8	1	9	1	8	1	8	1.2	12	– 0.3	– 20.0	1.5	14.3
Ringu	8	56	8	64	7	56	6	54	5.8	46.4	5	40	8	80	0	0.0	24	42.9
Sana	4	28	4	32	4	32	3	27	3	24	3	24	4	40	0	0.0	12	42.9
Khata	1.5	11	1.5	12	1	8	1	9	1	8	1	8	1	10	– 0.5	– 33.3	– 0.5	– 4.8
Gaila Malla	2.5	18	2	16	2	16	1.8	16.2	2	16	1.9	15.2	2	20	– 0.5	– 20.0	2.5	14.3
Patharkot	4	28	4	32	4	32	4	36	3	24	3	24	4	40	0	0.0	12	42.9
Gaila Talla	2.5	18	2.5	20	2	16	2	18	2	16	1.5	12	2	20	– 0.5	– 20.0	2.5	14.3
Madkot	1	7	1	8	1	8	1	9	0.5	4	0.5	4	0.8	8	– 0.2	– 20.0	1	14.3
Dobri Narki	3	21	2.5	20	1.5	12	1.5	13.5	1	8	1	8	1	10	– 2	– 66.7	– 11	– 52.4
Waiga	1	7	1	8	1	8	1	9	1	8	1	8	1.5	15	0.5	50.0	8	114.3
Rapti	1.5	11	1.5	12	1	8	1	9	1	8	1	8	2	20	0.5	33.3	9.5	90.5
Ropar	1	7	1	8	0.9	7.2	0.9	8.1	0.8	6.4	0.8	6.4	0.8	8	– 0.2	– 20.0	1	14.3
Walthi	3	21	3	24	2.5	20	1.5	13.5	1	8	1	8	1	10	– 2	– 66.7	– 11	– 52.4

(continued)

Table 11.6 (continued)

Villages	Collection in Kg (2016)	Total Income in Rs (Lakhs)	Collection in kg (2017)	Total Income in Rs (Lakhs)	Collection in kg (2018)	Total Income in Rs (Lakhs)	Collection in kg (2019)	Total Income in Rs (Lakhs)	Collection in kg (2020)	Total Income in Rs (Lakhs)	Collection in kg (2021)	Total Income in Rs (Lakhs)	Collection in kg (2022)	Total Income in Rs (Lakhs)	Change in collection (in kg)	Change in the collection (%)	Change in Income in RS (Lakhs)	Change income %
Dolma	2	14	1.5	12	1.5	12	1.5	13.5	1	8	1	8	1	10	– 1	– 50.0	– 4	– 28.6
Total	59.5	417	55.5	444	48.4	387.2	38.7	348.3	35.4	283.2	33.6	268.8	43.3	433	– 16.2	– 27.2	16.5	4.0

Source Field Survey (2022)
Keedajadi is valuable only in the Mandakini river valley
RS 7 lakh (US$ 10,421.31) per kg in 2016, RS 8 lakh (US$ 12,290.67) per kg in 2017, RS 8 lakh (US$11,694.20) per kg in 2018, RS 9 lakh (US$ 12,784.02) per kg in 2019, 8 RS lakh (US$ 10,791.53) per kg in 2020, RS 8 lakh (US$ 10,924.48) per kg in 2021, and RS 10 lakh (US$ 12,759.98) per kg in 2022

Table 11.7 Utilization of income earned from *Keedajadi*

S. No.	Utilization	Total no. of families	No. of families (%)
1	Taxi	9	1.94
2	Personal vehicle (bikes-30, and car-4)	34	7.34
3	Shop	27	5.83
4	Restaurant	14	3.02
5	Higher education (321 persons)	262	56.8
6	Animal husbandry (12 family—1010 sheep's and goats, 3 family—6 mules, 11 family—13 cows, 5 family—5 buffalo	31	6.69
7	House construction (in the village)	308	66.52
8	House construction (out of village)	32	6.9

Source Field Survey (2022)

Table 11.8 Keedajadi collection and income of Mandakini Valley

Year	Total collection (in Kg)	Total income (in Rs. Lakh)	Cost in Rs.& US$/Kg
2016	59.5	417.0	7 Lakh/US$10,421.3
2017	55.5	444.0	8 lakh/US$ 12,290.67
2018	48.4	387.2	8 lakh/US$ 11,694.20
2019	38.7	348.3	9 lakh/US$ 12,784.02
2020	35.4	283.2	8 lakh/US$ 10,791.53
2021	33.6	268.8	8 lakh/US$ 10,924.48
2022	43.3	433.0	10 lakh/US$ 12,759.98

Source Primary data collected during a field visit

The total collection is decreasing from 2016 to 2021. However, in 2022 it increased. In 2020–21 because of the COVID pandemic, the stay duration of people in Bugyals decreased resulting in less collection of Keedajadi. The cost of the Keedajadi is increasing yearly so total income is increasing though production is decreasing. A study in Mandakini Valley shows that a major portion of the income from Keedajadi collection is used by 308 (66.52%) families on building construction. Other investments are 262 (56.8%) in the higher education of children, 31 (6.69%) in animal husbandry, 27 (5.88%) in the opening of shops, and 14 (3.02%) in the hotel business.

11.6 Conclusion

The exploitation of Keedajadi started in the year 2001 in the Mandakini River valley area which transformed the economic condition of the people in the area. People go to Bugyals (meadows) for a short time between April and June (during the melting of snow cover) every year for Keedajadi collection. Due to the effect of Keedajadi, there has been an increase in the number of families in this valley, income has also increased, but the amount of production of *Keedajadi* is continuously decreasing. After the collection of *Keedajadi*, the income received from its marketing is used by most families for building construction, educating children, starting their employment-oriented businesses, and meeting their day-to-day needs.

Acknowledgements The authors are highly obelized and grateful for the information provided by the elderly persons of all villages and the cattle herders of the area. We are also thankful for providing the facilities of the available infrastructure in the Department of Geography, Kumaun University, Nainital. We extend thanks to the U.S. Geological Survey for providing satellite data. We would also like to acknowledge Google for Google Earth Pro data. Mr. Masoom Reza, Assistant Professor (C) is duly acknowledged for helping in the formatting of the article.

References

Adhikar MK (2008) The diversity of Cordycepioid fungi (Ascomycotiona: Clavicipitales) reported from Nepal. Bull Dept Plant Resour 30:1–8

Cannon PF, Hywel-Jones NL, Maczey N, Norbu L, Tshitila S, Dup T, Lhendup P (2009) Steps towards sustainable harvest of *Ophiocordyceps sinensis* in Bhutan. Biodivers Conserv 18:2263–2281. https://doi.org/10.1007/s10531-009-9587-5

Caplins L, Halvorson SJ (2017) Collecting *Ophicordyceps sinensis*: an emerging livelihood strategy in the Garhwal, Indian Himalaya. J Mountain Sci 14(2):390–402. https://doi.org/10.10007/s11629-016-3892-8. https://www.researchgate.net/publication/313321558

Census of India (2011) Census Data 2011 Office of the Registrar General & Census Commissioner India. Ministry of Home Affairs Government of India http://censusindia.gov.in

Dai Y, Wu C, Yuan F, Wang Y, Huang L, Chen Z, Zeng W, Wang Y, Yang Z, Zeng P, Lemetti, P, Mo X, Yu H (2020) Evolutionary biogeography on *Ophiocordyceps sinensis*: an indicator of molecular Phylogeny to geochronological and ecological exchanges. Geosci Front 11:807–820. https://doi.org/10.1016/j.gsf.2019.09.001

Devkota S (2008) Distribution and status of highland mushroom: a study from Dolpa, Nepal. J Nat Hist Museum 23:51–59

Devkota S (2010) *Ophicordyceps sinensis* (Yarsagumba) from Nepal, Himalaya: status, threats and management strategies. In: Grassland Monitoring and Supervision Centre, Chinese Ministry of Agriculture (eds). Cordyceps sinensis resources and environment. LanZhou University Press, Gansu, pp 91–108

Garbyal SS, Aggarwal KK, Babu CR (2004) Impact of cordyceps sinensis in the rural economy of interior villages of Dharchula subdivision of Kumaon Himalayas and its implications in the society. Indian J Tradit Knowl 3(2):182–186

Gurung M (2019) Socio-economic impact of cordyceps collection on cordyceps collectors in sephu Gewog of Bhutan. J Innov Sci Res Technol 4(10):763–770

He J, Smith-Hall C, Zhou W, Zhou W, Wang Y, Fan B (2022) Uncovering caterpillar fungus (*Ophicordyceps sinensis*) consumption pattern and linking them to conservation intervention. Conserv Sci Pract 4(8):e12759. https://doi.org/10.1111/csp2.12759

Kuniyal CP, Sundriyal RC (2013) Conservation salvage of Cordyceps sinensis collection in the Himalayan Mountains is neglected. Ecosyst Serv 3:e40–e43

Laha A, Badala R, Hussain AS (2018) Earning a livelihood from Himalayan caterpillar fungus in Kumaon Himalaya. Mountain Res Dev 38(4):323–331

Li Y, Wang XL, Jiao L, Jiang Y, Li H, Jiang SP, Lhosumtseiring N, Fu SZ, Dong CH, Zhan Y, Yao YJ (2011) A survey of the geographic distribution of *Ophiocordyceps sinensis*. J Microbiol 49(6):913–919

Negi CS, Koranga PR, Ghinga HS (2006) Yartsa Gumba (cordyceps sinensis): a call for its sustainable exploitation. Int J Sust Dev World 13(6):165–172. https://doi.org/10.1080/13504500609469669

Negi CS, Pant M, Joshi P, Bohra S (2014) Yartsagnbu (*Ophicordyceps sinensis*) [Berk] GH. Singh et al. The Issue of its Sustainability. Curr Sci 107(5):882–887

Negi CS, Joshi P, Bohra S (2015) Rapid vulnerability assessment of Yartsa gunbu (*Ophicordyceps sinensis* [berk.] in Pithoragarh district Uttarakhand, State, India. Mountain Res Dev 35(4):382–391. https://bioone.org/journals/mountain

Negi CS, Pant M, Bohra S (2016) Conserving the caterpillar fungus [*Ophicordyceps sinensis* (Berk.): a case study of habitat ecology and sustainability in District Pithoragarh, Western Himalaya, India. Biodiv Conserv 8(9):196–205. https://doi.org/10.5897/IJBC2014.0742, https://www.academijournals.org/IjBC

Negi SV, Rana SK, Giri L, Rawal RS (2020) Caterpillar fungus in the Himalayas, current standing and future possibilities. G.B. Pant National Institute of Himalayan Environment, Kosi-Katarmal, Almora Uttarakhand, India

Olsen CS, Larsen HO (2003) Alpine medicinal plant trade and Himalayan mountain livelihood strategies. Geogr J 169(3):243–254. https://doi.org/10.1111/1475-4959.0008. https://www.Jstor.org/stable/3451450

Pradhan BK, Sharma G, Subba, B, Chettri S, Chettri A (2021) Distribution harvesting, and trade of Yartsa Gunbu (*Ophicordyceps sinensis*) in the Sikkim Himalaya, India. Mountain Res Dev 40(2). https://doi.org/10.1659/MRD-JOURNAL-D-19-00039.1

Sharma S (2004) Trade of *Cordyceps sinensis* from high altitudes of the Indian Himalaya: conservation and biotechnological priorities. Curr Sci 86(12):1614–1619

Shrestha B, Sung JM (2005) Notes on cordyceps species collected from the Central Region of Nepal. Mycobiology 33:235–239

Sanjukta S, Karati D, Priyadarshin IR, Dua TK, Paul P, Ranabir S, Gouranga N (2023) *Cordyceps sinensis* (Yarsagumba): pharmacological properties of a mushroom. Pharmacol Res Modern Chin Med 8:100294. https://doi.org/10.1016/j.prmcm.2023.100294

Sen S, Karati D, Priyadarshini R, Dua TK, Paul P, Sahu R, Nandi G (2023) Cordyceps Sinensis (Yarsagumba): pharmacological properties of a mushroom. Pharmacol Res Mod Chin Med 8:100294. https://doi.org/10.1016/j.prmcm.2023.100294

Shrestha B, Zhang W, Zhang Y, Liu X (2010) What is the Chinese caterpillar fungus *Ophicordyceps sinensis* (Ophiocordycipitaceae)? Mycology 1:228–236. https://doi.org/10.1080/21501203.2010.536791. https://www.tandfonline.com/loi/tmyc.20

Stone R (2008) Last stand for the body snatcher of the Himalayas? Science 322(5905):1182

Shrestha UB, Bawa KS (2014) Economic contribution of Chinese caterpillar fungus to the livelihoods of mountain communities in Nepal. Biol Cons 177:194–202. https://doi.org/10.1016/j.biocon.2014.06.019

Subedi A, Hamal S, Joshi A (2023) A review on *Ophiocordyceps sinensis*. Sci Herit J (GWS) 7(2):44–48. https://doi.org/10.26480/gws.02.2023.44.48

Thapa BB, Panthi S, Rai RK, Shrestha UB, Aryal A, Shrestha S, Shrestha B (2014) An assessment of Yarsagumba (*Ophicordyceps sinensis*) collection in Dhorpatan Hunting Reserve, Nepal.

J Mountain Sci 11:555–562. https://doi.org/10.10007/s11629-013-2692-7. https://www.researchgate.net/publication/236951240
Wei Y, Zhang L, Wang J, Wang W, Niyati N, Gou Y, Wang X (2021) Chinese caterpillar fungus (*Ophiocordyceps sinensis*) in China: current distribution, trading, and futures under climate change and overexploitation. Sci Total Environ 755:142548
Winkler D (2008) Yartsa gunbu (*Cordyceps sinensis*)and the fungal commodification of Tibet's rural economy. Econ Bot 62:291–305. https://doi.org/10.1007/s12231-008-9038-3
Winkler D (2009) Caterpillar fungus (*Ophicordyceps sinensis*) production and sustainability on the Tibetan plateau and in the Himalayas. Asian Med 5(2):291–316. https://doi.org/10.1163/157342109X56.8829
Yadav PK, Saha S, Mishra AK, Kapoor M, Kaneria M (2016) Conserving *Ophiocordyceps sinensis* in the Nanda Devi Biosphere Reserve, India. CICADA, Dehradun
Yang ZL (2020) *Ophiocordyceps sinensis* (amended version of 2020 assessment) IUCN red list of threatened species 2020: e.T58514773A179197748

Chapter 12
Impact of Climate Change on the Customary Life of the Brokpa Community of Arunachal Pradesh; Northeastern Himalayan Region

Nandini C. Singh

Abstract Climate change has been a persistent worldwide phenomenon throughout recorded history, but the rate of change has accelerated in the last few years. Since temperatures have been rising recently, most land areas have experienced an increase in hot days and seasonal variations, including heat waves. According to NOAA scientists (National Oceanic and Atmospheric Administration) 2020 ranked as the second hottest year on record for the planet effectively tying with 2016 that has been recorded as the one of the warmest years ever. The land and ocean surface temperatures in the Northern Hemisphere were 2.30 °F (1.28 °C) above average, the highest in 141 years. Massive wildfires and increased heat-related disease spread are therefore becoming signs of an increasingly worrying scenario that portends a difficult present and a dismal future. These extreme heat waves creating consequential water scarcity are initiating. Global conditions leading to predicted outcomes such as extreme and frequent storms, flooding, and droughts posing immense threat to the existence of humanity in general and indigenous groups' in particular by endangering their survival through diminishing the natural habitats that serve as their primary means of subsistence. One such notable example are the Northeastern states of India, whose inhabitants uniquely reflect the wide amalgamation of physical and cultural diversity truly showcasing India's physical and cultural multiracial nature. Among these states, Arunachal Pradesh stands out as occupying the largest geographical area and being bestowed with picturesque physical diverseness and divine serenity with rare flora and fauna forming an ecological hotspot. The state is predominantly inhabited by Scheduled Tribes having 26 major and 106 minor tribes. Among these tribes, the Brokpas who are also referred as highlanders are the only pastoral nomadic tribe found in Northeastern India and through their livelihood and lifestyle depict the intricate man-nature interaction and adaptation in most natural form. The Brokpas draw their origin as a subgroup of the Monpas who are located in the Tawang and West Kameng districts forming the western most part of the state. Being confined to the higher altitudes, and given their tradition of herding of yaks the local animal found

N. C. Singh (✉)
Department of Geography, Rajiv Gandhi University, Rono Hills, Arunachal Pradesh, India
e-mail: nandini.singh@rgu.ac.in

S. C. Rai (ed.), *Himalaya: Mountains of Destiny*,
https://doi.org/10.1007/978-3-031-94905-0_12

in these high altitude as a profession for livelihood, the Brokpas are dependent upon climate as a crucial factor in the traditional way of life of herding. As nomads the Brokpas migrate seasonally to find suitable grazing land for their livestock a tradition being followed for generations. However, recent significant changes in the climate in this region of the Himalayas have begun to directly affect the Brokpas by means of disappearing pastures and shortening of cold periods as well as the rising temperature in this high altitude areas. This has started affecting the economic and socio-cultural practices, as well as the generation old livelihood pattern and the associated lifestyle of this nomadic tribe. This Present chapter attempts to understand and explain these shifting aspects of the Brokpas' way of life in the Himalayan region of Arunachal Pradesh.

Keywords Brokpa · Climate change · Culture · Dimension · Livelihood · Tradition · Yak

12.1 Introduction

Climate change refers to long-term shifts in temperatures and weather patterns. These shifts are often reasoned as the natural cycle of climate phenomena. But since the beginning of the twentieth century, the pace and pattern of this change have been abrupt and fast, the justification of which could not be restricted to the singularity of nature alone but brought out very aptly the human activities as the main driver of change, primarily due to the burning of fossil fuels (like coal, oil, and gas), which produces heat-trapping gases. Climate change is the substantial variation in average weather patterns over several decades or more that results in warmer, wetter, or drier circumstances (World Bank 2021). Although the phenomenon of climate change has been happening on Earth for all of the recorded time, the changes have accelerated recently. It has been noted that the current temperature rise, which has mostly occurred since the middle of the twentieth century, is happening faster than it has in the past (Turrentine and Denchak 2021). According to NOAA, National Centres for Environmental Information, Annual 2023 Global Climate Report (2024), the year 2023 was the warmest with average land and ocean surface temperature at 1.18 °C since the beginning of global records in 1850 which is above the twentieth century average of 13.9 °C. It is such incidences of rising temperatures, intense precipitations, floods, heatwaves, and droughts occurring across the globe, that communities, livelihoods, and infrastructure are greatly affected (IPCC 2014). These changes impact at the local, national, and global levels, affecting human societies and natural ecosystems through the nature of accessibility of, natural resources, its usage and conservation. Such effects are more visible in developing and underdeveloped countries where mitigation of such climatic hazards is yet to be improvised at a mass scale due to lack of entrepreneurship, viable economic sustenance, resources availability and accessibility. However, it can not be ignored that even within these countries, there are different and isolated zones inhabited by indigenous tribes who have been affected the most as they continue to occupy heterogeneous areas and have survived by relying totally on the weather patterns for their traditional way of livelihood, living conditions customs, and rituals. These situations has influenced and have created harsh conditions for their survival and growth on one hand, but has simultaneously created unique cultural heritage and tradition marking their diversity and ability to adopt and build in an intricate relation with the surrounding nature forming their habitat.

It is therefore obvious that such indigenous tribes who inhabit diverse and geographically difficult physical areas are impacted the most by the phenomena of climate change by any shift in the pattern of the daily to yearly weather conditions in their areas of habitation in the form of either rising temperature, rainfall or snow, or all. The reason can be best understood in terms of understanding their livelihoods or traditional practices for survival is directly depended upon the natural surroundings and provides for their sustenance ranging from daily food requirements in the form of edible herbs/shrubs, roots to generating economic resources from the forest products along with the rained primordial agriculture practices, fishing and even hunting. Therefore any change in the climatic pattern affects and threatens their survival and well-being (USPA 2024). India stands out as one of the unique examples that exhibit varied physical diversity stretching from the highest of the mountain ranges to the coasts of the southern seas and from the creeks of the west to the swamps of the East. This vast variegated landscape incorporates in itself a mosaic of the diverse social milieu. In its easternmost part lies the Northeastern states which reflect a similar picture with diverse physical and cultural backgrounds. Arunachal Pradesh is one of the biggest states in geographical terms of NE and is enormously significant both in the context of geopolitics and the physical boundary of not only the region but for the nation as a whole, barricading the extreme northeast. Popularly known as the land of the rising Sun, the state is consecrated with nature's bounty of shimmering diversity of rare flora and fauna. These diversities have been sustained by equally diverse climatic zones varying from the Alpine to subtropical in conglomeration with the vast expanse of varying landscapes starting from plains, and valleys to the lofty ranges of high Himalayas. Like its varied topography, the state of Arunachal Pradesh reflects distinct diversity in its population base too having being inhabited by 26 major tribes and 106 minor tribes forming the predominant share of schedule tribes of the population of the state. Distributed into different physical zones, these tribes exhibit the intricated picture of adaptation to the given natural setup and consequently reflect a mosaic of speckled cultural practices often emanating from their customary pattern of livelihood practised over generations. Of the major tribes, the Monpas of Tawang district shows extreme resilience in adapting to their heterogeneous local habitat in terms of their livelihoods and cultural practices. The Monpas are scattered over a geographic zone encompassing the west Kameng and Tawang districts. Predominantly Buddhist, the Monpas have a cultural affinity with the Tibetan people across the border and are different from other tribes of the state.

The most demarcating factor of the Monpa tribe is the presence of prominent subgroups known as the Brokpas the only transhumant pastoralist community inhabiting the Tawang and West Kameng districts of the state. Transhumance is the regular movements of herds between fixed points to exploit the seasonal availability of pasture (Blench 2001). The term Brokpa comes from the combination of two words, i.e. Brok meaning 'pasture or grazing land' and Pa meaning 'People or community'. The word therefore refers to the herdsman/people/community that dwells in the pastures and rears livestock (Norbu 2015; Dorjee 2015), explaining the occupational profession practised by the Brokpa community whose primary livelihood is revolves around Yak rearing and as such their livelihood pattern and practices can be compared to the Gaddis and Bakarwal of the Northern Himalayas. The Brokpas of Arunachal Pradesh are confined to the higher altitudes of the Arunachal Himalayas given their tradition of herding yaks as a profession for livelihood and practising vertical movements between established points. They usually move to the higher altitudes with their flocks in the summer in search of grazing land and move downhill during winter, depending primarily on their livestock which includes yaks and their hybrid. It is obvious that in such circumstances, climate plays a key role in maintaining the traditional livelihood system of herding as the Brokpas move seasonally in search of grazing land (Tsering et al. 2010). This study attempts to assess how with the changing times, the yak herder tribe of the Monpa community is struggling to catch up with the changing economic and social impact which poses a grave challenge to the age-old cultural occupation of grazing yaks in hostile geo-climatic conditions in a fixed shifting pattern of the pastoralist tribe in the face of climate change.

Within this background, the current chapter has attempted to explore the emerging changes by setting before itself the following two objectives: (i) to assess and comprehend the impact of climate change on the upper reaches of the Arunachal Himalayan region vis-a-vis sustainability of the pastoral lifestyle of the Brokpa community inhabiting these areas, and (ii) to assess the emerging trajectory of socio-cultural, economic, and lifestyle change among the Brokpa community of Arunachal at large.

12.2 Study Area

Arunachal Pradesh covers a geographical area of 83,743 km^2, which roughly accounts for 2.55 of the total area of India. It shares an inter-state boundary with Assam and Nagaland and an international boundary with Bhutan, Myanmar, and China's Tibet Autonomous Region. As per the 2011 census, it has a total population of 1,383,727, of which 1,066,358 are in rural areas and 317,369 are in urban areas. The state is divided into 26 districts. Among these, Tawang and West Kameng districts are located in the westernmost part of the state. Tawang has an area of 2172 km^2 and is situated at a height of 3500 m above sea level. West Kameng district has an average elevation of 2415 m with an area of 2172 km^2 (Fig. 12.1).

The Brokpas are inhabited in the westernmost part of Arunachal Pradesh which shares international boundaries with Bhutan and Tibet (China). Mostly their habitation is found to be scattered in the northern and northwestern higher altitude in the areas of West Kameng and Tawang districts. The study area is situated between 27°25′ to 28°01′ N latitudes and 91°31′ to 92°40′ E longitudes (Fig. 12.1). The two districts

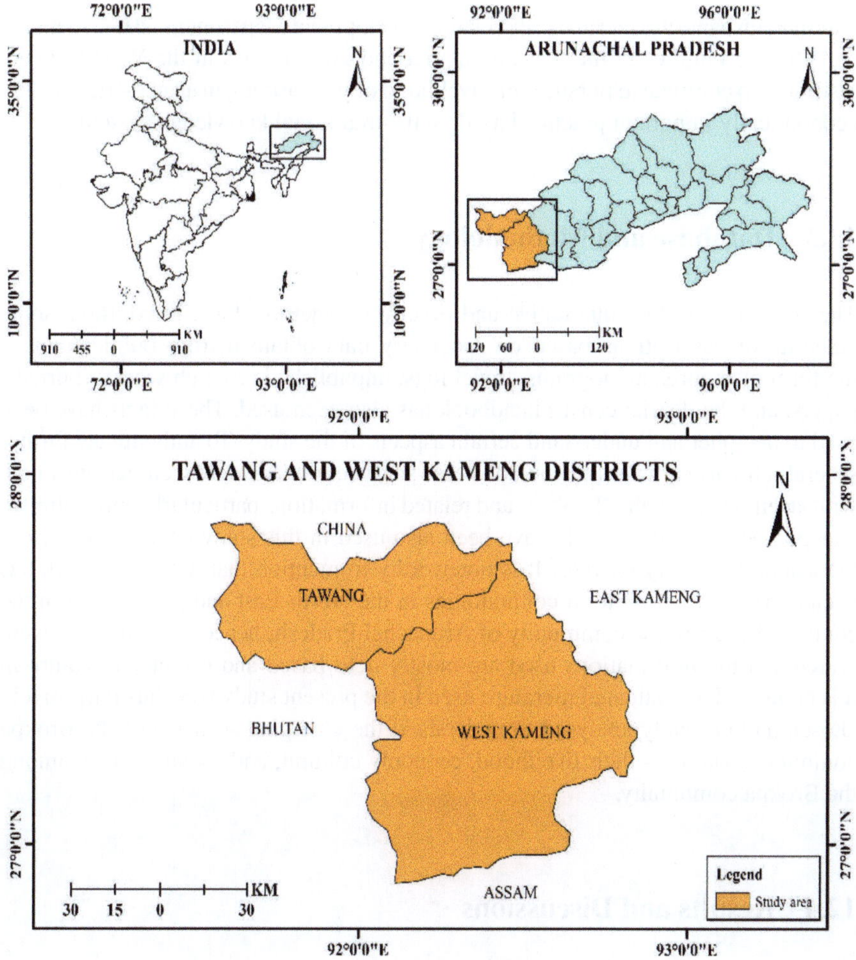

Fig. 12.1 Map showing the study area—West Kameng District

combinedly cover an area of 9594 km which is about 11.45% of the total geographical area of the state. The total population of the district stood at 33,924 which has a share of 9.67% of the total population of the state with a population density of only 16 as compared to 172 persons per km of the state. The literacy rate of the district is 65% with a wide gender gap representing 72.1% male and 54.77% female literacy rates, respectively. The sex ratio of the district shows maximal imbalance with a ratio of 760 females per 1000 males (Census 2011). The present district came under the effective Indian administration on February 12, 1951, when Major Bob Khating, the then Assistant Political Officer, re-established India's sovereignty over the Tawang and Monpa settlements of West Kameng district. As stated earlier, the Brokpas live sparsely in the highland regions of Tawang and West Kameng districts. As per the

Animal Husbandry, Veterinary and Dairy Development Department, 2019, there are 384 Brokpa families in the Tawang district and 260 families in the West Kameng district who continue to pursue their livelihood of yak farming in an eco-friendly and economically important practice based on the traditional knowledge system.

12.3 Database and Methodology

The study is based on qualitative and descriptive methodology, in addition, some quantitative descriptions based on secondary data obtained from publications in the form of reports, monograms, travel logs, unpublished research works, journals, papers, and the district census handbook has also been used. These facts have been used to decipher and understand certain aspects of the study. To authenticate further several scholarly articles that included case study references were included and cited, with an emphasis on the 'Brokpa' and related informations particularly about climate change. Descriptive methods have been also used in this study drawing data from the available literary sources. It is noteworthy to mention that not much work has been carried out on pastoral communities in the North-East and particularly in the context of the Brokpa community of Arunachal Pradesh; hence, relevant data were scarce and the informations used are mostly field based and primary in nature in this context. The evaluated literature used in the present study was thus purposively chosen and critically analysed to understand the emerging scenario of the Brokpa community via-a-vis their livelihood, economy cultural, and social, change among the Brokpa community.

12.4 Results and Discussions

Geographically, Arunachal Pradesh is the largest state in the North Eastern Himalayan Region of India, covering an area of 8.37 million hectares and boasting a rich physical and cultural diversity. The state is custodian of 23.52% of total flowering plants of India (Hegde 2002), including around 4500 species of angiosperm and 550 species of orchids, and is also regarded as nature's repository of medicinal plants (Haridassan 1989), where around 500 medicinal plants were identified during the preliminary survey. Arunachal Pradesh possesses India's second-highest level of genetic resources (SAPCC 2011). The fauna diversity includes 85 species of mammals and 760 species of birds (SAPCC 2011). In continuation with this diversity, the area also exhibits close adoptive human habitats and their dependency on this varied natural landscape. It is thus obvious that with such verse diversity and immense natural bounty, the climate factor plays an extremely significant role in the context of both substance and prospects of the fragile ecosystem. Any deviation therefore in terms of climate change is going to have an intricate impact both from the perspective of physical and human existence. It is alarming to mention that such trends have already started appearing and affecting the delicate balance of this resource-rich region in the form of changing patterns of flora on one hand and crops and cropping patterns of the communities who are dependent upon the surrounding environment for their livelihood on the other.

The State Climate Change Cell of Arunachal Pradesh carried out a study analysing IMD gridded data from 1980 to 2019 to ascertain the state's temperature trend. The findings showed that the state's annual mean temperature increased by 0.015 °C years. This attests to the state's temperature rising by 0.59 °C during the previous 40 years. The maximum temperature rose at the rate of 0.034 °C annually and the minimum temperature increased at the rate of 0.017 °C per year during this timeframe. Nonetheless, this shift in climatic conditions will seem particularly concerning and prevalent in mountainous areas due to their status as the most environmentally fragile regions globally (Tshe-ring et al. 2009). In addition to being home to a variety of plant and animal species, these areas also provide a habitat for some of the world's most endangered species and unique communities that rely entirely on nature for their survival (Kollmair et al. 2005). Additionally, the Eastern Himalayan region is included in the Himalayan global biodiversity hotspot, with exclusive species and distinct tribal groups relying on the mountain ecosystem that is significantly impacted by the changing climatic conditions, particularly the increasing temperatures. Similar effect are becoming prominently visible not only in other parts of India like the higher Himalayas of Darjeeling in West Bengal and the North eastern states of Arunachal Pradesh, Assam, Manipur, Mizoram, Meghalaya, Nagaland and Tripura, but are spanning across Central Nepal, Tibet autonomous region of China and Bhutan, and Myanmar as well (Fig. 12.2).

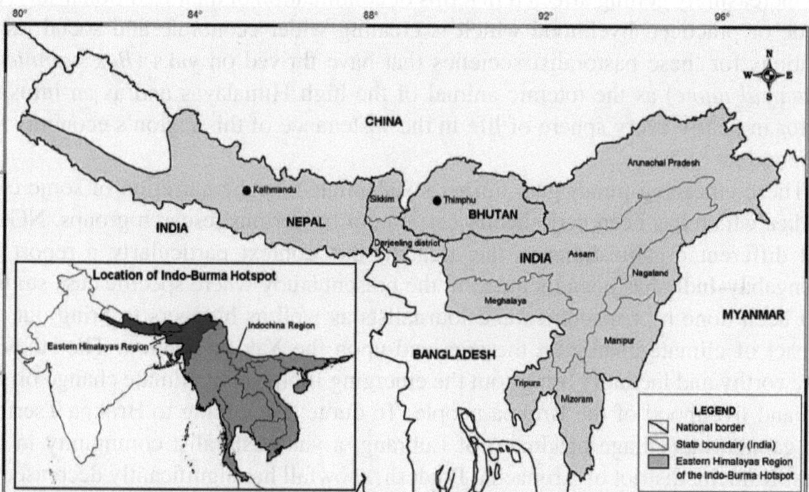

Fig. 12.2 Map showing the Eastern Himalayan Region. *Source* Critical Ecosystem Partnership Fund, Eastern Himalayan Region (2005)

The highland of Arunachal, Tawang, and West Kameng districts of Arunachal Pradesh is also seeing a remarkable shift in the physical and human aspects in terms of their changing livelihood pattern. In the light of this, the current chapter concentrates on the Brokpas, the state's highlanders, and the only transhumant pastoralist group who usually live in sparse settlements reflecting a harmonious relationship with the environment and rely on the forest's resources for their sustenance (Kumari and Rimung 2022). Their main source of sustenance consists primarily of yaks and yak-related products which they prepare by utilizing their milk, meat, fur, skin, and dung as essential resources. The Brokpas as transhumance pastoralists are characterized by mobility patterns at an altitude of 3000–4500 m with yak and sheep to find resources such as high mountain pastures covering different habitat types and ranges of forests (Dorjee 2015; Tsering et al. 2010). The most significant factor of highland pastoralists is based on pasture availability which relies on the climatic condition of the rearing tracks, and any change in these conditions is bound to cause a total shift of livelihood patterns affecting not only the economy but also the socio-cultural fabric as well as the development and adaptation pattern and process of the inhabitants in the given context.

The fluctuating climatic conditions in Arunachal Pradesh therefore tend to pose an immense threat to the very livelihood of these highlanders. Therefore, it becomes necessary to highlight and attempt to mitigate this problems being faced by the Brokpas, who are traditional yak herders, as any change in their pastoral grounds is bound to effect adjusting to newer herding breeds and would throw up the challenge of departure from the traditional practices of yak herding.

Thus, it becomes evident that of late with the emerging trends and effect of climate change, the yak herders, the Brokpas of the high Himalayas in Arunachal Pradesh are increasingly getting affected in terms of continuation of their traditional survival mode or practiced livelihood which is creating wider economic and social ramifications for these pastoralist societies that have thrived on yaks (*Bos grunniens, botanical name)* as the totemic animal of the high Himalayas and as an infusing factor in nearly every sphere of life in the sustenance of this region's economy for generations.

These emerging trends gets further substantiated by the narration of some case studies which has been periodically carried out by various research groups, NGOS and different organizations in the area. In this context particularly a report by Mongabay-India has been included in the present study where specific case studies had been done by researchers and journalists as well as bloggers to bring out the impact of climate change in the area and upon the Yak population. The study is noteworthy and factually brings out the emerging footprint of climate change on the life and livelihood of the Brokpa people. To quote "According to Brokpa Tsering, the gaonburha (village headman) of Lubrang, a yak pastoralist community in the West Kameng district of Arunachal Pradesh, snowfall has significantly decreased in the area during the last couple of years. Two decades ago, at the time of the *Losar* festival, held in February, the villages were mostly covered in at least two feet of deep snow. But now no snow is to be found around at that time of the year" (Bhattacharya 2020).

The sign of rising temperature is also apparent in the form of different kinds of growth of flora and fauna which were little known to the area a decade ago. One such effecting factor is the appearance of the pest leech which are otherwise typically found in the tropical forest are now becoming part of these high altitudes grazing grounds. It is important to highlight that during the last few years by the onset of summer, the entire landscape gets infested by leeches making it difficult for the grazing animals. These pests which are typically found in the warm foothills were unknown in these areas of high altitude and cold weather conditions. Such pests are becoming a threat to the yaks and the young calves particularly affecting their grazing grounds which in turn have started affecting the health and longevity of the yaks and the livelihood dependency of the yak herders, the Brokpa's.

Besides these, there are other perceptible signs too with a change in the temperature range the nature and growth pattern of the flora is marking a difference in the area. Prominent among these are the upslope growth and spread of warm-climatic plants like the rhododendron that is encroaching upon pastures which are the source of yaks feed, particularly the local variety Paisang leaves (*Quercus griffithii;* a kind of oak) which is considered as yak's favourite winter food. Additionally, the rising temperature is leading to a fall in the water level and also drying up of high-altitude wetlands (HAW) or Jheels (in local terms) which have been traditionally the sole source of water for yaks and yak herders moving around the mountain top pastures (Fig. 12.3).

Fig. 12.3 Brokpa settlement and grazing grounds and daily chores of a Brokpa household of Lubrang in the West Kameng district of Arunachal Pradesh. *Photo source* https://india.mongabay.com/2020/01/climate-change-threatens-traditional-way-of-life-of-brokpa-h

In the face of these difficulties, the number of Brokpa families practising the transhumant yak-rearing system has started to drastically plunge within two decades falling from a total of 800 households in 1994 to around 200 households by 2014 in Arunachal Pradesh (Norbu 2015). Along with these of-late an emerging trend in these remote areas has started to take shape in the form of exposure to the audio-visual media particularly social media bringing in consumerism in the form of consumer durables making inroads in these areas particularly among the new generation. Once considered one of the richest communities among the Monpas, the nomadic tribe, who by their nature of livelihood and cultural tradition preferred to live in isolated locations up in the highlands in the Arunachal Himalayas are now facing an existential threats both from outside in the form of climate change on one hand and from the changing cultural trends from within their social structure on the other. It is therefore not unusual to observe migration and declining trend of the age-old occupation of grazing yaks as preferred livelihood option particularly among young generation of these pastoralist tribes and the rising tendency and inclination to change from their traditional yak herding to white collar jobs, and different fields of business to earn fast money is becoming apparent among the new generation of the Brokpas of the area.

In conclusion, the findings of the study brings out very pertinently the fact that highlanders in the isolated region of the Arunachal Himalayas have not been spared from the threat posed by climate change, which is emerging as the most critical problem facing human societies worldwide today. This in turn has brought in a major challenge to the Brokpas' long-standing cultural occupation of grazing yaks in fixed shifting patterns is the widening temperature range and altitude. The climate of Arunachal Pradesh can be categorized as tropical, subtropical, temperate, and alpine based on altitude levels where tropical climate extends over 80–900 m above mean sea level, subtropical climate falls from 900 to 1800 m, above mean sea level, temperate climate lies between 1800 and 3500 above mean sea level and alpine zone range from 3500 m above mean sea level. The Brokpas rely on the mountain pastures above 3000 m for herding. However, as the grazing fields are shrinking due to changes in local weather conditions creating a scarcity of alpine fodder upon which the Yaks survive thereby creating a resonating challenge to the traditional Brokpas of Tawang district, which mostly inhabit the last village of Arunachal Pradesh along the Indo-China border. Given the present situation, many of the Brokpa families today tend to climb to higher altitudes as their yaks are unable to sustain in the warm climatic conditions in the lower altitudes which were till recently favourable habitat for the yaks and their herders. what complicates the situation further is such climbs run the risk of venturing in to the marked security area or near the border as many of the traditional herders are unaware of the international demarcation and the nature of boundary drawn they move unknowingly with their yaks to these areas in search of higher pastures and accidently venture towards the Line of Actual Control (LAC)) and often end up entering the restricted area of the Army facing trespassing charges for security reasons (Sangno 2023).

The following graphs, which clearly show the emerging trends of temperature and precipitation change in Tawang Town over the last 24 years, from 1979 to 2023,

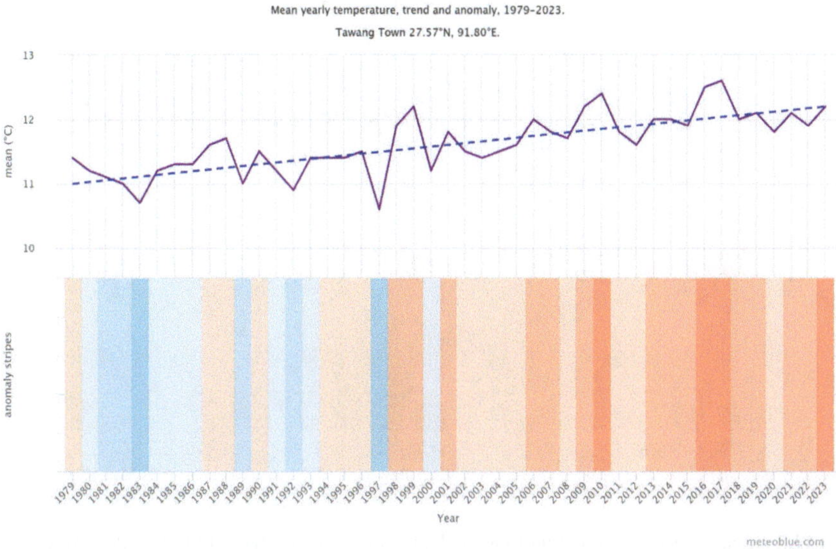

Fig. 12.4 Mean yearly temperature trend and anomaly. *Source* Metablue.com: https://www.meteoblue.com/en/weather/week/tawang-town_india_1254832

make it clear that these nomadic herdsmen are facing a serious threat from climate change. The data of the graph has been compiled and sourced from the report of the climate change assessment cell of Arunachal Pradesh (Fig. 12.4).

The top graph shows an estimate of the mean annual temperature for the larger region of Tawang Town. The dashed blue line is the linear climate change trend. When the trend line is going up from left to right, the temperature trend is positive and it is getting warmer in Tawang Town due to climate change. When horizontal, no clear trend is seen, and if it is going down, conditions in Tawang Town are becoming colder over time. In the lower part, the graph shows the warming stripes. Each coloured stripe represents the average temperature for a year—blue for colder and red for warmer years (Fig. 12.5).

The top graph shows an estimate of mean total precipitation for the larger region of Tawang Town. The dashed blue line is the linear climate change trend. When the trend line is going up from left to right, the precipitation trend is positive and it is getting wetter in Tawang Town due to climate change. When it is horizontal, no clear trend is seen, and if it is going down, conditions are becoming drier in Tawang Town over time. In the lower part, the graph shows the precipitation stripes. Each coloured stripe represents the total precipitation of a year—green for wetter and brown for drier years.

Though in both the graphs, sharp changing trends of the climate become evident over the Tawang region only, it can be very well assumed and assessed that similar emerging trends are also having similar patterns in the higher reaches of the Arunachal Himalayas, affecting the Brokpa habitat, and is consequently changing the economic

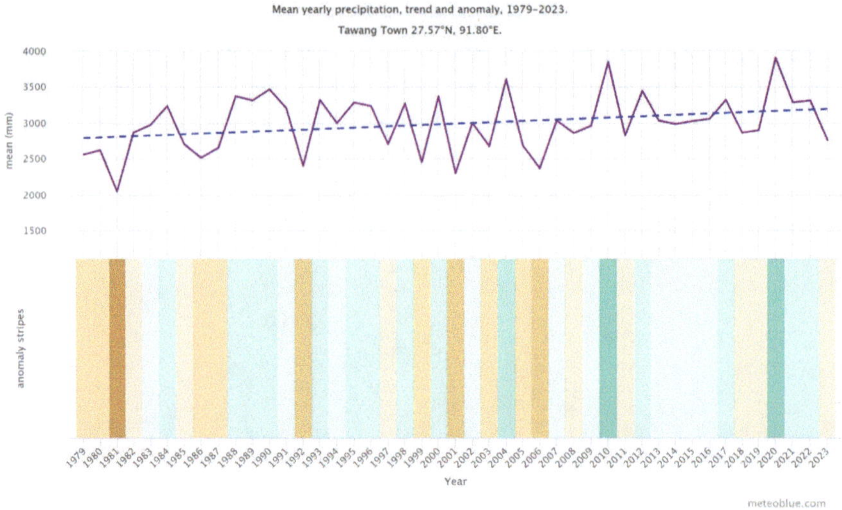

Fig. 12.5 Mean yearly precipitation trend and anomaly. *Source* Mebo.com: https://www.meteoblue.com/en/weather/week/tawang-town_india_1254832

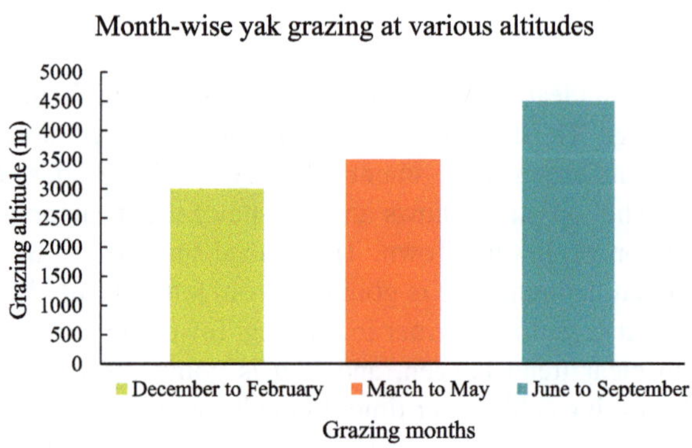

Fig. 12.6 Month-wise yak grazing at various altitudes. *Source* Singh et al. (2018)

and cultural heritage of the Brokpas at large. Further, the effect of global warming on the traditional practices of the Brokpas is becoming a serious matter of concern as the yaks which are the primary source of their livelihood are dying frequently being affected by the increasing heat and prolonged summer seasons (Ajum 2023; Bhattacharya 2020). Less availability and access to the pasture land of high mountains for herding further accentuates the situation. Such emerging trends can be authenticated by referring to the state climate change cell report, the study of which was initiated from 1980 to 2019, and the collected data depicts that the West Kameng district started to experience a decrease in annual rainfall at a rate of 37.44 mm per year. This has impacted the grazing patterns of the yaks and the movement of the Yak herders. Figure 12.6 given provides an idea of the nature of altitude change of the yak and the herdsmen in terms of their migration patterns.

In the traditional seasonal migratory system, during the four months from June to September, yak grazing usually takes place in the alpine pastures located at an altitude of more than 4500 m above sea level. Between March and May, yak herders migrate with the yaks to high pastures, and between October and November, they return, to the grazing grounds in mid-hills at elevations ranging from 3500 to 4500 m above sea level. The period from December during the beginning of the winter season to the end of February is dedicated to moving between grazing grounds below 3000 m above sea level.

Biologically the yaks are cold adaptive animals with an ideal temperature of − 10 °C and thrive at an altitude of 2000–5000 m above sea level (Bhattacharya 2017). With temperature being one of the most significant environmental factors along with limited water resources, the health, welfare, and productivity efficiency of yaks is emerging as a potential problem. The alterations in metabolic processes and physiological responses to temperature rise may be detrimental to the health and production of yaks (ICAR-National Research Centre on Yak 2022, State Climate Change Cell-Arunachal Pradesh, n.d.). They can withstand temperatures of − 35 °C but struggle if the temperature rises above 12–13 °C (Mukherjee 2024). As its production is regulated by weather patterns (Berger et al. 2015), therefore changing climatic conditions such as temperature rise consequently affect the livelihood system of the Brokpa community which has traditionally been dependent on yaks since time immemorial. This indicates the hitches among the Brokpas in continuing their traditional livelihood practices because of the emerging climate change conditions, forcing them to shift and adopt other sources of livelihood, consequently posing a transition among the Brokpa community to preserve their ancient traditions and culture. The overall situation becomes more forbidding with the assessment of the rural habitat and disaster vulnerability of the state to climate change, with a sizable proportion of the population being below the poverty line, having less access to drinking water, electricity, sanitation, and housing facilities (State Climate Change Cell-Arunachal Pradesh, n.d.).

12.5 Conclusion

Even though the changing climate is a global problem, certain communities have suffered more than others. For the tribal community as a whole, this is particularly true, as their way of life depends on a complex relationships with the environment. In addition to anthropogenetic activities like urban sprawl and shrinking agricultural land, the new trends of climate change—such as unprecedented disasters, rising temperatures, and drying water sources are becoming cause of a situation which will eventually lead to the loss of traditions and alter the ancient heritage if mitigations are not taken into account. The Brokpas, highlanders, or yak herders of Arunachal Pradesh are one such group that stands out as a community that continues to practise transhumant way of life. They are entirely reliant on yak herding and manufacture a variety of goods that serve as their primary source of income. Growing challenges, such as the low yield of traditional dairy products on the one hand and the desire for a different living situation on the other, are causing younger generations to abandon the traditional practice of rearing yaks, a trend that can be attributed to as a result of globalization. Thus, the persistence of the Brokpa culture is becoming a more restricted phenomenon, impacted by the erratic weather condition and uncertainty leading to migration and adoption of other occupation among the Brokpa Community.

The emerging situation therefore needs a balance between modernity and continuity. While it is true that change is inevitable, what matters is the pace and measured degree for a sustainable growth without changing the ecosystem and the social fabric of the area. The younger Brokpa generation needs to be made aware about the value of maintaining their culture, particularly as Highlanders of Arunachal Pradesh, their cultural significance holds immense value as no other tribe has such practices. Yak herding among the Brokpas is an inherited occupation that is passed on to the next generation. They continue to raise yaks using antiquated techniques passed down from a tradition that is deeply ingrained in their culture, which makes them unique in themselves. However, the concern is the impending threat of a dwindling Brokpa population in the area and conserving and continuing their traditions.

While the changing physical and cultural dynamics and the concern of the Brokpa herders of possibility of losing their traditional way of life looms at large what becomes very pertinent to mention is the changing social dimensions among the Brokpas and the new trend of gender relations. While this is an accepted and proven fact that in the face of changing climatic condition, women in the rural areas be farmers or herders bear the impact of risk of losing their means of subsistence and access to resources of fuel fodder and water to run the family hearth the most and consequently are burdened with managing the household on one hand and support in the labour for the economic works as well indicating gendered division of labour created more as social directives and practice and not based upon the biological need. But it is noteworthy to mention that in the traditional social structure of the Brokpas there was no practice of gendered division of labour. But of late gender dynamics are now changing with impact of climate change and changing livelihood choices. As yak-rearing is appearing to become more and more difficult to sustain both from the physical and social aspects, a good number of the Brokpa families are opting to settling down as farmers, traders, and daily wage labourers, and are getting exposed to consumerism and money economy, many for the first time ever. One can not deny that as part of the formal monetized economy that is affecting the world in its own shuttle way, a change undoubtedly inevitable but what needs to be understood and planned is the ways and means to sustain their traditional cultural practices along with these shifts so that the rich traditional and cultural heritage and the distinct uniqueness and identities of the Brokpa community of Arunachal Pradesh do not disappear into oblivion.

References

Azad AK, Jha SK, Maiti S, Kadian KS (2023) Brokpa's organic yak farming in Arunachal Pradesh: an ethnographic study. Int J Curr Microbiol App Sci 12(4):230–234. https://doi.org/10.20546/ijcmas.2023.1204.026

Ajum B (2023) Brokpa tradition dying slowly. The Arunachal Times. https://arunachaltimes.in/index.php/2023/03/13/brokpa-tradition-dying-slowly/

Bhattacharya BK (2017) The Brokpa Yaks: a dying breed? The Diplomat. https://thediplomat.com/2017/12/the-brokpa-yaks-a-dying-breed/

Bhattacharya BK (2020) Climate change threatens the traditional way of life of Brokpa herders in Arunachal Pradesh. https://india.mongabay.com/2020/01/climate-change-threatens-traditional-way-of-life-of-brokpa-herders-in-arunachal-pradesh/

Berger J, Schaller GB, Cheng E, Kang A, Krebs M, Li L, Hebblewhite M (2015) Legacies of past exploitation and climate affect mammalian sexes differently on the roof of the world the case of wild yaks. Sci Rep 5(1):8676

Blench R (2001) You can't go home again: Pastoralism in the new millennium. ODI, London

Census of India (2011) National population register & socio-economic survey

Critical Ecosystem Partnership Fund (2005) Eastern Himalayas Region. WWF-US, Asia Program

Das PJ, Kour A, Deori S, Begum SS, Pukhrambam M, Maiti S, Sivalingam J, Paul V, Sarkar M (2022) Characterization of *Arunachali* Yak: a roadmap for pastoral sustainability of yaks in India. Sustainability 14(19):12655. https://doi.org/10.3390/su141912655

Developmental Statistics of Arunachal Pradesh (2020) Directorate of economics & statistics government of Arunachal Pradesh, Itanagar

Dorjee T (2015) Yak pastoralist (Brokpa) of Arunachal Pradesh: mobility and institutional arrangement regarding regulation of seasonal use. Int J Innov Res Dev 4(9):27–33

Haridassan M (1989) Agroforestry systems for the cerrado region of central Brazil: potentials and constraints. J Meteorol Agrofor. An e-publication by the World Agroforestry Centre

Hegde SN (2002) Orchid Wealth of India. Indian J Nat Sci B63(3):229–244

ICAR-National Research Centre on Yak (2022) Annual report. Dirang, Arunachal Pradesh, India

International Panel on Climate Change (2014) Climate Change 2014: synthesis report. In: Core Writing Team, Pachauri RK, Meyer LA (eds) Contribution of Working Groups I, II, and III to the Fifth Assessment Report of the Intergovernmental Panel on Climate Change. IPCC, Geneva, Switzerland, 151pp

Kalita K (2024) Arunachal's cross-breed Dzomo is a promising choice for yak herders. Times of India, Guwahati. https://timesofindia.indiatimes.com/city/guwahati/arunachals-cross-breed-dzomo-a-promising-choice-for-yak-herders/articleshow/111097370.cms

Kollmair M, Gurung GS, Hurni K, Maselli D (2005) Mountains: special places to be protected? An analysis of worldwide nature conservation efforts in mountains. Int J Biodiv Sci Manage 1(4):181–189

Kumari M, Rimung T (2022) The Brokpas of the Forest. Sanctuary Cub 42(7)

Livestock Census Report (2012) Department of Animal Husbandry, Dairying and Fisheries, Ministry of Agriculture, Government of India, Krishi Bhawan, New Delhi

Livestock Census Report (2019) Department of Animal Husbandry, Dairying and Fisheries, Ministry of Agriculture, Government of India, Krishi Bhawan, New Delhi

Meteoblue, Awindy.com Company https://www.meteoblue.com/en/weather/week/tawang-town_india_1254832

Mukherjee R (2024) The happy days are now just nostalgia. United Nations Development Programme. https://www.undp.org/india/happy-days-are-now-just-nostalgia

NOAA National Centers for Environmental Information (2024) Annual 2023 Global Climate Report, Retrieved on 22 June 2024 from https://www.ncei.noaa.gov/access/monitoring/monthly-report/global/202313

Norbu L (2015) A geographical analysis of transhumance among the Monpa Tribe of Arunachal Pradesh

Sangno A (2023) Brokpas is caught between modernity and marginalization. The Arunachal Times.
SAPCC (2011) Sikkim action plan on climate change (2012–2030). Government of Sikkim
Singh RK (2009) Indigenous knowledge of yak breeding and management by the Brokpa Community in Eastern Himalaya, Arunachal Pradesh
Singh RK, Sureja AK, Maiti S, Tsering D (2018) Grazing and rangeland management: trans-human adaptations by Brokpa community in fragile ecosystems of Arunachal Pradesh
State Climate Change Cell-Arunachal Pradesh (n.d.) Climate Change Risk Assessment in the Indian Himalayan Region, Arunachal Pradesh
Tsering R, Bora L, Ando K, Kosaka Y (2010) The Brokpa and Their Social Development: the Work of M/s Dunkarpa Welfare Association at Dirang Circle of West Kameng District, Arunachal Pradesh, India 11:191–195
Tshe-ring K, Eklabya S, Nakul C, Arun S (2010) Climate change vulnerability of mountain ecosystems in the Eastern Himalayas. Synthesis report
Tshe-ring L et al (2009) Yak husbandry in Bhutan. Unpublished report of the Animal Husbandry Department, Thimpu, Bhutan
Turrentine J, Denchak M (2021) What is climate change? The lowdown on the earth's central environmental threat. Natural Resources Defense Council, New York, United States. https://www.nrdc.org/stories/what-climate-change#whatis
United States Environmental Protection Agency (2024) Climate impacts on society. https://19january2017snapshot.epa.gov/climate-impacts/climate-impacts-society_.html
World Bank Group (2021) What is climate change? Climate change knowledge portal. For development practitioners and policy makers. https://climateknowledgeportal.worldbank.org/overview

Chapter 13
Springshed Management: A Present-Day Need to Revive the Life Line of Mountain Inhabitants

Ajay Vashisht

Abstract Recognizing the importance of springs, adopting practices that could revive the degraded biodiversity by shielding the mountainous ecosystem is crucial to ensure its long-term existence and the well-being of millions of inhabitants. The essential steps required to reach the cause(s) of springshed deterioration are compiled with suitable examples. Based on the exact cause, relevant engineering measures or revival strategies applicable to the selected springshed are finalized. Therefore, in this chapter, the pressing need for adopting the springshed management is justified for protecting and rejuvenating the springs.

Keywords Indian Himalayan Region · Springshed management · Engineering measures for springs' revival · Spring rejuvenation

13.1 Introduction

Millions of inhabitants in the mountain ranges across India depend upon springs. Springs are the visible appearance of groundwater on the earth's surface. Springs exist in the Indian Himalayan Region, *Sahyadri* mountain range of western *ghats*, eastern *ghats*, and *Satpura* and *Vidhyas* mountain ranges of central India. Springs may be considered the epicenter for the sustainability of rivers, wetlands, and lakes that are further essential for the survival of biodiversity, agriculture, and industry. For the past century, all the regions have been facing the wrath of climatic extremes. Other than climate change, deforestation and un-scientifical developments in these fragile ecosystems have created an imbalance. Springs are affected majorly. Either these are dried up and become seasonal, or their discharge rates are reduced drastically. Therefore, springshed management is the present-day need by which the springs

A. Vashisht (✉)
Department of Irrigation and Drainage Engineering, College of Agricultural Engineering and Post Harvest Technology, CAU (Imphal), Ranipool, Sikkim, India
e-mail: akvashisht74@yahoo.com; akvashisht74@gmail.com

S. C. Rai (ed.), *Himalaya: Mountains of Destiny*,
https://doi.org/10.1007/978-3-031-94905-0_13

will be safeguarded and rejuvenated by incorporating scientific studies and ancient wisdom.

The occurrence of springs is a natural phenomenon in which water from the elevated saturated formations emerges at some localized locations. In previous literature, it is mentioned that the geology and topography of the region are the decisive factors responsible for the spring oozing out locations (Dar et al. 2022). However, other than the topography of the region, it is the hydrogeology of the spring catchment that is mainly responsible for the existence of a spring. Though the majority of springs discharge at a low rate, however, their uninterrupted continuous supply makes them able to fulfill the needs of millions of inhabitants. The temperature of the spring water varies with the location of the spring catchment concerning the earth's core. There exist numerous hot springs in the Indian Himalayan region.

13.2 Spring Management

It is a natural human tendency to campground near a perennial source of water. However, with the increased population and consequently, the construction of buildings and roads (mostly unplanned) the catchments of these life-supporting springs are severely deteriorated. For the past half-century, various organizations (including state government, central government, and non-government organizations) have been making efforts to prepare databases of available natural resources so that these can be preserved. Their efforts have changed the scenario at a few locations. Still, we have to cover a long distance back to rejuvenate the complete scenario. For this purpose, the first step is to map the springs and to understand their catchments through reliable and lesser time-consuming methods.

Initially, the majority of the organizations concentrated their efforts on geotagging the springs by noting their latitudes and longitudes. With time, it was realized that geotagging does not provide sufficient information to be considered as a basis for making a policy for the region. Additional parameters were added in due course such as seasonal discharge rate, its quality in terms of electrical conductivity (EC), and pH. A few springs at some strategic locations were studied by various researchers in detail. However, the observed characteristics of such springs cannot be applied to other springs even if these are oozing out in the same region. Because of this reason, it is an extremely difficult and time-consuming task to collect the complete details of all the springs within a short period.

13.2.1 Classification of Springs

First of all it is necessary to understand the definition of spring. When the Earth oozes out water naturally from a specified location that takes its supply from the water stored in the catchment it is termed spring. Next, why there is a need to classify the springs?

All springs behave distinctively in terms of their annual discharging patterns as well as their oozing-out ways. Monitoring of springs is the first step to understanding their annual discharging behavior. How and when to monitor the spring so that maximum information can be exploited with minimum data is only possible if we have an understanding to classify the spring. A spring, if identified in the wrong classified group may lead to a waste of efforts that will be performed for its management. Moreover, classification is essential for recognizing its catchment contributing to its discharge rate. Springs can be classified into various groups based on the following characteristics:

(i) *Flow type*: Visually, water comes out of the earth in two ways, i.e., (i) intermittent shoot-out of non-gravitational springs; and (ii) free flow or gravitational springs. Non-gravitational springs are either volcanic springs or fissure springs and are usually thermal. The free flow can further be classified into (i) free flow dispersion gravitationally (usually termed 'dhara') and (ii) free flow accumulation due to lack of gravity outlet on the surface (usually termed 'naula').

(ii) *The average discharge rate of spring*: In a region, springs can be classified based on their mean discharge rate. However, such a type of classification cannot be applied in every region. For example, Meinzer (1923) classified the springs of the United States based on their mean discharge which cannot be applied to the Indian springs.

(iii) *Hydrogeological conditions of the spring catchment*: Depression springs are oozed out at the intersection of the ground surface and the water table. When a water-bearing formation overlies an impermeable or semi-permeable formation, water oozes out from the intersection point on the ground surface. Such a spring is termed a contact spring. The topographical intersection of water-bearing fractures or faults in the Earth's crust leads to the origin of fracture/fault springs. The springs oozing out from the limestone or dolomite rock formations are termed karst springs.

(iv) *Lithology*: As rainwater enters and comes in contact with the rock formation of the spring catchment, its composition changes, and the concentration of the composition depends upon the residence time of rainwater in the formation before oozing out as spring. Hence, springs can be easily classified based on the rock type of the spring catchment.

(v) *Flow continuation*: Springs may be perennial or seasonal. Certain literature uses the term 'intermittent' for the free-flowing (gravitational) springs. Technically, it is wrong to use this term for the springs since only gysers discharge intermittently.

(vi) *Temperature*: Based on the temperature of the water, springs can be classified as superheated, hot, warm, normal, and cold.

(vii) *Chemical composition*: The chemical composition of the water that oozes out of the spring orifice is a criterion for classifying and comparing the springs.

(viii) *Discharge variability ratio*: Other than exceptions, springs do not discharge steadily throughout the year. The available discharge rate of the spring during

Table 13.1 Discharging pattern of a perennial spring in Uttarakhand Himalayas

Year	Rainfall (mm)	Yield (l)	Maximum (l)	Minimum (l)	Maximum/Minimum ratio
1999	939.70	4,915,253	34,833.60	6912.00	5.04
2000	1334.40	6,494,314	43,632.00	8409.60	5.19
2001	719.10	2,924,539	12,355.20	6768.00	1.83
2002	1254.60	4,681,837	28,800.00	7862.40	3.66
2003	1173.80	3,931,206	26,582.40	6163.20	4.31
2004	1174.70	4,590,619	31,291.20	3873.60	8.08
2005	1386.90	6,904,546	42,984.00	7675.20	5.60
2006	897.00	3,556,373	28,800.00	5976.00	4.82

Source Vashisht and Sharma (2007)

the lean season is an important criterion for understanding its reliability. Alfaro and Wallace (1994) expressed the ratio of high flow that exceeded 10% of the time to low flow that exceeded 90% of the time and designated this ratio as the discharge variability ratio. Vashisht and Sharma (2007) have used a maximum-to-minimum flow ratio for expressing the reliability of a spring (Table 13.1). The ratio of the maximum annual yield to the minimum of a spring describes its discharge rate uniformity. The springs discharging uniformly present a lesser ratio of maximum to minimum yield and vice-versa.

As explained earlier the classification of spring gives an insight into various parameters that are responsible for the existence of spring in a region.

13.2.2 Mapping of Springs

Spring identification within the area under survey is the prime objective of the level I field survey. A discharge rate measuring device, GPS, and a camera are sufficient for such a type of survey. In the level I field survey, the latitude/longitude and altitude of the springs are monitored along with the springshed characteristics such as its slope and persisting vegetation. As an example, Table 13.2 presents the preliminary information collected during the spring identification before starting the intensive data collection.

In the level II survey, physio-chemical, biotic, and isotopic characteristics of the spring water are assessed. Moreover, this information leads us to understand the recharge environment of the catchment, the residence time of infiltrated rainwater in it, and ultimately its lithology. A long-term monitoring of the selected springs comes under the level III survey.

Table 13.2 Springs and their general characteristics

S. No.	Name of spring	Latitude/ longitude	Altitude (m)	Slope	Type of forest trees/bushes	Springshed surface coverage
1	Kosi	N29°38.373′ E79°37.128′	1295	Steep	The thin density of Pine trees with thorny bushes and local grasses	A government institute covering a certain area of the springshed
2	Kosi Katarmal	N29°38.323′ E79°36.861′	1327	Gentle	Pine and Oak trees forest	A government institute covering a certain area of the springshed
3	Barsimi	N29°34.321′ E79°37.843′	1463	Gentle	Pine trees at the upper catchment with bushes and local grasses at lower elevation	The upper catchment of the springshed is under dense step farming
4	Patwadangar	N29°20.359′ E79°26.810′	1525	Gentle	Forest trees like Pine and Oak. *Kalabansa, Bichchu ghas,* Lantana Cammara, Solanum Sapplings, and Parthenium Sapplings are the local grasses growing there	Approximately 50% of the catchment area is under building structures
5	Kalimath-B	N29°37.630′ E79°40.050′	1711	Steep	Dense pine forest without any local vegetation. The soil surface is covered with Pine leaves only	No

(continued)

Table 13.2 (continued)

S. No.	Name of spring	Latitude/ longitude	Altitude (m)	Slope	Type of forest trees/bushes	Springshed surface coverage
6	Kalimath-A	N29°38.827′ E79°40.115′	1715	Steep	Pine trees at the top of the catchment. Grasses are growing on the steep slopes	No
7	Chaubatia	N29°37.284′ E79°27.437′	2039	Gentle	Dense forest having Oak and Deodar trees	Military Cantonment covers the majority of the area without cutting the forest trees

Source All India Coordinated Research Project on Groundwater Utilization, Pantnagar Center (2008)

13.3 Data Collection

To investigate the response of spring to changes in the environment, data is collected. Usually, the parameters like spring discharge rate, pH, EC, and temperature are studied. Monitoring the discharge rate (daily, fortnightly, monthly, and seasonally) is the basic step for studying the behavior of springs. For intermittent monitoring, manual methods are preferred. However, continuous monitoring requires sophisticated devices. A simple bucket of known volume and a stopwatch are sufficient to measure the discharge rate of a free-flowing gravitational spring. However, for a spring without a gravity outlet, a different strategy is followed to measure the discharge rate. Though manual monitoring is a widely accepted practice it has certain limitations. Spring having very high discharge rates cannot be monitored accurately. Frequent monitoring of springs at different locations within a specified period is a labor-intensive task. Unfavorable weather conditions on the data monitoring day make the task difficult and the situation further worsens if spring is remotely accessed.

To overcome the above-mentioned difficulties faced in manual monitoring, instrumentation was required. Various organizations installed flumes and other devices along with sensors to monitor the spring discharge. In this direction, a continuous monitoring device was conceptualized, designed, fabricated, and installed in the field (Fig. 13.1). Different portions of the springshed contribute to spring discharge incorporating different time lags. Frequent short-term variations in the discharge rate of the spring occur in response to intermittent rainfall events specifically during the pre-monsoon, post-monsoon, and during winter months. Manual data collection (even one time a day) may fail to monitor this short-term variation in discharge rate. In the present-day scenario, it is recommended to monitor the spring discharge continuously to avoid any skipping of such short-term events.

Fig. 13.1 Automatic spring discharge measuring device patented by Ajay Vashisht. *Source* https://patents.google.com/patent/WO2023026294A1?

Continuous monitoring of rainfall is an essential parameter while managing the springshed. Other than rainfall, temperature, humidity, wind speed, and wind direction are other important parameters frequently required to perform various studies. For this purpose, an automatic weather station should be installed as per the recommendations of IMD.

Considerable instrumentation of the springshed warrants regular visits to the site for inspection. Data should be collected systematically and compiled neatly. Any unusual event (like an earthquake, excess runoff, or landslide) that happened in or near the data collection site should be noted in detail. Such information sometimes becomes essential during data analyses and while concluding. Before concluding analyzed data should be interpreted with utmost care and using different techniques/ methods.

13.3.1 Methodology of Springshed Management

A springshed may be defined as a catchment area from where rainwater after infiltrating leads to the spring orifice(s). Springshed management is an approach that ensures reliable water supply from the springs and restoration of the ecosystem. Since this approach is based on scientific and social principles, the involvement of inhabitants is a must. The methodology of springshed management is developed based on the field experience gained by various government and non-government organizations. Originally, this methodology was developed by the Central Himalayan Rural Action Group (i.e., CHIRAG) and it includes nine-step interventions. On this proposed methodology, a project was initiated in the year 2008 in which the Pune-based not-for-profit organization named Advanced Center for Water Resources Development and Management (i.e., ACWADAM) was also a technical partner. It becomes possible to further refine this methodology only when it is applied in the field. Finally, an eight-step procedure was finalized (Table 13.3) that was accepted by NITI Aayog as protocol (NITI Aayog 2018). Firstly, the Rural Development Department, Government of Sikkim applied the springshed methodology under the name of the 'Dhara Vikas Program'. Later on, NITI Aayog recommended this methodology for the whole country.

As mentioned previously springshed management is a people-centric approach and it encompasses the compulsion to involve residents of the spring catchment for the sake of developing a sense of ownness. *The mismanagement practices of deteriorating and degrading an ecosystem knowingly or unknowingly do not warrant any qualification, expertise, or involvement of the inhabitants or government agencies.* On the contrary, the practices to revoke the complete ecosystem warrant the implementation of engineering techniques with the expertise-involvement of inhabitants and government agencies after carefully considering the hydrological and social aspects of the spring catchment.

13.4 Revival of Springs Through Agronomical and Engineering Measures

It is proposed in the literature that a combination of agronomical practices and engineering interventions are required to be implemented in springsheds. However, if the spring catchment area contributing supply to springs is lying in the restricted area only engineering measures are to be adapted for springshed management. These measures aim to enhance the recharging of rainwater by increasing the time of concentration of the rainwater in contact with the infiltrating surface and to check the soil erosion.

The grasses like broom grass and vetiver can control soil erosion and enhance the infiltration rate of trapped runoff water. That is why the plantation of these grasses at specific locations is termed 'vegetative measures'. Therefore, these grasses are planted on the bunds of the fields or water courses. Other than these two, nature has blessed several other plant species in different regions. After maturity, these grasses have commercial value. Correct selection and establishment of any of such grass species for a springshed is a task that can be completed after surveying the local market. In hilly terrain, these vegetative measures trap the sediments flowing with the runoff and settle down it. The effectiveness of vegetative measures on sloppy lands is excellent and these are cost-effective and long lasting. Recently, geotextiles have gained popularity as an excellent soil conservation measure. Contour cultivation and the use of mulches are other field practices that need to be adapted for better soil and water conservation in the springshed.

The slope, soil characteristics, land use land cover of the spring catchment along with rainfall and hydrogeological conditions are the factors responsible for the selection of suitable engineering measures. The rectangular-shaped staggered contour trenches are mainly used as recharging structures in springshed (Fig. 13.2). The purpose is to break the sloppy surface, collect the runoff water, facilitate infiltration

Table 13.3 The eight-step methodology

Steps	Sub-steps	Ultimate goal
First: Comprehensive mapping of springs and springshed	1. Background information on the identified spring catchment area is collected 2. A reconnaissance survey of the springshed is completed 3. Springs are mapped, and aimed data is collected 4. The springshed area is delineated	– Delineation of water tower – Comprehensive map of springs
Second: Setting up a data monitoring system	1. Data collection concerning why we are collecting, who will collect, from where to collect, what to collect, and how to collect 2. Compilation of collected data and its management 3. Development of any software or mobile application after analyzing the data 4. Sharing of results with the community	– Installation of rain gauge station – Compilation of hydrometeorological data in Excel
Third: Understanding social and governance aspects	1. Existing institutions and management systems are to be analyzed using questionnaire surveys, group discussions focused on selected issues, interviews with the key informants, and communication and dialogue with the community and public policymakers	– Management of springshed and springs by the lead community – Questionnaire tool
Fourth: Development of a hydrogeological map	1. Obtain a geological map of the area from the concerned department 2. Observe geology, adjoining topography, and landscape during the transect walk. Note latitude, longitude, elevation, spring location, geological observations, and other site-specific measurements 3. Create a base map on a convenient platform	– Compilation of hydrogeological data in Excel format – Development of a base map of the area
Fifth: Creating a conceptual hydrogeological layout of springshed	1. Create a geological map based on the transect walk 2. Preparation of a draft of the cross-sectional layout	– Development of the geological map of spring and springshed – Development of a cross-sectional layout

(continued)

Table 13.3 (continued)

Steps	Sub-steps	Ultimate goal
Sixth: Classifying spring types, identify mountain saturated formation and recharge area	1. Identify spring(s) and formations of the spring catchment 2. Delineate recharge area	– Example of spring types – Outline of recharge area
Seventh: Developing springshed management protocols	1. Hydrogeological inventory for springsheds 2. Negotiable and non-negotiable land use and land cover change 3. Institutional mechanism 4. Conservation and intervention, measures of recharge and discharge area 5. Develop operational and maintenance guidelines	– Revival activities using voluntary labor – Recharge structures
Eighth: Measuring the impact of spring revival	1. Impact study 2. Continuous monitoring	– Comparison between the positive change

Source ICIMOD, Kathmandu, Nepal

by increasing the time of concentration of collected runoff, and ultimately trap the sediments. In low-rainfall regions, continuous contour trenches are preferred over staggered trenches.

Terracing in hilly areas is a well-recognized practice to control runoff and conserve soil moisture. Depending upon the crop tolerance limit, the bund height of the terraces can be increased to store more rainwater and allow it to infiltrate and recharge the spring catchment. Moreover, the excess water from one terrace is not allowed to drain but is guided to move downstream terrace only (Fig. 13.3). In hilly terrain, the occurrence of natural depressions on sloppy land is a common phenomenon. After

(a) Snapped near Almora, Uttarakhand

(b) Snapped near Resum (or Poison) Lake, Rinchenpong, Sikkim

Fig. 13.2 A view of staggered contour trenches at different locations suitable modifications, these depressions attain the capability to store huge quantities of water to promote percolation (Fig. 13.4).

Fig. 13.3 Terrace cultivation with raised bunds. Snapped near Ranipool, Sikkim

Fig. 13.4 A modified natural depression in the hilly terrain of Ranichauri village, Uttarakhand

Other soil and water conservation measures such as brushwood check dam and loose boulder check dam are site-specific interventions for water management and control of sediment loss.

13.5 Importance of Hydrological Analyses for Springshed Management

The Greater Himalayas are fulfilling the water demand of Asian countries and are considered the water tower of Asia. From the past century, climatic variability in the Himalayan region has become acute due to global warming and as a result, the region is experiencing extreme hydrological events such as floods and droughts. Probably, the rainfall pattern has responded greatly to this climatic variability as the low-intensity longer-duration rainfall spells have changed to high-intensity smaller-duration spells, and the average number of rainy days in the winter months has reduced drastically. It is a well-known fact that the infiltration opportunity time for rainwater received from extreme events is considerably less, or in other words, lesser recharging of the formations in the springshed responsible for feeding the springs. With every coming year, the average flow rate of perennial springs is declining and the springs fed by smaller catchments have become seasonal or completely dried.

Various measures have already been explained earlier to cope with this situation. How the spring is responding to the adapted water management practice is only possible when we perform hydrological analyses of the data monitored before and after adopting measures. Moreover, various criteria are available in the literature for judging the impact of adopting measures (Vashisht 2017). These are (i) the time-lag of rainwater from the infiltration location to the spring outlet; (ii) the quantity of rainwater stored in the formation(s) (Vashisht and Bam 2013); (iii) the ratio of maximum flow rate to minimum flow rate for every hydrological year; and (iv) by predicting the discharge rate of the spring and its yield for the hydrological year using artificial intelligence or analytical techniques (Vashisht 2016).

13.6 Capacity Building of the Stakeholders

The springshed management program is aimed revive the springs and revoking the degraded ecosystem of the hilly regions of India. It is mentioned earlier that other than the geohydrologists, groundwater engineers, ecologists, and social scientists, stakeholders are the backbone of this proposed program. There are three aspects for which the various functionaries involved in the springshed management program work, i.e., concept, approach, and strategies. The core of the springshed management program indicates that it is a blend of traditional knowledge/practices with present-day technologies to rejuvenate the whole biodiversity. The stakeholders know their traditional knowledge and customs, however, if we want to revive the whole system comparatively within a shorter period, they should have an awareness of modern technologies being used in the springshed. It is necessary to prepare the training material in the local language and all the methodologies should be written in layman's wording for better understanding. As far as possible, certain thumb rules need to be explained in the training manual to understand the changes that occur after adopting springshed management measures.

13.7 Conclusion

In this chapter, the eight-point springshed management methodology has been explained with relevant studies. All aspects of this methodology are discussed minutely to understand their roles in the way to manage the springshed. A few constraints of this methodology are also discussed. Understanding the importance of the monitoring of intensive temporal hydrological data of the springs in remote locations, it has been justified its monitoring by developing suitable devices. A comparison of the extent of information that can be retrieved between the data collected through manual efforts intermittently and continuous monitoring of the springs through automatic devices is presented. In this regard, an automatic spring discharge measuring device recently developed and patented is presented in this chapter. Ultimately aim is to analyze the data for the determination of the factors responsible for the deterioration of springs and based on these, location-specific engineering measures are proposed.

References

Alfaro C, Wallace M (1994) Origin and classification of springs and historical review with current applications. Environ Geol 24:112–124

Dar T, Rai N, Kumar S (2022) Distinguishing mountain front and mountain block recharge in an intermontane basin of the Himalayan region. Groundwater 60(4):488–495

Meinzer OE (1923) Outline of ground-water hydrology, with definitions. Technical report. US Govt. Print. Off

NITI Aayog (2018) Report of working group i: inventory and revival of springs in the Himalayas for water security. https://niti.gov.in/writereaddata/files/document_publication/doc1.pdf

Vashisht AK (2016) Development of an analytical function for optimizing the capacity of spring water storage structure. Sustain Water Resour Manage 2:337–352

Vashisht AK (2017) Use of mathematical techniques for determining the reliability of springs. In: Development of Water Resources in India. Springer International Publishing, pp 113–124

Vashisht AK, Bam B (2013) Formulating the spring discharge function for the recession period by analyzing its recession curve: a case study of the Ranichauri spring (India). J Earth Syst Sci 122:1313–1323

Vashisht AK, Sharma HC (2007) Study on hydrological behaviour of a natural spring. Curr Sci 93(6):837–840

Part III
Geo-environmental Hazards

Chapter 14
Geospatial Analysis of Landslide Susceptibility Along Kangan to Kargil Highway (NH1) in the Western Himalayas Using Heuristic and Statistical Techniques

Mohd Taqi, Mohammad Ali, Wani Suhail Ahmad, and Saleha Jamal

Abstract The present study conducts a detailed landslide susceptibility assessment on the NH-1 National Highway between Kangan and Kargil in Ladakh, a high-altitude region showcasing diverse geological formations. In this study, the weight overlay method was employed to evaluate the landslide causative factors and generate landslide susceptibility maps (LSMs). The study area was categorized into four susceptibility zones: low (1.71%), moderate (24.4%), high (59.58%), and very high (14.2%). The aspect analysis revealed that regions such as Kangan, Sonamarg, and Zoji La, with north and northeast-facing slopes, are more susceptible to landslides compared to regions with south and southwest-facing slopes like Drass and Kargil. Higher elevations (over 3200 m) from Kangan to Pandras were more prone to landslides than lower elevations (around 3000 m) from Drass to Kargil. Proximity to roads and streams was another critical factor, with areas closer to the road (< 300 m) from Zoji La to Pandras being more susceptible. Most landslides (25–30) were identified between Pandras and Kangan, and 10–15 between Drass and Kargil. The largest landslide (0.65 km^2) was between Zero Point and Gumri, while the smallest (6056 m^2) was in Hardass. The regions near Zoji La were more susceptible to landslides compared to Drass and Kargil, as evidenced by Google Earth imagery. The final Landslide Susceptibility Map offers a critical tool for future planning and disaster mitigation efforts in the region.

Keywords Landslide susceptibility mapping · Weight overlay analysis · Kargil · Remote sensing and GIS

M. Taqi · M. Ali · W. S. Ahmad
Department of Geography, University of Ladakh, Kargil, Ladakh, India

S. Jamal (✉)
Department of Geography, Aligarh Muslim University, Aligarh, India
e-mail: salehajm@gmail.com

S. C. Rai (ed.), *Himalaya: Mountains of Destiny*,
https://doi.org/10.1007/978-3-031-94905-0_14

14.1 Introduction

Landslides are one of the most destructive natural hazards, posing significant risks to both human life and infrastructure. The increasing frequency and intensity of landslides, driven by factors such as climate change, deforestation, and urban expansion, underscore the urgent need for effective risk assessment and management strategies. They can occur in various environments, from steep mountain ranges to coastal cliffs and even underwater as submarine landslides. While gravity is the primary driving force, additional factors such as heavy rainfall, earthquakes, and human activities like construction and mining can trigger landslides. Human activities, particularly urban sprawl and resource exploitation, often exacerbate landslides. Deforestation and land degradation reduce soil stabilization by vegetation, while climate change increases the frequency of extreme weather events, further contributing to landslide risks. Mitigation strategies are essential to reduce the human impact of these natural disasters.

Geographic information systems (GIS) have emerged as a powerful tool in the field of natural hazard assessment and management. GIS technology enables the integration, analysis, and visualization of spatial data, facilitating a comprehensive understanding of the factors contributing to landslide susceptibility (Ganaie et al. 2021; Jamal and Ahmad 2020; Jamal et al. 2022; Ahmad et al. 2024a, 2024b; Behera et al. 2023). Landslide hazard analysis and mapping can provide useful information for catastrophic loss reduction, and assist in the development of guidelines for sustainable land use planning. "*Cardenas 1984*" reported evidence on the exhaustive use of GIS in conjunction with uncertainty modelling tools for landslide mapping. Remote sensing techniques are also highly employed for landslide hazard assessment and analysis.

Landslides are one of the most common geological hazards in many parts of the world (Shahabi et al. 2012). To minimize losses incurred by landslides, it is essential to have a comprehensive understanding of their causative factors which are useful for assessing landslide susceptibility of an area. The identification of landslide-prone regions is relevant for carrying out quicker and safer mitigation programmes, as well as future development planning of the area (Saadatkhah et al. 2015). National Highway 1 from Kangan to Kargil lies at the northwestern Himalayan Mountain range, which is one of the tectonically most active zones on earth. Among the various land degradation processes prevalent in the Himalayas, landslides are one of the most significant phenomena (Ahmad and Joshi 2010) as this region is tectonically very unstable with rugged topography, unstable geological structures, soft and fragile rocks, common earthquakes, along with heavy landslides are among the most damaging natural hazards in mountainous regions. The study assessed landslide susceptibility along National Highway 1D, connecting Kashmir Valley to Ladakh, using a frequency ratio method within a GIS framework. Factors like slope, rainfall, and proximity to fault zones were analyzed to create susceptibility maps, identifying that 42% of the area falls under high or very high-risk zones. Key vulnerable regions include Matayin, Gumri, Pandras, and Sonamarg, with significant landslide threats identified, especially near Zojila Pass and Hardas's region (Nanda et al. 2021).

This study of landslide susceptibility was conducted on the NH-1 National Highway between Kangan and Kargil in the union territory of Ladakh, which was part of Jammu and Kashmir until August 5, 2019. This high-altitude region, spanning 8 km from NH-1, showcases diverse geological formations. Starting from Kangan in the Kashmir Valley, the route traverses the geological splendour of the Himalayas, formed from ancient sedimentary and metamorphic rocks. The journey crosses the rugged Zoji La pass, characterized by crystalline metamorphic formations, and descends into the Suru Valley, featuring sedimentary rocks like limestone, sandstone, and shale. These formations reveal the region's geological history, shaped over millions of years. Digital elevation models (DEMs), sourced from missions like SRTM and ASTER, provide crucial elevation data for slope analysis. Slope and aspect maps derived from DEMs are key factors in assessing landslide susceptibility. Geological maps, available from organizations such as the USGS and GSI, offer information on rock types, fault conditions, and structural features. Soil maps, sourced from the FAO and national databases, provide additional data on soil types and properties.

14.2 Study Area

The study of landslide analysis was made on the National Highway-1 (NH-1) between Kangan to Kargil, connecting the Kashmir valley with Ladakh region. The union territory of Ladakh region was a part of erstwhile Jammu and Kashmir before the 5th of August 2019. The study area was made on a high-altitude region of the Kangan-Kargil National Highway, with an influence area of 8 km from National Highway-1 as shown in (Fig. 14.1).

The comprehensive information about the study area from Kangan to Kargil can be concluded in two major aspects or characteristics, that is the physical characteristics and the second one is human characteristics which have been discussed as follows.

14.2.1 Physical Characteristics

The region between Kangan and Kargil, located in the Indian Union Territory of Ladakh, is known for its rugged and varied terrain. Kangan sits at an elevation of approximately 2800 m (9200 feet) above sea level. It is characterized by lush meadows, dense forests, and stunning mountain vistas, surrounded by snow-capped peaks and alpine meadows. The climate in Kangan is temperate, with heavy snowfall in winter and mild summers, creating a lush, green landscape during the summer months. The vegetation here includes dense coniferous forests, alpine meadows, and a variety of wildflowers. The Sindh River flows through Sonamarg, adding to its

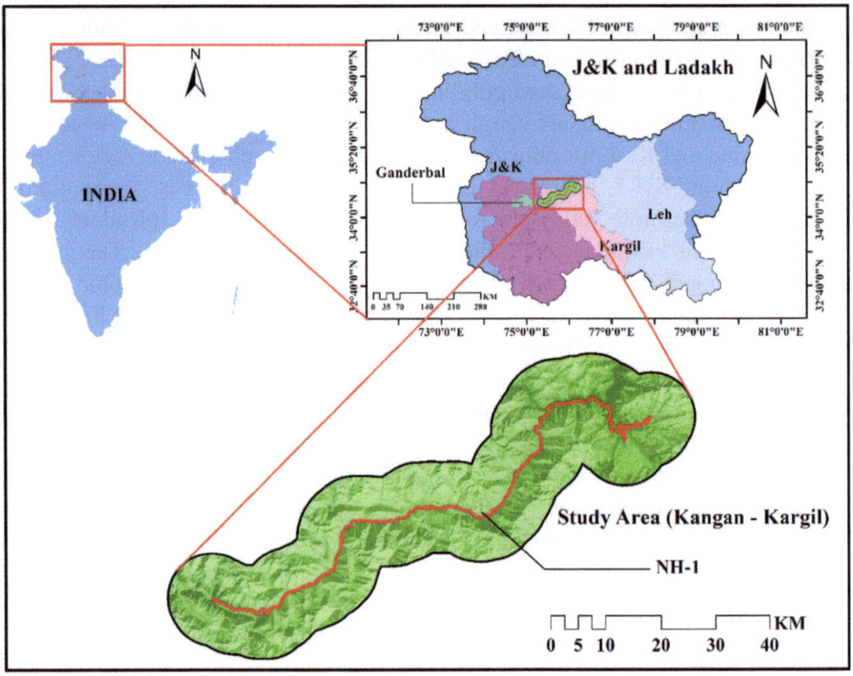

Fig. 14.1 Study area map, NH-1 (Kangan—Kargil)

verdant scenery. The journey from Kangan to Kargil involves significant elevation changes and crossing several mountain passes. One of the most notable features along this route is the Zoji La Pass, which sits at an elevation of approximately 3528 m (11,575 feet) above sea level. Zoji La Pass is a high mountain pass known for its steep inclines, rugged terrain, and narrow roads. It is one of the highest passes in the region and is often snow-covered, making it treacherous during the winter months. The climate here is harsh and unpredictable, with heavy snowfall in winter and cool temperatures in summer. Dras, a small town along this route, is situated at an elevation of approximately 3280 m (10,760 feet) above sea level, known as the second coldest inhabited place on Earth. The terrain in Kargil is less forested compared to Kangan of Kashmir side, with more barren and rocky landscapes. Kargil experiences a cold semi-arid climate, with very cold winters and relatively warm summers. The vegetation is limited, with some agriculture possible along the river valleys where irrigation is feasible. The landscape here is more arid compared to the lush greenery of Kangan (Fig. 14.2).

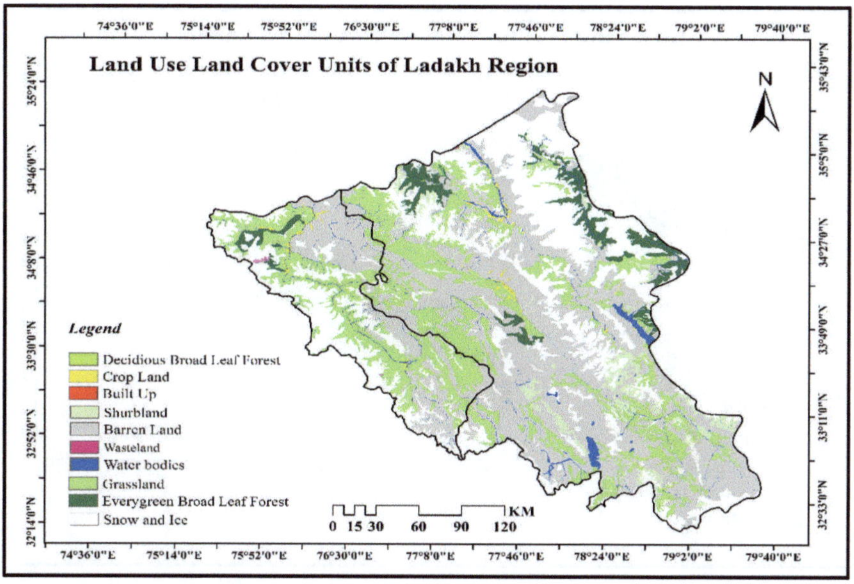

Fig. 14.2 Land use/cover map of Ladakh region. *Source* Prepared by Author/researcher using ArcGIS 10.8.2, with data from land use and land cover (LULC) classification products at 100-m resolution for India at decadal intervals for the data were derived from Landsat 4 and 5

14.2.2 Physiographic Features

The regions of Kangan and Sonamarg being a part of Kashmir valley are the parts of the greater Himalayas with major metamorphic rock formations with unconsolidated debris due its higher exposure to the physical and chemical weathering within the region, which makes the region more susceptible to land sliding, on the other hand, regions of Kargil are situated on Trans Himalayans and the side within south-western part of Tibet have a major granitic plutonic rock formation of what is called *"Ladakh batholiths"* which are high resistive to weathering and erosion. The region from Sonamarg to Kargil in the Indian Union Territory of Ladakh is characterized by diverse and dramatic physiographic features, reflecting the complex geology and varied climatic conditions of the greater Himalayan range; some important regions are Sonamarg and Kangan are situated at approximately 2800 m (9200 feet) above sea level that lies in the Sindh Valley. The region is home to several glaciers, which feed into the Sindh and Suru rivers. These glaciers and rivers have shaped the valleys and provide essential water resources, supporting both the natural ecosystem and human habitation. The area encompasses a range of climate zones, from the temperate, forested valley of Sonamarg to the harsh, cold semi-arid climate of Kargil. This climatic variation is a direct result of the significant elevation changes and the geographic barrier posed by the Zoji La Pass. The vegetation varies from dense coniferous forests and alpine meadows in Sonamarg to sparse, hardy shrubs and

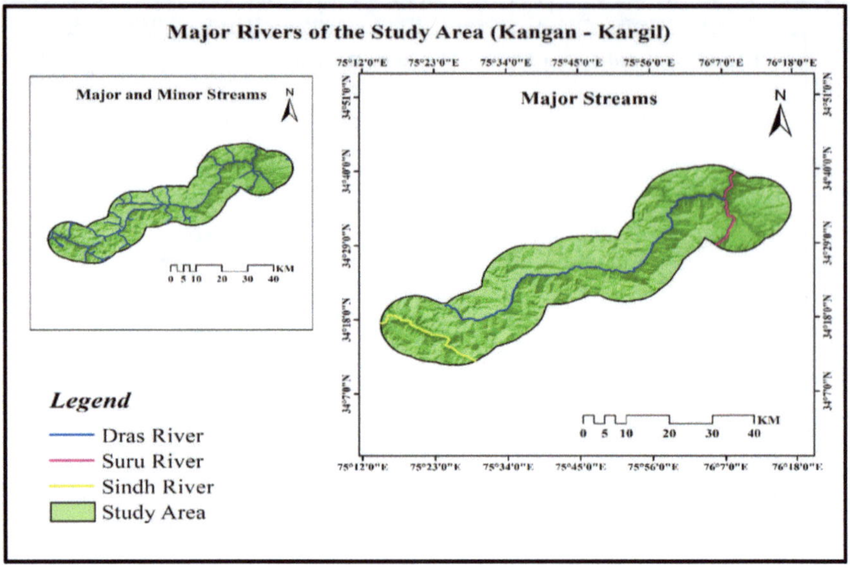

Fig. 14.3 Major rivers of the study area (Kangan—Kargil). *Sources* Prepared by author/researcher using ArcGIS 10.8.2

grasses in Dras and Kargil. The Sindh River is one of the major rivers flowing through Sonamarg. Originating from the Machoi Glacier near the Zoji La Pass, it traverses the lush Sindh Valley. The Dras River is another significant river in this region, flowing through the Dras Valley. It originates from the snowmelt and glaciers in the surrounding mountains. The Dras River's flow is crucial for the local agriculture and sustenance of the sparse vegetation in this harsh, cold environment (Fig. 14.3).

14.2.3 Lithology and Climate

The region from Kangan to Kargil in the Indian Union Territory of Ladakh is characterized by a diverse and complex geological history, resulting in varied soil types and rock formations. Starting with Kangan, situated at approximately 2800 m (9200 feet) above sea level, the soil here is typically loamy with a high organic content due to the dense vegetation cover. This soil is rich in humus, well-drained, and capable of retaining moisture, making it fertile and suitable for various plant species. The lithology in Kangan is dominated by sedimentary rocks such as sandstone and shale, which are often interbedded. The area also features limestone formations, contributing to the region's karst landscapes. Additionally, glacial deposits are significant due to past and present glacial activity, creating moraines and outwash plains. Moving towards Zoji La Pass, the soil is sparse and poorly developed due to the high altitude, steep slopes, and harsh climatic conditions. It is rocky and coarse, with a significant amount of gravel and sand, and has a low organic content and minimal vegetation cover, resulting in reduced soil fertility. Moving towards Zoji La Pass, the soil is sparse and poorly developed due to the high altitude, steep slopes, and harsh climatic conditions. It is rocky and coarse, with a significant amount of gravel and sand, and has a low organic content and minimal vegetation cover, resulting in reduced soil fertility. The lithology here is characterized by high-grade metamorphic rocks such as schists and gneisses, formed under extreme pressure and temperature conditions. Phyllites, which are fine-grained metamorphic rocks, can also be found, indicating lower-grade metamorphism.

Additionally, granitoids, including granite, contribute to the rugged terrain and cliffs. In the Dras Valley, the soil is thin, rocky, and poorly developed, consisting mainly of glacial till and alluvial deposits. In Kargil, the soil is generally arid and less fertile compared to Kangan. It consists mainly of coarse sands and gravels with low organic content. Limited vegetation results in minimal humus formation, and irrigation is often required for agriculture. The lithology in Kargil includes sedimentary rocks such as *sandstone, shale, and conglomerates* (Fig. 14.4) often showing signs of past fluvial processes. Volcanic rocks, including basalt, indicate past volcanic activity in the region. High-grade metamorphic rocks like schists and gneisses are also present due to regional metamorphism. The climate from Kangan to Kargil varies significantly due to the changing altitude and geographical features along the route. Starting from Kangan, which lies at an altitude of about 2800 m (9186 feet), the climate is typically alpine. Summers in Kangan are mild and pleasant, with temperatures ranging from 10 to 25 °C (50–77 °F), and the region experiences moderate rainfall, making it lush and green. This period is ideal for tourism and trekking. Winters, from November to March, are extremely cold with heavy snowfall, and temperatures can drop to − 10 to − 15 °C (14–5 °F).

14.2.4 Geological Settings

From Kangan to Kargil reach through a landscape with its geological formations and lithological diversity. As we leave Kangan, nestled in the Kashmir Valley, the formation begins the geological splendour of the Himalayas. These mountains, formed primarily from ancient sedimentary and metamorphic rocks, bear witness to millions of years of geological processes. Through northwest towards Kargil, the route crosses through the rugged terrain of the Zoji La pass. Here, the rocks predominantly consist of crystalline metamorphic formations, marked by their rugged textures and often exhibiting signs of intense geological activity over eons and epochs. Further along the journey, as the road descends into the Suru Valley, the landscape shifts once again. This region is characterized by a mix of sedimentary rocks, including limestone, sandstone, shale, conglomerates, and breccias formed under different environmental conditions millions of years ago. These formations tell us stories of ancient seas and geological events that have shaped the present-day topography. The terrain of Kargil is arid and mountainous, with stark cliffs and rock faces dominating the scenery. "The

Fig. 14.4 Shale and sandstone depositions in the Kargil-Akchamal sedimentary formation. *Source* Researcher field Observation image

Himalayas can broadly be classified into four major geo-tectonic units" (Ganser 1964; Valdiya 1984), which from south to north are; Siwalik's or sub-Himalaya separated from the Indo-Gangetic plains by the Himalayan Frontal Fault (HFF) in south. Lesser Himalaya is separated from the Siwalik by the Main Boundary Thrust (MBT). Higher Himalaya or Himadri is separated from the Lesser Himalaya by the Main Central Thrust (MCT). Trans Himalaya or the Tethyan zone which is north of the Higher Himalaya. "The geology and tectonics of the north-west Himalaya: regional perspectives and constraints on the evolution of the Nanga Parbat syntaxis" *Journal of the Geological Society* (Streule et al. 2010). Trans-Himalayan Batholith consists of the old Eurasian crust overlain by shallow sea (platform) sediments deposited before the Late Cretaceous and intruded/overlain by Late Cretaceous to Eocene magmatic rocks of the Trans-Himalayan Batholith. This diverse geological framework underpins the varied landscapes and ecosystems observed across this region. Sedimentary rocks such as sandstones, mudstones, conglomerates, and limestones result from the consolidation of sand, mud, pebbles/sand, shell debris, and lime mud which were originally deposited in river beds, glacier moraines, deserts dunes (as in the Shayok Valley near Hundar), lakes, and oceans. The Himalayan range properly started to form during the Late Jurassic (160 Ma), when the oceanic crust and sediments attached

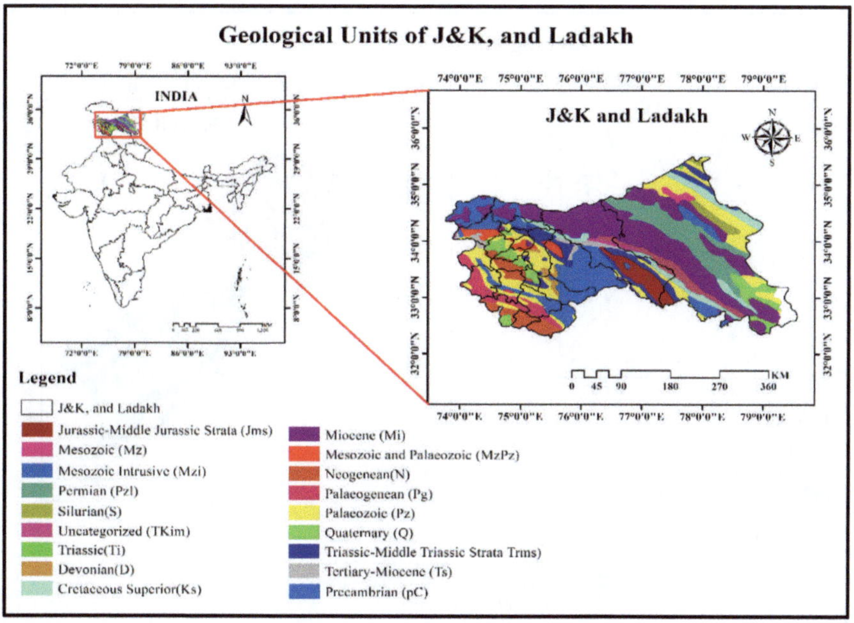

Fig. 14.5 Geological units of J&K, and Ladakh. *Source* Prepared by author/researcher using ArcGIS 10.8.2

to the northern edge of India started to plunge under Eurasia. At approximately the same time and on a more local scale an island arc with its attached basins developed in the area between Kohistan and Leh (the so-called Kohistan-Ladakh Arc), to the West where the collision with Eurasia was less direct (Figs. 14.5 and 14.6).

The Ladakh Batholith is but a part of the much larger Trans-Himalayan Batholith which extends from Kohistan to Gangdese in southern Tibet. Its chemistry and mineralogy are similar to those of the South and North American batholiths, which are located in a similar geological environment. Regions from Kangan to Kargil in J&K and Ladakh feature a geologically rich and diverse terrain along the strategic Srinagar-Leh Highway (NH1). Starting from Kangan, which lies at the foot of the Zojila Pass, the area primarily features metamorphic rocks such as schists and gneisses. These rocks have undergone significant transformation due to high pressure and temperature. Additionally, metamorphic rocks like slates and phyllite can be found, indicating ancient depositional environments that have been altered by regional metamorphism. When we are moving towards the Zojila Pass, the rocks are typically phyllites and quartzites, formed under moderate metamorphic conditions.

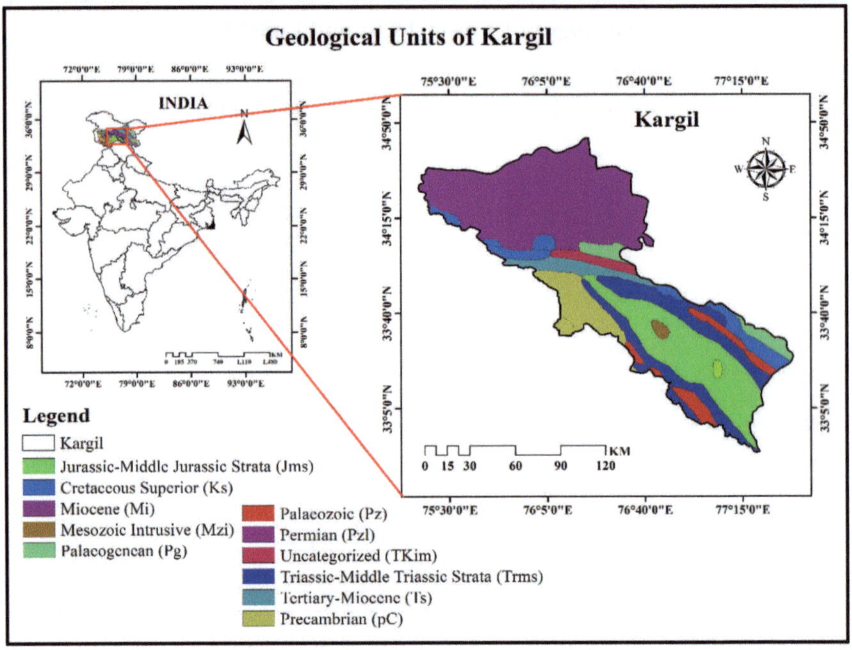

Fig. 14.6 Geological units of Kargil District. *Source* Prepared by author/researcher using ArcGIS 10.8.2

14.3 Data Sources and Methodology

The data was derived from Landsat 4 and 5 Thematic Mapper (TM), Enhanced Thematic Mapper Plus (ETM+), and Multispectral (MSS) data, India Remote Sensing satellites (IRS) Resource sat Linear Imaging Self-Scanning Sensor-1 or III (LISS-I, LISS-III) data, ground truth surveys, and visual interpretation. A mosaic of Landsat-9 was created and also achieved from Shuttle Radar Topography Mission Digital elevation Model (SRTM-DEM) for elevation, slope, and aspect map preparations. Also, Landsat-8 and Landsat-9 are used for normalized difference vegetation index and model preparations. Other than these sources some other data were used for outline mapping and others for which the data used from OSM, MapInfo, and Survey of India. Also, for the preparation of soil classes, the data were collected from FAO soil data that is Food and Agricultural Organization of United Nations (Digital Soil Map of the World- Geonet work). For the preparation of precipitation and rainfall maps, the sources were collected from CRU-TS version 4.08 data variable precipitation-2023. Global Landslide Catalogue Downloadable Products were used for the preparation of Landslide Inventory Mapping. The Cooperative Open Online Landslide Repository (COOLR) is a worldwide database of landslide events from NASA, scientists, and citizen scientists. That can be from the COOLR catalogue as a file geodatabase (.gdb), shapefiles (.shp), or comma-separated values (.csv) (Table 14.1).

For the mapping of the vegetation index, Normalized Difference Vegetation Index-NDVI was used, as we have the data from Landsat—9, the following methods were applied for the formulation and mapping of vegetation patterns within the study area as follows:

$$\text{NDVI} = (\text{Band 5}-\text{Band 4})/(\text{Band 5} + \text{Band 4})$$

The value should range between − 1 to + 1, and the value for the current final output of the study area was; higher; + 0.87 and lower; − 0.93.

For precipitation and rainfall units' calculation, the following band formulas and methods were applied, while data collected from CRU-TS, for the year 2023 comprising of the last three years' data for the study area of NH-1 (Kangan—Kargil) as calculated from below methods:

$$\text{Total number of bands} = 36(\text{In a buffer of 8 km within the study area})$$
$$1 \text{ year} = 12 \text{ months} = 12 \text{ bands}$$

To find the number of years in the calculation of average annual rainfall using interpolation, we need to divide 12 by the total number of bands as follows:

$$36/12 = 3 \text{ years.}$$

Therefore, the last 3 years' average annual rainfall values need to be calculated as shown in Table 14.2.

Also, for the preparation and framework of Landslide Inventory Mapping, both Google Earth Imagery mapping and NASA's COOLR shapefile were used. Landslide Inventory mapping with 25% of the data is 11 Landslides collected from NASA Cooperative Open Landslide Repository (COOLR) database shapefiles, but the rest

Table 14.1 Different geospatial data used in landslide susceptibility mapping

Period	Satellite	Sensor	Spatial resolution
05/02/2022	Landsat-9	Operational Land Imager 2 (OLI-2): Thermal Infrared Sensor 2 (TIRS-2)	30 m for visible, NIR, and shortwave infrared bands 15 m for the panchromatic band 100 m for thermal infrared bands
02/06/2023	ESRI Sentinels-2	Multispectral Instrument (MSI)	10 and 20 m respectively
03/04/2022	Landsat-8	Operational Land Imager (OLI) Thermal Infrared Sensor (TIRS)	(OLI):30 m for visible, near-infrared, and shortwave infrared bands (TIRS):100 m for thermal infrared bands

Table 14.2 Showing the formulation and calculation of years for rainfall map from several bands

Years	Months	Number of bands
2021	12	1–12
2022	12	13–24
2023	12	25–36

So, to calculate the average annual rainfall from the raster calculator = $\frac{\text{SUM layer}}{3\text{years}}$

75% are nearly 34 landslides were framework and identified from Google Earth Imagery of a total of 45 landslides were identified within the study area as shown in following flowchart of Fig. 14.7.

The main methodology used is based on both qualitative and quantitative analysis of the landslide susceptibility model, qualitative in the sense that based on the researcher's imagery observation and some literature review helped in proper weightage which is introduced in the weight overlay section of the final map, also

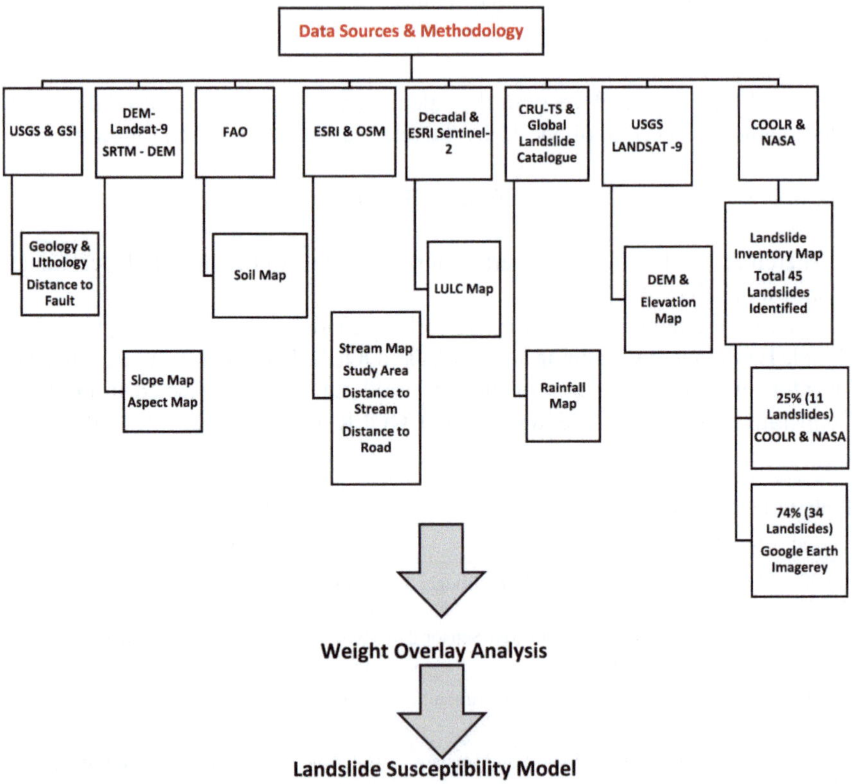

Fig. 14.7 General flowchart for the methodology used in the study of landslide susceptibility mapping (*NH1-Kangan—Kargil*)

quantitative analyses were done while making the susceptibility model, the comprehensive illustration is shown in Fig. 14.7, first of all, the parameters of studying landslides were determined and analyzed which could be slope, aspect, lithology, distance to road, distance to stream, land use land cover, rainfall or precipitation units, elevation, contour units, distance to fault, soil classes, and landslide inventory map were prepared for which data are collected from different sources, like USGS, GSI, FAO, and ESRI, etc., as mentioned earlier. Statistically integrating weights and overlaying the thematic maps generated from these conditioning factors in a GIS platform, specific landslide susceptibility maps can be developed (Nanda et al. 2021). Factors used as input data comprise distance to road, slope angle, lithology, soil, rainfall, aspect, land use land cover, distance to faults, distance to streams, and elevation (Sahabi et al. 2012). Total parameters and units taken into consideration were in total of 15 units which include; aspect unit, contour, distance to fault, distance to road, distance to stream, elevation unit, geological parameters, lithological units, LULC units, NDVI units, precipitation or rainfall units, slope units, soil classes, and landslide inventory mapping, for which the various data sources have been illustrated in above flowchart of Fig. 14.7. Among all the parameters and units some important that are 7 units were taken as the further examination under heuristic and statistical analysis which would result in the overlay of all these parameters by providing with suitable weights from expert and researcher, which finally forms the landslide susceptibility mapping by weight overlay method, these 7 units were aspect, distance to road, distance to stream, elevation model, LULC, precipitation units, and the most important slope classes were taken into consideration. Although landslide inventory mapping and other rest units were helped in qualitative and quantitative analysis of the study. On these 7 units of landslide susceptibility mapping a proper weight system was introduced based on expert knowledge and experience to assign weights to different factors contributing to landslide susceptibility. This approach involves, knowledge-based weighting, experts assign weights to various factors like slope, geology, land use, etc., based on their perceived importance in causing landslides (Shit et al. 2016). The weight system of overlay is a simple method, direct, and adequate methodology for the evaluation of land-susceptible zones and landslide areas (Ahmed et al. 2015). Under this methodology, each of the 7 unit classes, based on several classes they were provided with suitable weights which were purely based on observation from Google Earth imagery and literature reviews. These weights were considered into total influencing factors of percentages ranging from (0–100), which were further converted into its corresponding normalized factors ranges from (0–1) by using a suitable normalization formula, as shown in Table 14.3. The reason behind converting it into a normalized factor was to use the values in formulae for the calculation of *the landslide susceptibility index* which could be further converted back into its percentage value as follows:

$$S(x, y) = \frac{1}{3} \times N_1(x, y) + \frac{1}{3} \times N_2(x, y) + \frac{1}{3} \times N_3(x, y) + \frac{1}{3} \times N_4(x, y) + \frac{1}{3} \times N_5(x, y)$$

$$+\frac{1}{3}\times N_6(x,y)+\frac{1}{3}N_7(x,y)$$

where

$S(x,y)$ = susceptibility index.

(x,y) is the location within the study area of units.

$\{N_1 \ldots N_7\}$ are normalized factors of different units which range from (0 to 1).

N_1 = Normalized factor for LULC

N_2 = Normalized factor for rainfall

N_3 = Normalized factor for aspect

N_4 = Normalized factor for distance to stream

N_5 = Normalized factor for distance to road

N_6 = Normalized factor for elevation

N_7 = Normalized factor for slope.

Formula for calculation of normalized factors:

$$N=\frac{z-x_{\min}}{x_{\max}-x_{\min}}$$

where

N = Normalised factor, $\{N_1 \ldots N_7\}$

Z = Weights of each unit

$X_{\min}$ = Minimum weight = 0.03

$X_{\max}$ = Maximum weight = 0.3.

By using the above illustration and normalized factors from a table following calculations were done for the formulation of the susceptibility index as follows:

Table 14.3 Various percentages of weights and their corresponding normalized factors of all seven units used for calculation of landslide susceptibility index

Parameters/units	Influence (%)	Weights (0–1)	Normalized factors
LULC	25	0.25	0.81
Rainfall	20	0.2	0.62
Aspect	7	0.07	0.14
Distance to stream	5	0.05	0.07
Distance to road	3	0.03	0.0
Elevation	10	0.1	0.25
Slope	30	0.3	1

$$
\begin{aligned}
S(x,\ y) &= \frac{1}{3} \times 0.81 + \frac{1}{3} \times 0.62 + \frac{1}{3} \times 0.14 \\
&\quad + \frac{1}{3} \times 0.07 + \frac{1}{3} \times 0 + \frac{1}{3} \times 0.25 + \frac{1}{3} \times 1 \\
&= 0.2 + 0.2 + 0.0 + 0.0 + 0.0 + 0.0 + 0.3 \\
&= 0.7 \text{ which is in between } (0 - 1) \\
&= 0.7 \times 100 \\
&= 70\% \\
&\therefore \text{Total Landslide Susceptibility Index} \\
&\quad \text{Value within the study area} = 0\%.
\end{aligned}
$$

The second approach for the calculation of the susceptibility index is as follows with the same normalized factors:

$$
\text{Susceptibility Index } S(x,\ y) = \sum_{i=1}^{n} Wi \times Ni
$$

where

n = number of factors = 7

Wi= weight of each factor

Ni = normalized factor.

$$
\begin{aligned}
S(x,\ y) &= 0.3 \times 1 + 0.25 \times 0.81 + 0.2 \times 0.62 + 0.1 \times 0.25 \\
&\quad + 0.07 \times 0.14 + 0.05 \times 0.07 + 0.03 \times 0 \\
&= 0.3 + 0.2025 + 0.124 + 0.025 + 0.0098 + 0.0035 + 0 \\
&= 0.6648 \sim 0.7 \\
&= 0.7 \text{ which is in between } (0 - 1) \\
&= 0.7 \times 100 = 70\%
\end{aligned}
$$

Low susceptibility index	0–0.3	0–30%
Moderate susceptibility index	0.3–0.7	30–70%
High susceptibility index	0.7–1	70–100%

The findings of the above illustration have been discussed in the next result and discussion part. In conclusion, these are not the only methods considered under study but along with these, some qualitative analyses were also done using Google Earth Imagery, Landslide inventory from which Landslide Inventory mapping with 25% of data that are 11 landslides were collected from NASA Cooperative Open Landslide Repository (COOLR) database shapefiles, but rest 75% that are nearly 34 landslides were framework and identified from Google Earth Imagery of total 45 landslides were identified within the study area as shown in above flowchart of Fig. 14.7.

14.4 Results and Discussions

The entire result and findings can be described from its various conditioning factors and units that helped in the final understanding of landslide susceptibility model, which could be understandable from its 15 units of study ranging from aspect, contour, distance to fault, distance to stream, distance to road, elevation, geology, lithology, LULC, NDVI, rainfall classes, slope, soil classes, landslide inventory map, and finally landslide susceptibility model. Among all these, 7 units of landslide susceptibility mapping which are aspect, distance to road, distance to stream, elevation, LULC, rainfall, and slope were given proper weight systems that are introduced based on expert knowledge and experience to assign weights to different factors contributing to landslide susceptibility. "This approach involves, Knowledge-Based Weighting, Experts assign weights to various factors like slope, geology, land use, etc., based on their perceived importance in causing landslides", Pravat Kumar et al. (2016). "The weight system of overlay is a simple method, direct, and adequate methodology for evaluation of land susceptible zones and Landslide areas", Erener and Uzgun (2008): Ahmed et al. (2024a). Findings, results, and discussions of all these parameters and units are discussed as follows.

14.4.1 Aspect or Face of Slope

An aspect of slope refers to the direction of the compass that a slope faces. It is the direction in which the slope is inclined relative to the cardinal points of north, south, east, and west. North and east facing slopes are usually more susceptible to landslides, this is because these slopes receive less direct sunlight, leading to slower drying of the soil after rainfall. Southwest-facing slopes are usually less susceptible when compared with northeast-facing slopes. Coming to the current unit of the study reveals that the regions such as Kangan, Sonamarg, and Zojila have north and northeast aspect of slope, also the terrain around Drass and Chanigound have south and southwest aspect of slope. The regions of Kargil majorly have southwest, south, and west aspect of slope face which is shown in Fig. 14.8.

14.4.2 Elevation and Contour

Higher elevation and contour would directly influence greater susceptibility of landslide occurrences, the current elevation units have been divided into four classes

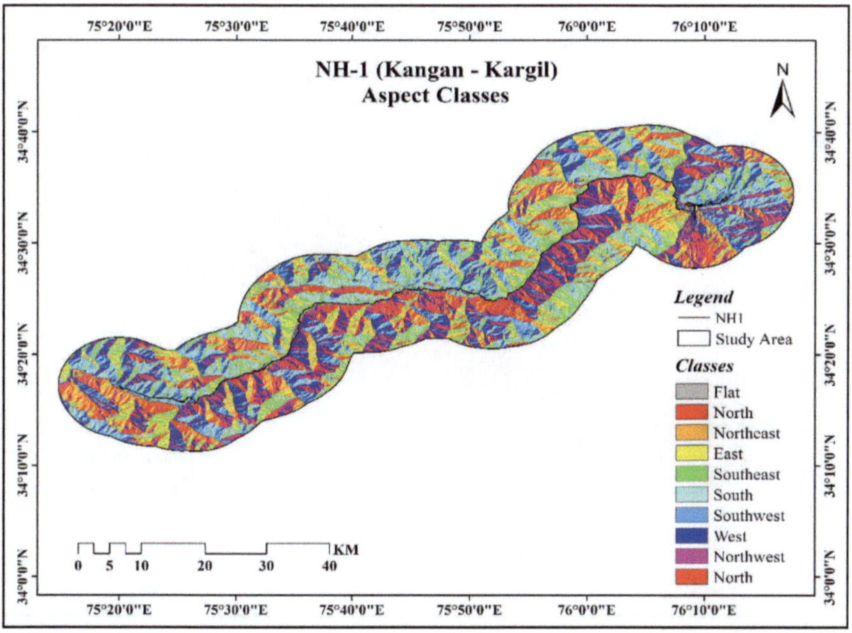

Fig. 14.8 Aspect classes of the study area NH-1 (Kangan to Kargil). *Sources* USGS Landsat-9, SRTM-DEM data using ArcGIS 10.8.2

2547–3341 m, 3341–3878 m, and 3878–4396 m. 4396–5445 m, respectively, among which the regions within the Kangan side and Kargil regions have an average elevation of 2547–3341, also the regions near Matayin and Zojila pass have an elevation greater than 3500 m from mean sea level. The similar results from contour units, where Kangan and Kargil have less than 3000 m of height, while Zojila pass has more than 3000 m of height (Figs. 14.9 and 14.10).

14.4.3 Distance to Fault

The distance to a fault affects landslide susceptibility because areas closer to faults are more likely to experience ground shaking during earthquakes, which can destabilize slopes and trigger landslides. Faults often have fractured and weakened rock, reducing slope stability. Additionally, faults can alter local groundwater flow, leading to increased water infiltration and slope saturation, further raising landslide risk. Thus, proximity to a fault generally means higher landslide susceptibility due to seismic activity and geological conditions. In the current study, two major fault lines were found within the Drass valley, from on an average coordinate of 34°25′41.45″ north, and 75°40′21.43″ east the highway comes in 2.5 km within the distance of fault line less than 1000 m and 3.7 km within the distance 1000–2000 m of fault

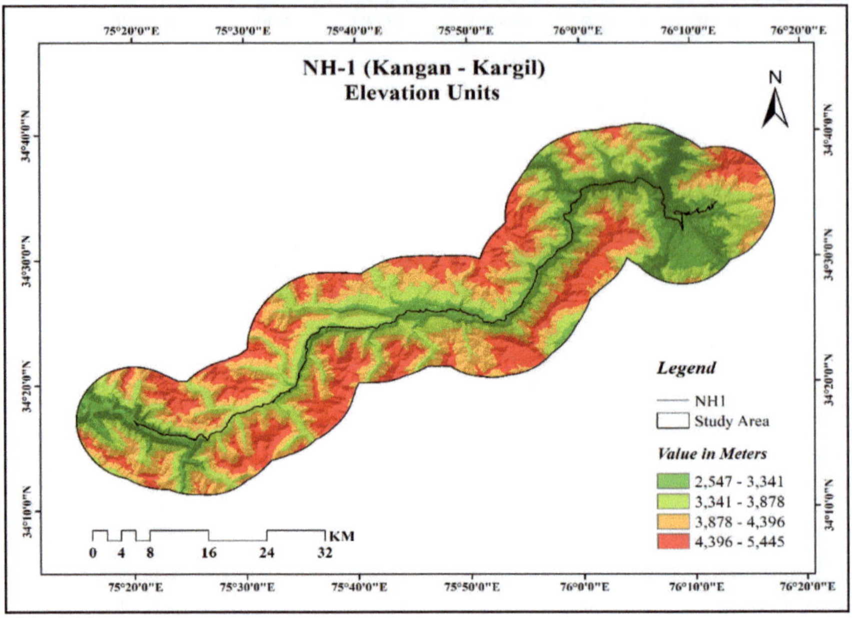

Fig. 14.9 Elevation units of the study area NH-1 (Kangan to Kargil). *Source* USGS, Landsat-9, DEM, Geological Survey of India

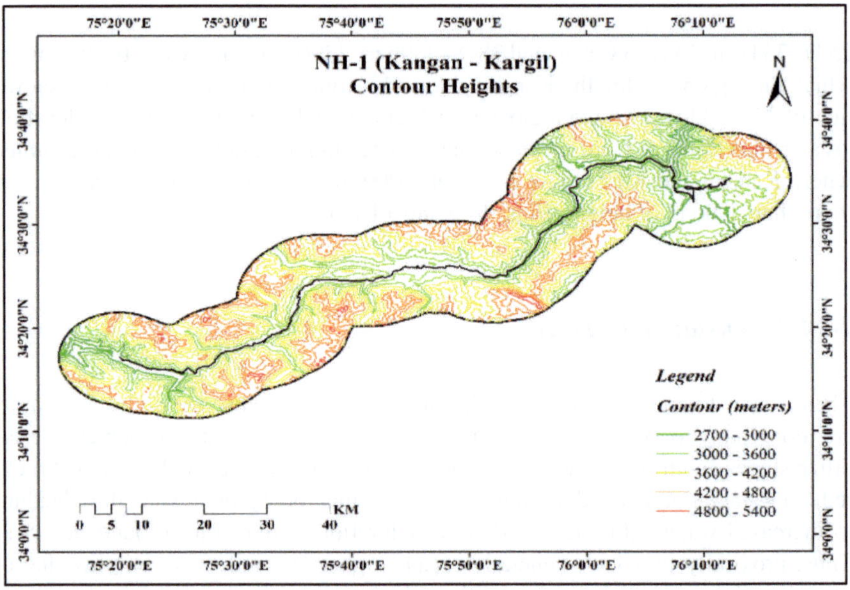

Fig. 14.10 Contour heights of the study area NH-1 (Kangan to Kargil). *Source* USGS, Landsat-9, DEM, Geological Survey of India

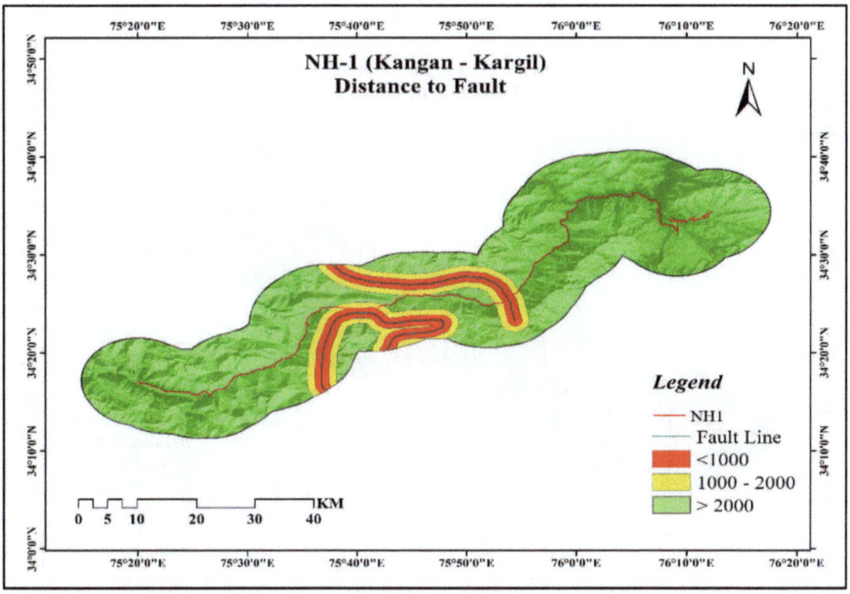

Fig. 14.11 Distance to fault units of the study area NH-1 (Kangan to Kargil). *Source* Data collected from USGS and ESRI (Environmental System Research Institute)

line. Similarly, within the Chokiyal region of coordinates 34º25′41.45″ north and 75º53′52.78″ east of 2.74 km road distance comes in less than 1000 km of fault line. Also, 2.3 km of the road distance within the same coordinate comes under 1000–2000 m of the fault line of the Chokiyal region. Although there could be the presence of some smaller fault lines within the study area only two major fault lines were recognized in Drass Valley (Fig. 14.11).

14.4.4 Distance to Road

The distance to a road affects landslide susceptibility because roads often involve cutting into slopes and altering natural terrain, which can destabilize the ground. Areas close to roads may experience increased water runoff due to paved surfaces, leading to greater soil saturation and reduced slope stability. Road construction and maintenance can also lead to vibrations and disturbances that weaken the soil and rock. Consequently, proximity to roads typically increases the risk of landslides due to these modifications and disturbances to the natural landscape. The road distance is divided into five classes 0–300 m, 300–1000 m, 1000–2000 m, 2000–3500 m, and 3500–4000 m respectively, with the help of Google Earth imagery observed that the distance from the road is greater over the Kargil side and getting narrower and lesser when approaching towards Zojila side (Fig. 14.12).

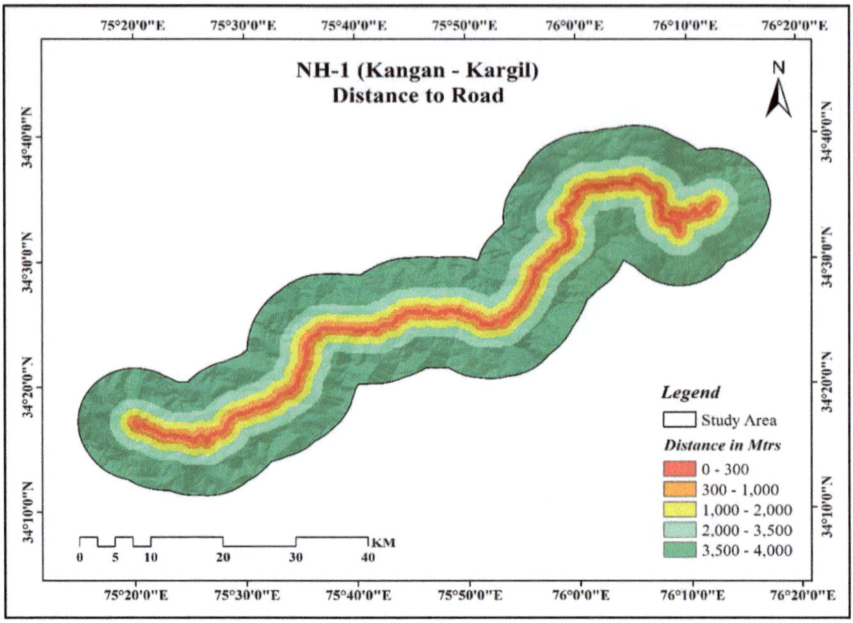

Fig. 14.12 Distance to road units of the study area NH-1 (Kangan to Kargil). *Source* Data collected from USGS and ESRI (Environmental System Research Institute)

14.4.5 Distance to Stream

The distance to a stream affects landslide susceptibility because areas closer to streams are more prone to soil erosion and undercutting slopes by flowing water, which can destabilize the ground. Stream banks and adjacent slopes often experience higher moisture levels, increasing soil saturation and reducing cohesion, which makes them more susceptible to landslides. Additionally, streams can alter groundwater flow patterns, contributing to slope instability. Therefore, proximity to a stream generally elevates landslide risk due to increased erosion, moisture, and potential undercutting of slopes. Thus from Kangan along Zojila to Drass is more proximity to rivers and streams as compared to regions closer to Kargil region. A total of 11 sample distances from the stream were collected from Google Earth imagery and the average distance is 0.5 km which is 500 m for every 10 km with the study area (Fig. 14.13).

14.4.6 Geology and Lithology

Geology and lithology significantly influence landslide susceptibility by determining the strength and stability of the Earth's materials. Different rock types, such as hard granite versus soft shale, vary in their resistance to weathering and

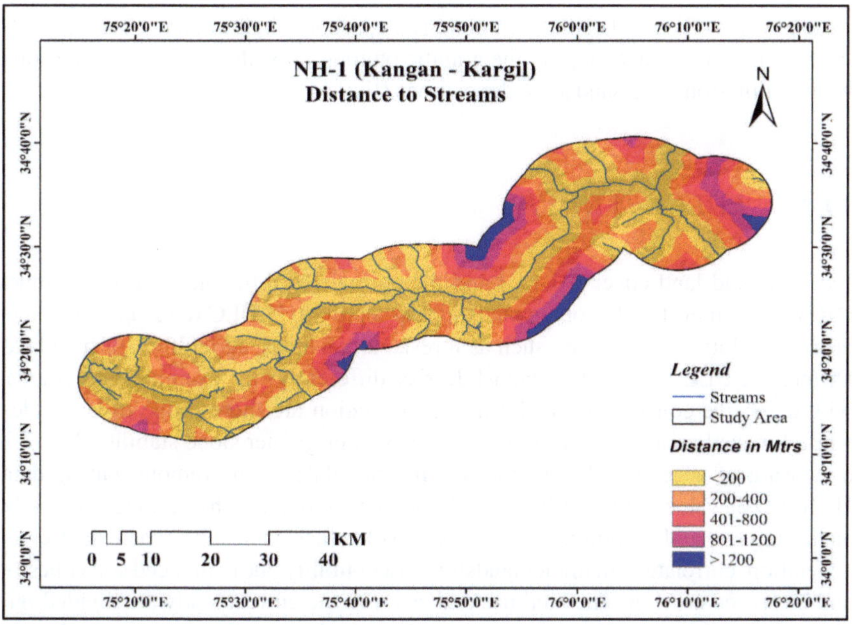

Fig. 14.13 Distance to stream units of the study area NH-1 (Kangan to Kargil). *Source* USGS, ESRI, and OSM data used

erosion.*"The geological composition, particularly the presence of sedimentary formations and significant thrusts such as the Main Boundary Thrust formation, exacerbate the region's susceptibility to recurrent landslides, posing a significant threat to this mountainous region*" (Sharma et al. 2020). Geological structures like faults and folds create zones of weakness, making slopes more prone to sliding. Soil composition, derived from the underlying rock, affects stability, with clay-rich and loose granular soils being more susceptible to landslides. The degree of weathering and erosion also impacts stability, as highly weathered rocks and eroded slopes are more prone to failure. Permeability influences drainage, with poorly draining soils increasing water saturation and landslide risk. Additionally, areas with a history of landslides often indicate inherent geological instability. Four types of geological formations were recognized within the study area which are majorly Miocene formations, followed by Cretaceous superior, Mesozoic formation, and Triassic-Middle Triassic formations. The regions closer to Kangan, Sonamarg, and Zojila have Cretaceous, Triassic, and Mesozoic formations, and entire regions from Drass to Kargil have Miocene formations. The geological formation of the Cretaceous, Mesozoic, and Triassic usually have sedimentary and metamorphic rocks, and the formations within the Miocene majorly have granitic batholiths and basaltic rocks, from a field visit to the regions of Kargil side also have sedimentary rock formations but in less when compared with the Cretaceous formation of Kangan, Sonamrg, and Zojila side. Similarly, the lithological units of study also reveal the same as the regions within the Zojila side have more shale, schist, phyllites, and quartzite beds, the Kargil regions have majorly the granitic plutons, amygdaloidal basalt, and some bedded limestone and sandstone depositions (Fig. 14.14).

14.4.7 LULC and NDVI

Land use and land cover (LULC) and the normalized difference vegetation index (NDVI) are important factors in landslide susceptibility. LULC refers to how land is utilized and its surface cover, such as forests, urban areas, agriculture, or bare land. Different LULC types affect landslide risk differently; for example, urban areas with extensive construction and reduced vegetation are more prone to landslides, while forested areas with strong root systems offer greater slope stability. NDVI is a measure of vegetation health and density, calculated using remote sensing data. High NDVI values indicate healthy, dense vegetation, which stabilizes slopes by anchoring soil and absorbing water. Low NDVI values, indicating sparse or stressed vegetation, correlate with higher landslide susceptibility due to reduced soil cohesion and increased erosion. The land use and cover in the current study are divided into classes such as waterbodies, vegetation, agriculture, built-up, snow cover, shrubs, and barren land. The major regions are coming under barren land, with more built-up area in Kargil and Drass region along NH-1, Kangan side usually has less built-up. Similarly, the regions within the Sonamarg have more evergreen vegetation and plantations as compared to others, but this would gradually decrease within the Zojila pass. Similarly, the vegetation profile from NDVI also reveals a greater vegetation index (0.7–0.8) within the Kangan region, Drass, and Kargil regions, while the regions along Matayin, Mushko, Pandrass, and Chanigound have less (− 0.8) vegetation index as compared to others (Figs. 14.15 and 14.16).

14.4.8 Precipitation and Rainfall

Rainfall significantly influences landslide susceptibility by affecting soil moisture and stability. Intense or prolonged rainfall saturates the soil, increasing pore water pressure and reducing the friction and cohesion among soil particles, leading to slope failure. Heavy rainfall can also cause erosion, removing the supporting material from slopes and triggering landslides. Additionally, sudden, intense rainfall can result in rapid infiltration and water build-up, quickly destabilizing slopes. Therefore, areas experiencing frequent, heavy, or prolonged rainfall are at a higher risk of landslides due to the increased likelihood of soil saturation and erosion. "Climate change, particularly in mountainous regions, has exacerbated landslide occurrences, especially during the monsoon season, which experiences heavy rainfall" Tharshini Murthy et al. (2023). The rainfall unit in the study area has been divided into five

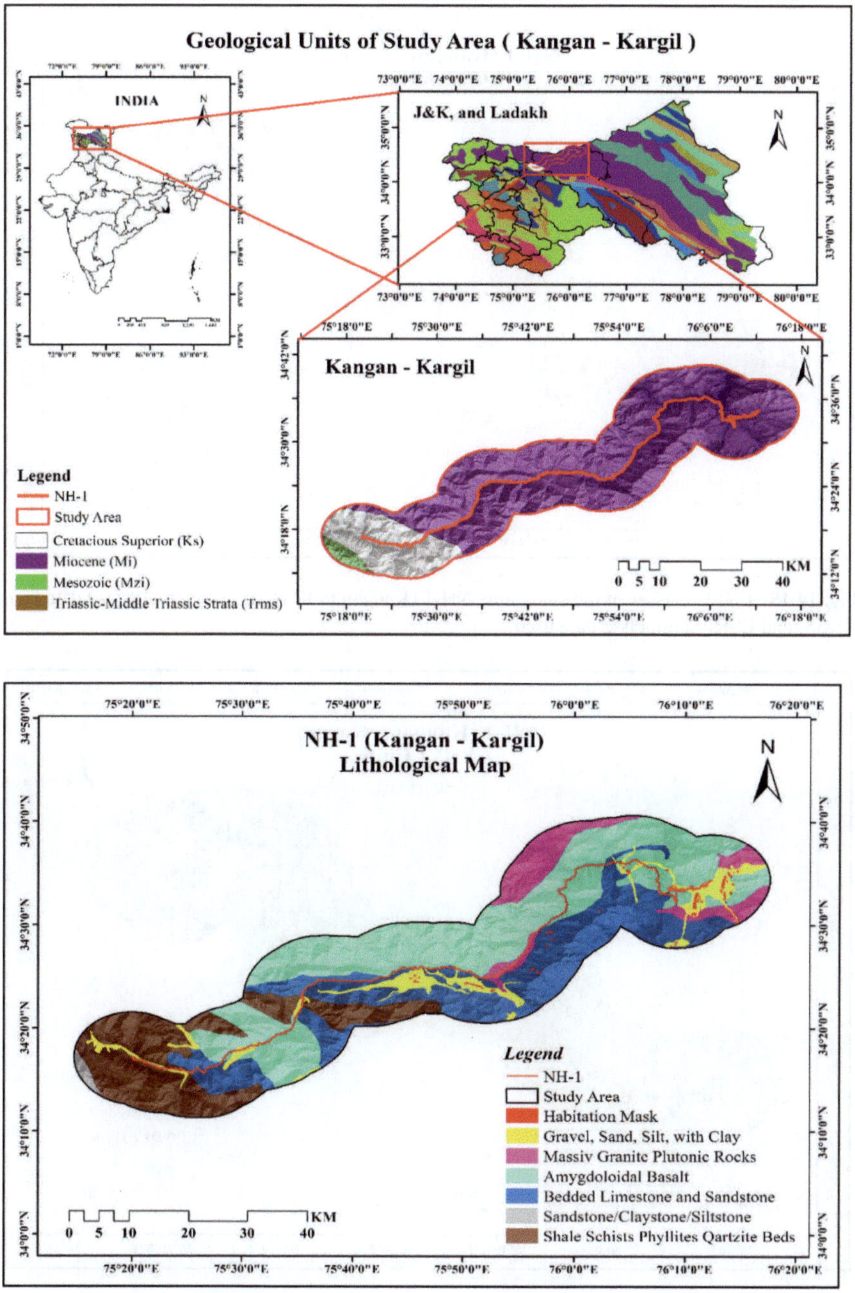

Fig. 14.14 Geological and lithological units of the study area (NH-1, Kangan—Kargil). *Source* USGS and ESRI data are used

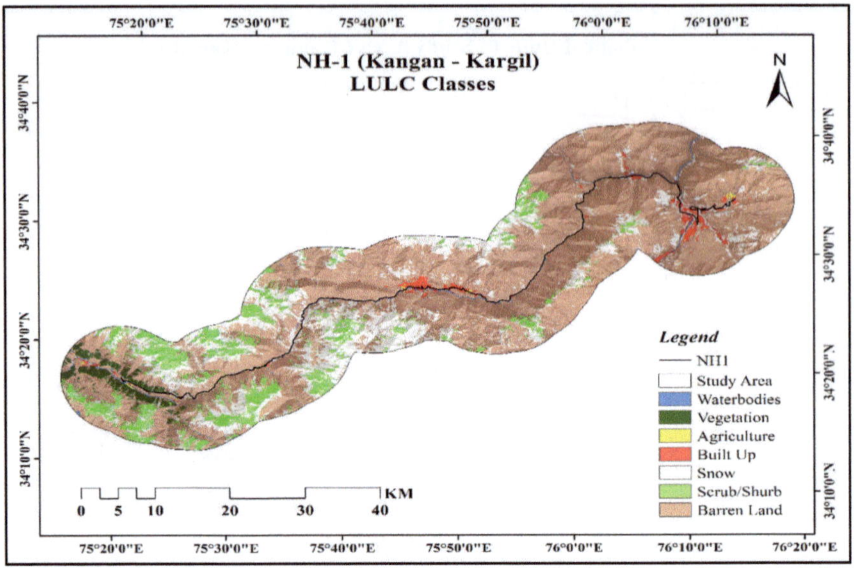

Fig. 14.15 LULC classes of the study area NH-1 (Kangan to Kargil). *Source* Landsat-9 data from USGS, and ESRI Sentinels-2 data used

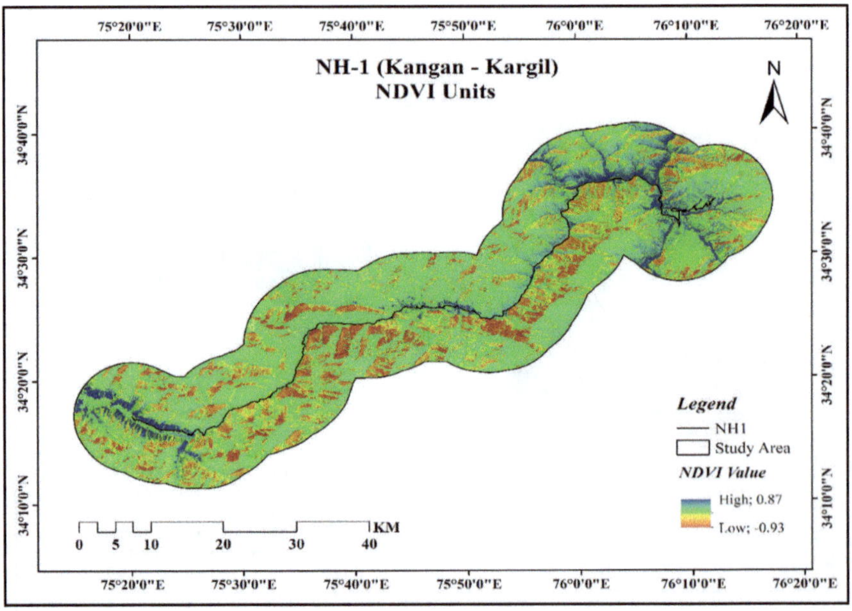

Fig. 14.16 NDVI classes of the study area NH-1 (Kangan to Kargil). *Source* Landsat-9 data from USGS, and ESRI Sentinels-2 data used

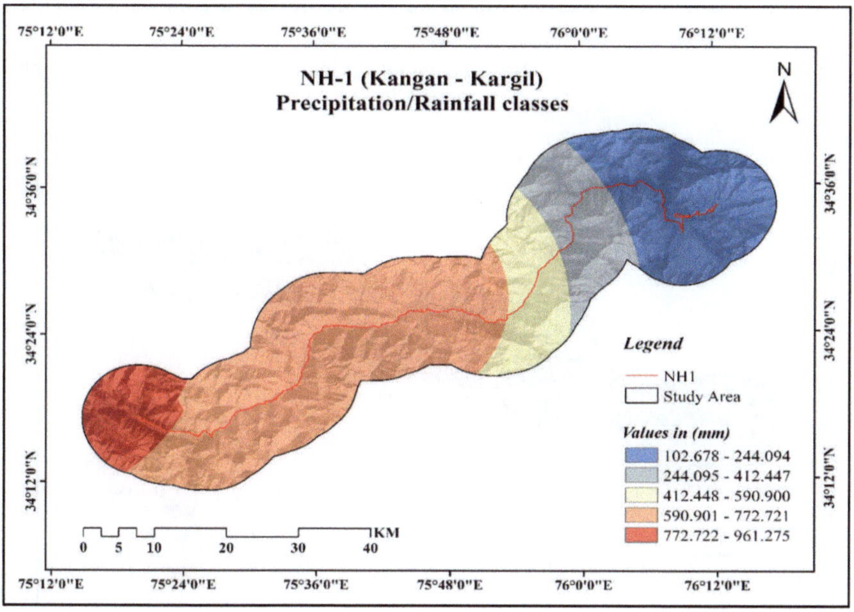

Fig. 14.17 Rainfall classes of the study area NH-1 (Kangan to Kargil). *Source* CRU-TS and global landslide catalogue data are used

classes with average low to high rainfall ranging from 102.678 to 961.275 mm annually. The regions from Kangan to Sonamarg usually have more than 772.722 mm of rainfall annually, from Sonamarg to Chokiyal the average annual rainfall is between 590.901 and 772.721 mm, the regions in between from Chokiyal to Kaksar the rainfall value is in between 412.448 and 590.900 mm annually, also the values from Kaksar, Chanigound to Kargil the rainfall is less than 412.448 mm annually (Fig. 14.17).

14.4.9 Slope Unit

Slope significantly affects landslide susceptibility. Steeper slopes are more prone to landslides because the gravitational force acting on the soil and rock materials is greater, increasing the likelihood of slope failure. Steep slopes also have less stability, as there is less material to resist the downward pull. Additionally, steep slopes often have higher rates of erosion, which can remove supporting material and further destabilize the slope. In contrast, gentler slopes have a lower risk of landslides due to reduced gravitational force and greater overall stability. Therefore, steeper slopes are generally more susceptible to landslides. The slope unit within the study area is divided into five classes of slope angles, which are 0°–15°, 15°–30°, 30°–45°, 45°–60°, and more than 60°. The regions within Kargil city region (elevation; 2702 m)

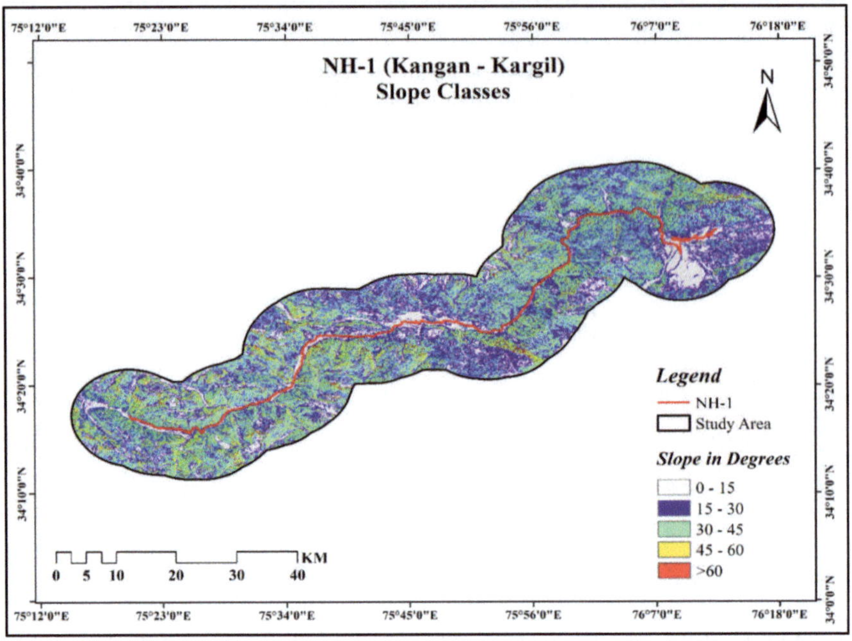

Fig. 14.18 Slope classes of the study area NH-1 (Kangan to Kargil). *Source* DEM, Landsat-9, and SRTM-DEM data are used

usually have slope less than 15°, from Hardas to Thangsgam of coordinates 34°26′8″ N and 75°37′14″ E with an elevation of 3193 m, for which the average slope unit is 30°–45°, also from Thangsgam to Pandrass with coordinates 34°24′43″ N and 75°37′14″ E with elevation of nearly 3334 m from mean sea level have slope units as 15°–18°, the slope units from Pandrass to Gumri (34°16′24″ N and 75°27′36″ E) with elevation of 3676 m has angle of average 30°–45°, and the regions from Gumri to Zojila (elevation 3687 m) have slope units as more than 45° which declines to 30° near Kangan side of Ganderbal region (Fig. 14.18).

14.4.10 Soil Classes

Soil classes significantly influence landslide susceptibility due to their varying physical properties. Clay soils, for example, can retain water and become very slippery when wet, reducing friction and increasing the likelihood of landslides. Sandy soils, while less cohesive, can quickly lose stability when saturated due to poor water retention. Silty soils, with fine particles, are also prone to losing cohesion when wet. On the other hand, loamy soils, which are a mix of sand, silt, and clay, generally provide better drainage and stability. Thus, soil composition, permeability, and cohesion are critical factors in determining landslide risk, with clay-rich and poorly draining

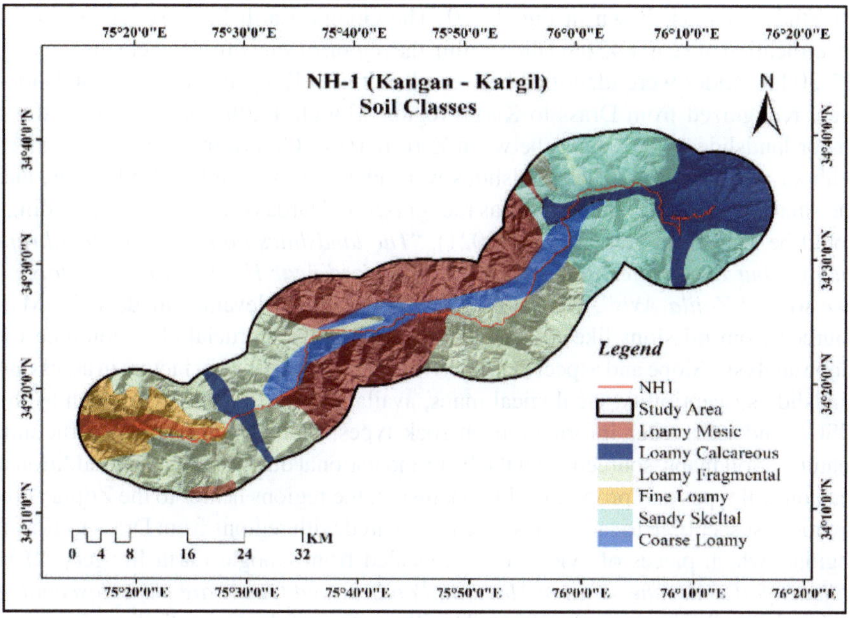

Fig. 14.19 Soil classes of the study area NH-1 (Kangan to Kargil). *Source* FAO data are used

soils being the most susceptible. The soil classes from the units also reveal that the regions from Kangan to Matayin usually have fine loamy, loamy fragmental type of soils. The regions from Matayin along with the Drass to Kaksar region have majorly loamy mesic and loamy fragmental type of soil, and regions from Kaksar to Kargil have sandy skeletal and loamy calcareous type soil units. As among soil types, the sandy skeletal are greater susceptible to landslides, followed by fine loamy, loamy fragmental, and loamy calcareous. The regions closer to Kangan have majorly fine loamy soils, while near the Kargil region, the soils are sandy skeletal, but major of NH-1 comes under loamy calcareous. This makes the regions of Kargil region less susceptible to landslides when compared with the Kangan side (Fig. 14.19).

Although above-mentioned finding up to Landslide Inventory units are the qualitative analysis of the study, as already mentioned in the methodology part both qualitative and quantitative analysis have been done in this study, the illustration of which is mentioned in the weight overlay analysis at the end of the chapter.

14.4.10.1 Landslide Inventory Unit Analysis

As earlier mentioned, the Landslide Inventory mapping with 25% of the data is 11 landslides collected from NASA Cooperative Open Landslide Repository (COOLR) database shapefiles, but the rest 75% are nearly 34 landslides were framework and identified from Google Earth Imagery of total 45 landslides were identified within the study area, as shown in Fig. 14.20. The Google Earth Imagery helped in the identification of nearly 45 landslide within the region of study from where the majorly 25–30 landslides were identified between Pandras to Kangan and 10–15 landslides were recognized from Drass to Kargil region. Google Earth imagery revealed the major landslide (~ 0.65 km^2) between Zero point to Gumri, major and large sizes with an area of 0.19 km^2 of landslides were identified within the Zojila Axis, and the smallest landslide (6056 m^2) was recognized in Hardass region. Similar findings could be seen from Nanda et al. (2021). "*The landslides do not show any abrupt change, but the smallest landslide was recognized near Hardas, while the largest one was at Zojila Axis*", Nanda et al. (2021). Digital elevation models (DEMs), sourced from missions like SRTM and ASTER, provide crucial elevation data for slope analysis. Slope and aspect maps derived from DEMs are key factors in assessing landslide susceptibility. Geological maps, available from organizations such as the USGS and GSI, offer information on rock types, fault conditions, and structural features. Soil maps, sourced from the FAO and national databases, provide additional data on soil types and properties. In conclusion, the regions nearer to the Zojila Axis are more susceptible to landslides when compared with regions from Drass to Kargil regions, which pieces of evidence are revealed from Google Earth Imagery. "*The Villages fall under the zones of Matayin, Pandras, and Gumri are more Susceptible to Landslides*" (Nanda et al. (2021). The illustration of the above findings is shown in Figs. 14.20 and 14.21 (Table 14.4).

As already mentioned, that, among all the 14 parameters and units, seven were selected for further examination through heuristic and statistical analysis. This process involved assigning appropriate weights to each parameter based on expert and researcher input, ultimately resulting in the creation of a Landslide Susceptibility Map using the weight overlay method. The seven units considered were aspect, road distance, distance to stream, elevation model, land use/land cover (LULC), precipitation units, and slope classes. Additionally, landslide inventory mapping and other units contributed to both qualitative and quantitative analysis of the study. A proper weighting system was implemented for these seven units of landslide susceptibility mapping. "*Experts used their knowledge and experience to assign weights to various factors such as slope, geology, and land use, based on their perceived importance in causing landslides*". "*The weight overlay system is a straightforward, direct, and adequate methodology for evaluating land susceptible zones and landslide areas*" (Ahmed et al. 2015). Each of the seven units was assigned suitable weights based on the number of classes, observations from Google Earth Imagery, and literature reviews. These weights were then converted into corresponding normalized factors, ranging from 0 to 1, using an appropriate normalization formula, as shown in Table 14.3. The susceptibility index [$S(x, y)$] which was calculated in the methodology section revealed a value of 0.7 and 70% and from second method also revealed a susceptibility value or index of nearly 0.7 and 70%, this results in the fact that the value ranges in between moderate susceptibility index to high susceptibility within the study area as shown above. Also, after the overlaying of all seven units together a final weight overlay map was prepared which was classified into 4 major classes of

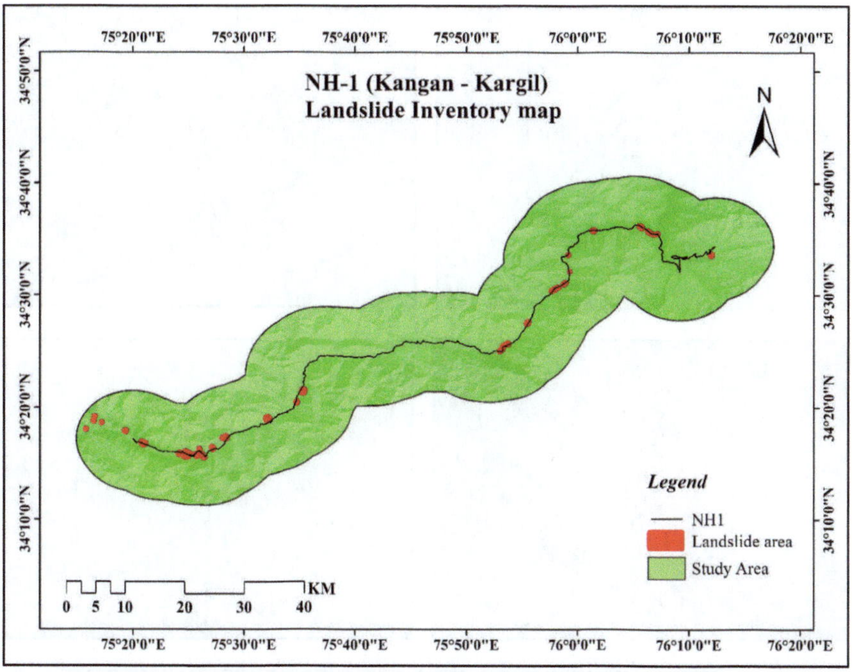

Fig. 14.20 Illustration of different 45 identified landslides within the study area of NH-1 (Kangan to Kargil). *Source* 25% of the data that are 11 landslides were collected from NASA Cooperative Open Landslide Repository (COOLR), the rest 75% are nearly 34 landslides were framework and identified from Google Earth Imagery

landslide susceptibility units which are: low susceptibility zones, moderate susceptibility zone, high susceptibility zone, and very high susceptibility zones, for which the area coverage is shown in Table 14.5 and weight assignment to respective parameters of overlay are shown in Table 14.6 as follows.

The information illustrated in Table 14.5 is derived from the process of overlaid mapping of seven parameters, which unveils that, the total size of cells in the entire study area is 220 × 220 is 48,400, also the entire landslide susceptibility unit is divided into low, moderate, high, and very high zones. The low susceptibility zone covers 32.67 km^2 which is 1.71% of the entire study area, the moderate susceptibility zone covers 467.931 km^2 which is 24.4% of the study area, the high susceptibility zone covers 1138.228 km^2 is 59.58% of the study area, also very high susceptibility zone covers 271.4272 km^2 that is 14.2% of the study area. Similar findings and divisions are made as; "*Landslide susceptibility units are divided into five zones as very high, high, moderate, low and very low zones*" (Nanda et al. 2021; Pravat Kumar et al. 2016). From the above illustration and analysis, we can conclude that the major landslide susceptibility area falls in between the moderate to high susceptibility zone making 83.98% of the entire study area, the similar result we have already seen from the susceptibility index earlier that major susceptibility units that are 70% falls in

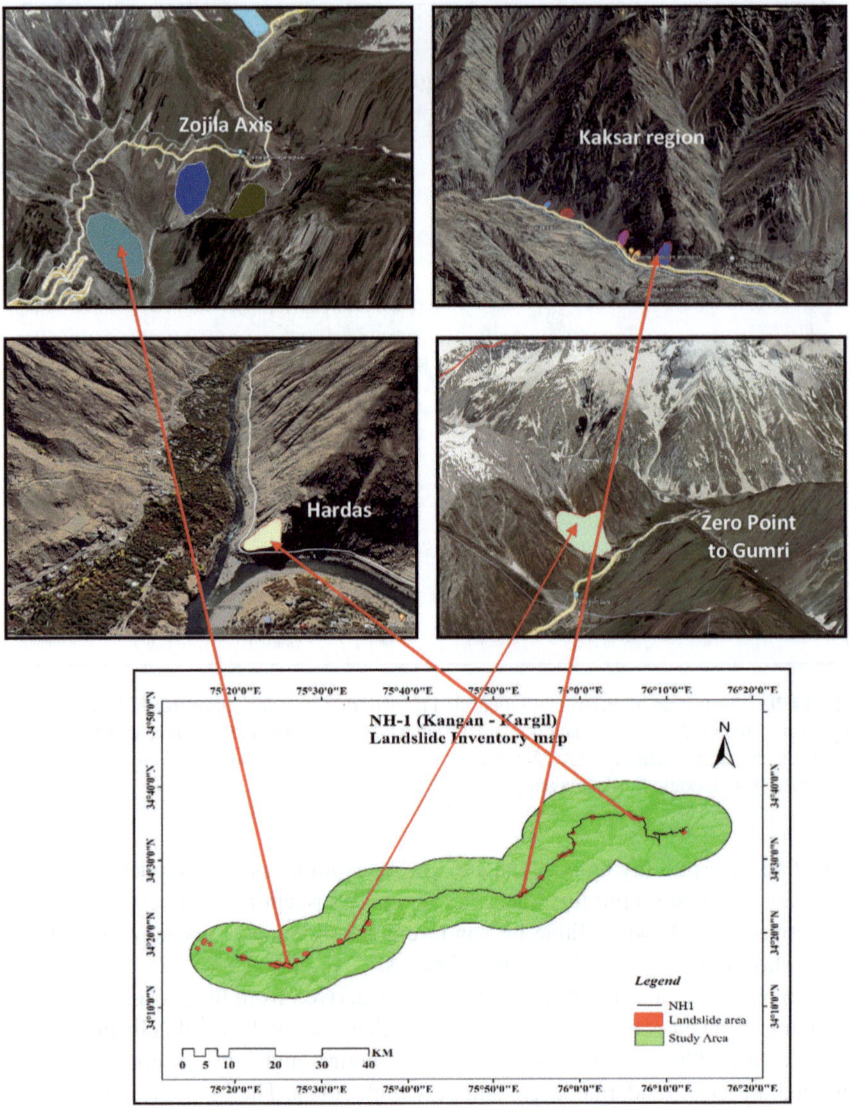

Fig. 14.21 Major and minor landslides within the vicinity of Zojila, Zero Point-Gumri, Kaksar, and Hardas's region, that are illustrated from Landslide Inventory Map prepared in ArcGIS 10.8.2. *Source* NASA Cooperative Open Landslide Repository (COOLR), and Google Earth Imagery

Table 14.4 Susceptibility index value and classes

Low susceptibility index	0–0.3	0–30%
Moderate susceptibility index	0.3–0.7	30–70%
High susceptibility index	0.7–1	70–100%

Table 14.5 Area coverage in m^2, km^2, and % age, for classes of weight overlay with 220 × 220 cell size of the study area

Cell size	Category or classes	Count value	Area in m^2	Area in km^2	Area in %
220 × 220	Low	675	32,670,000	32.67	1.71
	Moderate	9668	467,931,200	467.931	24.4
	High	23,517	1,138,222,800	1138.2228	59.58
	Very high	5608	271,427,200	271.4272	14.2
Total = 48,400		39,468	1,910,251,200	1910.251	100

between moderate to high susceptibility zones. Also, from landslide inventory unit reveals that the major susceptibility zone was in between Zojila to Pandras of regions along with Gumri and Matayin. As earlier mentioned, that from entire 45 landslides were mapped among which nearly 30 landslides that are 66.6% of landslides were identified between Zojila to Pandras, also rest nearly 15 landslides that are 33.3% of which were identified between Pandras to Kargil. Comparing the result with others; "*the major susceptibility zone falls under Very High and High Susceptibility Zones of 42% of the entire study area, also these regions include Zojila, Gumri, Matayin and Pandras*" (Nanda et al. 2021).

Table 14.6 shows the proper weights that are assigned to the respective parameters of the study that are taken for weight overlay analysis, based on the author's experience and visualization the weights were provided which are discussed in the methodology chapter.

The above formulation and framework helped in the preparation of the weight overlay map, and by calculating its corresponding normalized factors, the landslide susceptibility index was calculated which was nearly 70% in both methods and falls in between moderate to high susceptibility index value. The weight assigned margin of greater importance to slope 30%, followed by LULC 25%, rainfall 20%, erlevation 10%, aspect 7%, distance to stream 5%, and distance to road 3%. These weights and influencing factors finally helped in the preparation of the weight overlay land susceptibility map as follows with its discussions (Fig. 14.22).

The above final illustration of the susceptibility map unveils that the majority of landslides that is 24.4% of moderate 59.58% of high and 14.2% of very high susceptibility zones are present within the regions from Kangan to Pandras (Table 14.5 and Fig. 14.22). The remaining low value of 1.71% could be seen majorly from Drass to Kargil regions.

The discussion of both qualitative and quantitative analyses that we have done so far from aspect to final landslide susceptibility of very high, high, moderate, and low zones are comprehensively highlighted as follows:

Very High Susceptibility Zone: This zone of very high covers an area of 271.4272 km^2 which makes up 14.2% of the entire study area, these zones have a purely north aspect of slope face, elevation greater than 3500 m from mean sea

Table 14.6 Assignment of different weights and ranking to every parameter

Parameters	Ranks	Categories or classes	Weights individual	Influencing factor (%)	Weight from % (0–1)	Normalized factor
Slope	1st	0–15	1	30%	0.3	1
		15–30	3			
		30–45	6			
		45–60	8			
		> 60	12			
LULC	2nd	Water	6	25%	0.25	0.81
		Vegetation	0.5			
		Agriculture	4			
		Built-up	8			
		Snow	1			
		Shrub	0.5			
		Barren land	5			
Rainfall	3rd	102.67–244	1	20%	0.2	0.62
		244–412.4	1			
		412.4–590.9	4			
		590.9–772.7	6			
		772.7–961.2	8			
Elevation	4th	2547–3341	1	10%	0.1	0.25
		3341–3878	2			
		3878–4396	3			
		4396–5445	4			
Aspect	5th	N	1	7%	0.07	0.14
		NE	2			
		E	3			
		SE	0.4			
		S	0.1			
		SW	0.1			
		W	0.1			
		NW	0.3			
Distance to stream	6th	< 200	2	5%	0.05	0.07
		200–400	1			
		401–800	1			
		801–1200	0.5			
		> 1200	0.5			
Distance to road	7th	0–300	1	3%	0.03	0.0
		300–1000	0.8			
		1000–2000	0.5			
		2000–3500	0.4			
		3500–4000	0.3			

level, major to a medium fault line, more shale, schist, phyllites, and quartzite beds, a distance of proximity to the road is < 300 m very little vegetation cover, more than 772.722 mm rainfall annually, angle of slope has an average of 30º–45º, fine loamy and loamy fragmental type of soil is common. Some regions included in this type

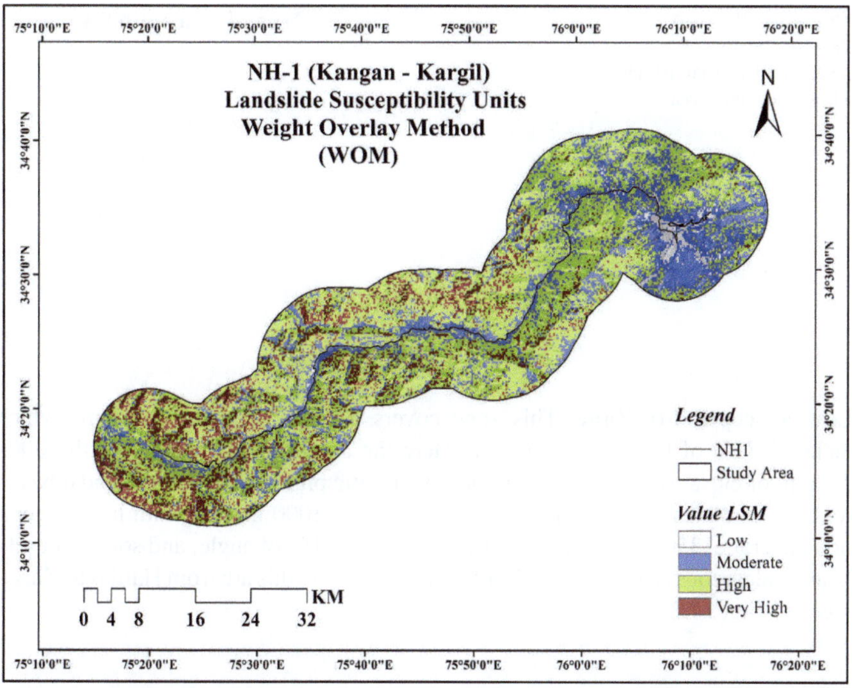

Fig. 14.22 Final landslide susceptibility map, by using weight overlay analysis from all seven parameters of thematic layer, i.e. slope, aspect, distance to road, distance to stream, elevation, LULC, and rainfall in ArcGIS 10.8.2

are Kangan, Sonamrg, Zojila, Zero Point, Gumri, and Matayin. "*Most of the villages that fall under this category are Sonamarg, Zojila, and Gumri*" (Nanda et al. 2021).

High Susceptibility Zone: This zone covers an area of about 1138.2 km^2 which makes 59.58% of the entire study area. Usually consists of North and East facing aspects, 3000 m elevation, medium fault lines, rocks such as schist, phyllites, and quartzite bed are common, distance of proximity to the road is 300–1000 m, less to medium vegetation cover, 412.448–590.900 mm annual rainfall, slope angle less than 30°, and loamy fragmental type of soil are common. Some regions that can be included here are; Kangan, Sonamarg, Drass Valley, Kaksar, and Chokiyal.

Moderate Susceptibility Zone: This zone covers an area of about 467.913 km^2 which makes up 24.4% of the study area. Consists of South East and South West aspects of the slope face, elevation of about 2547–3341 m, rock types are typically amygdaloidal basalt and some bedded limestone and sandstone depositions with minor granitic plutons. The distance of proximity to the road is 1000–2000 m. Its vegetation cover is moderate to high, with 412.448 m rainfall annually, 15º–30º of slope angle, loamy fragmental and loamy calcareous type of soil. Some regions included here are Chanigound and Hardas.

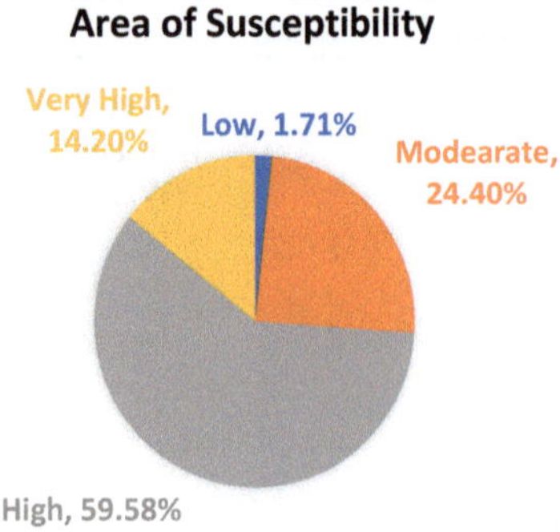

Fig. 14.23 Pie chart showing the area of susceptibility of landslides within the study area

Low Susceptibility Zone: This zone covers an area of about 3.67 km^2 which makes 1.71% of the total study area. Here the aspect is purely west, with 2000–2500 m of elevation, rock types are purely granitic plutons and amygdaloidal basalt rocks. The distance of proximity to the road is > 3000 m, also with high vegetation coverage, 350–400 mm rainfall, slopes are 10–15° of angle, and soil are purely loamy calcareous type of soils. The regions included in this are from Hardas to Kargil regions (Fig. 14.23).

14.5 Conclusion

Landslides are highly destructive natural hazards that pose significant risks to human life and infrastructure. This study of landslide susceptibility was conducted on the NH-1 National Highway between Kangan and Kargil in the union territory of Ladakh. The study area was categorized into low, moderate, high, and very high susceptibility zones. The aspect analysis revealed that regions such as Kangan, Sonamarg, and Zoji La, with north and northeast-facing slopes, are more susceptible to landslides compared to regions with south and southwest-facing slopes like Drass and Kargil. Villages near Matayin, Pandras, and Gumri are particularly vulnerable. The study's findings, illustrated in various figures and tables, demonstrate that 24.4% of the area is moderately susceptible, 59.58% highly susceptible, and 14.2% very highly susceptible to landslides, with only 1.71% of low susceptibility, primarily in regions from Drass to Kargil. The conclusions can be made from the fact that the landslide susceptibility model between Kangan to Kargil unveils the findings that majorly regions from Kangan to Pandras are more susceptible to landslides when compared with regions between Drass to Kargil. A landslide inventory map was prepared for the identification of landslides. It unveils 66.6% of landslides identified along Kangan to Pandras and nearly 33.3% along Drass to Kargil regions. In the end by overlaying the thematic layer in ArcGIS 10.8.2 a weight overlay analysis was applied, resulting in the formation of the final susceptibility map, which unveils the major of the landslide towards the Zojila side, also during overlay analysis two susceptibility index methodology was run on the parameters which also unveils nearly 70% of landslide susceptibility index in the study area. Although a comprehensive study has been used for this study despite these the study lacks some limitations such as more samples of landslides could be taken for better results, in the current study only 45 landslide samples were taken during landslide inventory mapping. Similarly, with the help of both primary and secondary data, the result would be more accurate and complete, although a short field visit was done just only to know the lithology and rock types on a few places like Chanigound and Akchamal side of Kargil.

To mitigate the risks posed by landslides in the Kangan to Kargil region, several strategies should be implemented. Early warning systems through the installation of advanced monitoring equipment such as ground-based radar and LiDAR can provide real-time data on slope movements and potential landslide triggers. Implementing robust land use planning and zoning regulations is crucial, especially in high and very high susceptibility zones. These regulations should restrict construction and development activities in vulnerable areas, thereby minimizing human exposure to landslide risks. Vegetation plays a critical role in stabilizing slopes, so reforestation and afforestation programmes should be prioritized in deforested and degraded areas. These programmes can help reinforce soil structure and reduce the likelihood of slope failures. Additionally, engineering solutions such as retaining walls, drainage systems, and slope reinforcement should be employed in particularly vulnerable locations to enhance slope stability. Public awareness campaigns and education programmes are also vital. These initiatives should inform local communities about the risks of landslides, safe practices, and emergency response strategies. Establishing and training local disaster response teams can further ensure a swift and effective response in the event of a landslide. Lastly, continuous collaboration between government agencies, local communities, and scientific institutions is essential to maintain an integrated approach to landslide risk management. Continuous monitoring and evaluation of implemented mitigation strategies are essential to ensure their effectiveness and to make necessary adjustments based on evolving conditions and new research findings. This iterative approach will help build resilience against landslides and protect both human lives and infrastructure in the long term. The events that happened in the past are the key ways to mitigate and overcome the landslide area from NH-1 Kangan to Kargil, these regions especially from Kangan to Pandras along with Gumri and Matayin need to take proper research and mitigation programmes in the future.

References

Abdo HG (2022) Assessment of landslide susceptibility zonation using frequency ratio and statistical index: a case study of Al-Fawar basin, Tartous, Syria. Int J Environ Sci Technol 19(4):2599–2618

Ahmed B (2015) Landslide susceptibility mapping using multi-criteria evaluation techniques in Chittagong Metropolitan Area, Bangladesh. Landslides 12(6):1077–1095

Ahmad R, Joshi MN (2010) Assessment of landslide susceptibility on land degradation processes in Chamoli and surrounding area using RS and GIS technique. Int Geoinf Res Dev J 1(3):1

Ahmad WS, Jamal S, Taqi M, El-Hamid HTA, Norboo J (2024a) Estimation of soil erosion and sediment yield concentrations in Dudhganga watershed of Kashmir Valley using RUSLE & SDR model. Environ Dev Sustain 26(1):215–238

Ahmad WS, Kaloop MR, Jamal S, Taqi M, Hu JW, Abd El-Hamid H (2024) An analysis of LULC changes for understanding the impact of anthropogenic activities on food security: a case study of Dudhganga watershed, India. Environ Monit Assess 196(1):105

Aiswarya L, Rema KP, Shyla J, Brijesh VK, Vaisakh V (2023) Landslide susceptibility mapping of Chaliyar River Basin by multivariate statistical model. J Exp Agric Int 45(12):23–35

Akgun A, Erkan O (2016) Landslide susceptibility mapping by geographical information system-based multivariate statistical and deterministic models: in an artificial reservoir area at Northern Turkey. Arab J Geosci 9:1–15

Aksoy H (2023) Determination of landslide susceptibility with Analytic Hierarchy Process (AHP) and the role of forest ecosystem services on landslide susceptibility. Environ Monit Assess 195(12):1525

Al-Sababhah N (2022) Development of landslide susceptibility mapping using GIS modeling in Jordan's Northern highlands. Environ Ecol Res 10(6):701–727

Batar AK, Watanabe T (2021) Landslide susceptibility mapping and assessment using geospatial platforms and weights of evidence (WoE) method in the Indian Himalayan Region: recent developments, gaps, and future directions. ISPRS Int J Geo Inf 10(3):114

Behera DK, Jamal S, Ahmad WS, Taqi M, Kumar R (2023) Estimation of soil erosion using RUSLE Model and GIS tools: a study of Chilika Lake, Odisha. J Geol Soc India 99(3):406–414

Biswakarma P, Joshi V (2021) GIS based bi-variate statistical study for the assessment of landslide susceptibility in the West Sikkim District of Sikkim Himalaya, India. J Ind Geol Congr 12(2):181–194

Dey S, Das S (2022, February) Comparative exploration of statistical techniques for landslide identification and zonation mapping and assessment: a critical review. In: International symposium on artificial intelligence. Springer Nature Switzerland, Cham, pp. 44–55

Ganaie TA, Jamal S, Ahmad WS (2021) Changing land use/land cover patterns and growing human population in Wular catchment of Kashmir Valley, India. GeoJournal 86:1589–1606

Ganser A (1964) Geology of the Himalaya. InterScience Publishers, New York, p 289

Gupta V, Singh K (2023, February) Landslide susceptibility mapping along Chandigarh-Shimla Road corridor NH-5, Himachal Pradesh, India. In: AIP conference proceedings. AIP Publishing, vol 2558, no 1

Gyawali P, Aryal YM, Tiwari A, Prajwol KC, Ansari K (2021) Landslide susceptibility assessment using bivariate statistical methods: a case study of Gulmi District, western Nepal. VW Eng Int 3(2):29–40

Hamid B, Massinissa B, Nabila G (2023) Landslide susceptibility mapping using GIS-based statistical and machine learning modeling in the city of Sidi Abdellah, Northern Algeria. Model Earth Syst Environ 9(2):2477–2500

Jamal S, Ahmad WS (2020) Assessing land use land cover dynamics of wetland ecosystems using Landsat satellite data. SN Appl Sci 2(11):1891

Jamal S, Ahmad WS, Ajmal U, Aaquib M, Ashif Ali M, Babor Ali M, Ahmed S (2022) An integrated approach for determining the anthropogenic stress responsible for degradation of a Ramsar Site-Wular Lake in Kashmir India. Mar Geodesy 45(4):407–434

Jari A, Khaddari A, Hajaj S, Bachaoui EM, Mohammedi S, Jellouli A, Barakat A et al (2023) Landslide susceptibility mapping using multi-criteria decision-making (MCDM), statistical, and machine learning models in the Aube Department, France. Earth 4(3):698–713

Kamp U, Growley BJ, Khattak GA, Owen LA (2008) GIS-based landslide susceptibility mapping for the 2005 Kashmir earthquake region. Geomorphology 101(4):631–642

Kanwal S, Atif S, Shafiq M (2017) GIS-based landslide susceptibility mapping of northern areas of Pakistan, a case study of Shigar and Shyok Basins. Geomat Nat Haz Risk 8(2):348–366

Imran K, Ashutosh K, Mansoor A, Inzamam UH, Asgher Md, Vishnu P (2022) Landslide susceptibility in the watershed of Ramban District, Jammu and Kashmir Himalaya

Lee S (2019) Current and future status of GIS-based landslide susceptibility mapping: a literature review. Korean J Rem Sens 35(1):179–193

Mallick J, Singh RK, AlAwadh MA, Islam S, Khan RA, Qureshi MN (2018) GIS-based landslide susceptibility evaluation using fuzzy-AHP multi-criteria decision-making techniques in the Abha Watershed, Saudi Arabia. Environ Earth Sci 77:1–25

Mansouri Daneshvar MR (2014) Landslide susceptibility zonation using analytical hierarchy process and GIS for the Bojnurd region, northeast of Iran. Landslides 11(6):1079–1091

Mir RA, Habib Z, Kumar A, Bhat NA (2024) Landslide susceptibility mapping and risk assessment using total estimated susceptibility values along NH44 in Jammu and Kashmir, Western Himalaya. Natural Hazards 1–40

Mowen X, Tetsuro ESAKI, Cheng Q, Lin J (2007) Spatial three-dimensional landslide susceptibility mapping tool and its applications. Earth Sci Front 14(6):73–84

Murthy T, Yusoff IM, Samsudin SH, Majid TA, Abir IA, Yong CH, Ismail MAM (2023) Geographical Information System (GIS)-based landslide susceptibility mapping in Malaysia: a review of past, current and future trends

Nanda AM, Hassan ZU, Ahmed P, Kanth TA (2021) Landslide susceptibility assessment of national highway 1D from Sonamarg to Kargil, Jammu and Kashmir, India using frequency ratio method. GeoJournal 86:2945–2956

Pal SC, Chowdhuri I (2019) GIS-based spatial prediction of landslide susceptibility using frequency ratio model of Lachung River basin, North Sikkim, India. SN Appl Sci 1:1–25

Roccati A, Paliaga G, Luino F, Faccini F, Turconi L (2021) GIS-based landslide susceptibility mapping for land use planning and risk assessment. Land 10(2):162

Saadatkhah N, Kassim A, Lee LM, Yunusa GH (2015) Quantitative hazard analysis for landslides in Hulu Kelang area, Malaysia. J Teknol (Sci & Eng) 73(1)

Saha A, Villuri VGK, Bhardwaj A (2020) Assessment and preparation of landslide susceptibility zonation map by geospatial method using remote sensing and GIS. In: Proceedings of the ACRS

Sarkar S, Kanungo DP (2017) GIS application in landslide susceptibility mapping of Indian Himalayas. GIS Landslide 211–219

Sengupta A, Nath SK (2022) GIS-based landslide susceptibility mapping in the eastern boundary zone of northeast India in compliance with Indo-Burmese subduction tectonics. In: Geospatial technology for environmental hazards: modelling and management in Asian countries, pp 19–37

Shahabi H, Ahmad BB, Khezri S (2012) Landslide susceptibility mapping using image satellite and GIS technology. Int J Res Eng Technol (IJRET) 1(6–2016)

Sharma A, Sur U, Singh P, Rai PK, Srivastava PK (2020) Probabilistic landslide hazard assessment using statistical information value (SIV) and GIS techniques: a case study of Himachal Pradesh, India. In: Techniques for disaster risk management and mitigation, pp 197–208

Shit PK, Bhunia GS, Maiti R (2016) Potential landslide susceptibility mapping using weighted overlay model (WOM). Model Earth Syst Environ 2:1–10

Singh AL, Jamal S, Ahmad WS (2024) Climate change vulnerabilities and adaptation understanding and addressing threats with insights for policy and practice. Springer Nature Switzerland, Cham

Singh P, Sharma A, Sur U, Rai PK (2021) Comparative landslide susceptibility assessment using statistical information value and index of entropy model in Bhanupali-Beri region, Himachal Pradesh, India. Environ Dev Sustain 23:5233–5250

SOMA AS. Land use change and landslide susceptibility assessment using GIS and multivariate quantitative predictive models for mountainous disaster mitigation in South Sulawesi Indonesia

Streule MJ, Strachan RA, Searle MP, Law RD, Law RD, Butler RWH, ... Strachan RA (2010) Comparing Tibet–Himalayan and Caledonian crustal architecture, evolution and mountain building processes Continental Tectonics and Mountain Building: The Legacy of Peach and Horne. Geol Soc 335:207–232

Tripathi G, Shakya A, Upadhyay RK, Singh SK, Kanga S, Pandey SK (2023) Landslide susceptibility mapping of Tehri reservoir region using geospatial approach. Climate change adaptation, risk management, and sustainable practices in the Himalaya. Springer International Publishing, Cham, pp 135–156

Tubog MV, Villahermosa RL, Perong JG (2023) Landslide susceptibility modelling derived from remote sensing, multi-criteria decision analysis, and GIS techniques: a case study in the Southeast Bohol Province, Philippines

Valdiya KS (1984) Aspects of tectonics. Focus on south-central Asia. Tata McGraw-Hill,New Delhi, p 319

Yadav M, Pal SK, Singh PK, Gupta N (2023) Landslide susceptibility zonation mapping using frequency ratio, information value model, and logistic regression model: a case study of Kohima District in Nagaland, India. Landslides: detection, prediction, and monitoring: technological developments. Springer International Publishing, Cham, pp 333–363

Zhou W, Minnick MD, Chen J, Garrett J, Acikalin E (2021) GIS-based landslide susceptibility analyses: case studies at different scales. Nat Hazard Rev 22(3):05021007

Chapter 15
Disaster Risk Reduction and Resilience in Mountain Areas: A Case Study of 2014 Floods of Kashmir Valley

Shamim Ahmad Shah, Peer Jeelani, and Sumira Mir

Abstract The 2014 floods in Kashmir Valley were among the most catastrophic natural disasters in the region's history. This chapter provides a comprehensive analysis of the flood's multifaceted impacts, the underlying risk factors, and proposed strategies for enhancing resilience against future disasters. The floods caused unprecedented economic losses, extensive damage to infrastructure, and significant disruption to the region's socioeconomic fabric. The agricultural sector, a cornerstone of Kashmir's economy, suffered substantial losses due to crop damage and soil erosion, while the tourism industry experienced a sharp decline due to the destruction of infrastructure. The chapter highlights the critical importance of flood risk management in the Kashmir Valley, given its unique topography and climatic conditions, which contribute to its susceptibility to natural disasters. Key risk factors include the region's location in a seismically active zone, rapid urbanization, deforestation, and inadequate drainage systems. The socio-political instability in the region further complicates disaster management efforts, hindering effective coordination and implementation of disaster risk reduction (DRR) initiatives. To better understand the determinants of flood risk and impact, a primary survey was conducted using Structural Equation Modelling (SEM). The survey revealed that exposure and vulnerability are key determinants of flood risk perception and impact. However, preparedness emerged as a crucial factor in mitigating these risks. Communities with better preparedness through evacuation plans, early warning systems, and community drills experienced lower perceived risk and were better able to cope with the impacts of the floods. Effective response and recovery efforts also played a significant role in reducing long-term negative effects, underscoring the importance of robust, well-coordinated response mechanisms. Based on the findings, the chapter advocates for a multifaceted approach to flood risk management in Kashmir Valley. Recommended strategies include strengthening flood defences and infrastructure,

---

S. A. Shah (✉) · P. Jeelani · S. Mir
Department of Geography and Disaster Management, University of Kashmir, Srinagar, Jammu & Kashmir, India
e-mail: shamimcrocus@gmail.com

S. C. Rai (ed.), *Himalaya: Mountains of Destiny*,
https://doi.org/10.1007/978-3-031-94905-0_15

implementing advanced early warning systems, enforcing stringent land-use planning and zoning regulations, enhancing community awareness and preparedness, and adopting ecosystem-based approaches to flood risk reduction.

Keywords 2014 floods · Kashmir Valley · DRR · SEM · Stakeholders

15.1 Introduction

The Kashmir Valley, nestled in the Himalayas, is no stranger to natural disasters, with its unique topography and climatic conditions making it particularly vulnerable to floods. The 2014 floods, however, stand out as one of the most catastrophic events in the region's history, causing widespread devastation. This chapter delves into the multifaceted impacts of the 2014 floods, focusing on economic losses, disruptions to education, loss of life, and property damage. Additionally, it underscores the importance of flood risk management and outlines strategies for building resilience against such calamities.

Kashmir Valley, often referred to as "Paradise on Earth", is a region of exceptional beauty nestled between the Greater Himalayas and the Pir Panjal Range (Jeelani et al. 2024; Jeelani and Shah 2023). The valley, with its picturesque landscapes, rich cultural heritage, and diverse ecology, has been a focal point of both admiration and conflict throughout history. The region is characterized by its temperate climate, lush forests, rivers, and lakes, making it a unique geographical entity within the Indian subcontinent (Jeelani et al. 2023; Jeelani and Shah 2024; Jeelani et al. 2025). Despite its natural beauty, Kashmir Valley is highly vulnerable to a range of natural hazards, including earthquakes, floods, landslides, and avalanches, making disaster risk management a critical concern for the region. The valley has been the subject of extensive research due to its strategic importance, complex socio-political history, and vulnerability to natural disasters. Several studies have explored the region's unique geography and the challenges it faces in terms of disaster risk management.

The valley's topography and climatic conditions contribute to its susceptibility to natural disasters. Bhat et al. (2021) highlighted that the region's location in a seismically active zone, coupled with its steep slopes and heavy rainfall, makes it prone to earthquakes, landslides, and floods. The catastrophic floods of 2014 serve as a stark reminder of the valley's vulnerability, with studies by Mir et al. (2024a, b) documenting the extensive economic losses, displacement, and infrastructural damage caused by this event. These studies underscore the need for improved flood management strategies and resilient infrastructure in the region.

The socio-political dynamics of Kashmir have further complicated disaster management efforts. The protracted conflict in the region has not only led to significant human suffering but has also hindered developmental initiatives, including those aimed at disaster risk reduction (DRR). According to Hakim (2020), the ongoing conflict has exacerbated the region's vulnerabilities by disrupting governance structures, limiting access to resources, and creating a pervasive sense of insecurity among the local population. This has made it challenging to implement effective disaster management policies and ensure community participation in DRR initiatives.

In recent years, there has been a growing recognition of the need to strengthen institutional frameworks to enhance disaster resilience in Kashmir. The work of Mir et al. (2024a, b) emphasizes the importance of integrating local knowledge with scientific expertise to develop context-specific DRR strategies. Their research advocates for the institutionalization of multi-stakeholder networks that include government agencies, non-governmental organizations, and local communities, as a means to build resilience against disasters. Additionally, the role of education in promoting disaster awareness and preparedness has been highlighted by Mir et al. (2023), who argues for the inclusion of DRR in the curriculum of educational institutions in disaster-prone areas like Kashmir.

15.2 The 2014 Floods: An Overview

In September 2014, unprecedented rainfall led to severe flooding in the Kashmir Valley. The Jhelum River, along with its tributaries, swelled beyond its capacity, inundating large parts of the valley. The floods were exacerbated by inadequate drainage systems, rapid urbanization, and deforestation, which hindered the natural absorption and drainage of water. The 2014 floods in Kashmir were among the most devastating natural disasters to strike the region in recent history, leaving an indelible mark on its socioeconomic fabric (Fig. 15.1). Triggered by unprecedented rainfall in early September, the floods resulted in widespread inundation across the Kashmir Valley, particularly in the districts of Srinagar, Anantnag, and Pulwama. The Jhelum River, which courses through the valley, breached its banks at multiple points, exacerbating the flood's severity. According to reports, more than 50% of Srinagar, the summer capital of Jammu and Kashmir, was submerged, paralyzing the city's infrastructure and displacing thousands of residents.

Fig. 15.1 Devastation of 2014 floods

The human toll of the disaster was significant, with official estimates citing around 300 deaths and over half a million people directly affected. The floods caused extensive damage to property, including the destruction of homes, schools, hospitals, and government buildings. Critical infrastructure such as roads, bridges, and communication networks was severely impacted, hampering rescue, and relief operations. The economic losses were staggering, with estimates suggesting damages running into billions of dollars, severely affecting the region's already fragile economy.

In addition to the immediate destruction, the 2014 floods had long-term implications for the region (Fig. 15.2). The agricultural sector, which is a vital part of Kashmir's economy, suffered substantial losses due to crop damage and soil erosion. The floods also highlighted the inadequacies in the region's disaster preparedness and response mechanisms. Studies have pointed to the lack of effective flood management infrastructure, such as adequate drainage systems and flood forecasting models, as contributing factors to the scale of the disaster (Romshoo and Rashid 2014). Furthermore, the socio-political instability in the region compounded the challenges of disaster management, as coordination between various stakeholders, including government agencies, non-governmental organizations, and local communities, was often lacking.

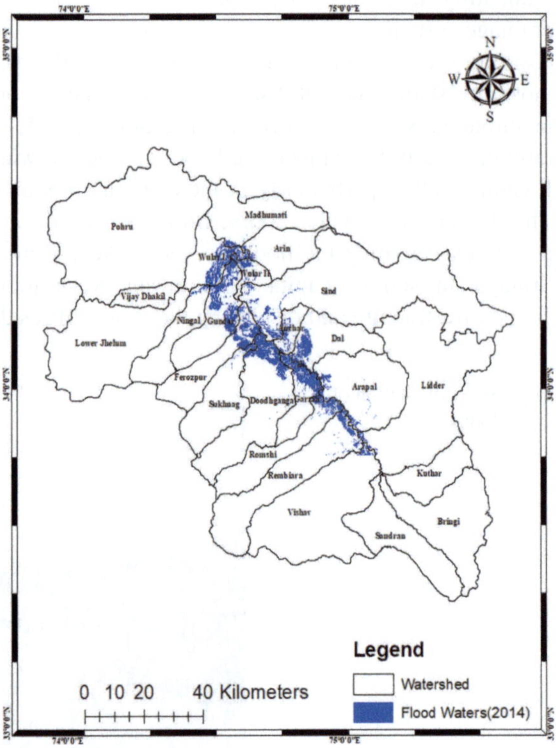

Fig. 15.2 Spatial extent of floodwaters across the watersheds in Central Kashmir

In the aftermath of the floods, there has been a concerted effort to improve the region's resilience to such disasters. Various studies have called for the integration of traditional knowledge with modern scientific approaches to flood management (Alam et al. 2018). The need for a comprehensive, multi-hazard approach to disaster risk reduction in the Kashmir Valley has also been emphasized, recognizing the region's vulnerability not only to floods but also to earthquakes, landslides, and other natural hazards.

However, the implementation of these measures has faced several challenges, including limited resources, political instability, and the ongoing conflict in the region. The 2014 floods serve as a critical case study for understanding the complexities of disaster management in conflict-prone areas and highlight the urgent need for sustainable and inclusive disaster risk reduction strategies.

The floods dealt a heavy blow to Kashmir's tourism industry, a critical component of the region's economy. The destruction of infrastructure, including hotels, houseboats, and transport facilities, led to a sharp decline in tourist arrivals, affecting businesses and livelihoods dependent on tourism. The floods caused significant disruptions to the education sector in Kashmir. Thousands of schools and educational institutions were either completely destroyed or severely damaged (Mir et al. 2024a, b). The immediate aftermath of the disaster saw educational activities come to a standstill as schools were either used as relief camps or rendered non-functional due to damage. The prolonged closure of schools resulted in a substantial loss of academic time for students. The floods struck at the beginning of the academic session, leading to interruptions that impacted the quality of education and student performance. Beyond the physical damage to schools, the floods had a profound psychosocial impact on students. Many children were traumatized by the disaster, and the disruption to their routine had long-term effects on their mental health and academic performance. Table 15.1 illustrates a list of major flood events in the Kashmir Valley, along with their interpretations.

15.3 Importance of Flood Risk Management

The 2014 floods underscored the urgent need for effective flood risk management in the Kashmir Valley. As the region remains prone to such disasters, building resilience through comprehensive flood risk strategies is crucial. Flood risk assessment is a critical component of disaster risk management, particularly in regions prone to flooding. It involves the systematic evaluation of potential flood hazards, vulnerabilities, and the capacity of communities and infrastructure to withstand flood events. The primary goal of flood risk assessment is to identify areas at risk, estimate potential impacts, and develop strategies to mitigate these risks. This process is essential for informed decision-making, enabling authorities to prioritize resources, design effective flood defences, and implement land-use planning measures that reduce the exposure of people and property to flood hazards.

Table 15.1 Some major floods in the Kashmir Valley

Flood	Interpretation of the Events
1900	Continuous rains led to widespread flooding throughout the Kashmir basin. This event caused riverbanks to breach, leading to prolonged inundation
1903	Srinagar City and surrounding areas were severely affected by heavy rainfall between July 11 and 23. Water levels rose but were 2.7 m lower than the catastrophic 1893 flood. Extensive crop loss and damage to infrastructure occurred
1905	Significant flood with a maximum peak flood discharge (MPFD) of 67,889 cusecs, affecting large areas in the valley. Resulted in crop and property damage
1909	The flood impacted the Kashmir Valley with an MPFD of 60,866 cusecs, leading to a loss of crops and livestock
1912	Major flood with MPFD of 46,820 cusecs. Water covered extensive areas of the valley, destroying crops and displacing people. Heavy losses reported
1928	Entire valley transformed into a water expanse. Severe impact on Srinagar, with 7000 dwellings marooned and a cholera epidemic following the flood that killed 4225 people
1950	Heavy rains led to breaches in the Jhelum basin, affecting over 112 km^2. More than 15,000 homes were damaged, and transportation was severely disrupted
1957	Prolonged monsoonal floods submerged much of the valley. Nearly 600 villages were affected, with Jhelum River overflow damaging crops and homes, which led to famine in some areas
1959	Four days of incessant rain caused flooding, submerging over 104 villages. MPFD reached 46,000 cusecs, with high financial losses in agriculture and public utilities
1973	The flood-affected 40 villages, primarily in Jammu and parts of Kashmir. Approximately 20% of the population in the affected areas was impacted, with financial losses estimated at INR 121.8 million
1992	One of the most devastating floods in the region, affecting both India and Pakistan. Over 60,000 people were impacted in north-western border districts. Landslides triggered by rainfall exacerbated the disaster
1996	Affected 29,600 km^2 in Kashmir, with flooding impacting 10,000 homes in Srinagar and surrounding areas. Water levels remained high for weeks, worsening damage
2006 (July–August)	Monsoonal rains caused widespread riverine flooding in the Jhelum and Chenab basins. Thousands were displaced, with 15,000 left homeless
2006 (September)	Flash floods added to the ongoing crisis, resulting in 19 fatalities. Nearly 800 people were injured, and significant property damage occurred
2014	One of the most destructive floods in recent history. Torrential monsoon rains resulted in riverine flooding across Kashmir, displacing 275,000 people and causing INR 16 billion in damages
2015 (July)	A cloudburst near Amarnath led to a flash flood in the Pahalgam area, with many tents washed away. The flood disrupted religious pilgrimage activities

(continued)

Table 15.1 (continued)

Flood	Interpretation of the Events
2015 (September)	Flash floods and continuous rain affected large parts of Kashmir Valley. Multiple landslides were triggered, leading to additional damage and casualties
2017	Flooding in Chadoora Village, Budgam District, affected 80,000 residents. Flash floods from heavy rain led to landslides, damaging homes and infrastructure

Source Ali et al. (2022)

Flood risk assessment is especially important in regions like the Kashmir Valley, where the topography, climate, and hydrological patterns contribute to a high flood risk. The devastating 2014 floods in Kashmir underscored the need for comprehensive flood risk assessment. The absence of accurate flood risk maps and inadequate forecasting systems were among the factors that exacerbated the disaster's impact. Effective flood risk assessments can help in anticipating such events, allowing for timely evacuations, the deployment of emergency services, and the protection of critical infrastructure.

Furthermore, flood risk assessments are vital for the sustainable development of flood-prone areas. By understanding the likelihood and potential severity of flood events, planners can avoid or limit development in high-risk zones, thereby minimizing future economic losses and ensuring the safety of communities. For instance, incorporating flood risk assessment into urban planning can prevent the construction of housing and infrastructure in floodplains, which are particularly vulnerable to flooding (Merz et al. 2010).

In addition to reducing immediate risks, flood risk assessments contribute to long-term resilience by informing the design of flood adaptation measures. These may include the construction of levees, flood barriers, and the restoration of natural floodplains, which can absorb excess water and reduce flood peaks. Additionally, flood risk assessments can guide the implementation of early warning systems, which are crucial for reducing casualties during flood events (Jha et al. 2012).

In summary, flood risk assessment is indispensable for reducing the human and economic toll of floods. It provides the foundation for proactive flood management strategies, supports sustainable land-use planning, and enhances the overall resilience of communities to flood hazards. As climate change continues to increase the frequency and intensity of extreme weather events, the importance of flood risk assessment in safeguarding lives and livelihoods cannot be overstated.

15.3.1 Understanding Flood Risk

Flood risk is a multifaceted concept that involves the likelihood of flood events occurring and the potential consequences of these events on people, property, and the environment. Understanding flood risk is essential for effective disaster risk management and involves analysing three main components: hazard, exposure, and vulnerability:

(A) Hazard

The hazard component refers to the natural occurrence of flooding itself, including the frequency, duration, and intensity of flood events. Hazards can arise from various sources, such as riverine flooding, flash floods, coastal flooding, and urban flooding. Each type of flood has different characteristics, with some being gradual and predictable, while others, like flash floods, can occur suddenly with little warning. Understanding the nature of the flood hazard is the first step in assessing flood risk, as it helps in predicting where and when floods are likely to occur (Ward et al. 2013).

(B) Exposure

Exposure involves identifying what is at risk from flooding, including people, buildings, infrastructure, and critical services. The degree of exposure is determined by the proximity of these elements to flood-prone areas, such as riverbanks or coastal zones. High levels of exposure increase the potential for significant damage and loss during a flood event. Urbanization, for instance, often leads to increased exposure as more people and assets are concentrated in areas susceptible to flooding (Winsemius et al. 2016).

(C) Vulnerability

Vulnerability refers to the susceptibility of exposed elements to suffer harm from a flood event. It encompasses both physical vulnerability—such as the structural integrity of buildings—and social vulnerability, which includes factors like socioeconomic status, age, and health of the affected population. Vulnerability is influenced by the capacity of individuals and communities to anticipate, cope with, and recover from flooding. For example, communities with poor infrastructure, limited access to resources, or inadequate flood defences are generally more vulnerable to the impacts of flooding (Fekete et al. 2010).

To fully understand flood risk, it is crucial to integrate these components—hazard, exposure, and vulnerability—into a comprehensive assessment. This integration allows for the identification of high-risk areas and populations, guiding the development of targeted mitigation strategies. For example, in regions with high flood hazards, reducing exposure through land-use planning and improving the resilience of vulnerable populations through education and infrastructure upgrades are key strategies. Moreover, understanding flood risk is an evolving process that must consider changes in climate patterns, land use, and demographic shifts. As climate change increases the frequency and severity of extreme weather events, the need for dynamic and adaptive flood risk management strategies becomes even more critical. Therefore, a deep understanding of flood risk enables decision-makers to implement proactive measures, reducing the potential for loss and enhancing the safety and resilience of communities.

15.3.2 Strategies for Flood Risk Management

The Kashmir Valley is highly susceptible to flooding due to its unique topography, climatic conditions, and the presence of several rivers and streams. The devastating floods of 2014 underscored the need for comprehensive flood risk management strategies in the region. To mitigate the impact of future floods and enhance the resilience of communities, a multi-pronged approach that combines structural and non-structural measures is essential.

Strengthening Flood Defences and Infrastructure

One of the most critical strategies for flood risk management in Kashmir Valley is the construction and maintenance of robust flood defences. This includes the development of levees, embankments, and floodwalls along vulnerable sections of the Jhelum River and its tributaries. Regular maintenance and dredging of river channels are also necessary to enhance their capacity to carry floodwaters without overflowing. Additionally, improving urban drainage systems, particularly in flood-prone areas like Srinagar, can help mitigate urban flooding by ensuring that excess rainwater is efficiently channelled away from populated areas.

Implementing Early Warning Systems

Effective early warning systems are crucial for minimizing the human and economic toll of floods. In Kashmir Valley, there is a need to develop and deploy advanced hydrological models that can predict flood events with greater accuracy. These models should be integrated with real-time data from weather stations and river gauges to provide timely warnings to at-risk communities. The establishment of a centralized flood monitoring and warning centre in the region could significantly improve response times and allow for coordinated evacuation efforts.

Land-Use Planning and Zoning Regulations

Proper land-use planning is a vital component of flood risk management. Restricting development in floodplains and other high-risk areas can greatly reduce exposure to flood hazards. In Kashmir Valley, there is a need for stringent enforcement of zoning regulations that prevent the construction of residential and commercial buildings in vulnerable areas. Additionally, existing settlements in flood-prone zones should be evaluated, and where feasible, relocation or elevation of structures should be considered. This approach not only minimizes risk but also preserves the natural floodplains, which can act as buffers during flood events.

Enhancing Community Awareness and Preparedness

Community-based disaster risk reduction (DRR) is an effective strategy for improving flood resilience. In Kashmir Valley, raising awareness about flood risks and promoting preparedness among local communities is essential. This can be achieved through educational campaigns, disaster preparedness drills, and the distribution of informational materials on flood safety measures. Encouraging community participation in DRR activities ensures that local knowledge is integrated into flood management strategies and empowers residents to take proactive steps in protecting themselves and their property.

Ecosystem-Based Approaches

Ecosystem-based approaches to flood risk management leverage natural processes to reduce flood risks. In the Kashmir Valley, restoring degraded wetlands and forests can enhance the region's natural ability to absorb and slow down floodwaters. Reforestation, especially in upstream areas, can reduce surface runoff and decrease the intensity of floods. Wetland restoration in floodplains can also provide additional water storage capacity during heavy rainfall events. These measures not only reduce flood risks but also offer co-benefits, such as biodiversity conservation and improved water quality.

Strengthening Institutional Frameworks and Coordination

Effective flood risk management requires strong institutional frameworks and coordination among various stakeholders. In Kashmir Valley, there is a need for better coordination between government agencies, non-governmental organizations, and local communities. Establishing a dedicated disaster management authority for the region, with a specific focus on flood risk management, could enhance the implementation of flood mitigation measures. Additionally, integrating disaster risk reduction into the region's development plans and ensuring that all relevant agencies are equipped with the necessary resources and expertise are crucial steps towards building resilience.

15.4 Conducting a Detailed Primary Survey Using SEM for Flood Risk Assessment

The primary survey aimed to gather data on the perceptions, experiences, and preparedness of individuals affected by the 2014 floods in Kashmir Valley. The survey was designed to understand the risk factors, vulnerabilities, and resilience strategies from the respondents. Total 100 respondents were selected from various flood-affected regions across the Kashmir Valley. The respondents included households, community leaders, local government officials, and members of civil society organizations.

15.4.1 Survey Instrument

A structured questionnaire was developed, covering key variables such as

Exposure to Flood Risk: Frequency of past flooding, proximity to water bodies, and flood awareness.

Vulnerability: Socioeconomic status, housing conditions, access to resources, and support networks.

Preparedness: Evacuation plans, early warning systems, and community drills.

Impact of the 2014 Floods: Economic losses, displacement, psychological effects, and recovery time.

Response and Recovery Efforts: Governmental and non-governmental aid, community response, and infrastructure restoration.

15.4.2 Structural Equation Modelling (SEM)

SEM was employed to analyse the relationships between the latent variables (e.g., exposure, vulnerability, preparedness) and their impact on flood risk perception and resilience.

Model Specification

Latent Variables:

Exposure (E): Measured by indicators such as past flood experience, distance from the river, and knowledge of flood risk.

Vulnerability (V): Measured by indicators such as income level, housing quality, and access to resources.

Preparedness (P): Measured by indicators such as possession of an evacuation plan, participation in drills, and knowledge of early warning systems.

Impact (I): Measured by indicators such as the extent of economic loss, property damage, and psychological trauma.

Response and Recovery (R): Measured by indicators such as the effectiveness of aid received, speed of recovery, and satisfaction with governmental response.

Hypotheses

H1: Higher exposure leads to a higher perception of flood risk.

H2: Greater vulnerability increases the negative impact of floods.

H3: Better preparedness reduces the perceived risk and negative impact.

H4: Effective response and recovery efforts mitigate the long-term effects of the flood.

H5: There is an interaction effect between preparedness and vulnerability on the overall impact.

Data Analysis: The survey data were analysed using SEM software (SmartPLS). The model was tested for fit indices, including the Chi-square test, Comparative Fit Index (CFI), Tucker–Lewis Index (TLI), and Root Mean Square Error of Approximation (RMSEA).

15.4.3 Scientific Results

Model Fit: The SEM model demonstrated a good fit, with the following indices:

Chi-square/df: 2.1 (acceptable if < 3).

CFI: 0.95 (values > 0.90 indicate a good fit).

TLI: 0.92 (values > 0.90 indicate a good fit).

RMSEA: 0.04 (values < 0.08 indicate a good fit).

Path Coefficients:

Exposure to Risk → Perceived Risk: $\beta = 0.65$, $p < 0.01$, indicating a significant positive relationship.

Vulnerability → Impact: $\beta = 0.72$, $p < 0.01$, indicating that higher vulnerability led to a greater negative impact.

Preparedness → Perceived Risk: $\beta = -0.50$, $p < 0.05$, showing that better preparedness reduced perceived risk.

Response and Recovery → Long-term Effects: $\beta = -0.40$, $p < 0.05$, indicating that effective response and recovery efforts mitigated the long-term negative effects.

Interaction Effect (Preparedness × Vulnerability → Impact): $\beta = -0.30$, $p < 0.05$, suggesting that preparedness moderated the effect of vulnerability on impact.

The results confirm that exposure and vulnerability are key determinants of flood risk perception and impact. Preparedness plays a crucial role in mitigating these risks, highlighting the importance of community education and early warning systems. The effectiveness of response and recovery efforts significantly influences the long-term resilience of affected populations.

The path diagram illustrates the relationships between the latent variables, along with the path coefficients showing the strength and direction of these relationships (Fig. 15.3). The SEM analysis provided valuable insights into the complex interactions between exposure, vulnerability, preparedness, and response efforts in the context of the 2014 Kashmir floods. The model underscores the importance of multi-dimensional risk assessments in disaster management. Community-based preparedness programmes should be strengthened to reduce vulnerability and mitigate flood impacts. Investing in technology and infrastructure for accurate and timely flood warnings can save lives and reduce economic losses. Special attention should be given to economically and socially vulnerable groups to enhance their resilience.

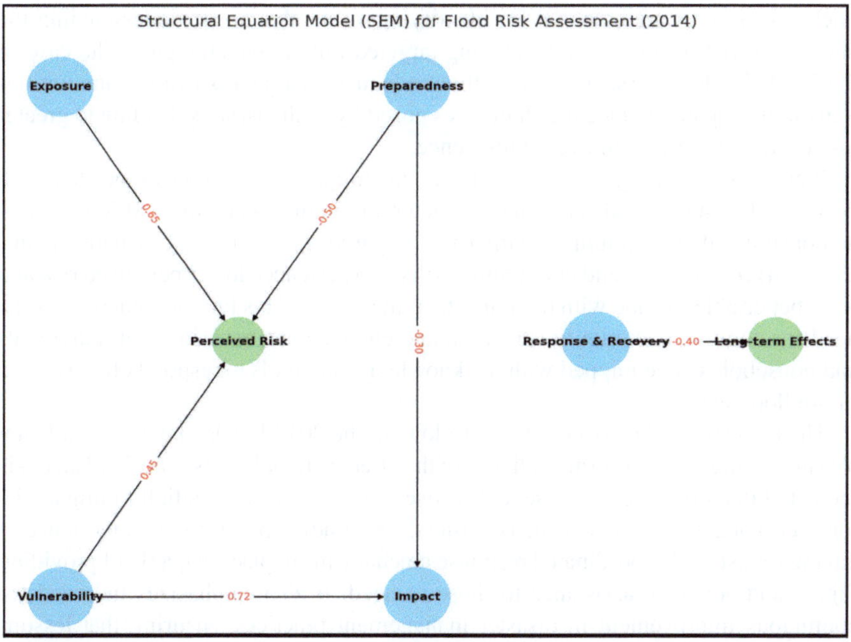

Fig. 15.3 SEM path diagram

Building more efficient and responsive mechanisms for disaster recovery can significantly reduce long-term negative impacts. This comprehensive survey and SEM analysis offer a scientific approach to understanding and managing flood risks in vulnerable regions, contributing to more resilient and informed disaster management practices.

15.5 Conclusion

The 2014 floods in Kashmir Valley served as a wake-up call, highlighting the region's vulnerability to natural disasters and the urgent need for comprehensive flood risk management. The floods inflicted severe economic damage, with losses spanning residential, commercial, and agricultural sectors. Infrastructure, including roads, bridges, and public utilities, was heavily damaged, leading to prolonged disruptions in transportation and essential services. The educational sector was particularly hard hit, with many schools closed for extended periods, affecting the academic progress of thousands of students. Beyond the material losses, the floods also had profound social and psychological impacts, displacing families, disrupting livelihoods, and causing long-term stress and trauma. The analysis highlights the critical importance of systematic flood risk assessment in disaster-prone regions like the Kashmir Valley.

Such assessments are essential for identifying vulnerable areas, understanding the underlying risk factors, and developing targeted mitigation strategies. The case of the 2014 floods underscores that without adequate risk assessments, communities remain ill-prepared to face the challenges posed by such disasters, leading to greater loss of life, property, and overall resilience.

Preparedness emerged as a key factor in mitigating the risks associated with flooding. The survey and subsequent Structural Equation Modelling (SEM) analysis demonstrated that communities with better preparedness—through evacuation plans, early warning systems, and community drills—experienced lower perceived risk and were better able to cope with the impacts of the floods. This finding underscores the need for ongoing community education and engagement, ensuring that individuals and households are equipped with the knowledge and tools to respond effectively to future flood events.

The response and recovery efforts following the 2014 floods played a significant role in shaping the long-term resilience of the affected populations. The SEM analysis indicated that effective response and recovery efforts can substantially mitigate the long-term negative effects of such disasters. This finding points to the importance of having robust, well-coordinated response mechanisms in place, capable of providing timely and adequate assistance to those in need. It also emphasizes the need for continuous improvement in disaster management practices, ensuring that lessons learned from past events are integrated into future strategies.

The analysis also highlighted the critical role of vulnerability in determining the impact of floods. Communities with higher vulnerability due to factors such as low socioeconomic status, poor housing conditions, and limited access to resources were disproportionately affected by the floods. This underscores the need for targeted interventions aimed at reducing vulnerability, particularly in marginalized and at-risk communities. Such interventions might include improving infrastructure, enhancing access to resources, and implementing policies that promote social equity and inclusion.

Based on the findings, several policy implications emerge. First, there is a need for comprehensive flood risk management strategies that integrate scientific risk assessments with community-based approaches. These strategies should focus on enhancing infrastructure resilience, improving early warning systems, and fostering community preparedness. Second, policies should prioritize the needs of the most vulnerable populations, ensuring that they are not left behind in disaster preparedness and recovery efforts. Third, there is a need for ongoing investment in disaster risk reduction (DRR) initiatives, including capacity-building programmes, research, and technological innovation.

Looking forward, the chapter suggests several areas for further research and action. There is a need to explore the long-term impacts of the 2014 floods on the socioeconomic fabric of the Kashmir Valley, including the psychological effects on survivors. Additionally, future research could focus on the effectiveness of different flood risk management strategies, providing evidence-based recommendations for policy and practice. Finally, there is a need for continuous monitoring and evaluation of flood risk management initiatives, ensuring that they are adaptive and responsive to the changing risk landscape. By learning from past experiences and implementing the lessons learned, there is an opportunity to build a more resilient and prepared future for the Kashmir Valley and other disaster-prone regions around the world.

References

Alam A, Bhat MS, Farooq H, Ahmad B, Ahmad S, Sheikh AH (2018) Flood risk assessment of Srinagar city in Jammu and Kashmir, India. Int J Disas Resil Built Environ 9(2):114–129

Ali N, Alam A, Bhat MS, Shah B (2022) Using historical data for developing a hazard and disaster profile of the Kashmir valley for the period 1900–2020. Nat Hazards 114(2):1609–1646

Bhat MS, Alam A, Sheikh H, Ali N (2021) Developing landslide hazard scenario of Kashmir Himalaya from the historical events

Fekete A, Damm M, Birkmann J (2010) Scales as a challenge for vulnerability assessment. Nat Hazards 55:729–747

Hakim MI (2020) Climate change & society in conflicted landscapes: Cases of Srinagar (Kashmir) and Jerusalem (West Bank, Palestine). Doctoral dissertation, University of Illinois at Urbana-Champaign

Jeelani P, Shah SA (2023) Tourism and the environmental components coexisting in the Himalayas: a trade-off between the two. J Qual Assur Hosp Tour 1–25

Jeelani P, Shah SA, Dar SN, Rashid H (2023) Sustainability constructs of mountain tourism development: the evaluation of stakeholders' perception using SUS-TAS. Environ Dev Sustain 25(8):8299–8317

Jeelani P, Rashid H, Shah SA, Ahad F, Gulzar S (2024) Sustainability in tourism demands a mode shift: appraisal of heritage tourism; a safe bet for developing regions. In: Natural resources forum. Blackwell Publishing Ltd, Oxford

Jeelani P, Ahad F, Shah SA, Rashid H (2025) Site suitability evaluation for nature-based tourism using gis and ahp: a case study of Kashmir Valley, India. Appl Geomat 1–15

Jeelani P, Shah SA (2024) Strategic insights for sustainable tourism development in Kashmir Valley: SWOT and QSPM analysis. Environ Dev 52:101092

Jha AK, Bloch R, Lamond J (2012) Cities and flooding: a guide to integrated urban flood risk management for the 21st century. World Bank Publications

Merz B, Hall J, Disse M, Schumann A (2010) Fluvial flood risk management in a changing world. Nat Hazard 10(3):509–527

Mir S, Shah SA, Bhat MS (2024a) Evaluation of disaster preparedness in educational infrastructure of remote geographies using analytical hierarchy process. Int J Disas Risk Reduc 108:104514

Mir S, Shah SA, Shafi Bhat M (2023) Disasters affecting education across time and space—a perspective from Kashmir. Environ Dev Sustain 1–19

Mir S, Shah SA, Bhat MS, Akhter S, Ahad F, Rashid H, Jeelani P (2024b) Disaster risk assessment of educational infrastructure in mountain geographies using PROMETHEE-II. Int J Disas Risk Reduc 107:104489

Romshoo SA, Rashid I (2014) Assessing the impacts of changing land cover and climate on Hokersar wetland in Indian Himalayas. Arab J Geosci 7:143–160

Ward PJ, Jongman B, Weiland FS, Bouwman A, van Beek R, Bierkens MF, Winsemius HC et al (2013) Assessing flood risk at the global scale: model setup, results, and sensitivity. Environ Res Lett 8(4):044019

Winsemius HC, Aerts JC, Van Beek LP, Bierkens MF, Bouwman A, Jongman B, Ward PJ et al (2016) Global drivers of future river flood risk. Nat Clim Chang 6(4):381–385

Chapter 16
A Holistic Overview of Glacial Lake Outburst Flood (GLOF) Hazard in Himachal Pradesh, India

Varun Joshi and Divya Singh

Abstract Glacial Lake Outburst Floods (GLOFs) are one of the most significant natural disasters in the fragile terrain of Himachal Pradesh, India. The state is facing a multitude of changes which include changes in the climate, i.e., increasing temperature and decreasing annual rainfall in the already varied geomorphology of the state. This results in the rapid melting of the glaciers which further results in the expansion of pre-existing glacial lakes and formation of new lakes in the study area. Along with that, the state is facing a surge of population rise which is further increasing the overall vulnerability of the state. Past events of GLOF within the state, i.e., the 2005, Kinnaur district GLOF event which was caused by the bursting of the Pareechu Lake. This event caused massive destruction within the district. Therefore, there is a need to study holistically the risk of GLOF within the study area. Emphasizing the intricacy of the risk factors involved, the interaction between faster glacier melt, climatic changes, and human activities like infrastructure development. This study focuses on providing a holistic overview of the GLOF disaster in Himachal Pradesh. It specifically discusses the glacial lakes in the state, the factors that lead to the formation of glacial lakes and the occurrence of GLOF, and potential mitigation strategies that can be used to lessen the adverse impacts of the disaster on society.

Keywords Glacial Lake Outburst Floods (GLOFs) · Himachal Pradesh · Climate change · Glacial melt · Mitigation strategies

---

V. Joshi (✉) · D. Singh
University School of Environment Management, Guru Gobind Singh Indraprastha University, New Delhi, India
e-mail: varunj63@gmail.com

D. Singh
e-mail: divyasngh951@gmail.com

S. C. Rai (ed.), *Himalaya: Mountains of Destiny*,
https://doi.org/10.1007/978-3-031-94905-0_16

16.1 Introduction

Glacial Lake Outburst Floods (GLOFs) are caused by the abrupt or unexpected release of vast volumes of water and debris from these lakes into a valley downstream, severely damaging natural resources, infrastructure, and human life (Taylor et al. 2023). Glacial lakes are generally formed due to the accumulation of meltwater in places exposed by glacial erosion, resulting in the formation of glacial lakes. This process is often caused by glacial retreat and recession (Buckel et al. 2018). Glaciers that are receding reveal topographic depressions that create proglacial lakes that are blocked by ice above and moraines that may have ice cores below (Kaushik et al. 2024). Due to the presence of such glacial lakes in the Himalayas has increased the risk of occurrence of GLOF thereby disturbing the socio and economic fabric of the area (Ives et al. 2010).

Himachal Pradesh in northern India is not an exception among high-altitude regions across the world to be increasingly vulnerable to Glacial Lake Outburst Floods (GLOFs). GLOFs, which are characterized by abrupt and catastrophic water releases from glacial lakes (Emmer 2017), are becoming more and more recognized as posing a serious threat to mountain infrastructure and human life. Few past events of GLOF within Himachal Pradesh have caused severe damage to both social and infrastructure sectors. The state's erratic history, which includes the unexpected emptying of some moraine-dammed lakes of the Shaune Garang glacier in 1981 and 1988 (Sangewar et al. 1999), also the transnational outburst floods from landslide-dammed lakes in 2000 and 2005 have previously caused considerable devastation in this geographic area indicating that the state is still vulnerable to the threat of GLOFs (Gupta and Sah 2008; Ruiz-Villanueva et al. 2016). The rising frequency and severity of floods in Himachal Pradesh (Allen et al. 2016), have been directly affected by the changes in the glacial melt patterns and climate change. This has led to worsening the condition and increasing the risk of flood in the state (Rinzin et al. 2023). A significant number of glacial lakes that are directly associated with the parent glaciers have been found within the state (Bhambri et al. 2018). The new formation of glacial lakes due to the frequent melting of the glaciers might be the major cause of the overall increasing threat of GLOF in Himachal Pradesh, India (Allen et al. 2016).

The state is situated in the northern part of the Himalayas which makes the topography of the area highly mountainous and the presence of an erratic climate causes the state to be vulnerable to the effects of GLOFs (Faizan et al. 2020). Past research indicates that rising global temperatures have hastened glacial melt in the Himalayan region, which substantially is involved in the creation and growth of glacial lakes (Bolch et al. 2012). This study focuses on providing a holistic overview of the GLOF disaster in Himachal Pradesh. It specifically discusses the glacial lakes in the state, the factors that lead to the formation of glacial lakes and the occurrence of GLOF, and potential mitigation strategies that can be used to lessen the adverse impacts of the disaster.

16.2 Study Area

Himachal Pradesh, a Himalayan state in northern India is located at 30°22′N and 33°12′N latitude and 75°47′E and 79°04′E longitude (Fig. 16.1), covering an area of approximately 53,673 km^2. The state is bordered by China to the east, Ladakh in the north, Uttarakhand in the south, Haryana in the south, and Punjab in the west. The state's topography is primarily mountainous with elevations ranging from 200 m to over 7000 m in the high Himalayas. The state comprises 12 districts, i.e., Bilaspur, Chamba, Hamirpur, Kangra, Kinnaur, Kullu, Lahaul-Spiti, Mandi, Shimla, Sirmaur, Solan, and Una. Himachal Pradesh, with its varied geography, is especially vulnerable to natural calamities (Chandel 2015).

Glacial Lake Outburst Floods (GLOFs) have been a source of concern in recent decades, as they are caused by the abrupt release of a considerable amount of water trapped in a glacial lake, regardless of the cause (Emmer 2017). The state's high-altitude regions are prone to GLOFs because of the existence of growing glacial lakes (Allen et al. 2016).

16.3 Number of Glacial Lakes in Himachal Pradesh

An overview of a few notable researches that comprise Himachal Pradesh's glacial lake inventory will be included in the section that follows. A total of 156 glacier lakes were mapped in the inventory of glacial lakes created by Survey of India (SOI) topographic sheets on a 1:50,000 scale and using LISS 3 satellite imagery data from 1999 to 2001(Bhagat et al. 2004). According to a study by Bhambri et al. (2018), high-resolution images (2011–2013) have been used to map 958 glacial lakes (size > 500 m) in Himachal Pradesh, which span an area of 9.6 ± 0.3 km. 345 of these were categorized as lakes as moraine-dammed lakes. According to this study the largest lake in Himachal Pradesh is located at the edge of the Samundra Tapu Glacier in the Chandra Basin and is classed as another moraine-dammed lake. The Miyar basin has the greatest number of glacier lakes (156), while the Siul basin has the fewest lakes (6).

The Himachal Pradesh Council for Science Technology-Environment (HIMCOSTE) Centre on Climate Change conducted a remote sensing-based study that found that the state's glacial lake count that could result in flash flooding nearly doubled from 562 in 2019 to 1048 in 2023 (India Today 2024). National Disaster Management Authority (NDMA) quoted 48 glacial lakes found in Himachal Pradesh are categorized as "high risk". Some studies cover the glacial lake inventory by river basin across the state. A study conducted by Vishwas et al. (2024) revealed 39 glacial lakes in the Kinnaur district, 10 of which were considered hazardous due to their size (> 0.1 km^2). Rongali et al. (2024) found that the size of Geepang Gath Glacier Lake expanded from 0.542 to 0.929 km^2 and Samudra Tapu glacial lake increased from 0.849 to 1.239 km^2 between 2005 and 2019. The formation of

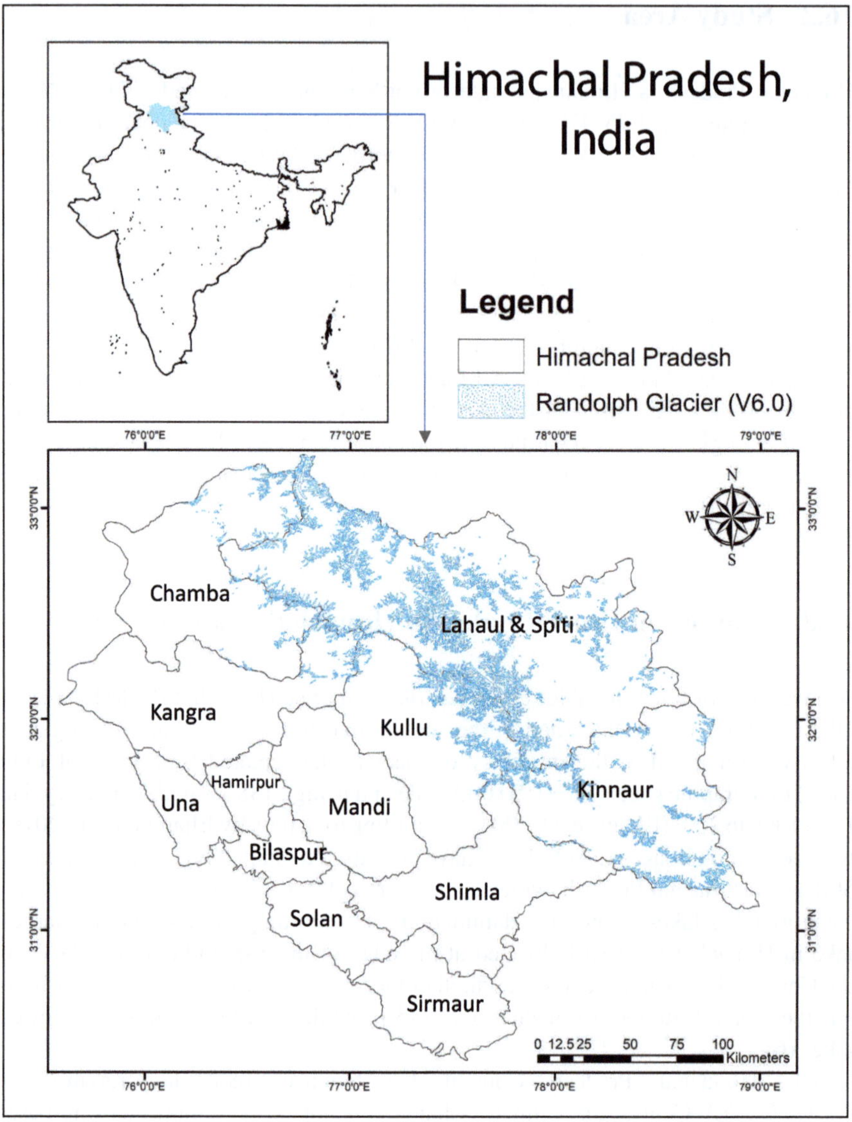

Fig. 16.1 Map of the study area

moraine-dammed lakes near the snouts of Samudra Tapu and Gepang Gath glaciers, as well as their fast growth, constitute a GLOF-related threat to downstream Chandra valley districts (Dhar et al. 2024). Such studies showcase the rising threat of GLOF in the downstream community due to GLOF. This suggests that the state might be susceptible to the menace of GLOF in the near future.

16.4 Factors Contributing to GLOF Hazards in Himachal Pradesh

The Himalayan region, especially Himachal Pradesh, is more susceptible to Glacial Lake Outburst Floods (GLOFs) due to a complex interaction between anthropogenic, geomorphological, and climate factors. To effectively manage the dangers connected with GLOFs and protect vulnerable communities and infrastructure, it is imperative to comprehend these contributing variables. The following contributing factors are analyzed in detail.

16.4.1 Climatic Factors

In comparison to the average worldwide increase in temperature rate of 0.74 °C per 100 years, geographic warming in the Himalayas has been reported to be around 0.15 and 0.60 °C per decade (Ahmed et al. 2021). Average temperature has risen and average precipitation has decreased in the Western Himalayas, according to a study by Lone and Jeelani (2024), which examined temperature and precipitation time series data collected from 1901 to 2022. The findings indicated a notable rise of 0.09 °C $year^{-1}$ from 1991 to 2022, with an average increase of 0.2 °C $decade^{-1}$ from 1901 to 2022, and a drop of 0.032 mm $year^{-1}$ in annual precipitation. Previous research has also addressed that, over the Western Himalayas, the trend toward increasing maximum temperature is greater than the trend toward growing minimum temperature yet, rising warm nights also indicate an increasing trend in minimum temperature (Dimri et al. 2021).

According to a study by Kaushik et al. (2020), annual precipitation in the Bhaga Basin, Himachal Pradesh is decreasing (– 2.724 $mmyr^{-1}$, 1951 to 2006) but annual temperature is increasing (0.027° Cyr^{-1}, 1961 to 2015). This is related to the loss in the overall area of glaciers within the Bhaga basin, which had decreased to 230.8 km^2 by 2017. According to an estimate, up to 25% of the world's mountain glaciers may disappear by 2050 and up to 50% by 2100 (Bajracharya et al. 2008). Due to the changing climate, Third Pole's glacial lakes have been growing lately. Proglacial lakes accounted for 83% of the increase in the glacial lake area, which was 22.68 ± 8.84 km^2 (3.03% relative area change) between 2018 and 2022 (Zhang et al. 2023). This impact of changing climate on the dynamics of glacial lakes is also impacting the lakes in Himachal Pradesh. It has also been reported the glaciers are retreating at a much faster rate in the Western Himalayas as compared to the Eastern and Central Himalayas (Sultan et al. 2022). Glacial area dynamics within Himachal Pradesh have been shown by a few studies such as Gara, Gor Garang, Shaune Garang, and Nagpo Tokpo Glaciers in the Satluj River Basin were researched by the Geological Survey of India (Vohra 1981), found an average retreat of 4.22 to 6.8 m/year. In the Chenab River Basin, the glaciers Bara Shigri, Chhota Shigri, Miyar, Hamtah, Nagpo Tokpo, Triloknath, and Sonapani receded between 6.81 and 29.78 m/year. The Bara Shigri

Glacier and the Chhota Shigri Glacier witnessed the greatest and lowest retreat, respectively (Bajracharya et al. 2008; Kulkarni et al. 2004).

In this context associated with changing climate, the most apparent and dynamic effects are the actual retreat of mountain glaciers and the growth of enormous glacial lakes (Bolch et al. 2012). A suitable example is Geepang Gath Lake as between 1989 and 2022, the glacial lake (Indus River Basin) at 4,068 m elevation in Himachal Pradesh, India, increased in size by 178%, from 36.49 to 101.30 hectares. The annual rate of increase is approximately 1.96 hectares (ISRO 2024). This indicates the changing climate which is experienced within the state might be one of the primary causes of the retreating of glaciers, which is associated with supplying meltwater to glacial lakes. Thus, increasing the area of the existing glacial lakes and also causing the formation of new glacial lakes within the study area. This is a point of concern as these glacial lakes if erupted can cause severe damage to downstream areas.

16.4.2 Geomorphological Factors

The risk of GLOF is also directly associated with the geological structure of Himachal Pradesh. Himachal Pradesh lies in the Western Himalayas region. The Himalayas consist of around 15,000 glaciers within its area; these glaciers are generally inaccessible (Rai et al. 2017). The state consists of 789 rock glaciers, covering an area of about 336.2 km^2 (Dash et al. 2024). The presence of glaciers is generally associated with the presence of glacial lakes in the area. These lakes if are located in steep valleys then the chances of their outburst and sudden release of water to the downstream area increase which is called GLOF. These lakes, which are mostly formed by loose, unconsolidated moraine debris, are extremely unstable and, if the lake is triggered, might severely flood the land downstream (Bendle 2024). Huggel et al. (2002) have stated that if the lake's slope of the moraine walls is greater than 10°, it could be the cause of the risk of glacial lake outburst flooding. Thus, one of the most important factors in assessing the risk of GLOFs is the stability of these moraine dams.

Himachal Pradesh is also a landslide-prone state which is also escalated by the ongoing anthropogenic activities within the state (Kahlon et al. 2014). Being a state that is prone to landslides increases the likelihood that a glacier lake ruptures, releasing massive volumes of water downstream, and creating a GLOF within the state. Since Pareechu Lake was dammed by the landslide the GLOF in the Kinnaur district in 2005 assumed that the lake erupted because of the landslide (Gupta and Sah 2008). The state has also a turbulent history marred by disasters such as earthquakes in locations such as Chamba, Kangra, Dharmashala, Kinnaur, and Sundernagar. The state is also an earthquake-prone state which also increases the chances of occurrence of GLOF in the area. An earthquake has the potential to cause landslides in locations with unstable terrain and to increase the magnitude of the flow, which could be harmful to the towns downstream (Chen et al. 2023).

16.4.3 Anthropogenic Factors

Increased anthropogenic activity, particularly a rise in built-up areas and population, renders already unstable terrain more vulnerable to disaster in many Himalayan states (Nagamani et al. 2024). According to a study on land-use/cover change (LULC) detection in Kullu, Himachal Pradesh by Parmar et al. (2022), the area under study that was built up surged between 2001 and 2017, but the area covered by glaciers decreased during the same period. Anthropogenic activities and rising air temperatures have been proposed to be the causes of the glacier's total area reduction. The decrease in the glacier area is a consequence of the melting of glaciers. The process of glacier melting channels water to the already-existing glacial lakes and also fosters space for the lakes to grow (Wang et al. 2015). This leads to the expansion of glacial lakes and also increases the number of glacial lakes present within the area. This is a rising concern due to the potential of dangerous glacial lakes to cause GLOF in the downstream community. This could lead to the loss of human life and infrastructural damage, disrupting the area's entire social and economic fabric. The central region of Kullu has a seven-fold higher likelihood of GLOF occurrence, according to a study by Allen et al. (2016). Additionally, the region that is downstream influenced by potential GLOF routes can be predicted to increase threefold. These findings result in a significant rise in the stipulated GLOF hazard stage from a "high" to "very high."

The rising population in the Himalayas is also a major concern as the accelerated persistence of changing climate is primarily due to human activities nowadays (Yıldırım et al. 2024). Human activities are directly associated with the changing climate which further results in more retreating of glaciers, formation of new lakes, expansion of pre-existing lakes, and therefore higher probability of occurrence of GLOF. A study conducted by Sultan et al. (2022) has mentioned that because of the ongoing human population rise in the Himalayan region, an increasing likelihood of vulnerability to natural events has just become recognized. India has immense hydroelectric potential, particularly in the Himalayas (Sharma and Kuniyal 2016). This resulted in the development of many hydropower plants in Himachal Pradesh. Few prominent hydropower plants in Himachal Pradesh include the Nathpa Jhakri Plant in the Satluj River (1500 MW); Bhakra Dam, Satluj River (1325 MW); Karcham Wangtoo Plant, Satluj River (1200 MW); Chamera Plant, Ravi River (540 MW); Rampur Project, Satluj River (420 MW). When the first power plant at Chabba, close to Shimla, was put into operation in 1912 to supply electricity to the town of Shimla, which was then the summer capital of British India, the development of hydropower in Himachal Pradesh officially began (Bansal 2005). In Himachal Pradesh, the Satluj (9866.55 MW), Beas (4527.90 MW), Chenab (2723.0 MW), Ravi (2226.75 MW), and Yamuna (602.52 MW) collectively possess a total identified hydropower potential of 20,670.12 MW, distributed throughout various phases of development in diverse sectors (Sharma and Kuniyal 2016). The presence of these hydropower projects that are proposed or currently functioning in the high-altitude areas in rivers that are fed by glaciers incurs the risk of GLOF which can not only destroy the hydropower infrastructure but change the yields of water and sediment and jeopardize downstream livelihoods (Schwanghart et al. 2016).

In addition to the expanding built-up area and population surge in the region, there is another likely factor-tourism-that contributes to the area's increased susceptibility to disasters such as GLOFs. Tourism is one of the most substantial sectors of Himachal Pradesh's economy. Shimla, Manali, Khajjiar, Dharmshala, McLeodganj, and other destinations rely heavily on tourism. Locals of Himachal Pradesh are dependent on these locations for their livelihood. According to a study by Chandel and Sharma (2024), the documented climatic changes may cause higher evapotranspiration and ablation, which could result in higher river discharge in the early summers. These changes are the most likely cause of disasters like GLOF, which have the potential to exacerbate societal and economic vulnerability.

16.5 Mitigation Strategies for GLOF Management

To predict the future GLOFs it becomes essential to monitor the glacial lakes. Risks are generally predicted and can be reduced by timely assessment of glacial lake sizes, their volume, moraine stability, and uncertain failure structures. Data on glacial lake dynamics can be obtained by using satellite remote sensing and also through field visits to the glacial lakes that are accessible. This will result in the creation of an updated and precise inventory of the glacial lakes within the area which can be further monitored to study the overall dynamics of the glacial lakes. The evolving climatic conditions can be monitored by the mapping of glacial lakes temporally which acts as an indicator of climate change (Ives et al. 2010). In a comprehensive risk assessment, the likelihood of GLOFs and their possible effects are assessed.

The generation of hazard maps that identify locations prone to GLOF and communities exposed to peril. This type of information is required to lessen the devastating effects of floods in the study area. These maps could be created using Geographic Information Systems (GIS), which include data on terrain, glacial lake characteristics, and socioeconomic data (Chen et al. 2022). The technique of mapping flood hazards is widely used to assess the probability of flooding and investigate flood features such as floodwater velocity and depth. This strategy has been greatly enhanced and incorporated into the planning and decision-making for policies in flood-prone areas during the past few decades (Mudashir et al. 2021). Improving the resilience of infrastructure against GLOFs is a crucial mitigation approach. In terms of the infrastructure development required for GLOF mitigation, (ICIMOD 2001) has proposed several mitigation infrastructures, including controlled breaching, the construction of an outlet control structure, pumping or siphoning water from the lake, and digging a tunnel through the moraine barrier or under an ice dam. These constructions will help to reduce the volume of glacial lakes. Check dam construction involves building little obstacles across gullies or streams to slow down water flow, and lessen erosion. Additionally, they lessen the risk of flooding and aid with runoff management (Sepehri et al. 2019). Another significant method for mitigation of the GLOF hazard might be the diversion of channels as it diverts floodwaters from locations that are susceptible by redirecting the channels into man-made canals or rivers. For effective flood control, additional structural elements like reservoirs and embankments (levees) are necessary in addition to check dams and diversion channels.

Riverbanks can be raised by embankments to help avoid spillover during floods, and reservoirs can hold extra water during peak flows and release it gradually to prevent flooding downstream (India WRIS 2024). The integration of GLOF management into national and state policies of the Himalayan states should become mandatory as it is very crucial. The National Disaster Management Authority (NDMA) has provided detailed guidelines for the management of GLOF which majorly focuses on all the aspects of the GLOF but it also consists of capacity development, awareness, and preparedness of GLOF. The awareness programs of the GLOF should include the creation of a village task force, the use of traditional art forms/traditional knowledge, and awareness through a Participatory Approach. Medical facilities, training of medical personnel, and creating awareness about drowning and its management should be a part of the plan, Involvement of all the stakeholders such as government and policymakers will all help to mitigate and properly manage the GLOF-like hazards in the Himalayan states. If any glacial lake is identified as vulnerable by the expert, then a real-time monitoring and early warning system could be installed in the area. So that in case any outburst takes place the information could be reached to concern and warning in the downstream could be issued.

16.6 Conclusion

This study concludes that there are a variety of probable factors such as altered climate, varied geomorphological features, and increasing anthropogenic activities that might be the reasons for increasing the vulnerability of GLOF risk in Himachal Pradesh. The probability of occurrence of GLOF within the area also increases due to the rapid melting of glaciers, which is associated with the formation of new glacial lakes and also expands the pre-existing glacial lakes. Since the area is constituted of fragile terrain a catastrophic GLOF event might be rendered more likely by the emergence of such unstable glacial lakes. Past events of GLOF within the state such as the 2005, Kinnaur district GLOF event which was caused by the bursting of the Pareechu Lake. This event caused massive destruction within the district. Such events act as a lesson to be learned and suggest future mitigation preparation strategies to prevent the state from future predicted GLOF events. Comprehensive risk assessment and management measures are thus imperative throughout the state. To address the issue of GLOF within the state effectively, a multi-faceted approach is required. The approach should include timely monitoring of the prominent or dangerous glacial lakes, awareness of the disaster within the community, and focus on infrastructure development. Enhancing early warning systems, stabilizing susceptible moraine dams, and striking a balance between development and environmental preservation should be the main goals of effective risk management. To better protect infrastructure and communities from the increasingly severe and frequent effects of GLOFs, stakeholders will benefit from a holistic strategy that considers both natural and human-caused factors.

Acknowledgements The second author is thankful to Guru Gobind Singh Indraprastha University for awarding the Short-Term Research Fellowship (STRF) to carry out this research. The authors are grateful to the University School of Environment Management, Guru Gobind Singh Indraprastha University, India, for providing space and facilities to carry out the research work.

References

Ahmed R, Wani GF, Ahmad ST, Sahana M, Singh H, Ahmed P (2021) A review of glacial lake expansion and associated glacial lake outburst floods in the Himalayan region. Earth Syst Environ 5(3):695–708. https://doi.org/10.1007/s41748-021-00230-9

Allen SK, Linsbauer A, Randhawa SS, Huggel C, Rana P, Kumari A (2016) Glacial lake outburst flood risk in Himachal Pradesh, India: an integrative and anticipatory approach considering current and future threats. Nat Hazards 84:1741–1763. https://doi.org/10.1007/s11069-016-2511-x

Bajracharya SR, Mool PK, Shrestha BR (2008) Global climate change and melting of Himalayan glaciers. In: Melting glaciers and rising sea levels: impacts and implications, pp 28–46

Bansal SP (2005) Private sector participation in hydropower development in Himachal Pradesh. In: Conference on hydropower projects—a perspective challenge, 20–22 April 2005, Shimla, India, pp V-37–V-44

Bendle J (2024) Glacial lake outburst floods (GLOFs). AntarcticGlaciers.org. https://www.antarcticglaciers.org/glacier-processes/glacial-lakes/glacial-lake-outburst-floods/

Bhagat RM, Kalia V, Sood C, Mool PK, Bhajracharya SR (2004) Inventory of glaciers and glacial lakes and the identification of potential glacial lake outburst floods (GLOFs) affected by global warming in the mountains of Himalayan Region. ICIMOD, Kathmandu, Nepal

Bhambri R, Misra A, Kumar A, Gupta AK, Verma A, Tiwari SK (2018) Glacier lake inventory of Himachal Pradesh. Himalayan Geol 39(1):1–32

Bolch T, Kulkarni A, Kääb A, Huggel C, Paul F, Cogley JG, Frey H, Kargel JS, Fujita K, Scheel M, Bajracharya S (2012) The state and fate of Himalayan glaciers. Science 336(6079):310–314

Buckel J, Otto JC, Prasicek G, Keuschnig M (2018) Glacial lakes in Austria-distribution and formation since the little ice age. Global Planet Change 164:39–51. https://doi.org/10.1016/j.gloplacha.2018.03.003

Chandel VB (2015) Geo-physical disasters in Himachal Pradesh, India: a spatial perspective. Int J Multidisciplinary Approach Stud 2(5):82–89

Chandel VB, Sharma MR (2024) Climate change and weather extremes in tourist towns: challenges for livelihood in Himachal Pradesh, India. In: Climate change, vulnerabilities and adaptation: understanding and addressing threats with insights for policy and practice. Cham, Springer Nature Switzerland, pp 61–78. https://doi.org/10.1007/978-3-031-49642-4_4

Chen H, Zhao J, Liang Q, Maharjan SB, Joshi SP (2022) Assessing the potential impact of glacial lake outburst floods on individual objects using a high-performance hydrodynamic model and open-source data. Sci Total Environ 806:151289. https://doi.org/10.1016/j.scitotenv.2021.151289

Chen N, Liu M, Allen S, Deng M, Khanal NR, Peng T, Tian S, Huggel C, Wu K, Rahman M, Somos-Valenzuela M (2023) Small outbursts into big disasters: earthquakes exacerbate climate-driven cascade processes of the glacial lakes failure in the Himalayas. Geomorphology 422:108539. https://doi.org/10.1016/j.geomorph.2022.108539

Dash A, Pradhan IP, Mahanta KK, Tiwari N, Shukla DP (2024) Comprehensive assessment of rock glaciers in the Himachal Himalayas: updated inventory and labelling. Progress Phys Geography: Earth Environ 48(4):571–594. https://doi.org/10.1177/03091333241263986

Dhar S, Rai SK, Kumar A, Kumar D (2024) Recession and development of moraine-dammed lakes in the Chandra Basin of Great Himalayan Range, district Lahaul and Spiti, Himachal Pradesh, India. In: Role of science and technology for sustainable future. Volume 1: Sustainable development: a primary goal. Singapore, Springer Nature, pp 223–241. https://doi.org/10.1007/978-981-97-0710-2_15

Dimri AP, Allen S, Huggel C, Mal S, Ballesteros-Cánovas JA, Rohrer M, Shukla A, Tiwari P, Maharana P, Bolch T, Thayyen RJ, Stoffel M, Pandey A (2021) Climate change, cryosphere and impacts in the Indian Himalayan Region. Curr Sci 120(5):774–790

Emmer A (2017) Glacier retreat and glacial lake outburst floods (GLOFs). In: Oxford research encyclopedia of natural hazard science. https://doi.org/10.1093/acrefore/9780199389407.013.275

Faizan OM, Rongali G, Tiwari KC (2020) Mapping of glacial lake outburst flood (GLOFs) using GIS and remote sensing for Lahul and Spiti District of Himachal Pradesh—India. In: 7th Annual international convention on sustainability in current scenario: challenges and solutions

Gupta V, Sah MP (2008) Impact of the trans-Himalayan landslide lake outburst flood (LLOF) in the Satluj catchment, Himachal Pradesh, India. Nat Hazards 45:379–390. https://doi.org/10.1007/s11069-007-9174-6

Huggel C, Kääb A, Haeberli W, Teysseire P, Paul F (2002) Remote sensing based assessment of hazards from glacier lake outbursts: a case study in the Swiss Alps. Can Geotech J 39:316–330. https://doi.org/10.1139/t01-099

India WRIS (2024) Flood management. https://indiawris.gov.in/wiki/doku.php?id=flood_management. Accessed on 19 Aug 2024

Indian Space Research Organisation (2024) Satellite insights on expanding glacial lakes in the Indian Himalayas. ISRO. https://www.isro.gov.in/Satellite_Insights_Expanding_Glacial_Lakes_Indian_Himalayas.html

Ives JD, Shrestha RB, Mool PK (2010) Formation of glacial lakes in the Hindu Kush-Himalayas and GLOF risk assessment. ICIMOD, Kathmandu, pp 10–11

Kahlon S, Chandel VB, Brar KK (2014) Landslides in Himalayan mountains: a study of Himachal Pradesh, India. Int J IT Eng Appl Sci Res 3(9):28–34

Kaushik S, Dharpure JK, Joshi PK, Ramanathan AL, Singh T (2020) Climate change drives glacier retreat in Bhaga basin located in Himachal Pradesh, India. Geocarto Int 35(11):1179–1198. https://doi.org/10.1080/10106049.2018.1557260

Kaushik S, Rafiq M, Dharpure JK, Howat I, Moortgat J, Joshi PK, Singh T, Dietz AJ (2024) Increasing risk of glacial lake outburst flood in Sikkim, Eastern Himalaya under climate warming. Remote Sens Appl: Soc Environ 36:101286. https://doi.org/10.1016/j.rsase.2024.101286

Kulkarni AV, Rathore BP, Suja A (2004) Monitoring of glacial mass balance in the Baspa basin using accumulation area ratio method. Curr Sci 86(1):101–106

Lone SA, Jeelani G (2024) Evaluating the potential impact of climate change on glacier dynamics in Western Himalayas, India. In: Climate crisis: adaptive approaches and sustainability.). Cham, Springer Nature Switzerland. https://doi.org/10.1007/978-3-031-44397-8_25

Mool PK, Wangda D, Bajracharya SR, Joshi SP, Kunzang K, Gurung DR (2001) Inventory of glaciers, glacial lakes, and glacial lake outburst floods: monitoring and early warning systems in the Hindu Kush Himalayan Region, Bhutan

Mudashir RB, Sabtu N, Abustan I, Balogun W (2021) Flood hazard mapping methods: a review. J Hydrol 603:126846. https://doi.org/10.1016/j.jhydrol.2021.126846

Nagamani K, Mishra AK, Meer MS, Das J (2024) Understanding flash flooding in the Himalayan Region: a case study. Sci Rep 14(1):7060. https://doi.org/10.1038/s41598-024-53535-w

Parmar S, Thakur PK, Chauhan M, Lata R (2022) Land use land cover change detection and its impact on land surface temperature of Malana Watershed Kullu, Himachal Pradesh, India. In: Singh RB, Kumar M, Tripathi DK (eds) Remote sensing and geographic information systems for policy decision support. Advances in geographical and environmental sciences. Springer, Singapore. https://doi.org/10.1007/978-981-16-7731-1_11

Rai PK, Mishra VN, Singh S, Prasad R, Nathawat MS (2017) Remote sensing-based study for evaluating the changes in glacial area: a case study from Himachal Pradesh, India. Earth Syst Environ 1:1–13. https://doi.org/10.1007/s41748-017-0001-2

Rinzin S, Zhang G, Sattar A, Wangchuk S, Allen SK, Dunning S, Peng M (2023) GLOF hazard, exposure, vulnerability, and risk assessment of potentially dangerous glacial lakes in the Bhutan Himalaya. J Hydrol 619:129311. https://doi.org/10.1016/j.jhydrol.2023.129311

Rongali G, Tiwari KC, Vishwas P (2024) Mapping of glacial lakes and glacial lake outburst flood in Lahul and Spiti District using remote sensing and GIS. In: Flood risk management: assessment and strategy. Singapore, Springer Nature, pp 49–68. https://doi.org/10.1007/978-981-97-2688-2_3

Ruiz-Villanueva V, Allen S, Arora M, Goel NK, Stoffel M (2016) Recent catastrophic landslide lake outburst floods in the Himalayan Mountain range. Prog Phys Geogr 41:3–28. https://doi.org/10.1177/0309133316658614

Saha B (2024) Flash flood risk in Himachal, Uttarakhand as glacial lakes in Himalayas rise: Report. India Today. https://www.indiatoday.in/environment/story/flash-flood-risk-himachal-uttarakhand-glacial-lakes-himalayas-rise-2601325-2024-09-17

Sangewar CV, Srivastava D, Singh RK (1999) Reservoir within the Shaune Garang glacier, district Kinnaur, Himachal Pradesh. In Abstract proceedings of the symposium on snow, ice and glaciers: a Himalayan perspective. Geological Survey of India

Schwanghart W, Worni R, Huggel C, Stoffel M, Korup O (2016) Uncertainty in the Himalayan energy–water nexus: estimating regional exposure to glacial lake outburst floods. Environ Res Lett 11(7):074005. https://doi.org/10.1088/1748-9326/11/7/074005

Sepehri M, Ildoromi AR, Malekinezhad H, Ghahramani A, Ekhtesasi MR, Cao C, Kiani-Harchegani M (2019) Assessment of check dams' role in flood hazard mapping in a semi-arid environment. Geomat Nat Haz Risk 10(1):2239–2256. https://doi.org/10.1080/19475705.2019.1692079

Sharma S, Kuniyal JC (2016) Hydropower development and policies in India: a case of Himachal Pradesh in the northwestern Himalaya, India. Energy Sources Part B 11(4):377–384. https://doi.org/10.1080/15567249.2011.633593

Sultan H, Zhan J, Rashid W, Chu X, Bohnett E (2022) Systematic review of multi-dimensional vulnerabilities in the Himalayas. Int J Environ Res Public Health 19(19):12177. https://doi.org/10.3390/ijerph191912177

Taylor C, Robinson TR, Dunning S, Rachel Carr J, Westoby M (2023) Glacial lake outburst floods threaten millions globally. Nat Commun 14(1):487. https://doi.org/10.1038/s41467-023-36033-x

Vishwas P, Tiwari KC, Rongali G, Vohra R (2024) Mapping of glacial lakes and glacial lake outburst flood in Kinnaur District, Himachal Pradesh using remote sensing and GIS. In: Advanced geospatial practices in natural environment resource management. IGI Global, pp 109–128

Vohra CP (1981) Note on recent glaciological expeditions in Himachal Pradesh

Wang W, Xiang Y, Gao Y, Lu A, Yao T (2015) Rapid expansion of glacial lakes caused by climate and glacier retreat in the Central Himalayas. Hydrol Process 29(6):859–874. https://doi.org/10.1002/hyp.10199

Yıldırım DÇ, Yıldırım S, Bostancı SH (2024) Anthropogenic impacts in the Himalayas and the sustainable development goals (SDGs). In: The Himalayas in the anthropocene: environment and development. Cham, Springer Nature Switzerland, pp 279–291. https://doi.org/10.1007/978-3-031-50101-2_12

Zhang T, Wang W, An B, Wei L (2023) Enhanced glacial lake activity threatens numerous communities and infrastructure in the third pole. Nat Commun 14(1):8250. https://doi.org/10.1038/s41467-023-44123-z

Chapter 17
Landslide Susceptibility Assessment Using Frequency Ratio Model in Kirti Nagar, Rudra Prayag, Uttarakhand, India

Prafull Singh, Ankita Das, and Kumar Ankit

Abstract Landslides are the most common natural hazard in the Himalayan regions triggered by many factors such as complex topography, bedrock lithological formation, climatic factors, soil characteristics, land use patterns, vegetation loss, hydrogeological characteristics, slope failure, and infrastructure development. In the recent past large number of landslide events were reported from the whole Himalayan region and pose a very significant threat not only to the socio-economic development in an area but also to the life and livelihood of the communities residing in mountainous regions. Geospatial technology has a great role in mapping the landslide vulnerable zones accurately and its management. In the present work total of 09 landslide-conditioning factors slope, elevation, aspect, geology geomorphology, TWI, SPI, LULC, and NDVI were evaluated and mapped through the use of satellite data and field data. Landslide-conditioning factors were properly checked and verified with the ground data and landslide inventory map of the area, thereafter all the contributing factors were assigned the suitable weight and ranking as per their importance in landslide occurrence using frequency ratio (FR), a statistical model to develop the susceptibility assessment. The developed susceptible zone re-classified into three zones low, moderate, and high. The classified zones were validated through the Google images and landslide inventory data of the area.

Keywords Frequency ratio · Landslide susceptibility · Kirti Nagar · Uttarakhand · India

P. Singh (✉) · A. Das · K. Ankit
Remote Sensing and Groundwater Modeling Laboratory, Department of Geology, Central University of South Bihar (CUSB), Gaya, 824236, India
e-mail: prafullsingh@cusb.ac.in

S. C. Rai (ed.), *Himalaya: Mountains of Destiny*,
https://doi.org/10.1007/978-3-031-94905-0_17

17.1 Introduction

The landscape of the planet is mostly shaped by natural forces, some of which are harmful to the people. These innate risks must be recognized, and avoided if possible, and their threat to assets and human life diminished. One of the recurring events in the last two decades is landslides triggered by neo-tectonic movements, earthquakes, heavy precipitation, changes in land use, deforestation, wrong agriculture practices, degradation of the natural drainage system, soil erosion and sedimentation, excess mining, construction of road, and other infrastructural development (Singh et al. 2023).

Landslide susceptibility analysis has been carried out to minimize the infrastructure and economic and save the lives of the people living in the susceptible region, and the maps and models created are also very useful for the awareness and capability building among the community and decision-makers (Demir 2019; Chen et al. 2014). Brabb (1984), Varners (1984), and Soeters and Van Westen (1996) defined landslide susceptibility as the prospect of the occurrence of a landslide in the terrain considering the geo-environmental, geological, and hydrological conditions. Landslide susceptibility assessment and mapping are considered one of the most primary level scientific investigations to understand the terrain conditions and develop suitable vulnerability maps which help the decision-makers and planners to develop disaster risk reduction policy (Erener et al. 2016).

In the recent past, a large number of studies on landslide hazard zonation have been published to identify the potentially vulnerable zones for the occurrence of landslides at the global level. Major studies are based on the use of multiple datasets and techniques such as multi-temporal satellite images, GIS techniques, and field data to understand the terrain conditions (Aleotti and Chowdhury 1999). Many statistical techniques were used to develop the models like the analytical hierarchy process and weighted linear combination, fuzzy AHP, frequency ratio (FR), weight of evidence model, multivariate, etc. Most of the methods are based on the weight of the various causative factors considered for the development of the landslide-susceptible models (Gorsevski et al. 2003; Clerici et al. 2006; Sarkar et al. 2013; Bai et al. 2015; Pal et al. 2019).

Examining the landslide hazards in mountainous regions such as the Himalayas is a complex task that involves studying these geo-environmental factors, and characterizing the triggering cause/s that are more influential in producing slope instabilities. Most of the landslides reported in the area include debris slide, debris flow, rock slide, and rock fall. The literature and previous records on landslides indicate that the majority of landslide events in the Himalayas were related to intense and prolonged monsoonal rainfall that leads to soil saturation and produces high transient pore pressure (Iverson 2000). Several reports and studies recognized that the impact of climate change affects the stability of natural and engineered slopes spatiotemporally. Scientists are continuously investigating the occurrence, magnitude, and frequency of landslide events under different scenarios according to climate model projections (Gariano and Guzzetti 2016; Tiranti and Chemonini 2019). Many other geo-environmental factors such as tectonics, seismicity, terrain complexity and topography, lithology, geology and structure, hydrology and hydrogeology, vegetation cover and density, soil characteristics, land use practices, and human activity can contribute to the occurrence of landslides (Pham et al. 2015; Zhang et al. 2020; Singh et al. 2023).

The fragile Himalayan landscape is highly susceptible to natural hazards, leading to increasing concern among scientists about current and future climate change impacts (Cruz et al. 2007). Large uncertainties in extreme precipitation have particularly been observed in the mountainous terrain such as the Himalayas which shows a positive correlation with the incidents of climate change. It is generally assumed that with an increase in extreme precipitation (heavy precipitation) there are more frequent instances of landslides provided all other geo-environmental factors remain constant. Over the past several decades, the precipitation extremes have shown a significant increase in many parts of the world, although such an increase differs seasonally and spatially. The changes in the Monsoon cycle in the Indian sub-continent show a direct correlation with the incident of climate change affecting the slope instability particularly due to heavier rainfall events and changes in the spatial distribution of precipitation. Hence, there is an urgent requirement for a thorough investigation at a finer resolution applying a reliable statistical technique and considering future climate change scenarios. This will help to identify the most sensitive areas that need immediate attention and the prediction of future landslides can be assimilated with future developmental planning and formulation of landslide mitigation strategies. The objective of the present work is to develop the landslide susceptibility of Kirti Nagar and Rudra Prayag using satellite data and GIS techniques.

17.2 Geographic, Geological, and Geomorphological Characteristics

The study area Kirti Nagar is a Nagar Panchayat of the district Tehri Garhwal Uttarakhand situated at 30°12′ to 30°15′ N Latitude and 78°45′–78°50′E Longitude (Fig. 17.1). It is bounded by the districts of Pauri Garhwal in the south, Uttarkashi in the North, Dehradun in the West, and Rudra prayag in the East. The study area is the part of lesser and higher Himalayas and highly dissected terrain and seismically very active zone traversed by many faults and the main central thrust (Sahoo et al. 2000).

Geologically Himalayan terrain classified under four major geological provinces: Siwalik Himalaya (SH), Lesser Himalaya (LH), Higher Himalaya (HH), and Indo-Tsangpo suture zone (ITSZ), and the area tectonics is defined by four north dipping thrust fault systems: the Himalayan frontal thrust (MFT), main boundary thrust (MBT), main central thrust (MCT), and south Tibetan detachment (STD) (Mandal et al. 2023). Geologically terrain comes under the Garhwal Group comprised of dolomites, slates, and quartzites, and the Central Crystalline Group comprises gneiss,

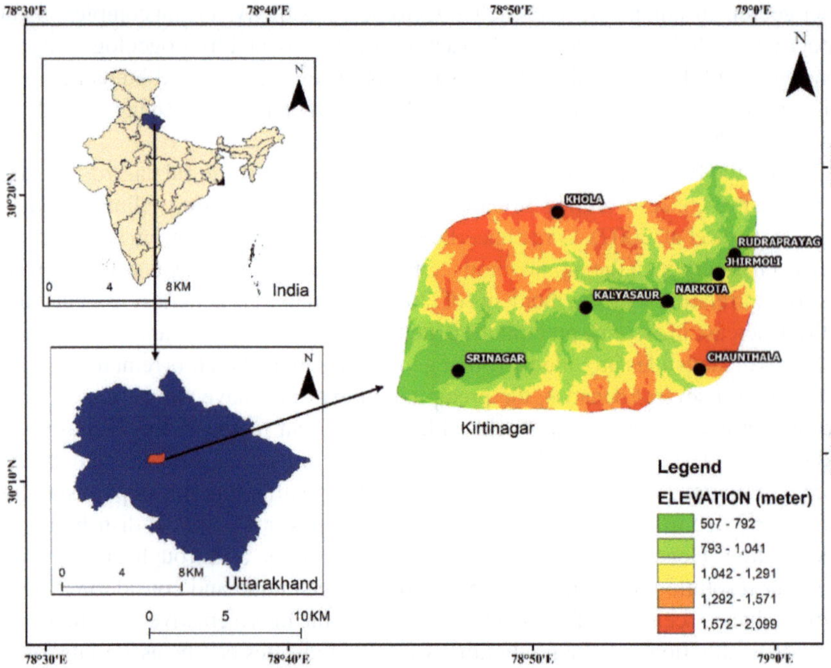

Fig. 17.1 Location map of the study area

schists, and metabasics. The formations are highly weathered and jointed. The Himalaya's geological features and frequent earthquakes are a result of the Indian plate's ongoing movement north and collision with the Eurasian plate. Active thrusts and faults accommodate the convergence, which is depicted in related geomorphic structures. The lower Himalaya, which is bordered by the main central thrust (MCT) in the north and the main boundary thrust (MBT) in the south, has rolling topography and less jagged mountains, indicating reasonably mature topography (Bhukosh 2021b).

Characterization and mapping of geomorphological features are very essential for landslide mapping and understanding the topography, depositional and erosional landforms, and their relationship with geological formations. The entire area is classified under eight geomorphic units on based image characteristics over satellite images and toposheets (Fig. 17.2a). The prominent landforms are the piedmont alluvial plain, lateral bar, point bar, ridge, valleys, highly dissected hills, valleys, terraces, rivers, etc. (Bhukosh 2021a and b). Geology and associated structural features are some of the important terrain characteristics supported for landslide assessment and management. Accurate geological mapping is one of the essential steps in landslide studies. In the proposed study area, the geological and lithological attributes were recognized using satellite data and already published geological. The geologically whole area is grouped under low-grade upper proterozoic phyllites, which merge with

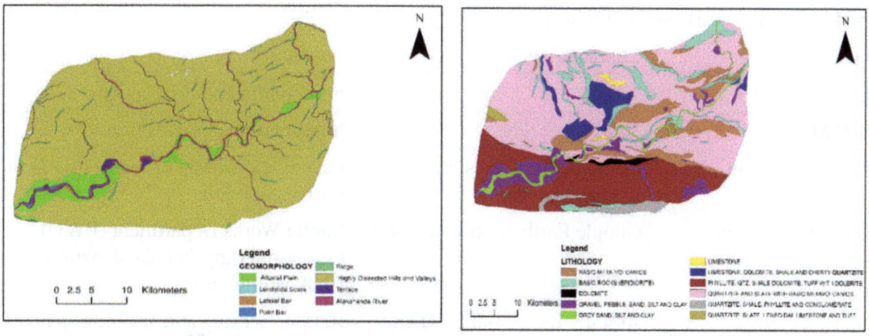

(a) (b)

Fig. 17.2 Geomorphological **a** and litholocal **b** units of the study area

the flaggy quartzites toward the North, Nagnithank-Berinag and Chamol formation, Chandpur-Amri formation that is mainly of phyllite, quartz, shale dolomite tuff with dolomite (Fig. 17.2b).

17.3 Datasets Used and Methodology

Any landslide assessment and vulnerability mapping required a systematic investigation of various landslide and environmental conditions supporting the landslide and other environmental issues in the terrain. The entire methodology and datasets used in the proposed study included satellite images, field data, and GIS modeling for developing the landslide susceptibility zones of the Kirti Nagar, Rudra Prayag, Uttarakhand, India. All the landslide contributing factors were generated through the use of satellite images using image processing and GIS software. The SRTM and Landsat 8 OLI data were used to extract all the thematic information (slope, elevation, aspect, geology geomorphology, TWI, SPI, LULC, and NDVI). The Bhukosh portal of the Geological Survey of India is used to verify and update the lithological and geomorphological features of the area and to find the landslide inventory information for validating and verifying the susceptible zones. The details of the dataset used and the general methodology followed to run the model and develop the landslide-susceptible zones are presented in Table 17.1 and Fig. 17.3.

The Frequency Ratio (FR) model is a statistical approach based on the practical spatial relation between each landslide-conditioning factor and the distribution of past landslides. This method is used for developing a spatial correlation between landslide location and landslide explanatory aspects (Lee 2005; Wang et al. 2019; Mohammady et al. 2012; Yilmaz 2009; Singh et al. 2023). FR values for each class of controlling factor were calculated based on their relationship importance and the same techniques were used for developing the landslide susceptibility map (Abedini

Table 17.1 Datasets used in the work

Data type	Characteristics	Source
Landsat-8 OLI	30 M	http://earthexplorer.usgs.gov
SRTM data	30 M	http://earthexplorer.usgs.gov
Geological map	Geological features, Landslide scar points, etc.	Geological Survey of India (http://bhukosh.gsi.gov.in)
Landslide inventory	Google Earth published reports	Public Works Department (PWD), GSI portal (http://bhukosh.gsi.gov.in/Bhukosh/Public)
Software used	ArcGIS 10.8.2 and ERDAS IMAGINE software, QGIS	Remote sensing and groundwater modeling laboratory, CUSB Gaya

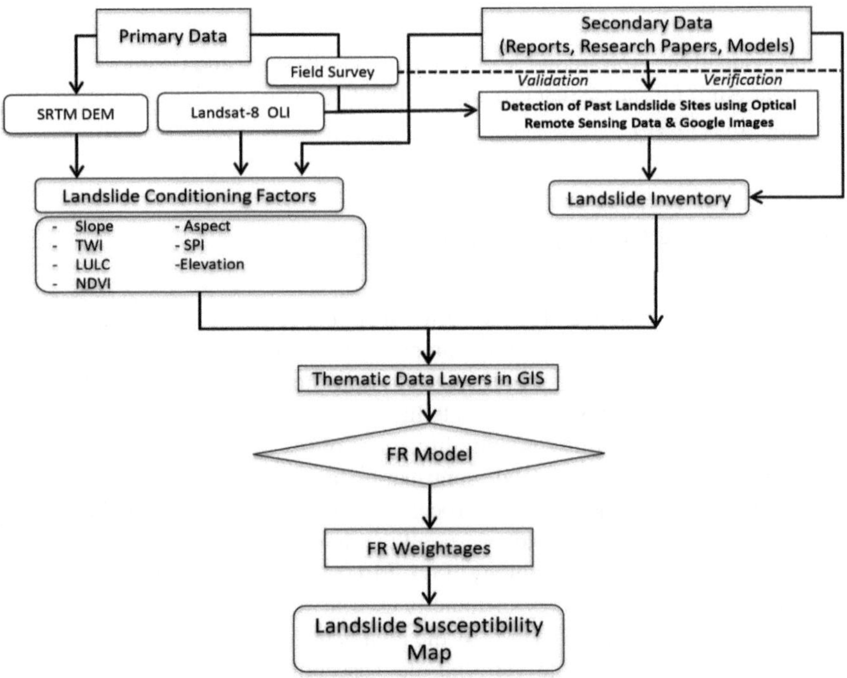

Fig. 17.3 Processing and modeling flowchart for landslide susceptibility mapping

and Tulabi 2018; Sur et al. 2021; Singh et al. 2023). In this method, landslide susceptibility prediction is considered the joint conditional probability based on the ratio of probability likelihood (Bonham-Carter 1994; Pardeshi et al. 2013). In this study, the thematic layers were developed at raster grids (150 m × 150 m), and subsequently, a binary layer of landslides was used by the input of dummy variables. Each thematic layer was then overlaid with the landslide's binary layer, the number of pixels under certain classes was copied into the Excel sheet, and the final correlation with the landslide under each geo-environmental variable was calculated. The resultant values were aggregated in Arc GIS using the lookup tool and finally integrated using map algebra to find out using the lookup tool and finally integrated using map algebra to find the LSI values ranging from low, moderate, high, and susceptible zones. Based on visual interpretation, the past landslide scars were identified using Google Earth satellite images and the Landsat-8 OLI.

$$FR_{ij} = \frac{N_p(S_{i,j}) / \sum_j N_p(S_{i,j})}{N_p(N_{i,j}) / \sum_j N_p(N_{i,j})} \tag{17.1}$$

where

N_{pix} (s_i) is the number of pixels in the entire study area having class *i*.

ΣN_{pix} (s_i) is the total number of pixels containing *i*.

Landslides in the study area ΣN_{pix} (ni) is the total pixels in the study area (Sur et al. 2020).

17.3.1 Development of Landslide-Conditioning Factor

A large number of work has been reported in the literature for landslide susceptibility zonation and their role in landslide prediction and mitigation plans. The accuracy of the developed models depends on the selection of appropriate conditioning factors that are equally important and given the priority. Landslides are initiated by the result of many factors alone and with combined effects such as geology, geomorphology, soil types, topography, climate, hydrology, tectonics, and many other human-induced activities (Sur and Singh 2020; Sur et al. 2022). In the present study, seven parameters such as slope, aspect, altitude, stream power index (SPI), topographic wetness index (TWI), normalized difference vegetation index, (NDVI), and land use/land cover map of the area to understand and map the landslide-susceptible zone. The data and methodology for the creation of all these landslide-contributing factors were discussed in the previous section.

Slope, Aspect, and Elevation

A digital elevation model (DEM) signifies the spatial variation in elevation. The slope, aspect, and elevation map were developed through the use of SRTM DEM with 30 m spatial resolution using ARC GIS software. Topographic features are considered as one of the important factors for landslide hazard zonation. The general slope of the area is grouped into five classes and ranges from < 1–13°, 14–23°, 24–31°, 32–40° to 41–76° degrees. The higher the slope and very low vegetation cover or barren are more prone to landslides. Aspect maps of any area provide important information about the slope facing, sunlight angle, wind direction, etc., and are crucial for understanding the erosional environment and landslide susceptibility mapping. The aspect map developed through the SRTM DEM grouped into six classes ranging from flat (<1–59), North (60–119), northeast and (120–179), south and southwest (180–239), east and southeast (240–299), and southeast (300–359). The elevation of the terrain directly affects the slide rate and is related to the landslide occurrence (Sur et al. 2022). It is known that the higher the elevation more the chance of sliding and landslide. In the present study, the elevation was categorized into 5 classes ranging from 507–792, 793–1041, 1042–1291, 1292–1571, and 1572–2099. The slope, aspect, and elevation map of the study area are shown in Fig. 17.4a, b, e.

Topographic Wetness Index (TWI)

Topographic wetness index (TWI) is another important terrain condition considered in landslide studies, which considers the slope gradient (in radians) and flow accumulation. It is also used to identify the hydrogeological nature of the catchment area and identify the moisture level. The TWI values in the study area range from 2.13 to 4.97, 4.97 to 6.71, 6.71 to 9.54, 9.54 to 13.49, and 13.49 to 22.24. TWI = Ln [CA/ Tan (slope) and presented in Fig. 17.4c.

Stream Power Index (SPI) The stream power index (SPI) is also an important hydrological property considered for landslide studies as it represents the streams' erosive power. It is observed that there is a positive correlation between the increase in the slope gradient, catchment area, and the erosive power of steams. The SPI values in the study area range from 0–1371, D1372–6512, 6513–16,109, 16,110–36,332, and 36,333–87,402, as given in Fig. 17.4d.

Land Use/Land Cover (LULC)

LULC is also considered one of the prime components for natural and human-induced landslides, land conversion due to large-scale infrastructural development largely causing the landslide events. LULC maps can be prepared through the use of satellite images with good spatial and temporal resolution. Vegetation cover contributes to managing slope failure. The LULC maps of the area were developed using Landsat-8 OLI image through the supervised classification and the entire area was grouped into the degraded forest, cropland, cropland, settlement, dense forest, barren rocky, and surface water bodies Fig. 17.4f.

Normalized Difference Vegetation Index (NDVI)

NDVI is an important biophysical parameter that represents the overall vegetation density of the landscape and is a very important factor to consider for landslide events. Observed a very positive correlation between vegetation density and landslide occurrence, the higher the density of vegetation cover, the lower the chance of landslide event and the lower the vegetation cover higher the landslide events in any area (Singh et al. 2023, 2021). NDVI for the study area is classified into five classes from lower to higher values and created by Landsat satellite images (Fig. 17.4g).

Landslide Inventory

Landslide inventory is an important input for understanding the previous landslide events and understanding the terrain, as well as verifying the developed landslide

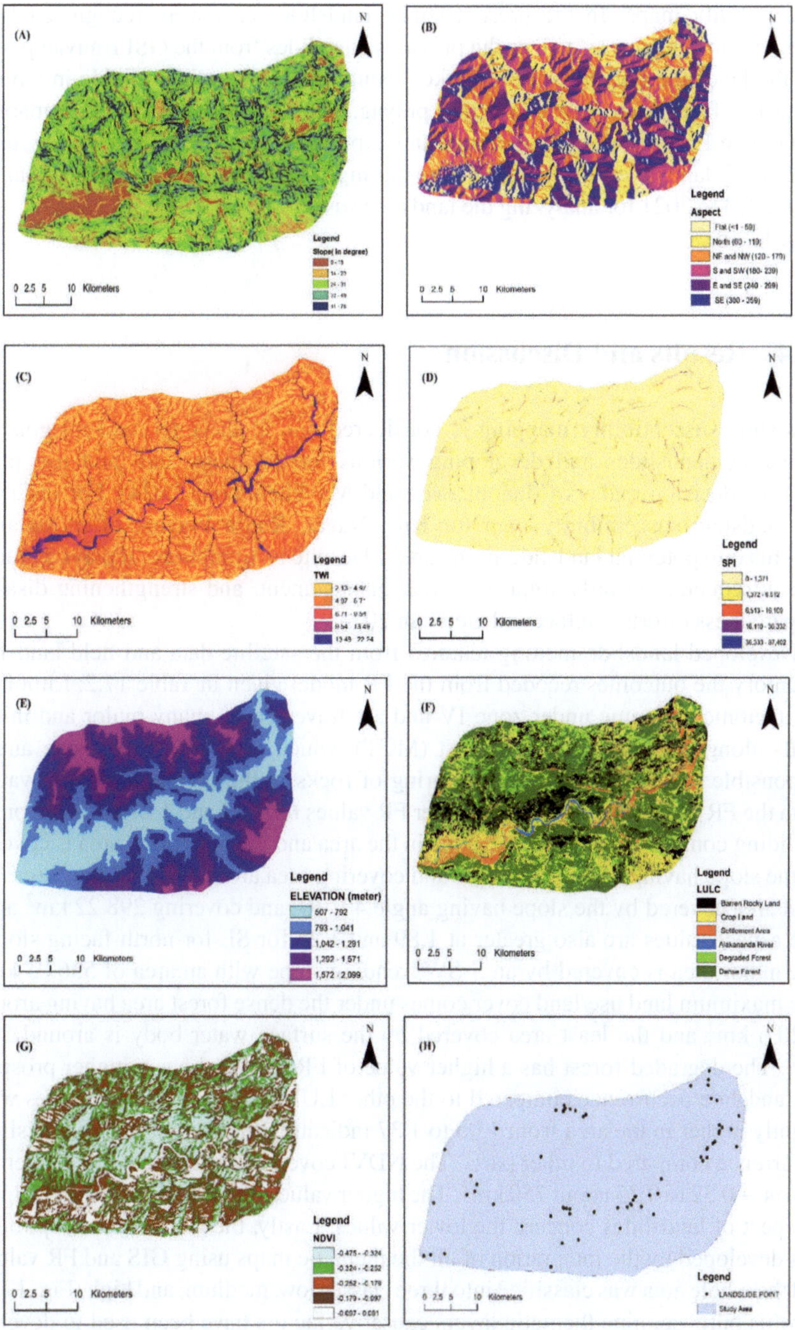

Fig. 17.4 Various landslide-conditioning factors such as slope, aspect, TWI, SPI, elevation, LULC, NDVI, and landslide inventory

susceptibility maps. In the present study, landslide scars were recognized from Google Earth images as well as the previous landslides from the GSI Bhuvan portal. All the landslide polygons were marked using digitizing tools available in Google Earth Pro from Kirti Nagar to Rudra prayag. The landslide locations are mapped on Google Earth as polygon features and exported to the Arc GIS shapefile and a total of 67 landslides were identified using high-resolution Google Earth Imagery from 2013 to 2021 for analyzing the landslide susceptibility and verification purpose (Fig. 17.4h).

17.4 Results and Discussion

Landslide susceptibility mapping is considered one of the important insights for mitigating landslides and developing various hazard maps and land use planning for decision-makers (Manopkawee and Mankhemthong 2024). For instance, the landslide susceptibility map for Kirti Nagar, Rudra Prayag, offers detailed insights into potential landslide-prone areas. Despite its limitations, it plays a crucial role in reducing hazards, improving risk management, and strengthening disaster preparedness efforts (Gulbet and Getahun 2024).

Developed landslide inciting features from the satellite data and field landslide inventory the outcomes recoded from the FR model given in Table 17.2. Lithology and tectonically come under zone IV and are traversed by many major and minor faults along with main central thrust (MCT), which is a very active zone and is responsible for the crushing and shearing of rocks in the region. The observation from the FR model shows that the higher FR values for 1.89 and 1.27 are more prone to sliding compared to rest of the slope in the area and the maximum area is covered by the slope having an angle 10–20° and covering area around 634.53 km^2 area, the least area covered by the slope having angle 40–76° and covering 298.22 km^2 area. The aspect values are also greater at 1.89 and 1.27 for SE for north-facing slopes. The major area is covered by an S-SW trending slope with an area of 536.05 km^2. The maximum land use/land cover comes under the dense forest area having around 1120.6 km^2 and the least area covered by the surface water body is around 29.8 km^2. The degraded forest has a higher value of FR 1.44 and has a higher prospect for landslide occurrence compared to the other LULC classes. The FR values were slightly higher in the area from 1.06 to 1.37 indicating the more prone to landslides occurrence compared to other parts. The NDVI covers the maximum area under the values − 0.32 to 0.25 about 750 km^2. The higher values of TWI and SPI with a higher prospect of landslides concern the lower values. Lastly, the landslide susceptibility was developed by the integration of all the thematic maps using GIS and FR values, and the whole area was classified into three classes low, medium, and high (Fig. 17.5). For this purpose, nine thematic layers causative factors have been used to develop. Further, the landslide inventories have been utilized to assess the success rate and validation of the results.

Table 17.2 Results derived from the analysis of the FR model

Factor	Unit	Class	Code	Pixel by class	Class %	LSD Pixel by Class	LSD %	FR
Slope	Degree	0–10	1	34,615	11%		0%	0.00
		10–20	2	70,504	22%	144	28%	1.27
		20–30	3	94,200	30%	159	31%	1.05
		30–40	4	81,360	26%	102	20%	0.78
		40–76	5	33,136	11%	101	20%	1.89
Aspect	Class	Flat	1	46,560	15%	208	37%	2.47
		North	2	48,562	15%	155	27%	1.76
		NE and NW	3	51,384	16%	62	11%	0.67
		S and SW	4	59,562	19%	20	4%	0.19
		E and SE	5	54,732	17%	20	4%	0.20
		SE	6	53,015	17%	103	18%	1.07
Altitude	Meter	507–792	1	60,795	19%	71	13%	0.65
		793–1041	2	78,546	25%	437	77%	3.10
		1042–1291	3	75,342	24%	60	11%	0.44
		1292–1571	4	68,284	22%		0%	0.00
		1572–2099	5	33,008	10%		0%	0.00
SPI	Ratio	0–5826	0	311,324	99%	565	99%	1.00
		5827–87,402	1	2491	1%	3	1%	0.67
TWI	Ratio	2.13–5.21	1	154,279	49%	305	54%	1.09
		5.22–7.34	2	115,075	37%	195	34%	0.94
		7.35–11.20	3	32,243	10%	45	8%	0.77
		11.21–22.24	4	12,218	4%	23	4%	1.04
NDVI	Ratio	– 0.475– – 0.324	1	56,316	18%	73	13%	0.72
		– 0.324– – 0.252	2	83,392	26%	134	24%	0.90
		– 0.252– – 0.179	3	68,918	22%	129	23%	1.05
		– 0.179– – 0.097	4	50,495	16%	121	21%	1.34
		– 0.097– – 0.091	5	57,350	18%	109	19%	1.06
LULC	Class	Alakananda river	0	3320	0%	0	0%	0.00
		Crop land	1	89,649	0%	136	32%	0.00
		Settlement	1	19,010	9%	24	6%	

(continued)

Table 17.2 (continued)

Factor	Unit	Class	Code	Pixel by class	Class %	LSD Pixel by Class	LSD %	FR
		Dense forest	1	124,518	56%	286	67%	1.19
		Degraded forest	1	34,707	16%	96	22%	1.44
		Barren rocky land	2	45,267	20%	24	6%	0.28

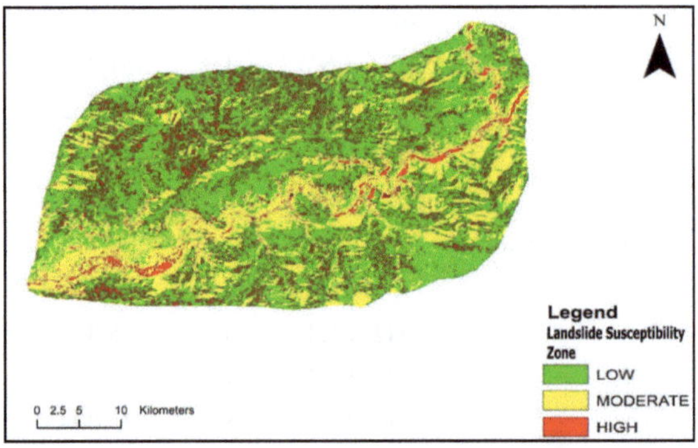

Fig. 17.5 Landslide susceptibility zonation map using FR model

17.5 Conclusion and Suggestions

In the present work, the integrated use of satellite images and the development of landslide-contributing factors provide a very important primary database for LSZ under changing climatic conditions and anthropogenic practices. The use of an advanced statistical model such as the frequency ratio (FR) approach to map the LSZ provides good accuracy after the available datasets (Achu et al. 2023). Further landslide investigation and development of landslide early warning systems required more scientific and multidisciplinary application through the use of AI, ANN, and other machine learning approaches to develop more accurate landslide susceptible maps (Rahaman et al. 2024). In this case study, the LSZ is divided into the low, medium, and high chance of landslide events through the use of a very limited database and suggested more scientific work for the terrain. In the Indian Himalayas, it is feasible to develop and employ trustworthy technologies for mapping landslides such as AI and machine learning including drone and LiDAR applications for developing high-resolution terrain data (Sur et al. 2022). The use of data-driven techniques like artificial intelligence, geotechnical investigation, and machine learning-based models may be investigated to increase the accuracy and usefulness of LSZ mapping.

References

Abedini M, Tulabi S (2018) Assessing LNRF, FR, and AHP models in landslide susceptibility mapping index: a comparative study of Nojian watershed in Lorestan province, Iran. Environ Earth Sci 77(11):426. https://doi.org/10.1007/s12665-018-7524-1

Achu A, Aju CD, Napoli M, Prakash P, Gopinath G, Shaji E, Chandra V (2023) Machine-learning based landslide susceptibility modelling with emphasis on uncertainty analysis. Geosci Front 14(6):101657. https://doi.org/10.1016/j.gsf.2023.101657

Aleotti P, Chowdhury R (1999) Landslide hazard assessment: summary review and new perspectives. Bull Eng Geol Env 58(1):21–44

Bai SB, Lu P, Wang J (2015) Landslide susceptibility assessment of the Youfang catchment using logistic regression. J Mt Sci 12(4):816–827

Bhukosh (2021b) Geological survey of India, the spatial data portal and a geophysical data repository, Lithological Data at https://www.gsi.gov.in/webcenter/portal/OCBIS

Bhukosh (2021a) Geological survey of India, the spatial data portal and a geophysical data repository; Earthquake Data at https://www.gsi.gov.in/webcenter/portal/OCBIS

Bonham-Carter GF (1994) In: Geographic information system for geoscientists: modelling with GIS. Pergamon Press, pp 398

Brabb EE (1984) Innovative approaches to landslide hazard and risk mapping. In: 4th International symposium on landslides, vol 1. Toronto, pp 307–324

Chen W, Li W, Hou E, Zhao Z, Deng N, Bai H et al (2014) Landslide susceptibility mapping based on GIS and information value model for the Chencang District of Baoji, China. Arab J Geosci 7:4499–4511

Clerici A, Perego S, Tellini C, Vescovi P (2006) A GIS-based automated procedure for landslide susceptibility mapping by the conditional analysis method: the Baganza Valley case study (Italian Northern Apennines). Environ Geol 50(7):941–961

Cruz RO, Harasawa H, Lal M, Ninh NH (2007) Asia. In: Parry ML, Canziani OF, Palutik JP of, van der Linden PJ, Hanson CE (eds) Climate change 2007: impacts, adaptation and vulnerability. Cambridge University Press, pp 469–506

Demir G (2019) GIS-based landslide susceptibility mapping for a part of the North Anatolian Fault Zone between Reşadiye and Koyulhisar (Turkey). CATENA 183:104211. https://doi.org/10.1016/j.catena.2019.104211

Erener A, Mutlu A, Sebnem Düzgün H (2016) A comparative study for landslide susceptibility mapping using GIS-based multi-criteria decision analysis (MCDA), logistic regression (LR), and association rule mining (ARM). Eng Geol 203:45–55

Gariano SL, Guzzetti F (2016) Landslides in a changing climate. Earth Sci Rev 162:227–252. https://doi.org/10.1016/j.earscirev.2016.08.011

Gorsevski PV, Gessler PE, Jankowski P (2003) Integrating a fuzzy k-means classification and a Bayesian approach for spatial prediction of landslide hazard. J Geogr Syst 5(3):223–225

Gulbet E, Getahun B (2024) Landslide susceptibility mapping using frequency ratio and analytical hierarch process method in Awabel Woreda, Ethiopia. Quaternary Sci Adv 100246. https://doi.org/10.1016/j.qsa.2024.100246

Iverson RM (2000) Landslide triggering by rain infiltration. Water Resour Res 36(7):1897–1910. https://doi.org/10.1029/2000WR900090

Lee S (2005) Application of logistic regression model and its validation for landslide susceptibility mapping using GIS and remote sensing data. Int J Remote Sens 26(7):1477–1491. https://doi.org/10.1080/01431160412331331012

Mandal P, Prathigadapa R, Srinivas D (2023) Evidence of structural segmentation of the Uttarakhand Himalaya and its implications for earthquake hazard. Sci Rep 13:2079. https://doi.org/10.1038/s41598-023-29432-z

Manopkawee P, Mankhemthong N (2024) Landslide susceptibility assessment using the frequency ratio model in the Mae Chan River Watershed, Northern Thailand. Quaternary Sci Adv 100263. https://doi.org/10.1016/j.qsa.2024.100263

Mohammady M, Pourghasemi HR, Pradhan B (2012) Landslide susceptibility mapping at Golestan Province, Iran: a comparison between frequency ratio, Dempster-Shafer, and weights-of-evidence models. J Afr Earth Sc 75:68–77. https://doi.org/10.1016/j.jseaes.2012.10.005

Pal SC, Chowdhuri I (2019) GIS-based spatial prediction of landslide susceptibility using frequency ratio model of Lachung River basin, North Sikkim, India. SN Appl Sci 1:416. https://doi.org/10.1007/s42452-019-0422-7

Pardeshi SD, Autade SE, Pardeshi SS (2013) Landslide hazard assessment: recent trends and techniques. Int J Geomat Geosci 4(3):497–513

Pham BT, Tien Bui D, Indra P, Dholakia M (2015) Landslide susceptibility assessment at a part of Uttarakhand Himalaya, India using GIS-based statistical approach of frequency ratio method. Int J Eng Res Technol 4(11):338–344

Rahaman A, Dondapati A, Gupta S, Raj R (2024) Leveraging artificial neural networks for robust landslide susceptibility mapping: a geospatial modeling approach in the ecologically sensitive Nilgiri district, Tamil Nadu. Geohazard Mech 2(4):258–269. https://doi.org/10.1016/j.ghm.2024.07.001

Sahoo PK, Kumar S, Singh RP (2000) Neotectonic studies of Ganga and Yamuna tear faults, NW Himalaya using remote sensing and GIS. Int J Remote Sens 21(3):499–518. https://doi.org/10.1080/014311600210713

Sarkar S, Roy AK, Martha TR (2013) Landslide susceptibility assessment using information value method in parts of the Darjeeling Himalayas. J Geol Soc India 82(4):351–362

Singh P, Sur U, Rai PK, Singh SK (2023) Landslide susceptibility prediction using frequency ratio model: a case study of Uttarakhand, Himalaya (India). Proc Indian Natl Sci Acad 89:1–15. https://doi.org/10.1007/s43538-023-00171-z

Singh P, Ankit S, Ujjwal S, Rai PK (2021) Comparative landslide susceptibility assessment using statistical information value and index of entropy model in Bhanupali-Beri region, Himachal Pradesh, India. Environ Developm Sustain 23:5233–5250. https://link.springer.com/article/10.1007/s10668-020-00811-0

Soeters R, van Westen CJ (1996) Slope instability recognition analysis and zonation. In: Turner AK, Schuster RL (eds) Landslide investigation and mitigation, Special Report, vol 247. National Research Council, Transportation Research Board, pp 129–177

Sur U, Singh P (2020) Landslide susceptibility indexing using geospatial and geostatistical techniques along Chakrata-Kalsi road corridor, India. J Indian National Cartographic Assoc 39:45–60

Sur U, Singh P, Meena SR (2020) Landslide susceptibility assessment in a lesser Himalayan road corridor (India) applying fuzzy AHP technique and earth-observation data. Geomat Nat Haz Risk 11(01):2176–2209. https://doi.org/10.1080/19475705.2020.1836038

Sur U, Singh P, Rai PK, Thakur JK (2021) Landslide probability mapping by considering fuzzy numerical risk factor (FNRF) and landscape change for road corridor of Uttarakhand, India. Environ Dev Sustain 23:13526–13554. https://doi.org/10.1007/s10668-021-01226-1

Sur U, Singh P, Meena SR, Singh TN (2022) Predicting landslides susceptible zones in the lesser Himalayas by ensemble of per pixel and object-based models. Remote Sens 14:1953.https://doi.org/10.3390/rs14081953

Tiranti D, Cremonini R (2019) Landslide hazard in a changing environment. Frontiers 1–200. https://doi.org/10.3389/feart.2019.00003

Varnes DJ (1984) Landslide hazard zonation: a review of principles and practice. UNESCO Press, Paris, pp 63

Wang Q, Guo Y, Li W, He J, Wu Z (2019) Predictive modeling of landslide hazards in Wen County, Northwestern China based on information value, weights-of-evidence, and certainty factor. Geomat Nat Haz Risk 10(1):820–835. https://doi.org/10.1080/19475705.2018.1549111

Yalcin A, Bulut F (2007) Landslide susceptibility mapping using GIS and digital photogrammetric techniques: a case study from Ardesen (NE-Turkey). Nat Hazards 42(3):451–465. https://doi.org/10.1007/s11069-006-9030-0

Yilmaz I (2009) Landslide susceptibility mapping using frequency ratio, logistic regression, artificial neural networks and their comparison: a case study from Kat landslides (Tokat—Turkey). Environ Monit Assess 157(1–4):277–293. https://doi.org/10.1016/j.cageo.2008.08.007
Zhang YX, Lan HX, Li LP, Wu YM, Chen JH, Tian NM (2020) Optimizing the frequency ratio method for landslide susceptibility assessment: a case study of the Caiyuan Basin in the southeast mountainous area of China. J Mt Sci 17(2):340–357

Chapter 18
Channel Dynamics and Slope Failures: A Case Study of Mahili Village Landslide Along the Beas River

Vijendra Kumar Pandey, Jeetesh Rai, Rupendra Singh, and Nitesh Verma

Abstract On July 10th, 2023, intense rainfall, combined with the dynamics of the river channel and slope erosion, triggered a massive landslide on the left bank of the Beas River. This landslide destroyed seven houses and partially damaged two others, along with agricultural lands and orchards. Landslides can be caused by various triggers and factors. To understand the mechanism behind this slope failure, we have analyzed maximum 1-day rainfall data for the upper Beas catchment and considered the river channel dynamics. The analysis revealed that erosion of the spur slopes by the river removed soil material, altering the stress distribution in the adjacent soil, which led to significant slope failures. These stress changes were substantial enough to disturb the static equilibrium that maintains slope stability, ultimately causing the landslide. Although river erosion was identified as a primary factor, the presence of a minor stream draining the slopes also increased infiltration. The combined effects of slope undercutting and increased infiltration contributed to the massive landslide. Additionally, channel migration analysis indicated that the river had shifted its course during the floods of 1995 and 2018, further eroding the slope base near Mahili village.

Keywords Channel dynamics · Slope erosion · Flood · River geomorphology · Landslide · Beas River

18.1 Introduction

The Himalayan region has long been severely affected by the devastating effects of extreme rainfall and subsequent landslides. These catastrophic events have had a profound impact, not only the loss of lives but also causing widespread destruction to economic infrastructure, thereby disrupting the lives of millions (Pandey

V. K. Pandey (✉) · J. Rai · N. Verma
Department of Geography, Kirori Mal College, University of Delhi, Delhi, India
e-mail: vijendrapandey@kmc.du.ac.in

R. Singh
CPS, School of Social Sciences, Jawaharlal Nehru University, New Delhi, India

S. C. Rai (ed.), *Himalaya: Mountains of Destiny*,
https://doi.org/10.1007/978-3-031-94905-0_18

2015; Pandey et al. 2022; Teja et al. 2019). Understanding the underlying mechanism inducing slope instability considering environmental fragility is very crucial to devise effective mitigation strategies (Pandey and Sharma 2023; Rai et al. 2024). The detailed analysis of channel migration-induced bank erosion and susceptibility of landslides has become very crucial in safeguarding the well-being of the region's inhabitants.

It is paramount to analyze the risk of landslides triggered by river dynamics and slope erosion during an extreme rainfall episode, preceded by the high-stage flow. The high-volume flow during the devastating floods induces channel migration and bank erosion, which in turn triggers massive slope failures (Ballesteros Cánovas et al. 2017; Bryan 2000; Dikshit et al. 2020; Dimri et al. 2017). Therefore, the understanding of the interactions between hydrological factors and slope instability is essential for developing effective risk assessment models that can predict the likelihood of landslide occurrences due to various environmental triggers, including heavy rainfall and riverbank erosion (Paringit et al. 2020). In addition, integrating advanced methodologies such as remote sensing and geographical information systems can enhance the analysis of slope stability and accurate identification of the areas at greater risk of landslides resulting from these dynamics (Paringit et al. 2020). The evaluation of landslide susceptibility requires a comprehensive approach that considers multiple factors, such as geotechnical data, geomorphological features, and hydrogeological assessments, to create models that reflect the complexities of slope behavior under varying environmental conditions (Devanand et al. 2023; Martino et al. 2020; Paringit et al. 2020; Xiao et al. 2020).

The failure mechanism of flash flood-induced landslides is a complex interplay of hydrodynamic forces, groundwater dynamics, and soil properties. Rainfall intensity, volume, and timing, as well as the antecedent conditions of the river and its drainage basin, can all contribute to the initiation of slope failures (Rinat et al. 2020). Hydrodynamic pressures exerted by the floodwaters, coupled with the rising groundwater table and increased pore pressures in saturated soils, can lead to the loss of soil cohesion and frictional strength, triggering the collapse of slopes and the mobilization of debris (Khajehei et al. 2020; Luo et al. 2020). While the mechanisms of flood-induced landslides in tropical regions are well-documented, the occurrence and underlying processes of such events in alpine valley settings, where the erosion of colluvium by successive floods may play a significant role, have been relatively understudied (Luo et al. 2020; Senouci et al. 2021).

Recent studies have explored innovative methods for assessing the suitability of nature-based solutions for landslide risk reduction (Devanand et al. 2023). These approaches highlight the importance of utilizing a combination of traditional geological assessments and contemporary data acquisition techniques, such as satellite remote sensing and LiDAR, to create detailed hazard and risk maps that can inform local authorities and stakeholders in their planning and resource allocation efforts (Devanand et al. 2023). This investigation involved field and drilling surveys to characterize the geomorphology and geological properties of the affected slope, alongside emergency monitoring to assess post-event stability and potential risk factors associated with future landslides triggered by ongoing hydrological conditions (Xiao et al.

2020). Furthermore, the development of geotechnical models that calculate failure probabilities and simulate potential run-out distances can significantly enhance our ability to predict landslide behavior, thereby enabling targeted risk mitigation strategies that address both immediate and long-term safety concerns within vulnerable regions (Devanand et al. 2023; Haninia et al. 2021; Zhang et al. 2021).

The existing research on river dynamics and slope erosion-triggered landslides emphasizes the need for a multifaceted approach that combines traditional geotechnical assessments with innovative data collection and modeling techniques. This comprehensive methodology not only aids in understanding the spatial distribution of landslide susceptibility but also highlights the critical role of technological advancements in enhancing data reliability and predictive accuracy, ultimately guiding effective land-use planning and disaster preparedness efforts in affected areas (Dikshit et al. 2020; Xiao et al. 2020). The rapid onset of heavy rainfall during flash flood events can overwhelm the capacity of riverbanks and slopes, leading to the destabilization and failure of soil and rock structures, resulting in devastating landslides that pose a significant threat to nearby settlements and infrastructure (Khajehei et al. 2020). These landslides are often characterized by high speeds, long travel distances, and the ability to accumulate large volumes of debris, which can then block and dam rivers, further exacerbating the flooding and hazardous conditions (Ding et al. 2022).

Climate change is expected to increase the frequency and intensity of extreme rainfall events, which will in turn heighten the risk of flash floods and associated landslides in many regions around the world. This is particularly true in developing communities where increased human encroachment into floodplains, lack of adequate flood response plans, and limited infrastructure and resources to mitigate the impacts of such hazards can amplify the vulnerability of local populations (Khajehei et al. 2020; Munyai et al. 2021; Yin et al. 2018). Notably, researchers have recognized the importance of incorporating climate variability into predictive models for landslide risk analysis. Understanding how changing weather patterns and extreme rainfall events contribute to slope instability is essential, particularly in regions prone to landslides, as studies have shown that integrating climate change factors can lead to more accurate hazard assessments and more effective management strategies to mitigate risks associated with slope failures (Dikshit et al. 2020; Paringit et al. 2020).

Effective strategies to predict, prevent, and respond to these cascading natural disasters must be grounded in a comprehensive understanding of the complex interactions between climate, hydrology, and geomorphology that govern the occurrence and behavior of these hazardous events (Khajehei et al. 2020; Munyai et al. 2021; Yin et al. 2018). Recent research underscores the need for a more location-specific and comprehensive approach to assessing the hazards and vulnerabilities associated with flash flood-induced landslides (Luo et al. 2020). Physically-based models that integrate hydrological, geotechnical, and geomorphological data can help identify the most susceptible areas and guide the development of early warning systems and risk mitigation strategies. By enhancing our understanding of the intricate interactions between climate, hydrology, and geomorphology, we can work toward building more resilient communities and infrastructure capable of withstanding the escalating threats posed by these cascading natural disasters (Khajehei et al. 2020; Martino et al.

2020). The present study focuses on analyzing channel dynamics with past floods using satellite images and generation of flood volume using the HecRAS modeling to understand the instability mechanism along the channels.

18.2 Study Area

Mahili village was hit by a catastrophic landslide on July 10th, 2023, which completely damaged 7 houses and partially damaged 2 others. According to the Census of India (2011), the village has 17 households, though 21 houses were reported during a field survey conducted by the villagers. It is located on the left bank of the Beas River, downstream of the Patlikuhal-Naggar bridge (Fig. 18.1). The village is part of the Naggar community development block in Kullu District, Himachal Pradesh. A detailed field investigation was carried out at the affected site to collect information about the damages caused by the landslide triggered by the July 2023 floods.

The upper catchment of the Beas River received heavy rainfall on July 9th and 10th, 2023, which caused a devastating flood in the downstream area. The high-stage flow disrupted the lives and livelihoods of the inhabitants. The village is situated on the left spur slopes, which are covered by dense vegetation and scrub. The area was also covered by sandy soil and loose regolith.

18.3 Materials and Methods

In the present study meteorological as the Statement on Climate for the State of Himachal Pradesh: 2022, published by Climate Monitoring and Prediction Group, Office of Climate Research and Services, India Meteorological Department, Pune; Rainfall Status Report (During 8th to 12th July 2023), published by Government of India, Ministry of Earth Sciences, India Meteorological Department Meteorological Centre, Shimla (HP), and Monthly Weather Report July 2023 published by Government of India, Ministry of Earth Sciences (MoES), India Meteorological Department Meteorological Centre, Shimla were used. Apart, from that satellite-based meteorological observations are available in the form of gridded data. Time Series, Area Averaged of 24-h accumulated precipitation (combined microwave-IR) estimate pattern from June 2000 to January 2024 (GPM_3IMERGDLv06), 24-h accumulated precipitation (combined microwave -IR) estimate July to September 2023 (GPM_3IMERGDLv06), (accessed from the https://giovanni.gsfc.nasa.gov/giovanni/) and Daily rainfall (in mm) trends for Kullu from 01-Jul-2023 to 31-Jul-2023 (IMD GRIDED data).

Digital elevation data was used for the physiographic analysis and of minor streams for the site. SRTM 1 arc-second digital terrain elevation global dataset was

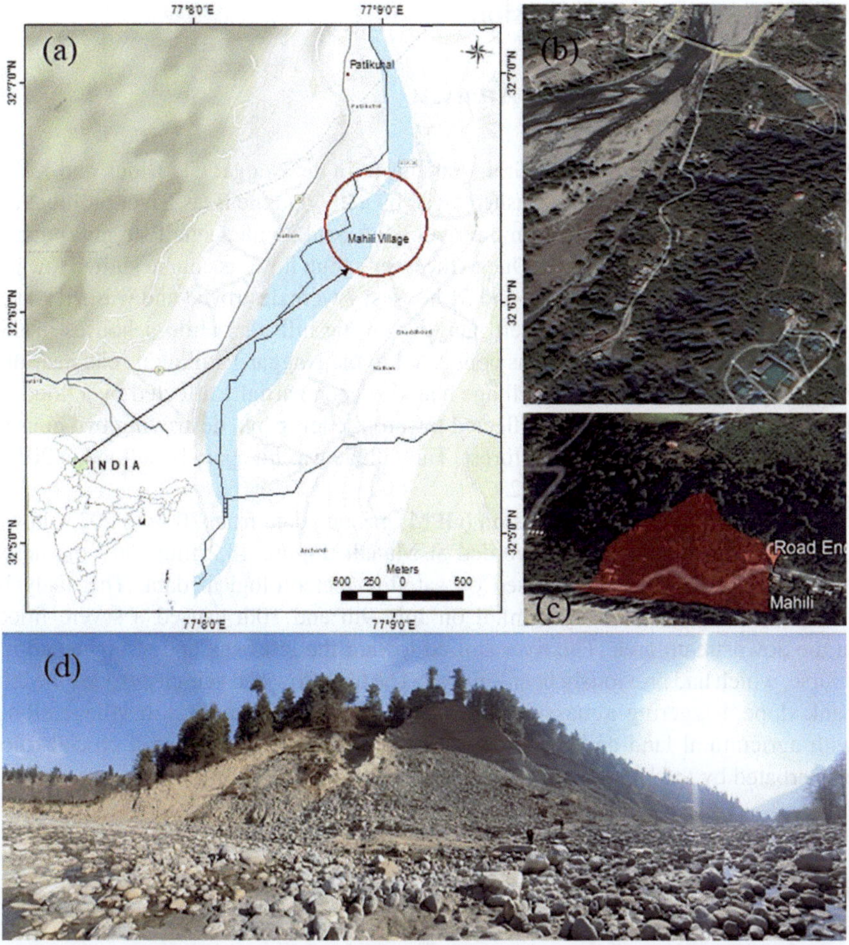

Fig. 18.1 **a** Location of the Mahili village on the left bank of Beas River near Patlikuhal-Naggar bridge, District Kullu, Himachal Pradesh, **b** Google image of April-2022 showing the road and houses of the village, **c** massive landslide area, and **d** photograph after 10th July 2023 of the landslide site

used (accessed from https://earthexplorer.usgs.gov/), accessed from the https://earthexplorer.usgs.gov/. The channel dynamics of the Beas River and land use have been analyzed using the LANDSAT series satellite images from 1980 to 2023. The images were analyzed using the QGIS 3.32.2 software using SCP plugins.

18.4 Results and Discussion

18.4.1 Rainfall Intensity Analysis

Mahili, a village under Nathan Gram Panchayat in the Naggar CD Block and Kullu Sub-Division was completely washed away by a flood in the Beas River on July 9th, 2023. The village was located on the river's left bank, with a small stream running through it and joining the Beas. Due to the river's high flow, its course shifted toward the left bank. The village had around 21 houses; 7 were destroyed and washed away, 2 were partially damaged and left hanging on the hillside, while 4 houses in the cluster remained safe. Mahili was connected to the Naggar-Patlikuhal road near the bridge by a concrete road. The village had also been partially affected by a flood in 1995 when the river similarly deflected toward the left bank, destroying two houses along with agricultural land and forest. The village was hit again by a flood in 2018, culminating in the disaster of 2023.

The Global Precipitation Mission (GPM) gridded data from 2000 to 2023 shows the 1-day maximum rainfall recorded at Manali. Figure 18.2 illustrates the daily rainfall estimates at Manali based on satellite meteorological data. The analysis indicates that high-intensity rainfall on July 9th and 10th caused a severe flood in the downstream area. The river shifted toward the left bank and straightened its course, which had previously been sinuous. The high flow and velocity eroded the left bank slope, triggering a massive landslide that washed away Mahili village along with agricultural land and orchards. The channel dynamics played a crucial role, exacerbated by the lack of bank protection measures (Table 18.1).

18.4.2 Channel Dynamics and Slope Erosion

The high-stage flow along the Beas River between July 9–11, 2023, caused significant disruption at various locations. On July 10th, a massive landslide occurred on the river's left bank, downstream of the Patlikuhal–Naggar bridge. Figure 18.3a shows the location of the bridge, which was restored after the flood. The road connecting Mahili village was completely washed away by the river. The cross-sectional profile indicates that the river flows as a braided channel downstream from the bridge (Fig. 18.3b) and has a wide river course (Fig. 18.3c). It was also observed that before the 1995 flood, the Beas River had deflected to the right side downstream of the bridge (Fig. 18.4a).

The river changes its course during the 2023 flood making its flow path straight and hitting the Mahili village. The subsequent lateral erosion triggered the landslide (Fig. 18.4a).

Figure 18.4b illustrates the channel morphology, highlighting features such as the water channel, point bar, and vegetated point bar within the riverbed. The longitudinal profile analysis of the channel, from the Sanjoin Nala confluence to the downstream end of the vegetated point bar, indicates that the channel bed has an average slope of

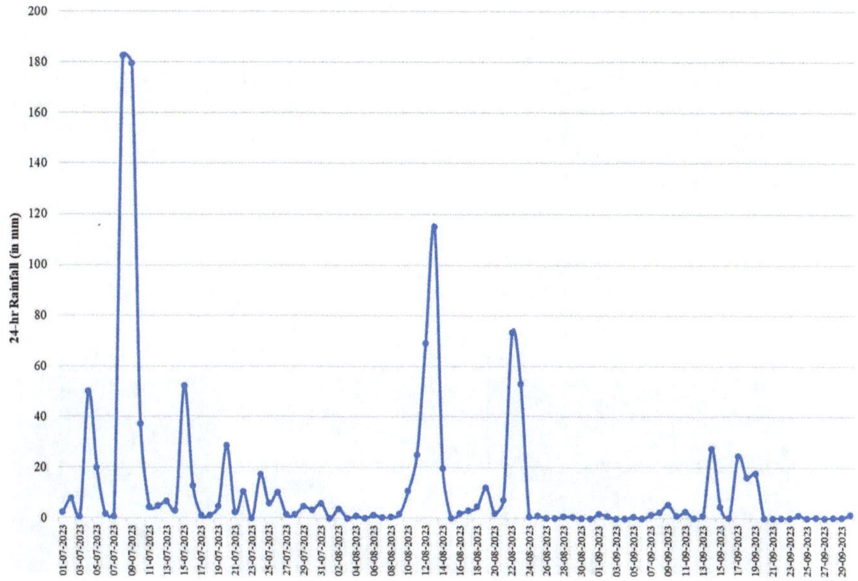

Fig. 18.2 1-day maximum rainfall data for July 2023 (Time Series, Area-Averaged of Daily accumulated precipitation (combined microwave-IR))

Table 18.1 1-day maximum rainfall records for the month of July 2023 in Himachal Pradesh

Station	District	New record		Previous record	
		Rainfall (in mm)	Date of July 2023	Rainfall (in mm)	Date
Manali	Kullu	131.3	9	105.1	09 July, 1971
Una	Una	228.5 (AWS)	9	224	22 July, 1927
Keylong	Lahaul & Spiti	83 (AWS)	9	78	28 July 1951
Rohru	Shimla	185	9	170	25 July 1966
Ghamroor	Kangra	166	9	164.8	19 July, 2021
Pachhad	Sirmaur	220	10	189.2	26 July, 1973
Nadaun	Hamirpur	160.5	9	146	30 July, 1996

Source Statement on Climate for the State of Himachal Pradesh: 2022, IMD

less than 2.4% (Fig. 18.4c). The gentle bed slope and the river's wide, sinuous course are key factors contributing to channel migration during high-stage flow. Satellite image analysis reveals that the river has altered its course during every major flood.

The river shifted its course to the left due to high-volume flow, eroding the slope material through lateral erosion (Fig. 18.5). This lateral erosion, combined with the excessive infiltration from prolonged rainfall, triggered a massive landslide. The

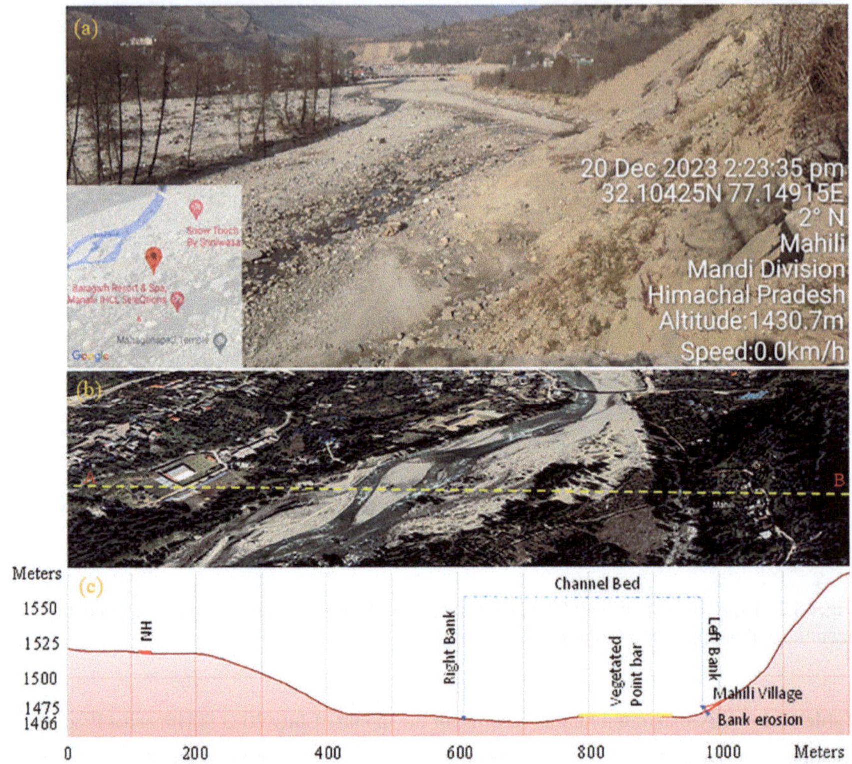

Fig. 18.3 Channel cross-section dynamics near the Mahili Village **a** landslide site on the left bank of Beas River, **b** Google image of May 2023 showing the houses and orchards, agriculture land, and **c** channel bed profile and side slopes

landslide displaced a significant portion of Mahili village, including houses, orchards, and agricultural land. A detailed field investigation of the site was conducted to understand the failure mechanism. It was observed that the crown of the landslide site was covered with dense vegetation, including trees and shrubs, while the middle section contained houses, roads, agricultural land, and orchards. The slope inclined approximately 47.5 degrees and was covered with regolith deposits. The upper slopes were used for cultivation and commercial purposes. Additionally, a minor stream drained from the landslide site into the Beas River.

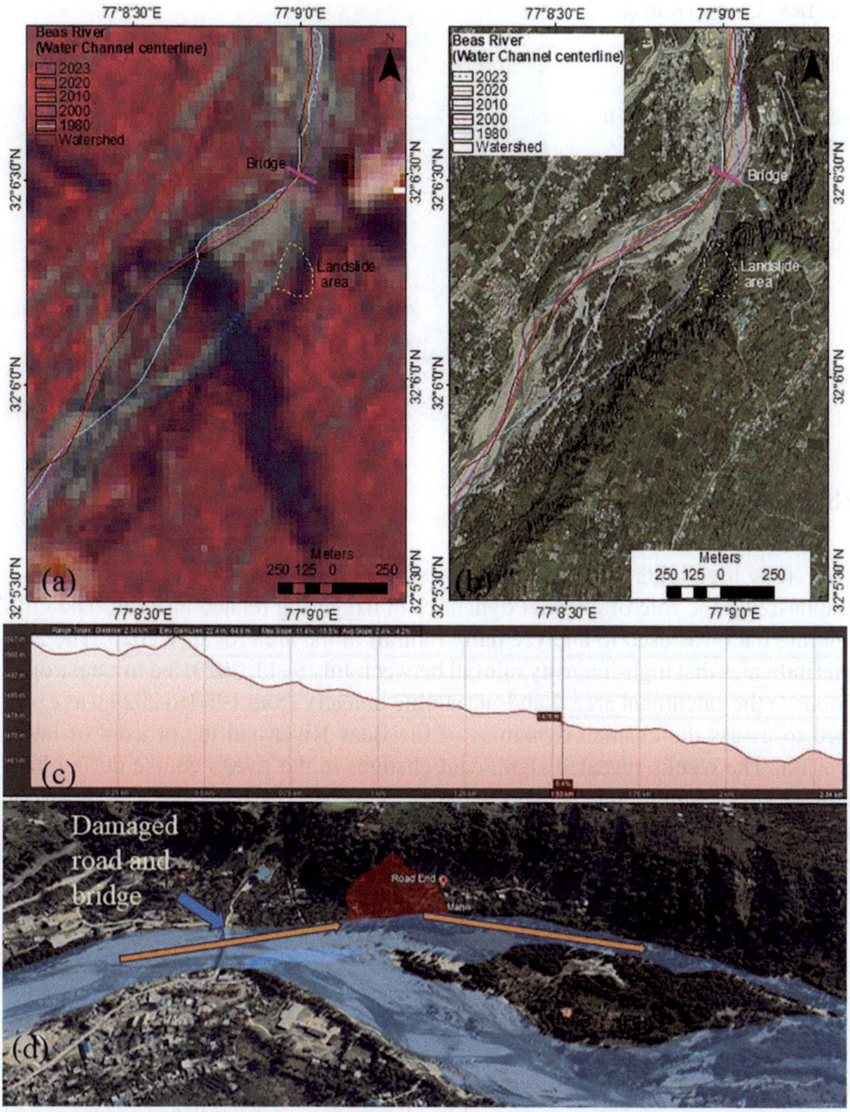

Fig. 18.4 **a** channel dynamics analysis using the Landsat series satellite imagery, **b** channel morphology and channel form on the Google images, **c** channel longitudinal profile and slope, and **d** change in the flood flow direction and associated damages and lateral erosion

Fig. 18.5 Channel path of the Beas River during the 10th July flood and associated lateral erosion induced massive landslide. It shows the traces of bed erosion and subsequent increase in channel depth, clearing of vegetated point bar, and bank erosion

18.5 Conclusions

This study highlights the recurring nature of floods in the Upper Beas River basin, emphasizing the role of channel dynamics in triggering related hazards. The GPM satellite data was used to analyze daily rainfall in the area for July 2023. The analysis indicates that high-intensity rainfall between July 9–11, 2023, led to catastrophic floods in the catchment area. Landsat satellite imagery from 1980 to 2023 was examined to assess the channel dynamics of the Beas River and the process of lateral erosion. The results revealed significant changes in the river's course during high-stage floods, resulting in extensive damage. Field investigations at the landslide site also suggest that the primary triggering factor for the massive landslide was the lateral erosion of the slope base and the river's channel shift. This was further exacerbated by the increased load due to rainwater infiltration.

The socioeconomic impact of the flood and landslide was profound, with significant losses in property and agricultural land for the affected families. Although government rescue and rehabilitation efforts were implemented, the scale of the disaster underscored the need for long-term measures for effective flood management and mitigation in the region. Community awareness and education programs are crucial for fostering a culture of preparedness and resilience. Strengthening disaster management frameworks and local capacities is essential to ensure a swift and effective response to future flood events and associated hazards in the area.

Acknowledgements The authors extend their gratitude and appreciation toward the Northern Regional Centre-Indian Council of Social Science Research (ICSSR), New Delhi, for their support and facilitation of this research endeavor under the Project Grant Sanctioned File No. NRC-ICSSR/RP/VKP/2022-23, dated 19.09.2023. The authors also extend their sincere thanks to the Principal, Kirori Mal College, University of Delhi, Delhi for all the necessary administrative support.

References

Ballesteros Cánovas JA, Trappmann D, Shekhar M, Bhattacharyya A, Stoffel M (2017) Regional flood-frequency reconstruction for Kullu district, Western Indian Himalayas. J Hydrol 546:140–149. https://doi.org/10.1016/j.jhydrol.2016.12.059

Bryan RB (2000) Soil erodibility and processes of water erosion on hillslope. Geomorphology 32:385–415. https://doi.org/10.1016/S0169-555X(99)00105-1

Devanand VB, Mubeen A, Vojinovic Z, Sanchez Torres A, Paliaga G, Abdullah AF, Leitão JP, Manojlovic N, Fröhle P (2023) Innovative methods for mapping the suitability of nature-based solutions for landslide risk reduction. Land 12(7). https://doi.org/10.3390/land12071357

Dikshit A, Sarkar R, Pradhan B, Segoni S, Alamri AM (2020) Rainfall induced landslide studies in Indian Himalayan region: a critical review. Appl Sci (Switzerland) 10(7). https://doi.org/10.3390/app10072466

Dimri AP, Chevuturi A, Niyogi D, Thayyen RJ, Ray K, Tripathi SN, Pandey AK, Mohanty UC (2017) Cloudbursts in Indian Himalayas: a review. Earth Sci Rev 168:1–23. https://doi.org/10.1016/j.earscirev.2017.03.006

Ding Y, Liu X, Chen R (2022) Numerical simulation of alpine flash flood flow and sedimentation in gullies with large gradient variations. Front Environ Sci 10.https://doi.org/10.3389/fenvs.2022.858692

Haninia A, Nazarnejad H, Naja S, Pourghasemi HR (2021) Landslide susceptibility assessment and mapping using statistical and data mining models in Iran. https://doi.org/10.21203/rs.3.rs-239985/v1

Khajehei S, Ahmadalipour A, Shao W, Moradkhani H (2020) A place-based assessment of flash flood hazard and vulnerability in the contiguous United States. Scient Reports 10(1). https://doi.org/10.1038/s41598-019-57349-z

Luo G, Zhong Y, Yang Y (2020) Failure mechanism and mitigation measures of the G1002 electricity pylon landslide at the Jinping I hydropower station. Adv Civil Eng 1. https://doi.org/10.1155/2020/8820315

Martino S, Battaglia S, D'Alessandro F, Della Seta M, Esposito C, Martini G, Pallone F, Troiani F (2020) Earthquake-induced landslide scenarios for seismic microzonation: application to the Accumoli area (Rieti, Italy). Bull Earthq Eng 18(12):5655–5673. https://doi.org/10.1007/s10518-019-00589-1

Munyai RB, Chikoore H, Musyoki A, Chakwizira J, Muofhe TP, Xulu NG, Manyanya TC (2021) Vulnerability and adaptation to flood hazards in rural settlements of Limpopo Province. South Africa. Water 13(24):3490. https://doi.org/10.3390/w13243490

Pandey VK (2015) Predictive landslide susceptibility assessment using the weight-of-evidence method in north-central Garhwal Himalaya, India. Phys Geography 36(6):510–536. https://doi.org/10.1080/02723646.2015.1116050

Pandey VK, Kumar R, Singh R, Kumar R, Rai SC, Singh RP, Tripathi AK, Soni VK, Ali SN, Tamang D, Latief SU (2022) Catastrophic ice-debris flow in the Rishiganga River, Chamoli, Uttarakhand (India). Geomat Nat Haz Risk 13(1):289–309. https://doi.org/10.1080/19475705.2021.2023661

Pandey V, Sharma MC (2023) Sedimentological characteristics and their relationship with landsliding in the Bhilangana Basin, Garhwal Himalaya, India. Earth Sci Res J 26(4):345–358. https://doi.org/10.15446/esrj.v26n4.94720

Paringit MCR, Cutora MDL, Santiago EH, Adajar MAQ (2020) Assessment of landslide susceptibility: a case study of Carabao mountain in Baguio city. Int J Geomate 19(71):166–173. https://doi.org/10.21660/2020.71.9261

Rai SC, Pandey VK, Sharma KK, Sharma S (2024) Landslide susceptibility analysis in the Bhilangana Basin (India) using GIS-based machine learning methods. Geosyst Geoenviron3(2):100253. https://doi.org/10.1016/j.geogeo.2024.100253

Rinat Y, Marra F, Armon M, Metzger A, Levi Y, Khain P, Rosensaft M, Morin E (2020) Hydrometeorological analysis and forecasting of a 3-day flash-flood-triggering desert rainstorm. Natural Hazard Earth Syst Sci 189:1–35. https://doi.org/10.5194/nhess-2020-189

Senouci R, Taibi NE, Teodoro AC, Duarte L, Mansour H, Yahia Meddah R (2021) GIS-based expert knowledge for landslide susceptibility mapping (LSM): case of Mostaganem coast District, West of Algeria. Sustainability 13(2):630. https://doi.org/10.3390/su13020630

Teja TS, Dikshit A, Satyam N (2019) Determination of rainfall thresholds for landslide prediction using an algorithm-based approach: case study in the Darjeeling Himalayas, India. Geosciences (Switzerland) 9(7). https://doi.org/10.3390/geosciences9070302

Xiao L, Wang J, Zhu Y, Zhang J (2020) Quantitative risk analysis of a rainfall-induced complex landslide in Wanzhou County, three Gorges Reservoir, China. Int J Disaster Risk Sci 11(3):347–363. https://doi.org/10.1007/s13753-020-00257-y

Yin J, Gentine P, Zhou S, Sullivan SC, Wang R, Zhang Y, Guo S (2018) Large increase in global storm runoff extremes driven by climate and anthropogenic changes. Nature Commun 9(1). https://doi.org/10.1038/s41467-018-06765-2

Zhang Y, Ma H, Yu Z (2021) Application of the method for prediction of the failure location and time based on monitoring of a slope using synthetic aperture radar. Environ Earth Sci 80(21):706. https://doi.org/10.1007/s12665-021-09989-6

Chapter 19
Urban Risk Scenarios and Multi-hazard Interrelationships: An Example of Eastern Himalaya

Lakpa Tamang, Saurabh Singh Barfal, Akash Chowdhury, and Mayank Joshi

Abstract India is called a 'land of diversity' due to the presence of a range of diversified landscapes from snow-capped mountains, deserts, plains, plateaus, and coastal plains. Due to its diversified landscapes, it becomes vulnerable to various kinds of natural hazards. In the eastern part of India, a tiny Himalayan state of Sikkim represents a unique blend of diversified landscapes within its boundary. It has topographic features of low flat land to highly elevated Kanchenjunga peak within an aerial distance of 114 km. Due to the sharp gradient change, the region possesses different climate types from sub-tropical to alpine. The general slope of the state is from north to south, on which a steep slope nearly occupies 43% of the total area and is highly susceptible to hazards. In the past, Sikkim is experiencing various kinds of natural hazards like landslides, earthquakes, flash floods, avalanches, GLOF, etc., since ancient times. However, the increased intensity and frequency of these natural hazards indicate a trend pointing towards climate-induced hazards such as GLOF, avalanches, etc. Despite all these challenges prevalent in the state, it also offers opportunities for those who wish to explore natural resources in terms of biodiversity, landforms, and geological features. It serves as a resource hub for enthusiasts from diverse fields who can use and decipher the nature's signature.

Keywords Natural hazards · Earthquakes · GLOF · Khanchedonga · Himalaya

19.1 Introduction

India has a geographically extremely diversified landscape; varying from snow-capped mountain ranges to deserts, plains, hills, and plateaus with elevation variation ranging from 8586 m at Kanchenjunga Mountain to − 2.2 m at Alappuzha District

L. Tamang · S. S. Barfal · A. Chowdhury · M. Joshi (✉)
G. B. Pant National Institute of Himalayan Environment, Sikkim Regional Centre, Pangthang, Gangtok, India
e-mail: mayank.geo@gmail.com

S. C. Rai (ed.), *Himalaya: Mountains of Destiny*,
https://doi.org/10.1007/978-3-031-94905-0_19

of Kerala. This variation in geography leads to expose the country to different kinds of hazards (NDMA Report). For example, more than 58.6% region is vulnerable to moderate- to very high-intensity earthquakes, ~ 12% of the area is exposed to floods and river erosion; coastal areas are vulnerable to cyclones and tsunamis, ~ 68% fertile area is prone to drought, and its hill regions are prone to landslides and avalanches (NDMA Report).

The Himalayas, a 2400 km long arc that runs from west-northwest to east-southeast, have been susceptible to various kinds of hazards such as earthquakes, landslides, floods, cloudbursts, glacial lake outbursts, and avalanches. Such natural disasters impose adverse effects on the economy (Skoufias 2003; Toya and Skidmore 2007). Sikkim, a small eastern Himalayan hill state, is bound between the latitudes 27°04' to 28°07' N and longitudes 88°01'to 88°55' E. It shares three international boundaries: in the west with Nepal, in the north with the Tibetan plateau, and with Bhutan in the east (SSDMA 2018). The general structural slope of the state is from north to south, which is dissected vertically into two halves by the River Tista flowing in a southerly direction (Kanade et al. 2018). Darjeeling District of West Bengal marks its southern boundary. The State covers an area of 7096 km^2 spread over six districts Gangtok, Pakyong, Namchi, Mangan, Soreng, and Gyalshing with a minimum elevation of 251 m to a highest elevation of 8586 m (Bhasin et al. 2002). The highest elevated areas of the state are covered with thick glacier cover and the source of river Tista, Rangit, and its tributaries. Sikkim, having rugged topography, is characterized by deep ravines, river terraces, ancient and recent flood plains, and steep and gently sloping hills. It is transversed by various geo-tectonic elements such as faults, folds, etc. (Thakur et al. 2012; Paul et al. 2015).

Existence of steep topography, fragile geology, and a high amount of rainfall may lead to sequential activation of hazards (Joshi 2022). Considering geology and geomorphology, the occurrences of natural hazards in Sikkim are area-specific. Incidences such as avalanches and GLOF were mainly reported from the northern part due to the presence of glaciers and snow-bound areas. Similarly, the major epicentres of earthquakes are located along the cone of the MCT line passing through the southern portion, and the impact of flash floods is mainly felt along the river valleys of Tista, Rangit, and its tributaries. So, all these events indicate Sikkim is a good host for the natural hazards.

This hill state is very vulnerable to different natural hazards as it falls under the Seismic Zone IV and receives one of the highest annual rainfalls in the country (Thakur et al. 2012; Paul et al. 2015). The state is notably affected by natural hazards such as landslides, subsidence, earthquakes, cloudbursts, lightning, forest fires, and occasionally by avalanches (Tables 19.1 and 19.2). Therefore, the present study is conducted to understand the intensity and impact of dominant natural hazards in the Sikkim Himalaya over the past six decades.

Table 19.1 Location of landslides reported in Sikkim Himalaya

Year	Landslide occurrences
1962	Rang Rang, Vong-I slide
1966	Rang Rang, Richu bridge slide, new Vong slide
1967	North Sikkim
1968	Rang Rang, Richu bridge slide, Lukvir slide
1969	Teesta Valley (NH-31A)
1970	Lantak Khola, Richu Bridge slide
1972, 1980, 1981	Richu bridge slide
1973	Manul-1
1975	Chandmari, Tathangchen, Rang Rang slide
1978	Rang Rang
1982	Namnsa slide, Lachung-Yumthang road, Richu bridge slide, Rang Rang slide
1983	Manual I&II, Lantak khola
1984	Tathangchen
1990, 1995	Syari & Deorali area
1997	Development area, Tathangchen, Chandmari, Six Mile Gangtok, Rongey basti, Chongey basti, Burdang, Rongli, Bhusuk, Zero point
1998	Development area, Ranipool, Bakthang-Burtuk slide, Samdong Gurdum slide
2000	Gyalshing bazaar
2003	Manupur khola slide
2005	Rakdong, Chawang, Sardong, Manzing
2007	Rangpo-Melli road, Gangtok, Bagdhara satam slide, Gangtok, Amdogolai
2008	Marchak, Dikchu, Hingdam,
2009	Sombaria, Daramdin, JN marg, 17 mile, Kyoongaala, Gangtok
2010	Gangtok, 32mile and Topakhani, Tintek khola, Tingda, Pegong, Sumin, Kamarey, Padamche, Rongli, Singtham, Lower Namphiang, Tsomgo-Nathula, Yaksha, Lachung-Yumthang, Lachen-Gurudongmar, JN marg
2011	Mangan, Chungthang, Lachung, JN road, Bakthang, Malbasey, 20 mile Baghey khola, Likhu bhir (NH-31A), Khurong kewa dara, Tikjuk & Lower pelling, Sichey and JN road, Gyalshing
2015	6th Mile
2016	Mantam, 13th Mile
2017	Pakyong

Table 19.2 Details of various natural hazards that occurred in Sikkim Himalaya

Event	Year	Place
GLOF	2023	South Lhonak
Avalanche	2023	14th Mile Nathula road
Flash flood	1968	Teesta basin
	2001	Tir Khola
	2003, 2010	Places were not specified
	2012	Chungthang
	2013	Rolep
Hailstorm	1983, (6 T) 1989, 1992, 1996, 1998, (2 T) 2001, 2002, (3 T) 2003, 2005, 2006, 2007, 2008, 2009, (2 T) 2010	Pangthang
	1998, 2007, 2008, 2009, 2010	Kabi
	2017	Pendam
	2020, 2023,	Gangtok
Forest fire	2000 (53 *In*), 2001 (9 *In*), 2002 (14 *In*), 2003 (18 *In*), 2004 (25 *In*), 2005 (42 *In*), 2006 (58 *In*), 2007 (29 *In*), 2008 (39 *In*), 2009 (120 *In*), 2010 (34 *In*), 2011(64 *In*), 2012 (2 *In*), 2013 (74 *In*), 2014 (80 *In*), 2015 (10 *In*), 2016 (54 *In*), 2017 (64 *In*), 2018 (53 *In*), 2019 (23 *In*), 2020 (74 *In*)	Sikkim
Earthquake	1980	Not known
	2006	Near Ralang
	2007	Chuchen
	2008, 2011, 2014	Nepal–Sikkim border region
	2009, 2017	Gangtok
	2012	Chadey Sikkim
	2013	Sangtok
	2015	Karmatar
	2016, 2019	Mangan
	2018	Rangpo
	2023	Yuksom
	2024	Soreng

T means times and *In* means Incidence

19.2 Study Area

The present study is mainly carried out in the tiny Himalayan state of Sikkim. It is located in the North-Eastern flank of India extending between 27.08°N to 28.13°N Latitude and 88.08°E to 88.92°E Longitude (Fig. 19.1).

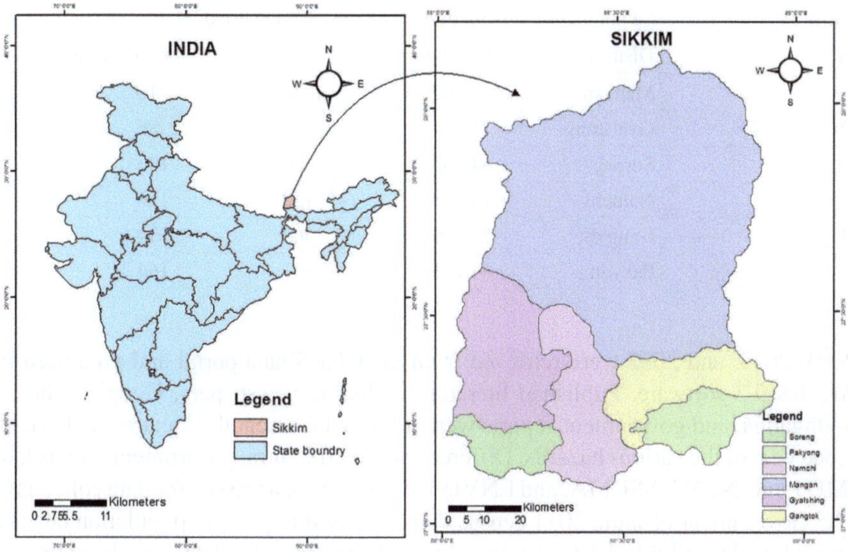

Fig. 19.1 Location map of the study area

Geographically, Sikkim State covers an area of 7096 km^2. The entire state extends over approximately 114 km from north to south and 64 km from east to west, with an altitudinal variation of 250 m to 8598 m (Subba 2008). The observable sharp change in gradient from south to north comprises nearly 43% of the state area under steep slopes along with rugged terrain. Due to sharp gradient changes, the southern portion of the state experiences a sub-tropical climate, whereas highly-elevated areas in the northern portion experiences an alpine climate with tundra vegetation cover (Aggarwal et al. 2017). The northern part of Sikkim falls in the Higher Himalayan Zone and the southern slope of the state falls in the Lesser Himalayan Zone. The study area experiences four main seasons: spring (March to April), summer (May to September), autumn (October to November), and winter (December to February). Despite rugged topography, the economy of the state mainly centres on agriculture (Joshi et al. 2023; Tamang and Joshi 2024) and allied activity, on which more than 60% of the population is engaged (Mishra et al. 2021). In general, high-elevated glaciers mountain tops are located towards the northern portion of the state, and low flood plain valleys are on the southern part.

19.3 Methodology

The present study is entirely based on primary field data and corroborated by the analysis of secondary data sources. The aspect map and slope map were prepared by ArcGIS 9.3, using the 30 m resolution SRTM data. The rainfall data for the years

Table 19.3 District-wise population distribution of Sikkim (2011 census)

S. No.	Districts	Population	Area km^2	Pop. Density
1	Mangan	43,709	4226	10
2	Gyalshing	71,675	836	86
3	Soreng	64,760	330	196
4	Namchi	146,850	750	196
5	Gangtok	219,450	563	390
6	Pakyong	64,133	391	164

2003, 2013, and 2022 were retrieved from the CHRS data portal and processed in ArcGIS 9.3 software. Published literature including newspapers, Google Scholar, Sodhganga, and government-reports were used to document the location and occurrence year of the various hazards. Different State and Central government portals like IMD Pune, NIDM, SSDMA, and ENVIS have also been accessed for data collection. The Government of India 2011 census data was used to plot the population density map of Sikkim (Table 19.3). To organize and represent the data, simple statistical techniques have been used.

19.4 Prominent Natural Hazard Occurrences in Sikkim

19.4.1 Landslides

Landslides are defined as the mass movement of rocks or debris generally along the slopes due to the influence of gravity (Joshi et al., 2018). Sikkim State receives its maximum precipitation during the monsoonal months, which occasionally causes cloudbursts. Landslides are very common in the state of Sikkim due to their high relief, weak lithology, higher precipitation, and active seismicity (Gupta et al. 2015), which causes a great loss of property as well as human lives (Table 19.4).

Generally, landslides are associated with all types of rocks and the Quaternary formation of Sikkim. The Daling group of rocks, particularly the Gorubathan formation with low-grade metamorphic rocks, is more prone to landslide than gneiss and schists of the Higher Himalaya. The Daling group of rocks has been attributed to landslides due to their severe shear distribution, due to loading and unloading during orogenesis and is highly susceptible to weathering. Slope instability is mainly associated with natural factors such as weak geology, high slope, frequent seismicity, and soil erosion and partially as a result of accelerated degradation (SSDMA 2018). These adverse conditions exacerbate the existing fragile, vulnerable, and multi-functional mountain ecosystem. The region experiences a lot of major landslides, which cause a lot of life loss in the region (Bhasin et al. 2002).

Table 19.4 Major disaster in Sikkim (landslides/flashflood)

Year	Location	Cause	No. of deceased	No. of injured	Remarks
1983	Sikkim	Landslide	63 people died	No information	Manual, North Sikkim
1995	Sikkim	Landslide	32 persons died, and 8 houses were completely damaged	No information	Gangtok
1997	Sikkim	Landslide	43 lives were lost, and 1300 houses were fully/partially damaged	No information	Gangtok and vicinity
2000	Sikkim	Landslide	5 lives were lost, and 28 families evacuated	No information	Gyalshing, West Sikkim
2005	Sikkim	Landslide	7 lives were lost, and 28 families evacuated	No information	Mangzing, South Sikkim
2011	Sikkim	Landslide	18 lives were lost, and 3 houses were fully damaged	No information	Gyalshing, West Sikkim
2012	Sikkim	Landslide/Flash Flood/Rock falls	41 lives were lost, 6 missing, and many houses were fully/partially damaged	20 injured	Throughout Sikkim

(*Source* SSDMA)

The dominant landslides in the region are mainly concentrated in the North, East, and West districts of Sikkim (SSDMA 2018). Sometimes, these major slides block the river and form temporary lakes. A similar lake formed due to a massive landslide in the Dzongu area, North Sikkim along the Kanaka River in the year 2016 (Sivasankar et al. 2021). In Gangtok, the capital of Sikkim, several active landslide zones were identified in different areas as 32 number, 3 miles, Chandmari, Tathangchen, Burdang, and 6 miles (Bhasin et al. 2023). Every year, several landslides incidence occur in 32-mile and 3-mile areas, which cause road blockages. The studies show that near urban areas, urbanization, and steep slope cutting for road construction are the key factors for landslides.

19.4.2 Land Subsidence

Land subsidence is another hazard, which state has faced in recent times. In the Gangtok area, there are some small land subsided areas (Sharma et al. 2022). This poses a risk to the people residing in the neighbouring area. In Gangtok, the capital of Sikkim, some areas like Chandmari, 16 mile, and Himkhola (Pangthang) are experiencing this kind of hazard. The construction cutting of underlying rocks in lower slopes facing the north-west direction of Gangtok ridge has posed serious threats to high- standing buildings, with emerging ripple of subsidence on the slope gradient.

19.4.3 Cloudburst

A cloudburst is defined as a sudden aggressive rainstorm falling within a short time limited to a small geographical area. Cloudburst is a very common phenomenon in Sikkim during the monsoonal seasons. Sometimes, it claimed lives also. Most of the time, the roads get blocked due to cloudbursts as loose material with rocks comes to the roadside (SSDMA 2018). In June 2019, 427 tourists were evacuated from North Sikkim after four days as they got stranded as a cloudburst damaged road infrastructure. Increasing intensity of rainfall is evident from the recent devastating Chatten landslide (North Sikkim), triggered by cloudburst in year 2025, which trapped 1600 tourists and resulted in 3 deaths.

19.4.4 Hailstorms

A hailstorm is a weather phenomenon, in which thunderstorms' severity outgrowths and balls or irregular-shaped lumps of ice start to fall with rain (IMD). A strong upward wind convention that causes the water droplets to go further upward into the freezing zone leads to the formation of hailstorms. The hailstorm is defined as severe when the size of the hail is larger than 1 inch (IMD). A hailstorm has also been reported from different places in Sikkim over the past years. Every year, it has become a major concern among the farming communities due to its potential to destroy the harvest. As per the people's observation, the size of the hailstone has increased over the years and has become unpredictable (Gulati 2015; Joshi et al. 2023).

19.4.5 Forest Fire

Forest fire is a common hazard in the forests of the Namchi District as it creates a threat to forest wealth as well as to fauna and flora, biodiversity, ecology, and the environment of a region. During autumn, when the temperature reaches high and months are without rain, the forests of Sikkim become littered with dry senescent leaves and twinges (SSDMA 2018). This could burst into flames, ignited by the slightest spark and lead to loss of vegetation cover and crops of that region. Lightning, high atmospheric temperatures, and dryness (low humidity) offer favourable circumstances for a fire to start naturally.

19.4.6 Earthquake

Himalayas, including Sikkim, is a young fold mountain chain that formed due to the collision of the Indian Plate with the Eurasian Plate (Thakur et al. 2020). Geologically, the area is dissected with many thrust faults as well as deep-seated lineaments (Paul et al. 2015) which cause repeated seismicity in the state. This put Sikkim under zone IV of the seismic zonation map of India (Thakur et al. 2012) and made it one of the seismically active regions with an expected PGA greater than 400 gal (SSDMA 2018). The 18th September 2011 earthquake with a magnitude of 6.8 was one such earthquakes that claimed many lives (Paul et al. 2015) generated several earthquake-induced landslides (Gupta et al. 2015; Joshi 2022), and propelled the construction of infrastructure in earthquake-resilient mode.

19.4.7 Glacial Lake Outburst Flood (GLOF)

Glacial lakes are formed by melting glaciers. An impulsive discharge of water and debris from glacial lakes is termed a Glacial Lake Outburst Flood (GLOF). Sikkim Himalaya is home to ~ 320 glacial lakes (Bajracharya et al. 2007; Agrawal et al. 2017). These lakes are expanding due to the fast rate of glacial melting. For example, the area of one of the lakes in the Gurudongmar Chho complex increased ~ 4 times from 1965 to 1989. Similarly, the Khanchung Chho, a moraine-dammed lake has grown ~ 1.5 times from 1965 to 1989. The increased size not only increases the potential for GOLF but also increases the risk factor (Agrawal et al. 2017). South Lhonak Lake had increased its size from 18 ha in 1976 to more than 126 ha in 2013 (SSCCC Report), which was breached in the 2023.

19.5 Results

19.5.1 Aspect Map of Sikkim

The aspect map is classified into ten categories to present information regarding the direction of the hill slopes and their steepness (Fig. 19.2).

As per the area calculation, flat land occupies 0.09% of the area, North-oriented area covers 11.01%, North-East-oriented covers 10.95%, East-oriented

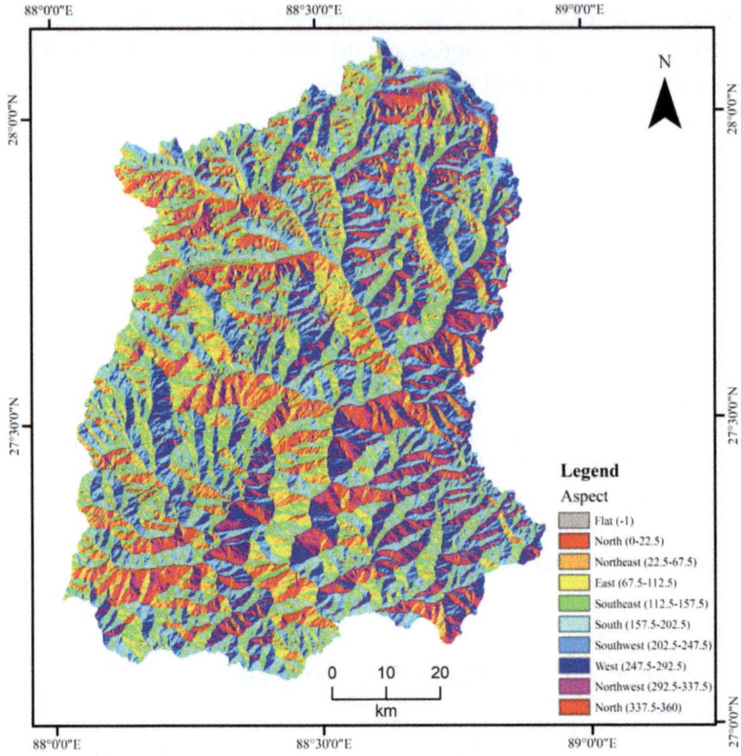

Fig. 19.2 Aspect map of Sikkim

covers 12.82%, South-East-oriented covers 13.95%, South-oriented covers 13.91%, South-West-oriented covers 13.77%, West-oriented covers 12.1%, and North-West-oriented covers 11.41% of area. However, in general South, the southeast and southwest-oriented areas lying between 112.5° and 247.5° cover the major portion of the state.

19.5.2 Slope Map of Sikkim

The slope map gives an idea about the degree of slope steepness of the entire state. Structurally, the entire state follows the slope gradient from north to south (Kanade et al. 2018). To represent the local area layout, the entire state is categorized into five categories based on slope degree (Fig. 19.3). Area having a degree of 0–15 is considered as gentle and covers an area of 16.39%; an area having a degree of 16–26 is considered as low, which covers an area of 27.14%; an area having a degree of 27–34 is considered as moderate, covering an area of 23.22%; the area having slope of 35–48 degree is considered as high and covers an area of 25.98%, and lastly area

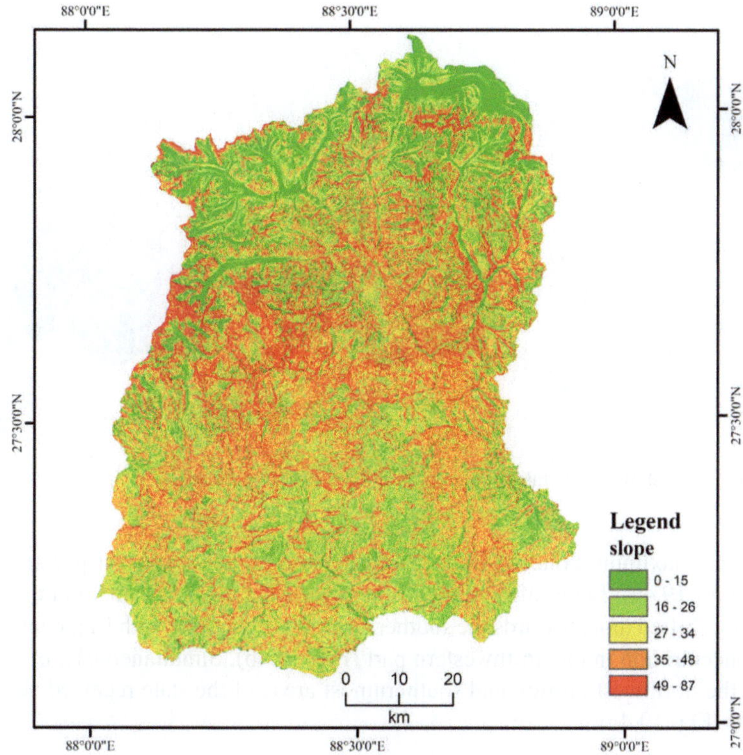

Fig. 19.3 Slope map of Sikkim

having a degree of 49–87 is considered as very high and covers an area of 7.27% only. A stable or low slope area covers the maximum portion of the state, and is immediately followed by a high steep slope area. The majority of common natural hazards (like landslides) are mostly confined to two categories of slope; high and very high slope areas. Few instances of landslides are also reported from moderate, low, and very low slope areas mainly due to unplanned developmental activities. 33.17% of (high and very high) slope area is very much susceptible to various natural hazards like landslides, forest fires, avalanches, etc.

19.5.3 Rainfall Distribution of Sikkim Past Two Decades (2003 to 2022)

The rainfall map in Fig. 19.4 indicates the changing rainfall pattern over two decades (2003 to 2022). The amount of rainfall received in different areas of Sikkim is classified into five categories—very high, high, moderate, low, and very low. In the year

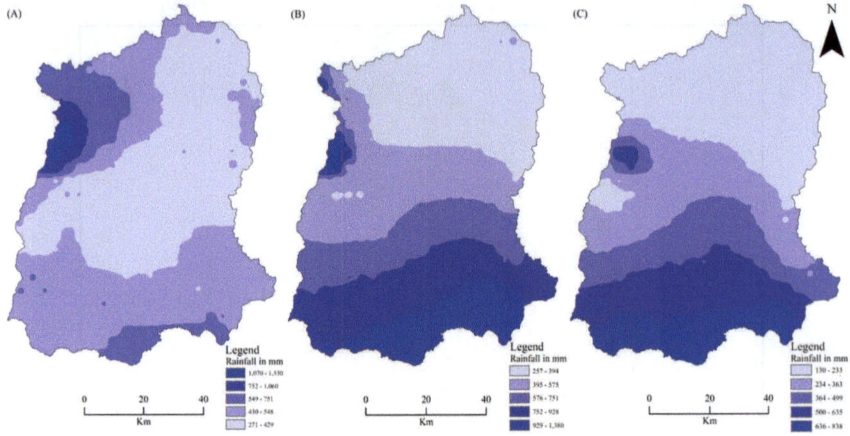

Fig. 19.4 Map showing the rainfall pattern of Sikkim over two decades

2002, the maximum concentration of rainfall was in the northwest portion of the state (Fig. 19.4a). Later, after one decade in the year 2013, the maximum rainfall intensity shifted more towards the southern portion of the state with fragmented rainfall concentration in the northwestern part (Fig. 19.4b). Simultaneously, in the year 2022, the northwest portion and southernmost areas of the state received very high rainfall (Fig. 19.4c).

19.5.4 *Natural Hazard Occurrences in Sikkim (1960–2024)*

Figures 19.5 and 19.6 indicate the major natural hazards and events of life loss from 1960 to 2024. Among these natural hazards, landslide is the dominant one, which prevails across the state. The concentration of microseismicity lies along the Main Central Thrust and various lineaments. The 2011 Sikkim earthquake of magnitude Mw 6.9 occurred along such a lineament (Sharma et al. 2013). Increasing incidences of forest fire are also reported from various parts of the state, particularly in the Namchi District. The impact of global climatic changes is also felt in Sikkim Himalaya in the form of unwelcomed hailstones, cloudbursts, GLOF, flash floods, and avalanches. Climatically triggered incidences are becoming more frequent, causing disruption and lots of damage to life and property. The distribution of these hazards indicates that most of the hazards have occurred in areas having slope degrees between 35° and 87° (Fig. 19.7). Hazards like flash floods and GLOF impacts have been felt in low-lying areas along river channels or hill base areas. However, in Sikkim, hazards operated in 35° to 87° slopes have claimed more lives than hazards triggered below this slope line.

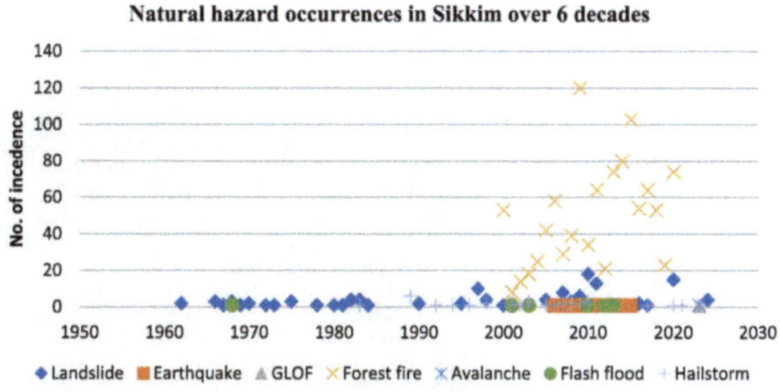

Fig. 19.5 Natural hazard occurrences in Sikkim over 6 decades

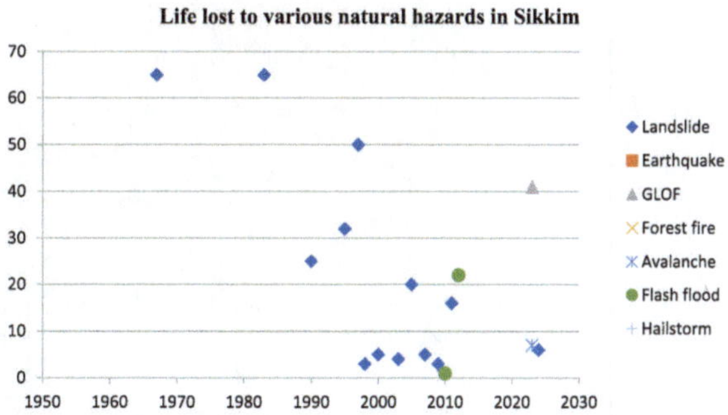

Fig. 19.6 Life lost to various natural hazards

19.5.5 Population Density and Natural Hazard Map

Figure 19.8 shows district-wise population density and the distribution of natural hazards. As per the 2011 census, out of the total population, 35.94% live in Gangtok, 24.05% live in Namchi, 11.74% live in Gyalshing, 10.61% live in Soreng, 10.50% live in Pakyong, and only 7.16% of the population lives in Mangan. The highest density of population is found in Gangtok District which is 390 persons per km^2, and the lowest density of population is observed in the Mangan District with only 10 persons per km^2. The hazard distribution over the population density map shows that most of the major hazards have been reported from the densely populated area of the Gangtok District. Several natural hazard events have been reported from this district only like hailstorms, avalanches, earthquakes, and landslides.

19.6 Discussion

Sikkim, being part of the active Himalayan belt, has witnessed several severe to moderate disasters or hazards in the past decades and will in the future as well. However, despite all the chaos experienced in the Sikkim Himalaya, the inhabited population has turned these challenges into opportunities for survival. For instance, having rugged topography and steep slopes is converted into agricultural land. On

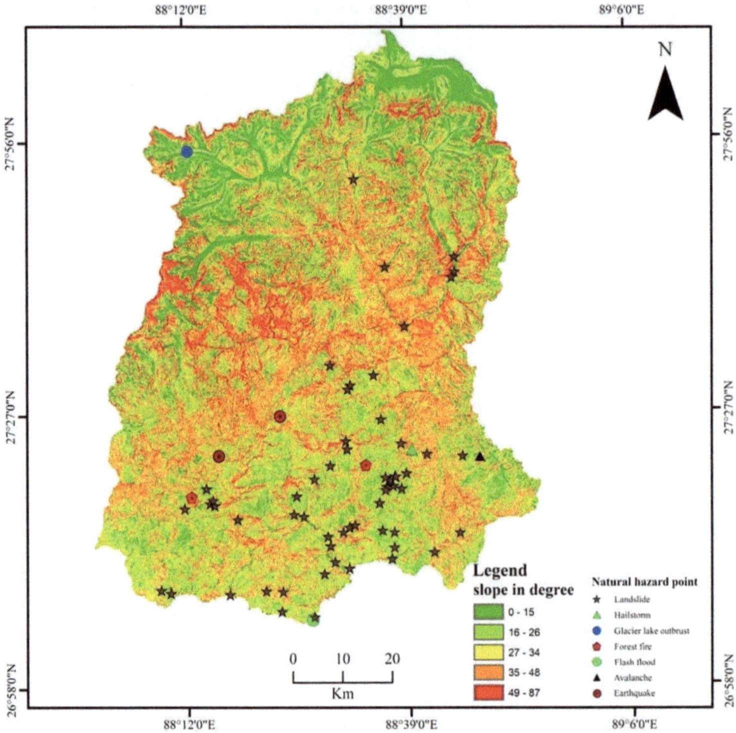

Fig. 19.7 Map showing natural hazard occurrences in Sikkim (1960–2024)

which more than 60% of the population is dependent (Mishra et al. 2021). The majority of natural hazards have occurred in the densely populated regions. The low flat lands, formed by river deposition, are in the southern part of the state like near Rangpo, Singtam, Jorethang area, etc.; whereas the highest peak like Kanchendzonga falls towards the northwest part of the state. So, for geomorphological studies, it offers a complete and unique landscape to study the entire process from origin (upliftment) to end (lowered landform), within a feasible distance. The processes involved in such activity can be easily monitored and recorded by an individual due to their existence within a local scale to the regional environment.

A diverse range of landforms can be found within the state, from glaciated mountain tops, glacier lakes, hanging valleys, deep valleys, river valleys, and flood plains. Different types of agents like glaciers, rivers, etc., are engaged in making such splendid landforms offering an opportunity for the landform enthusiast. Therefore, Sikkim holds a great scope for those who wish to study landforms and their related characteristics. Further, Sikkim has a small geographical expansion and experiences diverse climatic conditions from tropical, and temperate to alpine climates. It has supported diverse biological diversity and the development of various landforms. 40% of biological diversity components occurring in the Indian subcontinent can be

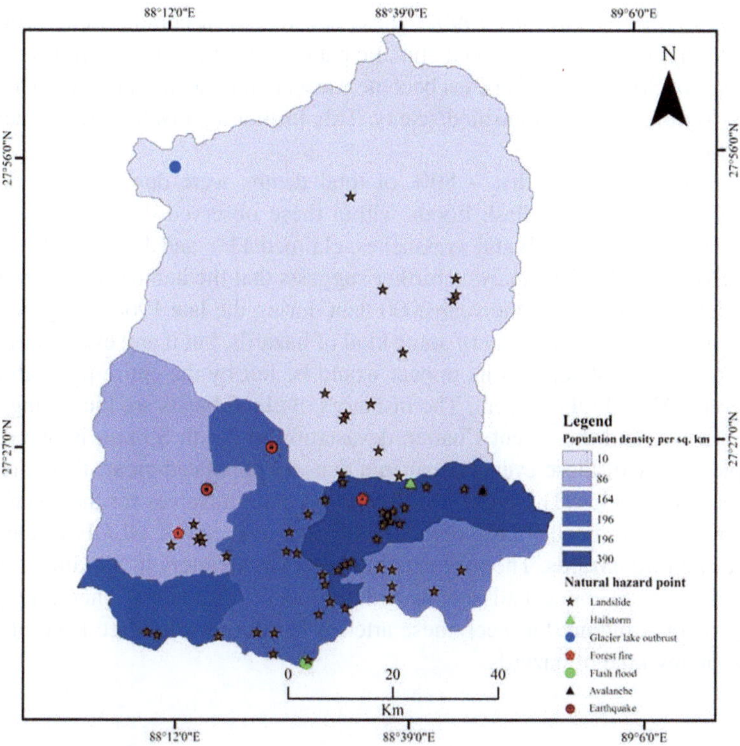

Fig. 19.8 Map showing district-wise population density and natural hazards

found in Sikkim only (Kharel and Bhutia 2013). It offers different climatic experiences with huge differences in flora and fauna within a short distance, which means Sikkim is a resource hub for climate experts, botanists, zoologists, and many experts.

Hazards are natural phenomena, which cannot be stopped but their impacts can be minimized. Apart from earthquakes and GLOFs, all the other hazards as landslides, subsidence, hailstorms, cloudbursts, and forest fires are of local scale. Thus, the intensity and magnitude of the hazards vary from place to place, as well as their impact. Sikkim, a mountainous state, experiences different natural hazards. Therefore, a better understanding as well as its characterization, followed by real-time monitoring, is an essential step for adopting mitigation strategies to prevent the adversities of any natural hazard. We documented the natural hazards, including landslides, earthquakes, GLOF, forest fires, avalanches, flash floods, and hailstorms, that occurred in the Sikkim Himalaya during the last six decades. Among these six natural hazards, landslides seem to be the more dominant type of hazard, and their occurrences are consistently recorded over the observed periods. From the year 1997 onwards, it is observed that the intensity of landslides is constantly rising due to earthquake and rainfall events. Several forest fire incidents were also raised after the year 2000. Climatically induced hazards like flash floods, hailstorms, and landslides occurred frequently after the 1990s. These hazards are becoming more prominent due to changing global or regional climatic patterns. Further, due to climate change, the winter of Sikkim Himalaya has become warmer, and the rainfall pattern has been shifted from regular to intensified/scanty. This further aggravated the frequency of forest fires.

Our findings suggest that ~ 80% of total deaths were due to landslides and 6.2% of deaths were by flash floods within these observed decades. The single-time events namely GLOF and avalanches, claimed 11% and 1.8% deaths, respectively. However, the data analysis further suggests that the number of deaths due to landslides decreased after the year 2000 than during the late 1960s. In general, the overall state is under the threat of some kind of hazards, but if any event is triggered in any part of the state, then its impact would be felt by the entire population like the recent 2023 GLOF incident. The instances of cloud bursts are increasing across the Himalaya, and the recent Chatten devastation in North Sikkim is one of such example. If any of these events occur in a densely populated areas, there are more chances of casualties. Therefore, we need to prepare ourselves for the hazards that are more likely to occur in the coming days like avalanches, GLOF, or any other climate-induced hazards. The road network serves as an artery in Sikkim, as it is the only active way to connect all regions with the rest of the country. Therefore, proper management is required to keep these arteries (road networks) free from blockage in case of any kind of hazards.

19.7 Conclusion

Various natural hazards from 1960 to 2024 have been considered for this study. The majority of natural hazards have occurred along the areas bearing slope between 35° and 87°. Among all the natural hazards, around 80% of death are due to landslides. A major shift in rainfall pattern is also observed just within two decades, which point towards new and intensified climatic hazards in the future. It has been found that the majority of hazards have been concentrated in the densely populated areas of Gangtok District, Soreng District, and Namchi District of Sikkim. Therefore, it indicates that such natural events have occurred not only due to natural processes but to a great extent due to our ignorance of developmental planning. We need to focus more on our developmental activities. If we properly plan our activities, then we are creating a safe environment not only for ourselves but for future generations as well. Likewise, if we make a proper plans today, then we are resilient to any kind of such natural hazards in the future.

References

Aggarwal S, Rai SC, Thakur PK, Emmer A (2017) Inventory and recently increasing GLOF susceptibility of glacial lakes in Sikkim, Eastern Himalaya. Geomorphology 295:39–54

Bajracharya SR, Mool PK, Shrestha BR (2007) Impact of climate change on Himalayan glaciers and glacial lakes: case studies on GLOF and associated hazards in Nepal and Bhutan, vol 12. International Centre for Integrated Mountain Development, Kathmandu

Bhasin R, Grimstad E, Larsen JO, Dhawan AK, Singh R, Verma SK, Venkatachalam K (2002) Landslide hazards and mitigation measures at Gangtok, Sikkim Himalaya. Eng Geol 64(4):351–368

Bhasin R, Aslan G, Dehls J (2023) Ground investigations and detection and monitoring of landslides using SAR interferometry in Gangtok, Sikkim Himalaya. Geohazards 4(1):25–39

Gulati V (2015) Understanding the vulnerability of rural livelihoods of East and South Sikkim across the different seasons (Doctoral dissertation, TERI University)

Gupta V, Mahajan AK, Thakur, VC (2015) A study on landslides triggered during Sikkim earthquake of September 18, 2011. Himalayan Geol 36(1):81–90

Gupta V, Chauhan N, Penna I, Hermanns R, Dehls J, Sengupta A, Bhasin RK (2022) Geomorphic evaluation of landslides along the Teesta river valley, Sikkim Himalaya, India. Geol J 57(2):611–621

Joshi M (2022) Co-seismic landslides in the Sikkim Himalaya during the 2011 Sikkim earthquake: lesson learned from the past and inference for the future. Geol J 57(12):5039–5060

Joshi M, Luitel K, Barfal SS, Kuniyal JC, Pande K (2023) Significance of indigenous knowledge systems in water conservation and management: a study from Sikkim Himalaya. In: Rai SC, Mishra PK (eds) Traditional ecological knowledge of resource management in Asia. Springer International Publishing, Cham, pp 159–174

Joshi M, Rajappan S, Rajan PP, Mathai J, Sankar G, Nandakumar V, Kumar VA (2018) Weathering controlled landslide in Deccan traps: insight from Mahabaleshwar, Maharashtra. J Geol Soc India 92(5):555–561

Kanade R, John R (2018) Topographical influence on recent deforestation and degradation in the Sikkim Himalaya in India; Implications for conservation of East Himalayan broadleaf forest. Appl Geogr 92:85–93

Kharel S, Bhutia JW (2013) Gazetteer of Sikkim. Gangtok, Government of Sikkim, Home Department

Mishra PK, Rai A, Abdelrahman K, Rai SC, Tiwari A (2021) Analysing challenges and strategies in land productivity in Sikkim Himalaya, India. Sustainability 13:11112

Parkash S (2011) Historical records of socio-economically significant landslides in India. J South Asia Disaster Stud 4(2):177–204

Paul H, Mitra S, Bhattacharya SN, Suresh G (2015) Active transverse faulting within underthrust Indian crust beneath the Sikkim Himalaya. Geophys J Int 201(2):1072–1083

Rai DC, Singhal V, Mondal G, Parool N, Pradhan T, Mitra K (2012) The M 6.9 Sikkim (India–Nepal Border) earthquake of 18 September 2011. Current Sci 1437–1446

Rainfall data site: https://chrsdata.eng.uci.edu/

Sharma ML, Sinvhal A, Singh Y, Maheshwari BK (2013) Damage survey report for Sikkim earthquake of 18 September 2011. Seismol Res Lett 84(1):49–56

Sharma A, Joshi M, Joshi R, Kumar D, Singh M, Kumar K, Barfal SS (2022) Effect of rapid urbanization on water quality: an experimental study from Indian Himalayan city, Gangtok. In: Facets of a smart city: computational and experimental techniques for sustainable urban development. Bentham Science Publishers, pp 76–91

Sivasankar T, Ghosh S, Joshi M (2021) Exploitation of optical and SAR amplitude imagery for landslide identification: a case study from Sikkim, Northeast India. Environ Monitor Assessm 193(7):386

Skoufias E (2003) Economic crises and natural disasters: coping strategies and policy implications. World Dev 31(7):1087–1102

SSDMA, Land Revenue & Disaster Management Department, Government of Sikkim (2018) Multi-hazard risk and vulnerability assessment of Gangtok municipal corporation, East Sikkim

Subba JR (2008) History, culture, and customs of Sikkim. Gyan Publishing House

Tamang L, Joshi M (2024) Transition in traditional cropping pattern led to engagement in alternative economic activity, accepted in climate change impact on Himalayan biodiversity. Springer International Publishing, Cham

Thakur VC, Mahajan AK, Gupta V (2012) Seismotectonics of 18 September 2011 Sikkim earthquake: a component of transcurrent deformation in eastern Himalaya. Himalayan Geol 33(1):89–96

Thakur VC, Joshi M, Suresh N (2020) Linking the Kangra piggy–back Basin with reactivation of the Jawalamukhi Thrust and erosion of Dhauladhar Range, Northwest Himalaya. Episodes J Int Geosci 43(1):335–345

Toya H, Skidmore M (2007) Economic development and the impacts of natural disasters. Econ Lett 94(1):20–25

Chapter 20
Characterization and Assessment of Landslide Hazard in Eastern Himalaya

Kapil Ghosh and Sunil Kumar De

Abstract The mountain terrains of the Sikkim Himalaya are geologically fragile and highly vulnerable to landslide hazards. This hazard has become a regular phenomenon during every monsoon season and earthquake events. The study focuses on assessing the landslide hazard and their characteristics. The hazard assessment was done through hazard zonation following the Bureau of Indian Standards (BIS in Criteria for earthquake resistant design of structures: general provisions and buildings 2002) guidelines, and the characterization of landslides was done by identifying the nature of the movement, materials, area, activity, and their association with land use/land cover and rainfall. Landslide points were collected through field surveys and the Geological Survey of India's report. Landslides hazard zonation mapping was done considering seven geo-environmental parameters. These spatial layers were integrated with the GIS using overlay techniques. It was observed that the landslide hazard map shows that very high and high-hazard areas are mainly found in the southern part of the district. The highest incidence of landslide per km^2 is recorded for the very high-hazard zone (73.67%). Characterization of landslides in the district shows that rock fall and sliding are the main mechanisms of movement and they are small to medium in size. Landslides are found even in dense forest areas which indicate the dominant control of geological structures in landslide occurrences.

Keywords Bureau of Indian Standard · Chungthang · Hazard zonation · Landslides · Sikkim Himalaya

K. Ghosh (✉)
Department of Geography, Diamond Harbour Women's University, Sarisha, West Bengal, India
e-mail: ghoshk.geo@gmail.com

S. K. De
Department of Geography, North Eastern Hill University (NEHU), Shillong, Meghalaya, India

S. C. Rai (ed.), *Himalaya: Mountains of Destiny*,
https://doi.org/10.1007/978-3-031-94905-0_20

20.1 Introduction

The landslide hazard undermines the overall physical-cultural setup in the North district, Sikkim, in the Eastern Himalayas. Occurrences of landslides in mountainous regions can be treated as cancer of hill slopes (Ghosh et al. 2012a, b). Huge damages happen almost every year due to the unexpected occurrence of landslides, and it is becoming very costly to tackle the associated problems with this hazard. A study shows that 95% of landslide disasters have been recorded in developing countries (Chung et al. 1995). It is a form of crustal displacement caused by weak geology, under the influence of different causative factors, like vulnerable geological structures, geo-tectonic activities, changing nature of river regime, high slope variation, deforestation, unscientific and unplanned uses of land as well as a huge amount of rainfall which acts as the triggering factors.

Despite various corrective and protective measures, the frequency and magnitude of landslides have superseded the controlling measures. Lack of proper understanding of the individual landslide character, mechanism of slide initiation as well as in-depth information about the landslide could be the reasons. Indeed, this type of hazard cannot be checked, but proper remedial measures can reduce the magnitude of occurrence and associated damages. A strong database and understanding of related geo-environmental factors are needed for this purpose. Thus, it is high time to prepare such a database and detailed landslide hazard zonation map to take proper necessary measures for controlling such hazards in the study area. Therefore, delineating landslide-prone areas through landslide hazard zonation mapping is the first step to managing this hazard (Ghosh et al. 2017).

Landslide hazard zonation, vulnerability, and risk analysis in India have also gained importance since 1980 because of the increasing frequency and magnitude of fatalities especially in the Himalayan region. A guideline for landslide hazard zonation mapping was prepared by the Bureau of Indian Standards (BIS 1998, 2002). They used the Landslide Hazard Evaluation Factor (LHEF) rating scheme and estimated the total hazard value for different facets of the slope. Pachauri and Pant (1992) proposed a systematic approach to landslide hazard mapping based on a weighted rating system following the guidelines. The weighted rating values of such classes were estimated by multiplying weightage and rating values.

Landslide zonation in the Himalayas has been attempted by numerous researchers (Majumder 1980; Valdiya 1985; Chatterjee 1986; Dave and Joshi 1987; Gairola and Shukla 1990; Sharan1992; Anbalangan et al. 1993; Bhattacharya and Sengupta 1996; Kanungo and Sarkar 2003; Joshi et al. 2003; Krishna et al. 2004; Bera 2005; De 2007). Most of the studies considered the preparation of different thematic maps depicting the various geological and geomorphological factors (Basu et al. 2009). Although many works have so far been carried out on landslide hazard zonation, characterization is very important for adopting remedial measures at a regional scale as well as a site-specific level. The present study focused on determining the landslide hazard zones for assessing the landslide scenario and identifying their inherent character including the triggering factor of rainfall for landslide occurrence.

20.2 Geographical Background of the Study Area

North Sikkim district is the largest district in the state of Sikkim and covers an area of 4296 km^2. The district lies between the latitude of 27°25′00″ N and 28°07′48″ N and longitudes of 88°03′00″ E and 88°53′00″E (Fig. 20.1). The Khangchendzonga Peak (8598 m) (third highest peak in the World) is the highest point of the district. The lowest elevation (512 m) is found at the extreme south point of the study area near Dikchu. The northern part of the district is mainly covered with snow, and it accounts for 949.34 km^2 (Source: IRS LISS-III data of 2001). The major river system—Teesta originated from the northern part of the district and the district is mainly drained by the Teesta river system. The district is the origin place of the major river of Sikkim—Tista. The major tributaries of Tista, i.e. Yumthang chhu, Lachung chhu, Lachen chhu, Zemu chhu, Goma chhu, Chhombo chhu, Tolung chhu, etc., have also originated from this district.

Geologically the entire district falls under the inner mountain range of the Himalayas. Four distinct longitudinal geo-tectonic-stratigraphic domains, i.e. Sub-Himalayan domain, the Lesser Himalayan domain, the Higher Himalayan domain, and the Tethyan belt are present in the district. About, 47.93% of the study area contained Kanchenjunga Gneiss/Darjeeling Gneiss (Undifferentiated). The two main structural elements, i.e. Main Frontal Thrust (MFT) and Main Boundary thrust (MBT), are passing over this district. The region falls under the seismic zone V, and a nos. of seismic activities are recorded in the thrust zones. The district is characterized

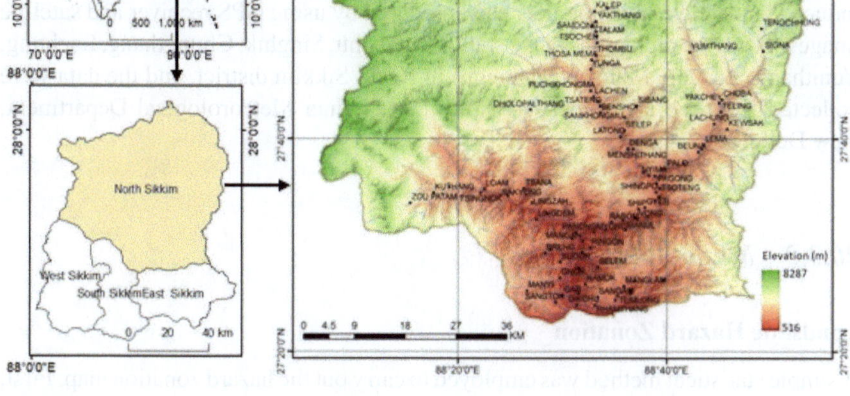

Fig. 20.1 Location map of the study area — (*Source* ElevationSRTM DEM)

by rugged topography, steep to very steep slopes, great variations in elevation, high ridges, and high-altitude passes. The maximum area of 2287 km^2 of the study area falls under a very steep slope (60–80°). Annual rainfall varies from 925 to 3389 mm, and it increased towards the southeast. The rainfall is seasonal.

Although it is the largest district of Sikkim, it has the least population (43,354 persons in the 2011 census) in Sikkim. The district accounted for 60% of the state, but it houses only 7.13% of the state's population. The population density is very low, and it is only 10 persons per km^2 (Census 2011). The decadal growth rate of the population was 5.67% in 2001–2011. There are 45 revenue blocks in the district, of which 5 are in the Chungthang sub-division and 40 are in the Mangan sub-division. The district comprises 8 forest blocks. Mangan, the district headquarters, is the only town of North district, Sikkim. Lepchas and the Bhutias are the dominant ethnic groups in the district.

20.3 Materials and Methods

20.3.1 Materials

Basic datasets were prepared to understand the landslide scenario in the study area. Morphometric maps, drainage maps, and sites of the old landslide scars have been prepared based on the Survey of India topographical sheets (77 D/8, 77 D/12, 77 D/16, 78 A/1, 78 A/5, 78 A/9, 78 A/13, 78 A/2, 78 A/6, 78 A/10, 78 A/14, 78 A/7, 78 A/11, 78 A/15 of 1:50,000 scale). The thematic maps were prepared using RS and GIS techniques. Satellite imageries (SRTM DEM, LISS III images of 1997 and 2009) were used to prepare relief maps and land use/land cover maps. Historical records of the landslides have been collected from topographical sheets, the Geological Survey of India's reports, and different published literature. Landslide points have also been marked with a handle GPS receiver during field surveys. Detailed earthquake-induced landslides' inventory has been prepared by using GPS receiver and satellite images. There are eight rain-gauge stations (Dikchu, Singhik, Chungthang, Lachung, Yumthang, Lachen, Mangan, Thanggu) in North Sikkim district, and the data were collected from the Hydromet Division of the Indian Meteorological Department, New Delhi.

20.3.2 Methodology

Landslide Hazard Zonation

A simple statistical method was employed to carry out the hazard zonation map. First, the landslide points were identified from toposheets, satellite images, and fields and the same was used for hazard zonation mapping. Identifying the hazard-prone areas

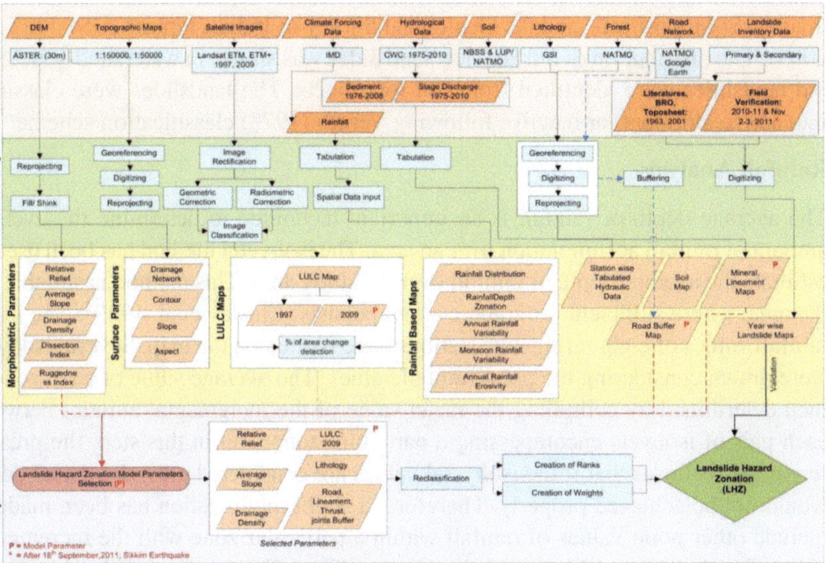

Fig. 20.2 Methodology for landslide hazard assessment

is the first step in disaster management. The hazard zonation technique can serve the purpose effectively. Landslides can occur as a consequence of interactions among various geo-environmental factors. The geo-environmental themes considered in the present study are given in a figure (Fig. 20.2).

The BIS guidelines (1998 and 2002) were followed to design and prepare the spatial database that was considered for the Landslide Hazard Zonation (LHZ). The causative factors were considered as the thematic layers, and they were prepared in the GIS platform. All the thematic layers (geology, geological structure, land use, relative relief, slope, drainage density, and road) were arranged according to their relative importance or probable influence on the landslide, and the weighting number (total 100%) was assigned based on multiple criteria decision-making techniques. Similarly, a rating was assigned for each class within a layer which ranges from 0 to 9. The relative score of each thematic unit in a theme was calculated by multiplying the weight of the theme with the rank of the thematic unit. Finally, the cumulative score of the landslide susceptibility index (LSI) based on the weight-rating method was calculated using multi-criteria decision techniques in GIS. The resultant map was classified into different (high, moderate, low, and very low) hazard zones. Finally, landslide points were superimposed on the hazard zonation map to verify the result.

Landslide Characteristics Analysis

Characterization of landslides is essential for the successful management of the hazard. A landslide inventory was prepared by considering attributes like their occurrence date, slide materials, movement involved with the slide, failure mechanisms, area/volume, depth, etc. The inventory was prepared from field surveys, satellite images, SoI toposheets, and the Geological Survey of India's reports (GSI 2007 and 2009). The distribution of old and new landslides was shown in watershed-wise. The old landslides were identified from SoI toposheets. The landslides were classified into active, dormant, or inactive following Vernes (1978) classification scheme.

Rainfall Analysis

The average depth of rainfall is an important technique to determine the average amount of rainfall accumulation over an area. The isohyetal method has been used to estimate the average depth of rainfall over the study area. This method is suitable for the area, where sufficient orographic effect persists (Linsley et al. 1949). The mean rainfall values of eight rain-gauge stations were plotted on the map. Several isohyets were drawn considering the point rainfall values. The average value of each zone is then determined by compiling the mean value of the incremental amount between each pair of isohyets encompassing a particular zone. But in this step, the present researcher believes that if the isohyetal values are considered only, the actual depth would not be achieved properly. Therefore, a slight modification has been made to include other point values of rainfall within a particular zone with the incremental value of each pair of isohyets of the same zone. The following formula has been used to estimate the average annual rainfall depth in the study area:

$$\bar{P} = \frac{1}{A}\sum_{j=1}^{J}\frac{A_j}{P_j},$$

where

$\bar{P}$ Average annual precipitation.
A $\sum_{j=1}^{J} A_j$, i.e. the total area computed by adding the individual zones.
P_j Average annual precipitation of the jth zone.
J Total number of zones.

20.4 Result and Discussion

20.4.1 Landslide Hazard Zonation (LHZ)

First, all spatial data layers were prepared in the same resolution and they were reprojected to a common project system (UTM) for integration in the GIS environment. Seven geo-environmental factors (Table 20.1) are considered for the landslide hazard zonation. Simple statistics of this additive normalized incidence are computed and used to classify the entire study area into five hazard classes: very high, high, moderate, low, and very low in Table 20.2. These semi-quantitative landslide hazard zones are colour coded to prepare a map depicting the hazard zones Fig. 20.3.

The landslide points were superimposed on the hazard maps, and the highest frequency of landslides per km^2 is observed for the very high-hazard zone (73.67%).

Table 20.1 Rating of the landslide hazard zonation in the north district, Sikkim

LHZ data layers	LHZ sub-data layers	Classes	Weightage (In %)	Rating
Geology	Upper pleistocene to holocene (Q_{12})	Sesela formation ($Q_{12}S$)	17%	1
	Triassic period (T)	Tso-Lhamo formation (Ts)		1
	Carboniferous to Permian (CP)	Lachi formation (CPl)		1
	Ordovician (O)	Everest limestone formation (Oe)		1
	Mesoproterozoic (Pt_2)	Lingtse Gneiss (Pt_2l)		5
	–	Everest pelite formation (Pte)		1
	–	Amphibole schist/ amphibolite (Pta)		1
	Daling group (Ptd)	Gorubathan formation (Ptdg)		6
	Central crystalline gneissic complex (CCGC)/Ptc	Kanchenjunga Gneiss/Darjeeling Gneiss (Ptck)		8
		Chungthang formation (Ptcc)		7
	Intrusive	(ϒ Cz) Tourmaline/ Biotite leucogranite, schroll Rock/ Pegmatite, Aplite (undifferentiated)		2
Geological structure	Thrust buffer	> 1000	5%	3
		500–1000		6
		< 500		9
	Lineament buffer	> 1000	5%	1
		500–1000		3
		< 500		5
	Joint buffer	> 1000	5%	3
		500–1000		6
		< 500		9
	Fault buffer	> 1000	5%	3
		500–1000		6
		< 500		9
Land use	Considered 2009 LISS III image	Very dense vegetation	17%	1
		Dense vegetation		2
		Sparse vegetation		3

(continued)

Table 20.1 (continued)

LHZ data layers	LHZ sub-data layers	Classes	Weightage (In %)	Rating
		Bare rocky land		5
		Steep bare rocky land		6
		Sediment deposit		4
Relative relief	Very high	> 2000	16%	9
	High	1500–2000		7
	Moderate	1000–1500		5
	Low	500–1000		3
	Very low	< 500		1
Slope	Very high	> 80°	15%	7
	High	60–80°		5
	Moderate	40–60°		3
	Low	20–40°		1
	Very low	< 20°		-
Drainage density	Very high	> 1.2	10%	9
	High	0.9–1.2		7
	Moderate	0.6–0.9		5
	Low	0.3–0.6		3
	Very low	< 0.3		1
Road network	Road buffer	> 1000	5%	3
		500–1000		6
		< 500		9
			100%	

Table 20.2 Segregation of landslide hazard zones with landslide incidences

Sl. No	LHZ	Landslide susceptibility index (LSI)	Area in km^2	No. of landslide	Frequency of landslide	% Frequency of landslide
1	Very high	5.4–6.9	9	7	0.778	73.67
2	High	4.8–5.3	146	13	0.089	8.43
3	Moderate	4.1–4.7	670	58	0.087	8.24
4	Low	3.4–4.0	1305	100	0.077	7.29
5	Very low	1.6–3.3	2166	53	0.025	2.37
Total			4296	231	1.056	100

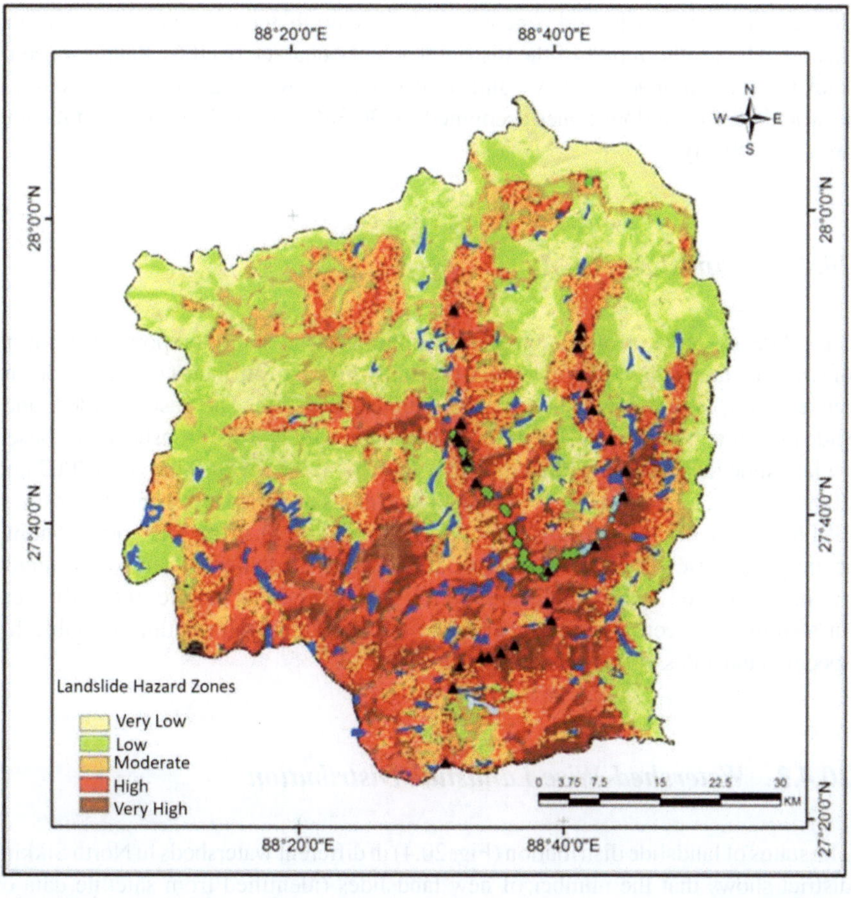

Fig. 20.3 Landslide hazard zonation (LHZ) with temporal landslide locations in the North District, Sikkim

The second highest frequency (8.43%) of landslide is found in the high-hazard zone. The lowest frequency (2.37%) was found in the very low hazard zone. The maximum number of landslides (100) is recorded from the thee Low hazard zone as the zone covers about 50% of the total area (Fig. 20.3).

The southern part of the district is mainly characterized by the presence of thrusts including MBT, the central crystalline gneiss, comparatively high to the very high slope (>30°), and high annual rainfall (>2500 mm). All these create a favourable condition for the occurrences of landslides in this area. A considerable number of landslides are found in the dense forest area as well as road sections. This indicates the dominant control of geological structure along with human interferences to the occurrences of the hazard. The final hazard zonation map shows that about 4.07 of the total study area falls under the very high to high-hazard zone. The moderate hazard zone is restricted within patches of the middle and southwestern part of the district. The very low hazard zone is found in the north and northwestern part of the district. The northern part of the district is mainly characterized by the presence of thick snow cover, glacial valleys, and comparatively low annual rainfall (<200 mm). Low and very hazardous zones accounted for 30.38% and 50.41% of the total study area respectively.

20.4.2 Landslide Inventory

Landslide inventory provides detailed information on the spatio-temporal distribution of recorded landslides. This inventory is usually linked to attributes on date of occurrence, types, failure mechanisms, area/volume, depth, etc. The past recorded landslide inventory has been prepared from Satellite images, and SoI toposheets and based on landslide information gathered by the Geological Survey of India (GSI 2007 and 2009). Sikkim Himalayas are dotted with landslides of varied types and dimensions. GSI has marked 347 landslide points in North Sikkim District. The corresponding textural database structure is developed through consultation of already available various published and unpublished reports and papers. For the present study area, an attempt has been made to incorporate the maximum information available for specific landslides within this database.

20.4.3 Watershed-Wise Landslide Distribution

The status of landslide distribution (Fig. 20.4) in different watersheds in North Sikkim district shows that the number of new landslides (identified from satellite data of (2002) is comparatively more than the old slides (identified from 1977 SoI toposheets of 1:50,000). The frequency of landslides in the last few years may be partially controlled by human interference along with the geo-environmental factors of the region. But, their frequencies are also influenced by climatic variability over the years.

20.4.4 Type of Movement and Materials

The reported landslides in the North Sikkim district show different types of movement and variations in the material. In a rockfall, rock material detaches from a steep slope or cliff and the materials can slide freely and extremely fast. Generally, rocky slopes of 40–90° are more prone to rock fall during earthquake shaking. From the study of earthquake-induced landslides, it has been found that rockfall is very common along road corridors. This type of movement accounted for about 10% of the total landslides in the district (Fig. 20.5).

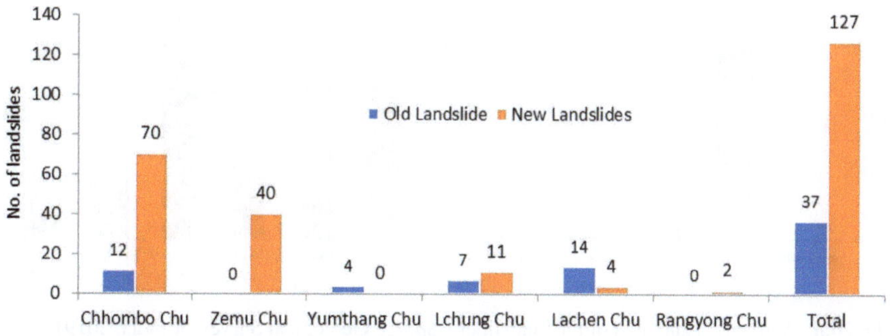

Fig. 20.4 Frequency of old and new landslides in various watersheds in the North District of Sikkim Himalaya. (*Source* Geological Survey of India, 2019)

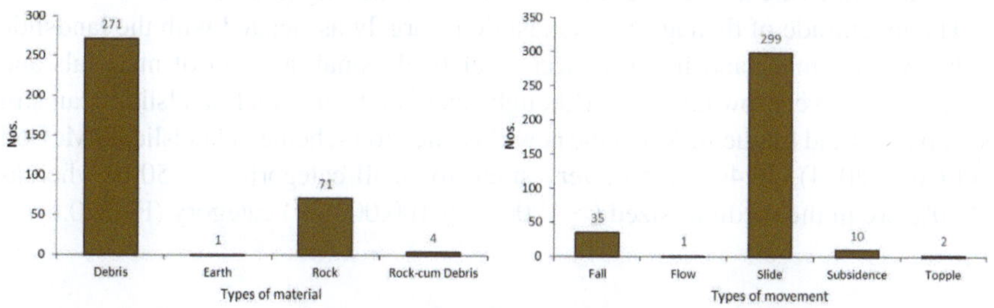

Fig. 20.5 Type of materials and movement involved in landslides. (*Source* Geological Survey of India, 2019)

Landslides are most likely to occur on moderately steep-to-steep slopes where unconsolidated colluvium overlies shallow soil/bedrock. The field study found that slides mainly occur during the monsoon season and are triggered by heavy rainfall. Most of the landslides in this district are slides (86%). Out of 347 landslides, only 10 were in this category, which is subsidence in nature. The materials of the landslides can be earth, debris, rock, and rock-cum debris. Seventy-eight % of the total landslides are characterized by debris material, while 20.46% of landslides consist of rock (Fig. 20.5). Rockfalls mainly occur in highly dissected hills.

20.4.5 Landslide Activity and Volume

Landslides can be classified as active, dormant, or inactive (Vernes 1978) based on the state of activity. Landslides generally occur when the shear stress on the slope material exceeds the material's shear strength. Several geo-environmental factors including slope, morphology, lithology, jointing and structure, and land cover/use influence the occurrence and reactivation of landslides. The triggering factors of

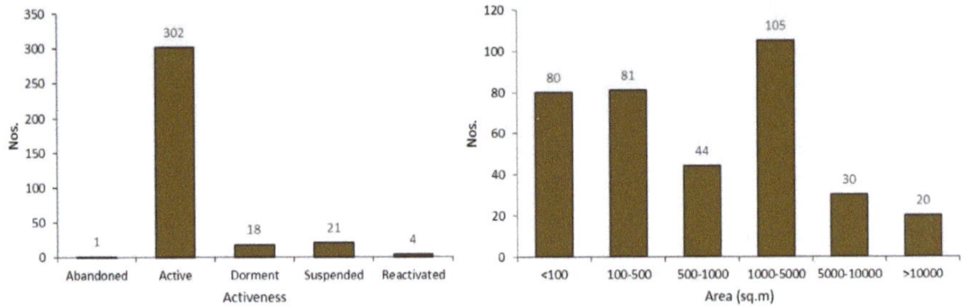

Fig. 20.6 Activeness and area of the landslides (*Source* Geological Survey of India, 2019)

rainfall and earthquakes can also influence them (Ghosh et al. 2012a, b). About 87% of the landslides are active, and they are currently moving (Fig. 20.6).

The magnitude of damage by a landslide is largely associated with the landslide body area. A small landslide produces a relatively small amount of materials and results in relatively low damage. Although, small and very small landslides can still be impactful and frequent. As per the new classification scheme of landslides (McColl and Cook 2024), 46.40% are in very small to small categories ($\geq$ 500), whereas 53.40% are in the medium-sized ($\geq$ 1000– $\geq$ 100,000 m^2) category (Fig. 20.6).

20.4.6 Landslide and Land Use/Land Cover

The undercutting of slopes due to heavy monsoon precipitation, deforestation, unscientific cultivation, and construction activities is known to combine with rather complex geological formations in the region to generate a variety of landslides. The landslides and their associated land use/land cover have been compiled mostly from GSI (2019) reports. The region is characterized by extensive tropical moist deciduous-semi-evergreen forest, sub-tropical broad-leaf hill forest, temperate forest, and mixed coniferous and fir forests. The northeastern side of the study area being a high-altitude zone has little forest cover. Alpine scrub also covers a very small portion of the total geographical area. The maximum portion of the northern side is covered by barren rocky land cover comprised of scree slopes and rocky outcrops with no vegetation at all. From Fig. 20.7, it is clear that the maximum landslides (32.24%) occurred in the moderate vegetation zone. Steep bare rocky land including snow cover constitutes 6.86% landslides. Twenty % of landslides have occurred in very dense vegetation, and it may be caused by heavy rainfall, local fracturing, and steeper slope morphometry.

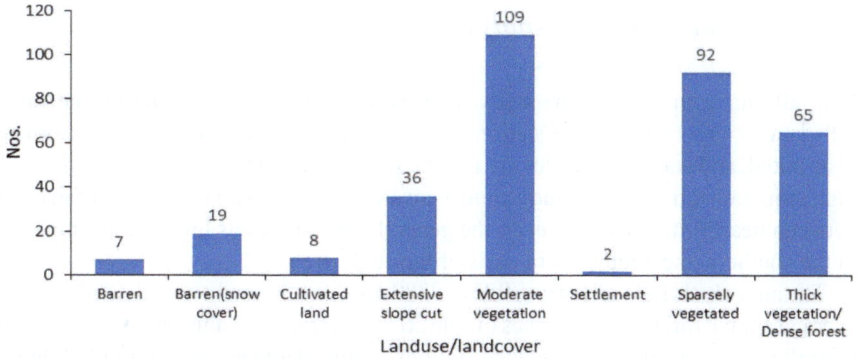

Fig. 20.7 Distribution of landslides and their associated land use/land cover (*Source* Compiled from the Geological Survey of India, 2019)

20.4.7 Rainfall and Landslide

Rainfall is one of the major triggering factors of landslides in the North district, Sikkim of Sikkim Himalaya (Ghosh and Ghosh 2017). The eastern Himalayan region including Sikkim receives the maximum rainfall every year in the country. Ayalew (1999) has mentioned the relationship between rainfall and landslides. Heavy precipitation (rainfall) can create hydrostatic pressure on the saturated materials and initiate certain types of mass movement. A close relationship between landslides and the amount of precipitation has also been observed by Tsereteli (in UNESCO/UNEP 1988). The considerable amount of rainfall increases the moisture content and creates pore water pressure in soils. Landslides are more likely to occur if high amounts of intense rainfall are preceded by a period of light but incessant precipitation (Ghosh and Ghosh 2017). The prolonged rainfall with low intensity can make the soil saturate. Thereafter, high-intensity rainfall can act as the triggering factor for the occurrences of landslides. The annual average rainfall (2010–2016) of the North Sikkim hydro-meteorological division is 3333.7 mm. The maximum rainfall occurs between June and September (monsoon), and it accounts for 64.1% of annual rainfall. High-intensity rainfall may result in some massive and devastating landslides in north Sikkim (Ghosh and De 2022).

The study revealed that 95.24% of landslides occurred under the influence of 10 days of antecedent rainfall. A similar result has been found in the case of the Lanta Khola landslide in North Sikkim. This landslide was initiated with normalized cumulative rainfall that exceeds 250 mm over at least 15 days. About 73% of the landslides in north Sikkim district were rainfall-induced landslides during 2010–2016.

20.4.8 Rainfall Depth Analysis

Rainfall characteristics of the study area were analysed by considering the rainfall data of eight rain-gauge stations. The rain-gauge stations are not uniformly distributed, and station distances are not the same. Due to the orographic barrier and high mountain terrain in the study area, there are significant variations in rainfall even between nearby stations. However, the general trend shows its higher concentration in the southern and southeastern parts of the study area.

The annual rainfall varies from 925 to 3389 mm, and the study was divided into ten average annual rainfall depth zones (Table 20.3). The rainfall value below the average rainfall of the considered rain-gauge stations is considered as the low rainfall depth zone. Lachen and Yumthang areas exhibit the highest rainfall depth zone (Fig. 20.8) The depth zone suddenly decreased towards the north and the very lowest rainfall depth is found towards the extreme northwestern part of the study area. The southern part of the study area also shows a very high depth zone with high annual rainfall. The depth zone IV has the highest area coverage (868.5 km^2). The depth zone I (< 1050) occupies the lowest area (111.7 km^2), and the average annual rainfall is 925 mm. The highest annual rainfall (3389 mm) is found in the depth zone of X. The frequency of landslides is less in the extreme northwestern part compared to the southern portion of the study area, because that area belongs to the very lowest rainfall depth zone.

Table 20.3 Rainfall depth zone with respective areas

Depth zone	Depth zone range	Average annual rainfall (P_j)	Area in km^2 A_j	$A_j \times P_j$
I	< 1050	925	111.7	103,322.50
II	1050–1300	1175	444.74	522,569.50
III	1300–1550	1425	484.86	690,925.50
IV	1550–1800	1690	868.5	1,467,765.00
V	1800–2050	1885	599.22	1,129,529.70
VI	2050–2300	2175	341.2	742,110.00
VII	2300–2550	2425	328.28	796,079.00
VIII	2550–2800	2710	447.35	1,212,318.50
IX	2800–3050	2925	273.12	798,876.00
X	> 3050	3389	397.03	1,345,534.67
Total area			4296.00	$\sum A_j \times P_j =$ 8,809,030.37

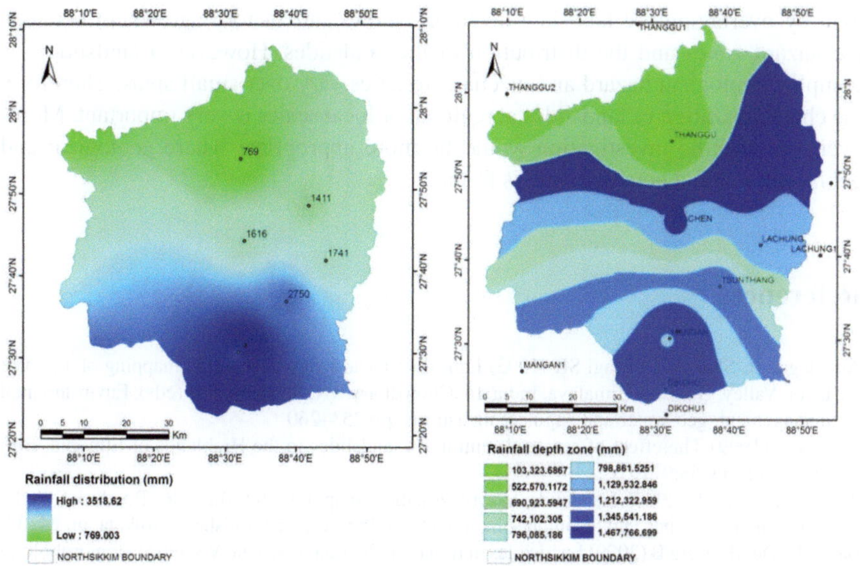

Fig. 20.8 Rainfall distribution map (left) and rainfall depth zone (right) of the North District, Sikkim. (*Source* Indian Meteorology Department, Meteorology Centre, Gangtok, Sikkim, 2019)

20.5 Conclusion

The present study shows that most of the northwestern glaciated areas in the study area are highly dissected and highly unconsolidated masses of debris waiting to be hurled down due to disturbance by any available natural agency like heavy precipitation at any point of time or by human interference. A considerable no. of landslides in this area indicates instability of the slopes. Several geological structures, e.g. faults, lineament, etc., also lead to the cutting, slide slumps, debris flows, etc., in the study area. These degradational activities are further assisted by the presence of steep slopes and softer rocks, particularly in the middle and southern parts of the North district, Sikkim. The spatial distribution of the Main Boundary Thrust (MBT) and Main Central Thrust (MCT-I & II) also accelerates the landslide hazard in the study area. Unscientific land use activities also lead to downcutting, slide slumps, debris flows, etc., in the study area.

Landslide hazard zonation mapping at the regional scale has become easier and more accurate through remote sensing and GIS, making it possible to integrate a large dataset with high location accuracy. The present study revealed that geomorphology, geology, and slope have greater control than the other factors for landslide occurrences in the north Sikkim district. In the recent past, the construction of roads in the mountainous region without consideration of local geological, morphological, and hydrological information and without factoring in the heavy monsoon rainfall increased the potentiality of landslides. Cross-verification of the landslide hazard map by overlaying the landslide inventory shows satisfactory agreement between the hazard zones and the distribution of the landslides. However, a landslide is a complex geological hazard and its characteristics vary over small areas. Therefore, the characterization of landslides at regional or local scales is very important. Moreover, site-specific investigation would be more appropriate where geological and geomorphological heterogeneity is found.

References

Anbalagan R, Sharma L, Tyagi SK (1993) Landslide hazard zonation (LHZ) mapping of a part of Doon Valley, Garhwal Himalaya, India. In: Chowdhury RN, Sivakumar M (eds), Environmental management, geo-water and engineering aspects, pp 253–260

Ayalew L (1999) The effect of seasonal rainfall on landslides in the Highlands of Ethiopia. Bull Eng Geol Env 58:9–19

Basu SR, Bera B (2007) Landslide hazard zonation mapping. In: Basu R, Bhaduri S (eds) Contemporary issues and techniques in geography. Progressive Publishers, Kolkata, pp 24–35

Basu SR, De SK, Bera B (2009) Landslides in India. In: Sharma, HS, Kale VS (eds) Geomorphology in India. Prayag Pustak Bhawan, Allahabad, pp 171–198

Bera B (2005) Assessment of landslide hazard: a case study of Gangtok town of the Sikkim Himalayas. In: Banerjee A et al. (eds) Combating disaster-perspectives in new millennium. ACB Publications, pp 45–149

Bhattacharya A, Sengupta CK (1996) Landslide hazard zonation along NH-31A from Sevok to Rangpo: a terrain evaluation study. J Eng Geol 27:1–40

BIS (1998) Guideline for preparation of landslide hazard zonation maps in Mountains Terains. Draft Proposal of the Bureau of Indian Standards 2:1–15

BIS (2002) Criteria for earthquake resistant design of structures: general provisions and buildings. Bureau of Indian Standards (BIS) IS 1893:2002 (Part-1)

Chatterjee B (1986) Final report on geological approach to the landslide hazard zonation in Sikkim-Himalaya. Rep Geol Surv Ind (Unpublished)

Chung CJF, Fabbri AG, Van Westen CJ (1995) Multivariate regression analysis for landslide hazard zonation. In: Carrara A, Guzzetti F (eds) Geographical information systems in assessing natural hazards, vol 5. Springer, Netherlands, pp 107–133. https://doi.org/10.1007/978-94-015-8404-3_7

Dave VKS, Joshi BC (1987) Landslide hazard zones in the Nayar Basin, Garhwal Himalayas. In: Proceedings of the 6th geological congress. Indian Geological Congress, vol 145. pp 1–37

De SK (2007) A study of landslides through susceptibility zoning in the hilly tract of the Balasan Basin of Darjiling Himalayas. IAG Regional Conference on Geomorphology, Kota Kinabalu, Malaysia, pp 58

Gairola BM, Shukla SD (1990) Preparation of landslide zonation map of U.P. Himalaya. Rec Geol Surv Ind 123(8):184–186

Ghosh GK, Ghosh K (2017) Spatio-temporal analysis of rainfall pattern in North District, Sikkim, India. Int Adv Res J Sci Eng Technol 6(5):87–92

Ghosh GK, Ghosh K, De SK, Rawat MS, Joshi V, Ayala IA (2012a) An inventory of recent (18th September 2011) earthquake triggered landslides in the Sikkim Himalaya. J Geo-Environ Observer 1(2):1–12

Ghosh K, Bandyopadhyay S, De SK (2012b) Landslide susceptibility zonation using weight rating based method in GIS: a case study in Dhalai District, Tripura. Indian J Geomorphol 17(2):127–145

Ghosh K, De SK (2022) Evaluation of hydro-geomorphic responses to climate change in North Sikkim District, Sikkim, India. In: Siddiqui AR, Sahay A (eds) Climate change, disaster and adaptations. Sustainable development goals series. Springer, Cham. https://doi.org/10.1007/978-3-030-91010-5_3

Ghosh K, Bandyopadhyay S, De SK (2017) A comparative evaluation of weight-rating and analytical hierarchical (AHP) for landslide susceptibility mapping in Dhalai District, Tripura. In: S. Hazra et al. (eds) Environment and earth observation. Springer International Publishing, pp 175–193

GSI (2007) Geology and mineral resources of Districts of Sikkim. Geological Survey of India, Eastern Region, pp 1–44

GSI (2009) Geological survey of India report. www.portal.gsi.gov.in

Gupta RP, Joshi BC (1990) Landslide hazard zoning using the GIS approach: a case study from the Ramganga Catchment, Himalayas. Eng Geol 28:119–131

Indian Meteorological Depattment (2019) Meteorology Centre, Gantgtok, Sikkim

Kanungo DP, Sarkar S (2003) Landslides and terrain parameters in Darjeeling Himalaya. Himalayan Geol 24(2):55–62

Krishna AP, Kumar N, Nag SK, Kundu S (2004) Landslide hazard susceptibility zonation (LHSZ) along roads: a case study from Sikkim Himalaya. In: Singh S, Sharma HS, De SK (eds) Geomorphology and environment. ACB Publications, Kolkata, pp 157–167

Linsley Jr. RK, Kohler MA, Paulhus JLH (1949) Applied hydrology. Mc. Graw Hill Book Company, USA, pp 689

Majumdar N (1980) Distribution and intensity of landslide processes in North-eastern India: a zonation map thereof. In: Proceedings of the international symposium on landslide. Sarita Prakashan, pp 3–14

McColl ST, Cook SJ (2024) A universal size classification system for landslides. Landslides 21:111–120. https://doi.org/10.1007/s10346-023-02131-6

Pachauri AK, Pant M (1992) Landslide hazard mapping based on geological attributes. Eng Geol 32:81–100

Sharan RB (1992) Landslide hazard zonation in parts of Beas valley, Himachal Pradesh. Rec Geol Surv Ind 123(8):182–183

UNESCO/UNEP (1988) Landslides and mudflows, vol 1. UNEP/UNESCO, Moscow, pp 236

Valdiya KS (1985) Accelerated erosion and landslide-prone zone in the Central Himalayan region. In: Singh JS (ed) Environmental regeneration in the Himalayas: concepts and strategies, Nainital. Central Himalayan Environmental Association and Gynodaya Prakashan, pp 12–38

Varnes DJ (1978) Slope movement types and processes. In: Schuster R, Krizek R (eds) Landslides, analysis and control. Transportation research board, National Academy of Sciences, Washington, DC, pp 11–33

Chapter 21
Data and Methods Used in Mapping, Monitoring, Management, and Mitigation for GLOF

Parmod Kumar and Swati Sharma

Abstract Glacial lake outburst floods (GLOFs) pose significant risks to communities and infrastructure in glacier-fed regions. Effective management and mitigation of GLOFs require a comprehensive approach involving advanced data collection and analytical methods. This study summarizes the data and methods employed in the mapping, monitoring, management, and mitigation of GLOFs. Mapping of glacial lakes and potential hazard zones utilizes a combination of remote sensing technologies, such as satellite imagery and aerial photography, alongside detailed topographic maps and digital elevation models (DEMs). These tools facilitate the identification and delineation of lake boundaries, glacier extents, and risk areas by integrating optical and radar data with ground-based surveys. Monitoring efforts focus on real-time tracking of glacial lakes and glaciers to detect early warning signs of potential outbursts. Key methods include hydrological monitoring with water level gauges and pressure sensors, glacier movement tracking using GPS stations and radar systems, and periodic remote sensing observations. These approaches collectively help in assessing lake stability and glacier dynamics, providing critical data for risk prediction. Early warning systems employing automated sensors and communication networks are also critical for providing timely alerts to at-risk populations. This study provides a comprehensive view of an integrated approach that combines advanced data collection, real-time monitoring, risk assessment, and targeted mitigation strategies essential for effectively managing and mitigating the risks associated with GLOFs.

Keywords DEM · GLOF · Remote sensing · GPS

P. Kumar
Amity Institute of Geoinformatics and Remote Sensing, Amity University, Noida, India

S. Sharma (✉)
Division of Research and Development, Lovely Professional University, Jalandhar 144411, Punjab, India
e-mail: swati.9238@gmail.com; swati.33384@lpu.co.in

S. C. Rai (ed.), *Himalaya: Mountains of Destiny*,
https://doi.org/10.1007/978-3-031-94905-0_21

21.1 Introduction

Recently published glacier inventory for the Indus, Ganga, and Brahmaputra basins using Indian Remote Sensing (IRS) satellite data have reported a total of 34,919 glaciers covering 75,779.10 km^2 of area in the Himalayas (Sharma et al. 2013). Glaciers depend on extremely particular climatic conditions to prevent the summertime loss of snow accumulation throughout the winter. A glacier is a mass of compacted, recrystallized ice on a surface that moves downward under gravity-driven pressure. Due to environmental and social impacts glacier melts and forms glacial lakes. A glacial lake is a body of water that is formed when a glacier melts and retreats, leaving behind a depression in the landscape that is filled with water. These glacial lakes act as ecosystems as well as cause of hazard in the form of glacial lake outburst flood hazard (GLOF). A GLOF is a sudden release of water from a glacial lake, typically caused by the failure of a natural dam made of ice or sediment. GLOFs can cause significant damage to downstream infrastructure, settlements, and ecosystems, and can result in loss of life. GLOFs can be difficult to predict but advances in remote sensing and monitoring technologies have improved early warning systems in some regions.

21.2 Characteristics and Formation of Glacial Lake

Glacial lakes are parts of glaciers that form mainly in depressions where some obstruction of sediments and moraine comes in the way of flowing meltwater from glaciers. They can be categorized as follows based on damming material.

(Mool et al. 2011) ICIMOD report describes the types of glacial lakes:

- Moraine-dammed lake—Lake dammed by moraine following glacial retreat.
  - End moraine-dammed lake
  - Lateral moraine-dammed lake (ice-free)
  - Lateral moraine-dammed lake (with ice)
  - Other moraine-dammed lakes
- Ice-dammed lake supra-glacial—Lakes dammed by glacier ice
  - Supra-glacial lake
  - Glacier ice-dammed lake
- Glacier erosion lake—Bodies of water that form because of an earlier glacial erosion process, which accumulate in depressions after the glacier has retreated or melted away
  - Cirque lake
  - Glacier through valley lake
  - Other glacier erosion lakes

- Other glacial lakes—Lakes formed in a glaciated valley, and fed by glacial melt, but damming material is not directly part of the glacial process.
  - Debris-dammed lake
  - Artificial lake
  - Other lakes fed by glacial melt.

21.3 Glacial Lakes in the Himalayan Region

Apart from the Arctic and Antarctic, the Himalayan region has the largest glaciarized area with glacial lakes. The report of ICIMOD creates the database of glacier and glacial lakes in Nepal and Bhutan. There are 3252 glaciers and 2323 glacial lakes that were identified in Nepal out of which 20 were considered as dangerous. 677 glaciers and 2674 glacial lakes were identified. In Bhutan, 24 glacial lakes were dangerous. Ives et al. (2010) describe the formation of glacial lakes in Hindu Kush Himalaya. There are more than 8000 glacial lakes that have been identified in this region among which 200 are potentially dangerous. Bajracharya and Mool (2009) stated that glaciers are retreating at a faster rate in the Mount Everest region resulting in the formation of glacial lakes. A study of a mapped glacier in the Himalayan region carried out on LISS III images of the years 2000/01/02 and 2010/11 which found that 87% of glaciers are stable, 12% showed retreat, and 1% glacier out of 2018 showed advancement of the snout. Glacial lake inventory has been carried out by SAC (ISRO) in the Himalayan region including Jammu and Kashmir, Himachal Pradesh, Uttarakhand, Sikkim, and Arunachal Pradesh using LISS III and LISS IV data on a 1:50,000 scale. There were 4699 glacial lakes that were mapped using different methods among which 1966 were in point file because of less area than the minimum mappable unit out of which 271 glacial lakes were in Himachal Pradesh (Panigrahy 2012). Spatiotemporal change of 763 glacial lakes is also analyzed by (Kumar et al. 2024a, b, c) using Landsat time series data from 1980 to 2017 in Himachal Pradesh. Glacial lakes have been mapped in Sutlej and Chenab basins by Randhawa et al. (2005). Bhambri et al. (2018) used LISS IV and Landsat data from years 2011, 2012, 2013, 2014, and 2015 for mapping of glacial lakes and found 958 glacial lakes in Himachal Pradesh.

21.4 Mechanism of GLOF

Glacial lake outburst floods (GLOFs) are catastrophic events that pose significant risks to communities, infrastructure, and ecosystems downstream of glacial lakes. In the Himalayas, the presence of numerous glaciers and glacial lakes makes the region particularly vulnerable to GLOFs.

21.4.1 The Major Causes of GLOFs Include

(i) Glacial Melting
(ii) Ice or Glacier Dam Failure
(iii) Landslides or Rockfalls
(iv) Earthquakes
(v) Volcanic Eruptions
(vi) Heavy Rainfall
(vii) Human Activities.

Each of these factors can contribute to the risk and severity of GLOFs, and often multiple factors are involved in a single event. Monitoring and managing these risks are crucial in areas prone to GLOFs, especially in regions experiencing significant environmental changes.

21.4.2 Risks and Hazards from GLOF

(i) Flooding
(ii) Loss of Life and Injury
(iii) Infrastructure Damage
(iv) Environmental Impact
(v) Economic Losses
(vi) Secondary Hazards
(vii) Displacement of Communities.

21.5 Historical Context of GLOF in the Indian Himalayan Region

The global glacial lake outburst flood database by (Veh et al. 2022; Lützow et al. 2022) encompasses data from more than 700 different sources including analyses of stream gages; satellite and aerial images; stratigraphy; tree rings; reports from local authorities; news outlets; workshop proceedings; social media accounts; and unpublished work. Additionally, they retrieved summary information regarding peak discharge, flood volume, and lake volume, as well as the methodology used to estimate these diagnostics, if possible. GLOF reported by RGI are 41 and ICIMOD reported 59 GLOF events in India. Source: http://glofs.geoecology.uni-potsdam.de/ and https://doi.org/https://doi.org/10.26066/RDS.1973283. Indian Himalayan region has witnessed several major GLOF events some of which are tabulated below (Table 21.1).

The recent one on October 3, 2024, the South Lhonak Lake in the Himalayan state of Sikkim ruptured, causing a glacial lake outburst flood (GLOF) that devastated four

Table 21.1 Major historical GLOF events in the Indian Himalayan Region

Indian Himalayan Region	1	77.7046	35.1806	1533	1 man died on the Murgo road; Kumdan road closed	Hewitt (2014)
	2	–	–	1971	13–16 fatalities	Ikeda et al. (2016)
	3	79.0815	30.7484	2013	Cause death of ~ 6000 people	Rastner et al. (2016)
	4	77.7046	35.1806	1835	Maximum devastation occurred in Lagzhung village (mid-Nubra) and Chorbat (Balti)	Bhambri et al. (2019)

districts. As a result, 77 more people are still unaccounted for and at least 42 people died. Scientists claim that the lake is still potentially dangerous more than a month after the incident. (The Hindu report). State-wise GLOF events are also given in Table 21.2 where Shyok has the most 25.

Table 21.2 GLOF events inventory recorded in different states and basins in the Indian Himalayan Region

Country	Province	River Basin	No. of GLOF	References
India	Arunachal Pradesh	Manas	1	Veh et al. (2022)
	Himachal Pradesh	Baspa	2	Sangewar et al. (2009)
	Kashmir	Nubra	5	Bazai et al. (2021)
	Ladakh	Chenab	4	Zheng et al. (2021)
		Indus	5	Zheng et al. (2021)
		Shyok	25	Zheng et al. (2021)
		Sutlej	1	Zheng et al. (2021)
		Upper Indus	4	Zheng et al. (2021)
		Zanskar	1	Zheng et al. (2021)
	Sikkim	Changme	2	Zheng et al. (2021)
		East Rathong	2	Zheng et al. (2021) Veh et al. (2019)
		Zemu	5	Zheng et al. (2021)
	Uttarakhand	Alaknanda	1	Rastner et al. (2016)
		Dhauliganga	1	Zheng et al. (2021)
India Total			59	

21.6 Data Used for GLOF Mapping and Monitoring

Modeling glacial lake outburst floods (GLOFs) requires a comprehensive array of datasets to accurately simulate the behavior and potential impacts of these events. This section outlines the key datasets used in GLOF modeling and their specific roles. Satellite remote sensing has revolutionized the way scientists observe and analyze glaciers and glacial lakes. High-resolution imagery from satellites allows for detailed assessments of glacier dynamics, glacier mass balance, and changes in glacial lakes over time.

21.6.1 Optical and Multispectral Imagery

Optical sensors on satellites are those that work in the visible band of the electromagnetic spectrum (EMS) and multispectral sensors (MSS) are the satellite sensors that work in the NIR and SWIR region of EMS (Table 21.3). Landsat and Sentinel-2 are mostly used datasets because of their free availability. The official website of these datasets is https://earthexplorer.usgs.gov/ and https://browser.dataspace.copernicus.eu/.

These datasets are used in identifying lake boundaries, monitoring changes in glacier size, and detecting new or expanding glacial lakes. These high-resolution datasets are also used in analyzing surrounding areas of glaciers and glacial lakes.

21.6.2 Synthetic Aperture Radar (SAR)

SAR systems, provide all-weather, day-and-night imaging capabilities. Sentinel-1 datasets (Table 21.4) are mostly used because of free availability and the website is https://search.asf.alaska.edu/#/.

These datasets are used in detecting changes in glacier flow, crevasses, and the stability of moraine dam structures. SAR interferometry provides insights into ground displacement. SAR also penetrates the snow which can help in finding buried lakes and also the polarimetric analysis (Kumar et al. 2024a, b).

21.6.3 Laser Altimetry

Spaceborne laser altimeters, like those on the ICESat and ICESat-2 missions (Table 21.5), measure the elevation of the Earth's surface with high precision. These

Table 21.3 Remote sensing satellites with panchromatic and multispectral sensors (Kumar and Bhagat 2019)

Satellite	Country (Year)	Sensors	Spatial resolution (m)	Revisit time (days)	Remarks
Resourcesat-2	India (2011)	4 MS - LISS IV	5.8	5	–
		4 MS - LISS III	23.5	5	
		3 MS - AWiFS	56	5	
Resourcesat-2A	India (2017)	4 MS - LISS IV	5.8	5	–
		4 MS - LISS III	23.5	5	
		3 MS - AWiFS	56	5	
Resourcesat-3 series	India (2020–2021)	5 MS	10/20	5	
Cartosat-1	India (2005)	PAN	2.5	5	Stereo
Cartosat-2A	India (2008)	PAN	< 1	4	–
Cartosat-2B	India (2010)	PAN	< 1	4	
Cartosat-2 series (2C, 2D, 2E & 2F)	India (2017 and 2018)	PAN	< 1	4	
Cartosat-3A	India (2020–21)	1 PAN 4 MS MIR VNIR & SWIR	0.25 0.50 5 30	Daily	Continuous
GISAT/GeoHR	India (2020–21)	6 VNIR	50	Daily 30 minutes	
IKONOS	USA (2000)	PAN	0.8	3–5	Stereo
		MS (4)	4	3–5	
Landsat-8	USA (2013)	OLI (8)	30	16	
		TIRS (2)	100		
		PAN	15		
SPOT-7	France (2014)	PAN	2.5	1–5	Stereo
		MS (3)	10/20		
Rapid Eye	Germany (2008)	MS (5)	6.5	1–5.5	
WorldView-3	USA (2014)	PAN	0.3	1–4.5	–
		MS (8)	1.2		
		MS-SWIR (8)	3.7		
Planet Labs	USA (2014)	PAN	3	–	–

(continued)

Table 21.3 (continued)

Satellite	Country (Year)	Sensors	Spatial resolution (m)	Revisit time (days)	Remarks
		MS (3)	5		
GeoEye-1	USA (2008)	PAN	0.46	3	
		MS (4)	1.84		
Sentinel 2A & 2B	EU (2015)	MS-VNIR (4)	10	5	–
		MS-VNIR (6)	20	5	
		MS-SWIR (3)	60	5	
Terra-ASTER	US (1999)	MS (14)	15/30	4–16	
		MS-IR (2)	90	4–16	
Terra-MODIS	US (1999)	MS-HIS (36)	250, 500 & 1000	Twice daily	–
Landsat-9	US (2023)	OLI (8)	30	16	–
		TIRS (2)	100		
		PAN	15		

Table 21.4 Microwave remote sensing satellites (Kumar and Bhagat 2019)

Satellite	Country (Year)	Sensors	Spatial resolution (m)	Revisit time (days)
RISAT-1	India (2012)	C-band SAR	1–50	
RISAT-1A/1B	India (2019–20)	C-band SAR	1–50	
ALOS-2	Japan (2014)	L-Band SAR	3	24
EnviSat	(EU (2002)	C-band SAR	28/50/950	35
Sentinel-1A&1B	EU (2014/ 2016)	C-band SAR	5/25	12
RADARSAT-2	Canada (2007)	C-band SAR	3/100	24
RADARSAT-3	Canada (2019)	C-band SAR	3/100	24
Terra-SAR-2	Germany (2007)	X-band SAR	1/16	11
TenDEM-X	Germany (2010)	X-band SAR	1/3	11
HJ-1C	China (2006)	C-band SAR	5/20	4
KOMPSAT-5	South Korea (2013)	C-band SAR	1/20	28
NISAR	India (2021)	L-Band SAR S-Band SAR	3–10	12

Table 21.5 LIDAR datasets used in GLOF analysis

Satellite	Country (Year)	Sensors	Spatial resolution (m)
ICEsat-2 (LiDAR)	US (2003)	2 HSI	70
ICEsat-1 (LiDAR)	US (2018)	a. HSI	10

measurements are crucial for calculating glacier thickness, and mass balance, monitoring changes in ice volume, and assessing the impact of glacier melt on sea levels. https://urs.earthdata.nasa.gov/.

21.6.4 Hydrological and Meteorological Data

Precipitation and Temperature Data

The website to download data is https://power.larc.nasa.gov/data-access-viewer/ and all datasets for SWAT hydrological modeling can be taken from https://swat.tamu.edu/data/india-dataset/.

Snow and Ice Data

This data is derived products in the form of snow cover datasets (e.g., MODIS Snow Cover), and ice mass balance data. These datasets are available on https://land.copernicus.eu/en/products/snow & https://earthexplorer.usgs.gov/ and https://modis.gsfc.nasa.gov/data/dataprod/mod10.php.

21.6.5 UAVs and Drone Technology

Unmanned aerial vehicles (UAVs), or drones, offer a flexible and cost-effective approach to studying glaciers and glacial lakes. UAVs equipped with high-resolution cameras and sensors can:

- Capture High-Resolution Imagery
- Conduct Close-up Surveys
- Monitor Glacier Dynamics.

21.6.6 Digital Elevation Models (DEMs)

DEMs, generated from various sources such as satellite imagery, aerial surveys, and ground-based measurements, provide detailed elevation data that is fundamental for studying glacier topography and glacial lakes.

- Topographic Analysis
- Volume Estimation
- Hydrological Assessment.

This data can be from different satellites like Shuttle Radar Topography Mission (SRTM), ASTER GDEM, Cartosat DEM LiDAR surveys, and DEM can be prepared using Sentinel 1 also. The website from where DEM can be downloaded for Aster GDEM is https://gdemdl.aster.jspacesystems.or.jp/index_en.html and for SRTM DEM https://earthexplorer.usgs.gov/.

21.6.7 Geographic Information Systems (GIS) Data Layers

GIS integrates various spatial data layers and provides tools for analyzing and visualizing complex geographical phenomena. In the context of glaciers and glacial lakes, GIS plays a pivotal role in:

- Mapping and Visualization
- Glacier Inventory and Analysis
- Hydrological Modeling.

The datasets included in this are:

21.7 Glacier Inventories

These datasets include glacier databases and national and regional inventories which will help document glacier locations, sizes, and characteristics. The main websites where this data can be available are:

- https://nsidc.org/data/glacier_inventory/browse.html
- https://www.glims.org/RGI/
- https://vedas.sac.gov.in/en/Himalayan_Glacier_Inventory_Atlas.html
- https://nsidc.org/data/nsidc-0770/versions/6.

Land Use and Land Cover Data

Landuse landcover can be taken from the website (https://livingatlas.arcgis.com/) which includes water, trees, flooded vegetation, crops, built area, bare ground, snow/ice, and clouds at 10 m resolution using Sentinel 2 images (Karra et al. 2021).

Curve number

Global Curve Number datasets are available on the website (https://figshare.com/articles/dataset/GCN250_global_curve_number_datasets_for_hydrologic_modeling_and_design/7756202) at a resolution of 250 m. This dataset is used in GLOF modeling.

Drainage and Basin

This dataset of river and drainage network is taken from open-source data which is overlaid on the ALOS PALSAR DEM (digital elevation model) data, downloaded from the Alaska vertex website, and mosaic together to a seamless raster. https://www.hydrosheds.org/products/hydrobasins#downloads

Soil Data

Soil data is used in modeling GLOF and sedimentation flow and in finding the GLOF-prone areas or landslide zones. Soil data is downloaded from the FAO website https://www.fao.org/soils-portal/data-hub/soil-maps-and-databases/faounesco-soil-map-of-the-world/en/.

21.7.1 Field Survey Data

Ground-Based Observations: These datasets include field surveys, geotechnical measurements, and manual observations which help in validating remote sensing data, assessing moraine dam integrity, and calibrating flood models (Kumar et al. 2024a, b). DGPS instruments play major role in field surveys.

Instrumentation Data

These datasets include lake level sensors, meteorological stations, and geotechnical sensors which provide early warning signals for potential GLOF events, informing flood risk assessments. https://www.data.gov.in/ministrydepartment/India%20Meteorological%20Department%20(IMD).

21.7.2 Modeled and Simulated Data

- Flood Simulation Models: These are used in predicting flood extents, estimating damage, and developing flood risk maps.
- Risk Assessment Models: This data is based on integrated risk assessment tools combining multiple datasets which helps in assessing vulnerability, planning for disaster preparedness, and informing mitigation strategies.
- Historical and Event Data: These datasets include historical records, case studies, and disaster reports which help in identifying patterns, understanding past triggers, and improving future risk assessments. http://glofs.geoecology.uni-potsdam.de/
- Event-Specific Data: These datasets include post-event surveys, and damage assessments which help in updating flood risk assessments and enhancing early warning systems. http://glofs.geoecology.uni-potsdam.de/.

21.7.2.1 Integration of Data and Modeling

The integration of data from various geospatial technologies allows for comprehensive modeling and analysis of glacier and glacial lake systems. Combined datasets enable researchers to:

- create 3D models
- simulate glacier dynamics
- assess climate impacts.

21.8 Software Used for GLOF

There are lots of remote sensing and GIS software available in the market where some major software are listed below:

- ArcGIS Map, ArcGIS Pro, Qgis, SNAP, Polsar Pro, Erdas, Envi, Global Mapper, and Ecognition are major remote sensing and GIS software for satellite images.
- AgiSoft, Pix4D, Global Mapper, Microstation, UAS Master, and Photomodeler are major software used for drone and LiDAR data and 3D work.
- Sql, MS Office for climate and hydrological data preparation.
- DAMBRK, SWAT, HEC-HMS, HEC-RAS, HEC-1, MIKE, and FLO 2D are major software packages used in hydrological modeling.

Google Earth Engine and some cloud-based platforms are also used nowadays for GIS tasks and also to download different datasets.

21.8.1 Methodologies Used for GLOF

21.8.1.1 Preprocessing of Satellite Images

Preprocessing the data involves removing noise, atmospheric effects, and other artifacts that could affect the accuracy of the results.

21.8.2 Glacial Lake Mapping

The preprocessed images can be processed (Fig. 21.1) for NDWI as suggested by (McFEETERS 1996). The NDWI is obtained by using the image processing software's modeler and the MNDWI is computed by using both SWIR bands of the LANDSAT 8 scene.

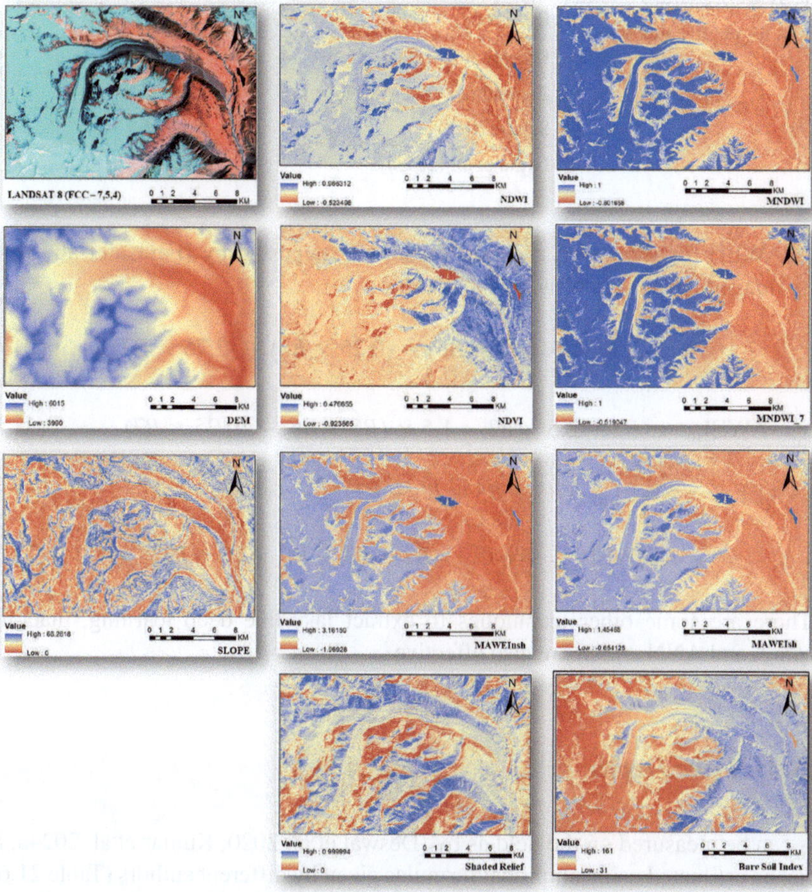

Fig. 21.1 Various indices applied on Landsat 8 FCC image: NDWI, NDWI, MNDWI, MAWEInsh, MAWEish, Bare Soil Inex and DEM, Slope, Shaded Relief on Aster GDEM

$$\text{NDWI} = (G - \text{NIR})/(G + \text{NIR}) \tag{21.1}$$

$$\text{MNDWI} = (G - \text{SWIR})/(G + \text{SWIR}) \tag{21.2}$$

$$\text{NDVI} = (\text{NIR} - R)/(\text{NIR} + R) \tag{21.3}$$

Normalized difference vegetation index (NDVI) suggested by (Rouse et al. 1974) and further evaluated by (Townshend and Justice 2007) and (Tucker and Sellers 1986) can also be used to differentiate the vegetation from the water body. There are two other indices named Advanced Water Extraction Index (AWEI) and Modified Advanced Water Extraction Index.

Table 21.6 Empirical formula for depth calculation

Reference	Formula
Huggel et al. (2002)	$D = 0.104\, A^{0.42}$
Budhathoki et al. (2011)	$D = 0.094\, A^{0.452}$
Sakai (2012)	$D_{\text{max}} = 95.665 \times A^{0.489}$
Fujita et al. (2013)	$D_{\text{m}} = 55\, A^{0.25}$

(MAWEI) for shadowed and non-shadowed suggested by (Feyisa et al. 2014) using image processing software's modeler.

$$\text{AWEI}_{\text{nsh}} = (4 \times (B3 - B6) - (0.25 \times B5) + (2.75 \times B7) \qquad (21.4)$$

$$\text{AWEI}_{\text{sh}} = (B2 + 2.5 \times B3 - 1.5 \times (B5 + B6) - (0.25 \times B7) \qquad (21.5)$$

s $\text{MAWEI}_{\text{nsh}} = \text{AWEI}_{\text{nsh}}/(B3 + B5 + B6 + B7)$ (6)

$$\text{MAWEI}_{\text{sh}} = \text{AWEI}_{\text{sh}}/(B2 + B3 + B5 + B6 + B7) \qquad (21.7)$$

There are some other techniques to extract lake-like deep learning, machine learning, and ANN which are also effective.

21.8.3 Lake Depth

This can be measured on the field as by (Deswal et al. 2020; Kumar et al. 2024a, b) or can be estimated using empirical formulas given by different authors (Table 21.6).

21.8.4 Peak Discharge, Lake Volume, Breach Width, and Breach Formation Time

Other than lake depth, more parameters are used for GLOF modeling. After the field survey of different GLOF events around the world, information has been calculated (Table 21.7) and analyzed, based on the relation among this information, a formula is derived for those parameters that cannot be measured directly using satellite images. Those parameters are listed below in the table with empirical formula.

Table 21.7 Peak discharge calculation formulas by various authors

Peak discharge calculation	
Walder and O'Connor (1997)	$Q = 0.0013\ \mathrm{PE}^{0.60}$
Huggel et al. (2002)	$Q_{\max} = 0.00077\ V^{1.017}$
Froehlich (1995)	$Q_{\mathrm{p}} = 0.607(V_{\mathrm{w}})^{0.295}(h_{\mathrm{w}})^{1.24}$
Volume Calculation	
Huggel et al. (2002)	$V = 0.104\ \mathrm{A}^{1.42}$
O'Connor et al. (2001)	$V = 3.114\ \mathrm{A} + 0{:}0001685\ A^2$
Fujita et al. (2013)	PFV = min [Hp; Dm] A
Breach width calculation	
Froehlich (2016)	$B_{\mathrm{ave}} = 0.27k_{\mathrm{M}}(V_{\mathrm{w}})^{1/3}$
Breach Formation time Calculation	
Froehlich (2016)	$T_{\mathrm{f}} = 63.2(V_{\mathrm{w}}/(gh^2{}_{\mathrm{b}}))^{0.5}/3600$

21.8.5 Vulnerability Criteria of Lake

Glacial lakes can pose a significant hazard when they become unstable, leading to outburst floods (GLOFs) that can cause damage to infrastructure, property, and loss of life downstream. Identifying potentially vulnerable glacial lakes is important for mitigating the risk of GLOFs. Here are some criteria (Table 21.8) that can be used to identify potentially vulnerable glacial lakes:

21.8.6 Discharge from Snowmelt

Field measurement: A wooden block or anything that matches the flow of water, can be taken to measure the flow of the open channel by measuring the distance traveled by it with water in a specific time.

Temperature-Based

For snowmelt, (Martinec 1975) has given this model and (Aggarwal et al. 2014) used the below formula which is based on the temperature lapse rate.

$$Q_{n+1} = \left[C_{S_n} a_n (T_n + \Delta T_n) S_n + C_{R_n} P_n\right] * A * (10000/86400) * \left(1 - K_{n+1}\right) + Q_n K_{n+1} \quad (21.8)$$

Energy Balance Method

Anderson (1976) calculated the temperature profile inside a snow cover using the energy balance method. In the western Tien Shan Mountains of Chi, (Hong et al. 1992) used the bulk transfer method to compute the energy balance over the snow surface to quantify the amount of snowmelt.

Table 21.8 Characteristics of lakes that are potentially vulnerable to GLOF

(1)	Lake and Glacier Characteristics		
	Increase in lake area per decade (Bolch et al. 2008)		
	> 100%	*50–100%*	***< 50%***
	Glacier Calving (Mergili and Schneider 2011)		
	Possible	***Not Possible***	
	Lake-glacier Distance		
	< 80 m	*80–600 m*	***> 600 m***
(2)	Dam Type (Huggel et al. 2004)		
	Ice	*Moraine*	***Bedrock***
	If moraine dam ⟶ 3, 4, 5 & 6; else ⟶ 5 & 6		
(3)	Dam Geometry (Huggel et al. 2004; Wang et al. 2013)		
	Moraine dam to width-to-height ratio		
	< 1	*1–2*	***> 2***
	The slope of the distal face of Moraine dam		
	> 20°	*3–20°*	***< 3°***
(4)	Freeboard (Worni et al. 2013)		
	< 5	*5–15*	***> 15***
(5)	Potential for lake impacts (Worni et al. 2013)		
	Ice or snow avalanche/debris flow/rock fall		
	Possible	***Not Possible***	

21.9 Results

Based on the parameters acquired from various data processing, mapping, discharge analysis, rainfall data measurements, lake volume and lake depth, etc., as mentioned in the above section, GLOF modeling can be performed which can be of different types like 1D, 2D, and 2.5D with sediment flow giving information on flood inundation and velocity of discharge data from GLOF events.

First of all, probability can be checked using the following equation given by (McKillop and Clague 2007):

$$P(Y=1)=\left\{1+\exp-\left[\alpha+\beta_1(\text{M\_hw})+\sum\beta_j\left(\text{Ice\_core}_j\right)\right.\right.$$
$$\left.\left.+B2(\text{Lk\_area})+\sum\beta_j\left(\text{Geology}_k\right)\right]\right\}^{-1}$$

Glacial lake mapped and GLOF in Uttarakhand by (Jain et al. 2012) and (Rana et al. 2013; Dobhal et al. 2013; Allen et al. 2016, and Bhambri et al. 2016) shows the Kedarnath disaster during 16–17 June 2013. Aggarwal et al. (2016) used a combined approach of remote sensing, GIS, and dam-break modeling in Sikkim Himalaya,

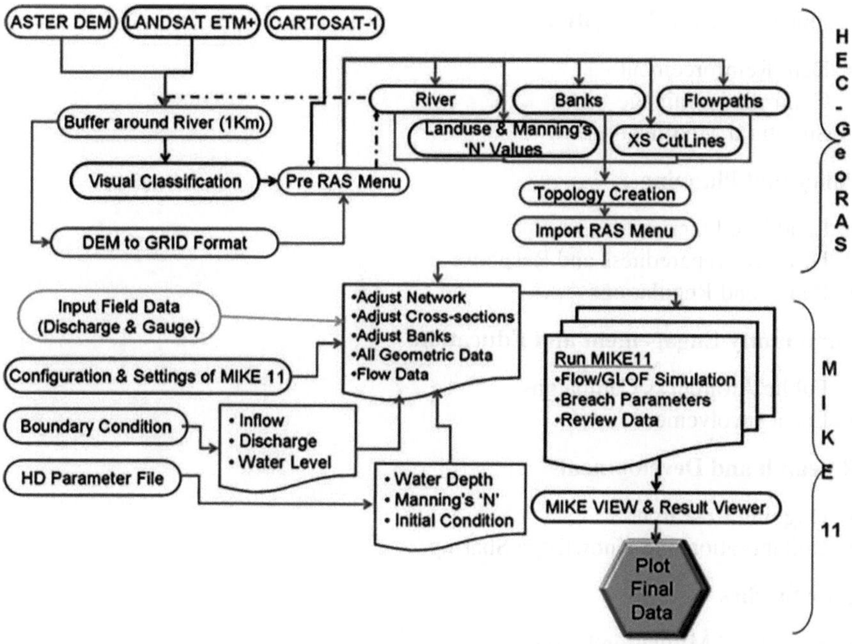

Fig. 21.2 GLOF modeling methodology figure taken from (Thakur et al. 2016)

whereas (Lamsal et al. 2016) assessed the conditions before and after the 1998 Tam Pokhari outburst in the Nepal Himalaya.

There are several studies based on software that are helpful in hazard assessment (Thakur et al. 2016) (Fig. 21.2) that use one-dimensional hydrodynamic modeling of GLOF in the Dhauliganga River. Vanderkimpen et al. (2009) used MIKE FLOOD and SOBEK 1D2D, Shrestha et al. (2010) used DAMBRK and HEC RAS flood modeling and downstream impact assessment in Sun Koshi River and Worni et al. (2013) used Mike on three lakes in the Himalayan region. Prakash and Nagarajan (2018) carried out a study in the Chandra Basin, Himachal Pradesh using time series data.

21.10 Early Warning Systems

- Automated Alerts
- Community Awareness
- Coordination and Communication

Preparedness and Mitigation

- Dam Reinforcement
- Controlled Drainage
- Structural Modifications

Policy and Planning

- Land Use Planning
- Disaster Preparedness and Response
- Policy and Regulations

Community Engagement and Education

- Public Awareness Campaigns
- Local Involvement

Research and Development

- Ongoing Research
- Collaboration and Knowledge Sharing

Case Studies

- Successful Mitigation Efforts
- Lessons Learned.

21.11 Discussion and Conclusion

In addressing the challenges posed by glacial lake outburst floods (GLOFs), a multi-faceted approach is essential. Mapping, monitoring, management, and mitigation efforts must be integrated and informed by diverse data sources and methods.

Mapping relies on remote sensing, topographic maps, and field surveys to provide a comprehensive understanding of glacial lakes and potential hazard areas. There are lots of datasets already available on different websites that can be used and analyzed, and researcher can create their datasets according to their need. **Monitoring** employs a combination of hydrological instruments, glacier movement sensors, and remote sensing to detect early warning signs of GLOFs. This can be done by closely observing the prone areas and lakes by using high-resolution satellite images or real-time monitoring using UAVs. Effective **management** strategies involve risk assessment models, community engagement, and infrastructure planning to prepare for and mitigate potential floods. Finally, **mitigation** efforts focus on engineering solutions, natural interventions, and early warning systems to reduce the impact of GLOFs and protect vulnerable communities.

By leveraging advanced technologies and adopting a holistic approach, it is possible to enhance our understanding of GLOFs and develop effective strategies to manage and mitigate their risks. Continuous research, data collection, and collaboration with local communities remain crucial in adapting to changing environmental conditions and safeguarding lives and property from the threats of GLOFs.

Acknowledgements The authors want to extend thanks to Amity University, Noida, for research facilities and logistic support.

References

Aggarwal A, Jain SK, Lohani AK, Jain N (2016) Glacial lake outburst flood risk assessment using combined approaches of remote sensing, GIS, and dam-break modelling. Geomat Nat Haz Risk 7(1):18–36. https://doi.org/10.1080/19475705.2013.862573

Aggarwal SP, Thakur PK, Nikam BR, Garg V (2014) Integrated approach for snowmelt run-off estimation using temperature index model, remote sensing and GIS. Curr Sci 106(3):397–407

Allen SK, Rastner P, Arora M, Huggel C, Stoffel M (2016) Lake outburst and debris flow disaster at Kedarnath, June 2013: hydrometeorological triggering and topographic predisposition. Landslides 13(6):1479–1491. https://doi.org/10.1007/S10346-015-0584-3

Anderson Eric A (1976) A point energy and mass balance model of a snow cover. NOAA Technical Report NWS, 19

Bajracharya SR, Mool P (2009) Glaciers, glacial lakes and glacial lake outburst floods in the Mount Everest region, Nepal. Ann Glaciol 50(53):81–86. https://doi.org/10.3189/172756410790595895

Bazai NA, Cui P, Carling PA, Wang H, Hassan J, Liu D, Zhang G, Jin W (2021) Increasing glacial lake outburst flood hazard in response to surge glaciers in the Karakoram. Earth Sci Rev 212:103432. https://doi.org/10.1016/J.EARSCIREV.2020.103432

Bhambri R, Hewitt K, Kawishwar P, Kumar A, Verma A, Snehmani Tiwari S, Misra A (2019) Ice-dams, outburst floods, and movement heterogeneity of glaciers, Karakoram. Global Planet Change 180:100–116. https://doi.org/10.1016/J.GLOPLACHA.2019.05.004

Bhambri R, Mehta M, Dobhal DP, Gupta AK, Pratap B, Kesarwani K, Verma A (2016) Devastation in the Kedarnath (Mandakini) Valley, Garhwal Himalaya, during 16–17 June 2013: a remote sensing and ground-based assessment. Nat Hazards 80(3):1801–1822. https://doi.org/10.1007/s11069-015-2033-y

Bhambri R, Misra A, Kumar A, Gupta AK, Verma A, Tiwari SK (2018) Glacier Lake inventory of Himachal Pradesh. Himalayan Geol 39(1):1–32. http://www.nrsc.gov.in/

Bolch T, Buchroithner MF, Peters J, Baessler M, Bajracharya S (2008) Identification of glacier motion and potentially dangerous glacial lakes in the Mt. Everest region/Nepal using spaceborne imagery. Natural Hazards Earth Syst Sci 8(6):1329–1340. https://doi.org/10.5194/nhess-8-1329-2008

Budhathoki KP, Bajracharya O, Pokharel B (2011) Assessment of Imja Glacier Lake outburst Flood (GLOF) risk in Dudh Koshi River Basin using remote sensing techniques. J Hydrol Meteorol 7(1):75–91. https://doi.org/10.3126/jhm.v7i1.5618

Deswal S, Sharma MC, Saini R, Dalal P, Kumar P (2020) Preliminary results of hybrid bathymetry and GLOF risk assessment for Neelkanth Lake, Lahaul Himalaya, India. Current Sci 119(9):1555

Dobhal DP, Gupta AK, Mehta M, Khandelwal DD (2013) Kedarnath disaster: facts and plausible causes. Curr Sci 105(2):171–174

Feyisa GL, Meilby H, Fensholt R, Proud SR (2014) Automated water extraction index: a new technique for surface water mapping using Landsat imagery. Remote Sensing Environ 140:23–35. https://doi.org/10.1016/j.rse.2013.08.029

Froehlich DC (1995) Embankment dam breach parameters revisited. Int Water Resources Eng Conf Proc 1:887–891

Froehlich DC (2016) Predicting peak discharge from gradually breached embankment dam. J Hydrologic Eng 21(11). https://doi.org/10.1061/(ASCE)HE.1943-5584.0001424

Fujita K, Sakai A, Takenaka S, Nuimura T, Surazakov AB, Sawagaki T, Yamanokuchi T (2013) Potential flood volume of Himalayan glacial lakes. Nat Hazard 13(7):1827–1839. https://doi.org/10.5194/nhess-13-1827-2013

Hewitt K (2014) Natural dams and outburst floods of the Karakoram Himalaya Monitoring Himalayan glaciers View project Large landslides in fluvial and glacial geomorphology, Karakoram Himalaya View project. https://www.researchgate.net/publication/238276365

Hong M, Zongchao L, Yifeng L (1992) Energy balance of a snow cover and simulation of snowmelt in the western Tien Shan mountains, China. Ann Glaciol 16:73–78. https://doi.org/10.3189/1992aog16-1-73-78

Huggel C, Haeberli W, Kääb A, Bieri D, Richardson S (2004) An assessment procedure for glacial hazards in the Swiss Alps. Can Geotech J 41(6):1068–1083. https://doi.org/10.1139/T04-053

Huggel C, Kääb A, Haeberli W, Teysseire P, Paul F (2002) Remote sensing based assessment of hazards from glacier lake outbursts: a case study in the Swiss Alps. Can Geotech J 39(2):316–330. https://doi.org/10.1139/t01-099

Ikeda N, Narama C, Gyalson S (2016) Knowledge sharing for disaster risk reduction: insights from a glacier Lake workshop in the Ladakh Region, Indian Himalayas. Mountain Res Dev 36(1):31–40. https://doi.org/10.1659/MRD-JOURNAL-D-15-00035.1

Ives JD, Shrestha RB, Mool PK (2010) Formation of Glacial Lakes in the Hindu Kush-Himalayas and GLOF risk assessment. https://doi.org/10.53055/ICIMOD.521

Jain SK, Lohani AK, Singh RD, Chaudhary A, Thakural LN (2012) Glacial lakes and glacial lake outburst flood in a Himalayan basin using remote sensing and GIS. Nat Hazards 62(3):887–899. https://doi.org/10.1007/s11069-012-0120-x

Karra K, Kontgis C, Statman-Weil Z, Mazzariello JC, Mathis M, Brumby SP (2021) Global land use/land cover with Sentinel 2 and deep learning. IEEE Int Geoscience Remote Sensing Symp IGARSS 2021:4704–4707. https://doi.org/10.1109/IGARSS47720.2021.9553499

Kumar P, Sharma S, Bahuguna IM, Singh SK, Deswal S (2024a) Polarimetric analysis of glacier surface features using RADARSAT-2 and Sentinel-2 data in part of Chandra Basin, Northwest India. Indian J Environ Protection 44(5):396–404. https://www.e-ijep.co.in/44-5-396-404/

Kumar P, Sharma S, Bahuguna IM, Deswal S, Sindhu P, Singh J (2024b) Validation of remote sensing derived ice-thickness with in-situ Lake Bathymetry in the recently vacated area of Gepang-gath glacier, Lahaul Himalaya, India. Geol Ecol Landscapes. https://doi.org/10.1080/24749508.2024.2392919

Kumar P, Sharma S, Bahuguna IM, and Partibha (2024c) Spatio-temporal change analysis of glacial lakes in Himachal Himalayas using geospatial technology. Int J Adv Technol Eng Exploration 11(117):1112–1128. https://doi.org/10.19101/IJATEE.2023.10102310

Kumar S, Bhagat V (2019) Remote sensing satellites for land applications: a review. Remote Sens Land 2(2):96–104. https://doi.org/10.21523/gcj1.18020203

Lamsal D, Sawagaki T, Watanabe T, Byers AC, McKinney DC (2016) An assessment of conditions before and after the 1998 Tam Pokhari outburst in the Nepal Himalaya and an evaluation of the future outburst hazard. Hydrol Process 30(5):676–691. https://doi.org/10.1002/hyp.10636

Lützow N, Veh G, Korup O (2023) A global database of historic glacier lake outburst floods. Earth Syst Sci Data. https://doi.org/10.5281/zenodo.7330345

Martinec J (1975) Snowmelt—runoff model for stream flow forecasts. Nord Hydrol 6:145–154

McFEETERS SK (1996) The use of the normalized difference water index (NDWI) in the delineation of open water features. Int J Remote Sens 17(7):1425–1432. https://doi.org/10.1080/01431169608948714

McKillop RJ, Clague JJ (2007) A procedure for making objective preliminary assessments of outburst flood hazard from moraine-dammed lakes in southwestern British Columbia. Nat Hazards 41(1):131–157. https://doi.org/10.1007/s11069-006-9028-7

Mergili M, Schneider JF (2011) Glacial lake outburst floods in the Pamir of Tajikistan: challenges in prediction and modelling. https://doi.org/10.5167/uzh-59538

Mool PK, Maskey PR, Koirala A, Joshi SP, Wu L, Shrestha AB, Eriksson M, Gurung B, Pokharel B, Khanal NR, Panthi S, Adhikari T, Kayastha RB, Ghimire P, Thapa R, Shrestha B, Shrestha S, Shrestha RB (2011) Glacial Lakes and Glacial Lake outburst floods in Nepal. ICIMOD, pp 1–109. 978-92-9115-193-6

O'Connor JE, Hardison III JH, Costa JE (2001) Debris flows from failures of Neoglacial-Age-Age Moraine Dams in the three sisters and Mount Jefferson Wilderness Areas, Oregon. In: USGS Professional Paper

Panigrahy S, Patel JG, Parihar JS (2012) National wetland atlas: high altitude lakes on India. In: Advances in remote sensing, vol 18, issue October. https://doi.org/10.1016/B978-0-444-52734-9.50009-8

Prakash C, Nagarajan R (2018) Glacial lake changes and outburst flood hazard in Chandra basin, North-Western Indian Himalaya. Geomat Nat Haz Risk 9(1):337–355. https://doi.org/10.1080/19475705.2018.1445663

Rana N, Singh S, Sundriyal YP, Juyal N (2013) Recent and past floods in the Alaknanda Valley: causes and consequences. Curr Sci 105(9):1209–1212

Randhawa SS, Sood RK, Rathore BP, Kulkarni AV (2005) Moraine-Dammed Lakes study in the Chenab and The Satluj River Basins using IRS data. J Indian Soc Remote Sens 33(2):285–290. https://doi.org/10.1007/BF02990047

Rouse Jr JW et al (1974) Metabolism of limonoids. Limonin D-ring lactone hydrolase activity in pseudomonas. Remote Sensing Centre Texas A (20):24–26. https://doi.org/10.1021/jf60203a024

Sakai A (2012) Glacial Lakes in the Himalayas: a review on formation and expansion processes. Glob Environ Res 16(2011):23–30

Sangewar CV, Shukla SP, Singh RK (2009) Inventory of the Himalayan glaciers : a contribution to the international hydrological programme|WorldCat.org. Geological Survey of India. https://www.worldcat.org/title/inventory-of-the-himalayan-glaciers-a-contribution-to-the-international-hydrological-programme/oclc/1102831424?referer=&ht=edition

Sharma AK, Singh SK, Kulkarni AV, Ajai (2013) Glacier inventory in Indus, Ganga and Brahmaputra basins of the Himalaya. Natl Acad Sci Lett 36(5):497–505. https://doi.org/10.1007/s40009-013-0167-6

Shrestha AB, Eriksson M, Mool P, Ghimire P, Mishra B, Khanal NR (2010) Glacial lake outburst flood risk assessment of Sun Koshi basin, Nepal. Geomatics Natural Hazards Risk 1(2):157–169. https://doi.org/10.1080/19475701003668968

Thakur PK, Aggarwal S, Aggarwal SP, Jain SK (2016) One-dimensional hydrodynamic modeling of GLOF and impact on hydropower projects in Dhauliganga River using remote sensing and GIS applications. Nat Hazards 83(2):1057–1075. https://doi.org/10.1007/s11069-016-2363-4

Townshend JRG, Justice CO (2007) Analysis of the dynamics of African vegetation using the normalized difference vegetation index 7(11):1435–1445. https://doi.org/10.1080/01431168608948946

Tucker CJ, Sellers PJ (1986) Satellite remote sensing of primary production. Int J Remote Sens 7(11):1395–1416. https://doi.org/10.1080/01431168608948944

Vanderkimpen P, Melger E, Peeters P (2009) Flood modeling for risk evaluation—a MIKE FLOOD vs . SOBEK 1D2D benchmark study. In: Flood risk management: research and practice, pp 77–84

Veh G, Korup O, von Specht S, Roessner S, Walz A (2019) Unchanged frequency of moraine-dammed glacial lake outburst floods in the Himalaya. Nat Clim Chang 9(5):379–383. https://doi.org/10.1038/s41558-019-0437-5

Veh G, Lützow N, Kharlamova V, Petrakov D, Hugonnet R, Korup O (2022) Trends, breaks, and biases in the frequency of reported Glacier Lake outburst floods. Earth's Future 10(3). https://doi.org/10.1029/2021EF002426

Walder JS, O'Connor JE (1997) Methods for predicting peak discharge of floods caused by failure of natural and constructed earthen dams. Water Resour Res 33(10):2337–2348. https://doi.org/10.1029/97WR01616

Wang X, Ding Y, Liu S, Jiang L, Wu K, Jiang Z, Guo W (2013) Changes of glacial lakes and implications in Tian Shan, central Asia, based on remote sensing data from 1990 to 2010. Environ Res Lett 8(4). https://doi.org/10.1088/1748-9326/8/4/044052

Worni R, Huggel C, Stoffel M (2013) Glacial lakes in the Indian Himalayas—from an area-wide glacial lake inventory to on-site and modeling based risk assessment of critical glacial lakes. In: Science of the total environment, pp 468–469. https://doi.org/10.1016/j.scitotenv.2012.11.043

Zheng G, Allen SK, Bao A, Ballesteros-Cánovas JA, Huss M, Zhang G, Li J, Yuan Y, Jiang L, Yu T, Chen W, Stoffel M (2021) Increasing risk of glacial lake outburst floods from future third pole deglaciation. Nat Clim Chang 11(5):411–417. https://doi.org/10.1038/s41558-021-01028-3

Part IV
Governance: Key for Environmental Sustainability

Chapter 22
Federal Discourse on Disaster Risk Funding in India: A Critical Reflection on the Concerns of Himalayan States and Union's Response

Rajendra Kumar Pandey, Sanjay Sharma, and Kalpana Agarhari

Abstract A self-plus shared government approach is offered by federalism theory. It harmonizes the disparities in geography, law, and society to create a unified political system. Federalism is best understood as the accommodation of various geographical concerns from a shared political platform, according to R.D. Dikshit and Ivo Duchacek, when viewed through the geographical lens of geographical variety. This theoretical dimension of geographical federalism has been applied while analyzing the concerns of the Himalayan states for differential treatment for disaster risk funding. Indian Himalayan Region (IHR) comprises 13 states and union territories and faces a variety of hazards and degrees of disaster vulnerability. Based on their geographical profile, vulnerability also increases due to poor availability of economic resources in comparison to the other non-Himalayan states. In addition, the economic cost of their contribution to providing a carbon sink for the industrial development of India. The IHR requests a new and higher level of funding for disaster risk management in light of all these variables. However, the current disaster risk funding that has evolved from the Second Finance Commission provides a small relief to the Himalayan states by reducing their contribution to the State Disaster Response Force (SDRF) from 25 to 10 percent but does not provide any other relief. The present research study primarily with empirical inputs from the region and available primary and secondary data applies a mixed-method approach to analyze the discourse between the Himalayan states and the central government for disaster risk funding from the theoretical perspective of federalism.

R. K. Pandey
Deshbandhu College, Delhi University, Delhi, India
e-mail: rkpandey.du@gmail.com

S. Sharma
Department of Politics & International Studies, Pondicherry University, Pondicherry, India
e-mail: sanjaysharma@pondiuni.ac.in

K. Agarhari (✉)
Department of Political Science, Kumaun University, Nainital, India
e-mail: kalpanaagrahari@kunainital.ac.in

S. C. Rai (ed.), *Himalaya: Mountains of Destiny*,
https://doi.org/10.1007/978-3-031-94905-0_22

Keywords Disaster risk funding · Indian Himalayan Region · Federal discourse · Disaster mitigation · Region-specific disasters

22.1 Introduction

The Himalayas extend from 28.5983° North to 83.9310° East. The Himalayas are spread over 600 km in width and around 2500 km in length. India has a large part of the Himalayas. The Indian Himalayan Region (IHR) encompasses 13 Indian States and Union Territories, including the former state of Jammu and Kashmir (now merged into two union territories of J&K and Ladakh), Himachal Pradesh, and Uttarakhand, which are considered western Himalayan states. Furthermore, the north-eastern states of Arunachal Pradesh, Assam, Meghalaya, Manipur, Mizoram, Sikkim, Nagaland, Tripura, and West Bengal (specifically the Darjeeling and Kalimpong regions) fall under the category of eastern Himalayan Region (NITI Aayog, n.d.). The IHR is home to about 50 million people and is characterized by a variety of demographics as well as diverse political, social, economic, and environmental systems. It divides South Asia from the northern portion of the Asian continent. The Himalayas are known as a "water tower of the world" and a lifeline due to their climatic regulating position; they nurture environmental biodiversity (Singh 2006). The Himalayan range is the youngest and most volatile in the global mountain system, making it highly susceptible to natural disasters. The Himalayan states have region-specific concerns and demand financial support from the Union government. Disaster risk funding in India has evolved from the Second Finance Commission's "Margin Money Scheme" to its present form of the "State Disaster Response and Management Fund" (SDRMF) and "National Disaster Response and Management Fund" (NDRMF). A close analysis of the disaster risk funding in Indian federalism reveals that states are grappling with many challenges like the poor quantity of funding, meager relief to region-specific disasters, which are not covered in the centrally notified list of disasters, funding not based on vulnerability mapping of the states, etc. States periodically put forth their demands to the union government for improvements in disaster risk funding. Instead, the union government claims that its role in disaster governance is largely supplementary as it acts when disaster is beyond the coping capacity of the states. In this federal discourse, the Himalayan states in India need special attention as these are highly vulnerable to disasters because of their geological conditions. The 6th Assessment Report of IPCC (2022) and Ministry of Earth Science Report on Assessment of Climate Change over Indian Region (2020) claim that soon, the Himalayan region will face increased numbers of climate change-related disasters, like flash floods, extreme weather conditions and melting of glaciers, etc. The Himalayan states are in double jeopardy from their geological-specific disasters. On the one hand, the Himalayan states are financially weak because of their geology, leading to large-scale migration and generating less revenue in comparison with the rest of the states. On the other hand, the Himalayan states are providing large green cover working as the non-incentivized carbon sink to the carbon produced by the industrial states. It is under this background that this study claims that demands of the Himalayan states for disaster risk funding need differential response and special attention from the Union government. The present chapter is an exploratory research work. It makes a modest attempt to deconstruct the political economy of disaster management in a broad framework of federal discourse with special reference to the demands of the Himalayan states. The present chapter is divided into four major segments: (i) Terrestrial theory of federalism, (ii) Disaster Financing in India: An Evolving Mechanisms, (iii) Concerns of the Indian Himalayan Region, and (iv) Response of the Union Government to the demands raised by The Indian Himalayan Region (IHR).

22.2 Terrestrial Theory of Federalism

Federalism is a theory of peace. It accommodates and reconciles the social, cultural, and spatial diversity into one political union. It is a political reflection of the existing diversity in society and it ensures that diversity can be retained with political coming together. There are multiple theoretical orientations of this federalism namely the theory of peace, sociological theory, and legal theory. However, the spatial or terrestrial theory of federalism needs special mention as federalism goes beyond the conventional areas of society and also demands the accommodation of terrestrial diversity in federalism.

Terrestrial theorist states that sheer recognition of social and cultural diversity is not sufficient to grasp the idea of federalism while the territories need to be converted into self-governing federal units. Hence, the requirement of territorialization should not be avoided. There are two important advantages of terrestrial federalism first administration can be decentralized and second through federal divisions frictions among the stakeholders can be reduced and federalism can become a source of peaceful association. There are two major contributors of the terrestrial theory of federalism namely R.D. Dikshit and Ivo D. Duchacek. Duchacek in his latest work "The Territorial Dimension of Politics" maintains that terrestrial indoctrination is an important concept as it is useful to understand the territorial distribution of power. In addition, he opines that all territories have different identities like symbols, anthems, personalities, and culture which helps in territorial socialization. He strongly believes terrestrial arrangements of federalism create multiple centers of power and governance that relieve the federal government and also assist in achieving the principle of self-governance. Finally, he also admires the fact that territorialization also results in an improved mobilization of resources, local planning, execution, and communication (Duchacek 1970). Nevertheless, territorialization of society and polity has certain benefits first interest articulation takes place within territorial divisions; second political disputes can be easily negotiated at the territorial levels; and third, it also enhances the decentralized democracy that is lacking in other political setups.

R.D. Dikshit, in Development in Political Geography, has also counted the advantages of the territorial indoctrination of federalism. He maintains that there are three major benefits of territorialization in federalism first it creates one territorial-based social, political, and economic autonomy to territories within the larger integrated federal territorial setup; second it also creates a sense of cultural locality as a different cultures come together in a territory creating a geography-based cultural identity; and third also provides a base for one of the federal requirements that are the dual government. In addition to counting these benefits, Dikshit has also made a distinction between the territorial division of federal societies with other forms of societies such as military societies, industrial societies, lower societies, and higher societies. He analyzed that while integration is achieved in these societies through compulsion, voluntary cooperation, and mechanical and organic societies territorial indoctrination provides the organization of societies (Dikshit 1997).

Institutionalization of federal diversities in the form of territorial indoctrination admits the territorial division of power. The territorial federal setup recognizes the fact that federal units have distinct identities and their denial and discomfort into merging into a single common identity. The territorial-based division of power also recognizes the fact that sovereignty is shared among all the territorial federal units and no one can impose on each other by demanding complete unity or secession from the federal union. In other words, both have a mutual interdependence and interact with the emotions of brotherhood. Hence, a federal constitution should be looked upon as a continuum of discourse among the federal governments and federal units and the territorial dimensions of federal diversity cannot be looked over (Duchacek 1970).

22.3 Disaster Financing in India: An Evolving Mechanism

Disaster Management is a complex activity that involves multiple responsibilities like preparedness, capacity building, rehabilitation, risk reduction, etc. To fulfill these responsibilities, a significant amount of resources is required. As a federal country, the center and state share financial resources in India. But, the responsibilities of the state are much more than the resources in hand, and that is why they have to ask for assistance and aid from the center, which in turn takes care of the majority of responsibilities like disaster management. Disaster financing in India has evolved over some time, which will be discussed in the following sections of this chapter.

22.3.1 Margin Money Scheme (MMS)

The "Margin Money Scheme" was the first of many schemes focusing on making disaster management more efficient and effective. The scheme was introduced by the "Second Finance Commission (1957–1962)." This scheme entailed a particular share of the annual revenues of a state to be kept aside, which can be utilized in times of natural disaster. The commission also recommended that all the states should set up a separate fund in which the fixed amount for each state can be deposited on an annual basis. Apart from the new scheme, the commission also suggested that states should make the ongoing schemes more effective by enhancing their scope. In case the balance in the fund, if not utilized, should be invested in cashable government securities (Reddy and Reddy 2018).

The scheme remained prominent in developing an effective disaster financing structure in India. The next few commissions till the ninth finance commission showed faith in the MMS scheme with some minor changes. For instance, the Fifth Finance Commission evaluated, based on the expenditure of states on disasters from 1957–58 to 1964–65, the amount of the margin money fund. Based on the review, the commission proceeded with a raise of twenty-five percent in the existing fund (Reddy and Reddy 2018).

22.3.2 Center's Contribution to the Margin Money Scheme

In case of a disaster, the responsibilities are shared by both the central and state governments. The central government also contributes a particular amount as relief expenditure to the Margin Money Scheme. In case a state is affected by a disaster, based on the compliance of the memorandum by the state, a central team visits the affected state. If the resources required to manage and mitigate the disaster exceed the state's margin fund, then the central government adds a supplementary share in the fund. Out of the entire expenditure made to deal with the disaster, the central government is responsible for providing 75 percent of the amount and the remaining 25 percent has to be managed by the state government from its financial resources. The central government's share further splits into the amount given as a loan, 67 percent, and the amount given as a grant, 33 percent (Sharma 2023).

22.3.3 Calamity Relief Fund/State Disaster Response Fund

The Ninth Finance Commission was set up in 1987. In its report, the commission recommended a major change in the disaster funding structure in India. The commission replaced the ongoing Margin Money Scheme with the "Climate Relief Fund (CRF)" to make disaster management more efficient (Salve 1989). The Margin Money Scheme was operational from the Second Finance Commission to the eighth finance commission. FC-II recommended that a margin from the state's annual revenue be added to a separate fund, which could be used in case of disaster. The new model, along with providing greater autonomy to the states, assigned the states greater responsibilities and aimed at making them more accountable. The state's capacity to deal with disaster was strengthened by providing more adequate means and better facilities which allowed the states to use the funds as per their needs. FC-IX calculated the CRF fund after considering an average upper limit of the expenditure that also included money margin, advance plan assistance, and special assistance from the union, along with the share accumulated by the state in ten years. After all calculations, the commission fixed the amount of Calamity Relief Fund at Rs. 804 crores per annum. Later on, the amount for the period 1995 to 2000 was fixed at Rs. 6304.27 crores by the Tenth Finance Commission (India, Ministry of Finance 2004).

The Calamity Relief Fund, part of the public account of the state, is utilized to meet the cost of expenditure incurred by "providing immediate relief to victims of the cyclone, drought, earthquake, fire, flood, tsunami, hailstorm, landslide, avalanche, cloud burst, and pest attack." As per the new structure, the center contributes 75 percent of the fund while the remaining 25 percent is managed by the respective state. The fund is managed by a state-level committee under the supervision of the Chief Secretary. While releasing the funds to the state, the committee considers past expenditures by the state on calamity relief measures. The scheme set the tone for disaster financing structure in the country after the Margin Money Scheme. However, the "Thirteenth Finance Commission" recommended the integration of the Climate Relief Fund with the "State Disaster Response Fund" of respective states. In deciding the finance of the SDRF, the Fourteenth Finance Commission followed the footsteps of previous commissions. It took into consideration the amount utilized, from 2006–07 to 2012–13, by the state on disaster relief to determine the SDRF for each state. Based on the evaluation of funds utilized from 2006–07 to 2012–13, the commission fixed an aggregate amount of Rs. 61,219 crores for the period 2015–20 combined for all SDRFs.

22.3.4 National Calamity Contingency Fund/National Disaster Response Fund

In case of disasters that are rare and severe, the Finance Commission, keeping in mind the resources that states have to deal with them, recommends a majority share in financial aid from the union. Disasters vary in terms of impact and damage they cause; thus, the seventh and eighth Finance Commissions stressed additional financial assistance from the union to the states that are affected by severe disasters. The issue of severe calamities and the mechanism to deal with them remained prominent for many years. The Ninth Finance Commission, to reduce pressure from the states, argued that severe disasters should be managed by the union government, and all the necessary expenditures should be met by the center (Salve 1989). Later on, the Tenth Finance Commission recommended a separate mechanism to deal with severe calamities. Rather than fixing a set of calamities as severe, it argued for a case-to-case analysis of calamities to measure the level of severity that would be based on the state's capability and resource availability to deal with the problem, level of relief assistance required and any other option available to the state (India, Ministry of Finance 2004). The commission further recommended the establishment of the "National Fund for Calamity Relief" that would be managed and supervised by the "National Calamity Relief Committee".

However, some of the major changes in the basic structure of the "National Calamity Contingency Fund" were introduced by the Eleventh Finance Commission. The salient features of NCCF are as follows:

1. The fund has a balance of Rs. 500 Crore, replenished through "National Calamity Contingency Duty" on consumer items like tobacco products, mobile phones, etc.
2. The fund is maintained within the public account of the "Government of India."
3. The Fund is supervised by a national-level committee that includes the Home Minister, Deputy Chairman of the Planning Commission, Agriculture Minister, and Finance Minister.
4. The fund can only be utilized for immediate relief activities and rehabilitation, not for the reconstruction of damaged infrastructure and assets (FC-XIII 2010).

The Thirteenth Finance Commission argued for merging the 'National Calamity Contingency Fund' with the 'National Disaster Response Fund' (India, Ministry of Finance 2010). Apart from the above-mentioned funds, Additional Central Assistance (ACA) is provided to the states to help them in post-disaster reconstruction works. A total sum of Rs. 645 crores was allotted to ten states in 2008–09 for post-disaster reconstruction. The union government supplements the cost of relief and reconstruction works through need-based applications of many centrally sponsored schemes like MNREGS and Indra Awas Yojana. The centrally sponsored schemes (CSS) are significant as they contribute around 10per cent of the cost of relief works. The Fourteenth Finance Commission recommended that the NDRF, wholly, should be financed by levying a cess on selected items. As a result, a total amount of Rs. 21,295.89 crores was collected through the "National Calamity Contingent Duty (NCCD)" from 2002 to 12, and, in the same period, a total sum of Rs. 23,346.92 crores was made available to the states affected by disasters. Apart from the amount collected through the NCCD, the government distributed Rs. 2051.03 crores through budgetary resources (India, Ministry of Finance 2015).

22.4 Concerns of the Indian Himalayan Region

In the past few decades, the vulnerability of this mountain region to natural disasters has increased manifold due to anthropocentric factors. The Himalayan states have very distinct features, not only in terms of their natural surroundings but also in terms of disaster risks. Recently, in February 2024, Mr. K.R. Suresh Reddy, Member of Parliament, sought the attention of the Rajya Sabha through his unstarred question no. 666, which queried the preparedness of the government because 59 percent of the Indian landmass is vulnerable to earthquakes (Government of India, Ministry of Earth Sciences 2024).

Disasters also have an impact on these states' tourism-based economies. On March 24, 2022, Dr. Santanu Sen asked in the Rajya Sabha about the effect of disasters on the tourism sector in Himalayan states such as Jammu and Kashmir and Uttarakhand (Government of India, Ministry of Tourism 2022). On March 16, 2023, Mr.

G. Kishan Reddy, the then Minister of the Ministry of Development of the North-Eastern Region, in response to a question in the Rajya Sabha related to the disaster vulnerability of Himalayan states, termed Assam "*the most vulnerable to climate change disasters among all Himalayan states.*" His response was based on the report titled "*Climate Vulnerability Assessment for the Indian Himalayan Region using a Common Framework*" *(2018–2019)*, submitted to the Department of Science and Technology, Government of India, and conducted by the Indian Institute of Science Bangalore, IIT Guwahati, and IIT Mandi. Under this project, a state-level vulnerability assessment for the IHR states was conducted using a Tier 1 methodology based on four broad indicators: "*socio-economic status (demographic status and health), the sensitivity of agricultural production, forest-dependent livelihoods, and access to information services and infrastructure.*" These indicators were chosen to highlight the inherent social, biophysical, and social systems of the twelve IHR states. The assessment reveals that Assam (0.72) and Mizoram (0.71) hold the highest vulnerability index, "*followed by Jammu and Kashmir (0.62), Manipur (0.59), Meghalaya and West Bengal (both 0.58), Nagaland (0.57), Himachal Pradesh and Tripura (0.51 each), Arunachal Pradesh (0.47), and Uttarakhand (0.45).*" With an index of 0.42, Sikkim is the least vulnerable state.

However, the report emphasizes that vulnerability is a relative measure, and there may be several inherent drivers of vulnerability that require attention. Sikkim and Uttarakhand ranked 2 (low vulnerability) "*on a vulnerability scale of very low, low, moderate, high, and very high; Arunachal Pradesh, Tripura, Himachal Pradesh, Nagaland, West Bengal, Meghalaya, Manipur, and Jammu and Kashmir ranked 3 (moderate vulnerability); and Mizoram and Assam ranked 4 (high vulnerability).*" Not a single state was classified as extremely low or extremely vulnerable (Department of Science and Technology 2018).

22.4.1 Drivers of Disaster Vulnerability in IHR States

The geographical, geological, and physical characteristics of IHR make it vulnerable to all major natural hazards, e.g., landslides, droughts, floods, earthquakes, cyclones, etc., as well as other human-induced disasters like forest and urban fires, flash floods, road accidents, and so on. For example, Assam, with its extensive river network, is highly susceptible to flooding and erosion. The Brahmaputra and Barak Rivers, with over 50 tributaries, cause annual flood devastation, resulting in an average annual loss of Rs. 200.00 crore (Government of Assam, n.d.). Similarly, the geographical and geological characteristics make Mizoram hypersensitive to all major natural and anthropogenic hazards, including floods, droughts, cyclones, landslides, and earthquakes, as 71percent of its total territory falls under very high to moderate hazard zones (Government of Mizoram 2020).

Likewise, the states of Jammu and Kashmir, and Uttarakhand are vulnerable due to their "*low road density, high migration rate, low livestock-to-human ratio, and low percentage area under horticulture crops.*" A spike in forest fire incidents in

Uttarakhand, for example, from 922 in 2002 to 41,600 in 2019, severely affects the forest structure and environment (Dimri and Farswan 2023) Ten people have died in Uttarakhand's continuing forest fire season as of June 13, 2024, when four frontline forest workers were burned alive in a wildfire (Azad and Nagarkoti 2024). Similar to these states, Manipur, Meghalaya, Sikkim, and other IHR states also embody a composite form of vulnerability, exacerbated by explicit socio-economic factors such as "*low per capita income and a low area under forests per 1000 households*" (Department of Science and Technology 2018).

However, the primary cause of concern is climate change, which is adversely affecting all Himalayan states equally. Numerous studies conducted worldwide have showcased strong causal evidence of climate change and disasters (IPCC 2023). Climate change forces all aspects of mountainous life, including socio-economic and biophysical systems like wetlands, rivers, mountains, forests, agriculture, and animal husbandry, to face its wrath. Another study cited by Anamika Barua elaborates that over the past five to six decades, there has been an increase in extreme warm weather events, a decrease in extreme cold events, and a rise in extreme weather and temperature-based indices (both minimum and maximum) in the Hindu Kush region, raising the high vulnerability of the IHR states (Barua et al. 2021).

22.4.2 Demands of the Himalayan States

The Indian Constitution does not explicitly name the government in charge of disaster management, and following Union List Entry no. 97, *any matter not enumerated in List II (state list) or List III (concurrent list)* falls under the jurisdiction of the Union government (Bakshi 2001). Even though it is the State government that primarily has to bear the disaster management responsibility, the central government plays a secondary and supportive role by providing necessary financial, technical, and material support. Even the huge burden of post-disaster relief, response, and rehabilitation also has to be borne by concerning State governments and district administration. It is important to note here that almost all of the budgetary allocations for the management of disasters come from the allocations given to the States annually for the five-year fiscal cycle. The Finance Commission (FC) of India, a constitutional body, proposes the allocation of these funds. (Bakshi 2001). Thus, the recommendations of the several FCs have played a key role in shaping the entire framework for financing disaster management in India. However, the relief-centric approach of FCs has been a major concern for states, which had been centered on the predominant principle that in the cases of calamities, providing immediate help for temporary livelihood to victims, is the only duty of the states and thus the financial allocation of states must be based only on this requirement. From the 2nd FC to the commencement of the 13th FC, this approach has been dominant in fund allocation to states, which ignored the greater role of states in providing compensation and restructuring the lost properties to the pre-disaster situation, and it was supposed to be executed by the States from their funds or with money from various Central government plan schemes. States have flagged their concern that this approach has hindered the strategic rationale for holistic disaster management and the measurement of resource gaps of the states to fill the gap (Chakrabarti 2009).

The augmentation of the State Disaster Response Fund (SDRF) allocation and the revision of its criteria have been among IHR states' key demands. Most states had demanded an increase in the State Disaster Response Fund (SDRF) allocation as well as a review of the allocation criteria to take into account risk and vulnerability parameters instead of the past expenditure-based criteria. The argument was that this approach would enable the states with higher disaster risks to get the appropriate funding for better preparedness. The Thirteenth Finance Commission considered these demands and finally allocated grants for capacity building, but it was derailed by the XIV Finance Commission. However, it is a matter of satisfaction that the XV Finance Commission, in its report for the five-year term (2021–22 to 2025–26), has embraced an unconventional methodology for deciding how to allocate funds for disaster management to states (Sharma 2022). This move represents a substantial change from the previous approach because it considers the three criteria demanded by states for fund allocation: *"capacity (as reflected through past expenditure), risk exposure (area and population), and hazard and vulnerability (disaster risk index)"* (Govt. of India, Ministry of Finance 2019).

The Himalayan states, which generate lower revenue mainly due to their geological fragility and climatic conditions, have raised several specific demands to mitigate the challenges of disasters faced by them. The states have reiterated many a time that disaster risk funding is scant to meet their needs, and they have appealed to successive Finance Commissions to expand the scope of disaster risk funding. In this context, the demands made by Himalayan states to various Finance Commissions, including the Fifteenth Finance Commission, underscore the need for enhanced financing and operations to mitigate the effects of disasters; however, many of these demands are the same raised on various occasions, but most of them awaiting to fructify.

The most prominent demand from Himalayan states is the modification of the existing 90:10 pattern of contribution to SDRF for IHR and North-Eastern states, as recommended by the XIIIth Finance Commission. They expect the center to shoulder the full financing of the SDRF, as their smaller economies make it difficult for them to make a matching contribution, thereby limiting their potential for disaster response (Govt. of India, Ministry of Finance 2019). However, the Union Government has been reluctant to accept this demand, mainly due to the fund crunch at its end, the principle of shared federal responsibilities, and the fear that it might encourage the over-reliance of states on central funds without accountability and lead to potential misuse.

Another major demand from the hilly and remote states reflects their suspicion of national-level parameters, which they term "one size fits all" criteria, not suitable for specific local needs. Therefore, states and State Disaster Management Authorities (SDMAs) must have more flexibility in disbursing funds to improve relief delivery (Govt. of India, Ministry of Finance 2019). Mizoram presented a memorandum to the Thirteenth FC, stating that the existing CRF/NCCF norms in northeast India were insufficient due to inaccessible terrain and climatic conditions, necessitating the consideration of separate norms (Chakrabarti 2009). This demand underscores the need to allow states to customize norms within certain guidelines to enhance the flexibility of disaster mechanisms.

These states have emphasized the need to regularly review and update the items and the qualifying parameters enlisted by the SDRF and NDRF, with input from disaster management experts and stakeholders, to ensure comprehensive coverage of all disaster-related expenses (Govt. of India, Ministry of Finance 2019). In 2009, Himachal Pradesh stressed that the hilly state is badly struggling to obtain reimbursement for expenditure on infrastructure damage, roads, and drinking water supply schemes and requested the FC to recommend a review of approved expenditure items in consultation with state governments, taking into account local needs under the Central Refund Scheme (CRF) (Chakrabarti 2009).

All the states, including IHR, have unanimously demanded the timely and transparent release of funds and assistance under the NDRF, as delayed funding and assistance in the past have been the main reasons for hampering prompt response and recovery efforts in times of disaster. This requires a better coordination process between state and central agencies (Govt. of India, Ministry of Finance 2019). Arunachal Pradesh specifically brought up this issue with the Thirteenth Finance Commission, and Mizoram requested the immediate release of assistance on an ad-hoc basis, following the recommendations of the State Level Team. It also suggested that the Inter-Ministerial Group finalize the assistance quantity based on the Central Team's assessment during their visit (Chakrabarti 2009).

Furthermore, these states have requested more resources for infrastructure restoration and recovery, which are significant parts of the post-disaster management strategy and require substantial funding. According to the memorandum presented to the Thirteenth FC, the Assam government stated that the state's losses outweighed the respite offered during 2007–08, and the state government was forced to spend an additional 198.62 crores of its funds. Manipur also stated that the estimated values of damaged assets are much higher than the scale of measurement prescribed for compensation (Government of Manipur 2020). No doubt increased CRF budget allocations and strict monitoring mechanisms may accelerate recovery and improve long-term resilience.

Uttarakhand gave a memorandum to the Thirteenth Finance Commission in 2009, drawing attention to the problem of natural disasters, including landslides and flash floods in hill states, which frequently result in the permanent destruction of mountainous habitations or make them unsuitable for human habitation. As a result, the rehabilitation of victims in alternative safe locations necessitates proper state funding arrangements. The proposal also recommended authorizing the state governments to use at least 1per cent of CRF funds for resource mobilization and expert hiring, to respond swiftly to disasters. It also demanded extending restoration work time limits in remote hilly areas from 45–60 days to 6 months due to difficult terrain and from 30–45 days to 3 months in plains (Chakrabarti 2009).

Likewise, the Tripura government has been consistently raising the issue of the Calamity Relief Fund guidelines that if funds deposited under the Public Account cannot be spent within the financial year, it necessitates a new release from the

Finance Department, causing accounting issues. Thus, simplifying the system could facilitate the easy flow of funding. Furthermore, Sikkim and Tripura both have raised the concern that under the CRF norms, states are bound to undertake only temporary and small restoration works, but such restoration works have no meaning during the monsoon. Thus, semi-permanent works should be permitted, and the guidelines may be modified to accommodate such expenditures, particularly in the states with hills (Chakrabarti 2009).

India, being the seventh largest country in the world, is susceptible to a wide range of calamities. However, the Disaster Risk List, by the Finance Commission, covers only 13 certain disasters, mostly of national character, and avoids state-specific disasters like lightning, snakebite, bamboo blossoming, man-wildlife conflicts, erosion of riverbanks and the sea coast, etc. Furthermore, the SDRMF's disaster risk funding does not adequately cover man-made disasters. As a result, states are never satisfied with the list of disasters provided by the Finance Commission and have been demanding the inclusion of local disasters, e.g., elephant attacks, lightning, heatwaves, and public health disasters like the pandemic, COVID-19, may qualify for funding assistance from the NDRF and SDRF, as well as the allocation for state-specific disasters, must be increased from 10percent to 25 percent (Govt. of India, Ministry of Finance 2019).

The West Bengal government has been insisting on recognizing erosion and lightning as natural calamities due to their annual damage to property and infrastructure. Uttarakhand has raised the issue of extreme cold conditions that often cause farmers to lose crops, necessitating the implementation of CRF norms to cover all disasters under the DM Act, of 2005 (Chakrabarti 2009).

Other key issues that IHR states presented to the XV FC included demands for funding for the Mitigation Fund by the Union, separate allocations for preparedness measures, and increased technical support to reduce reliance on the National Disaster Response Force and armed forces. They also demanded the restoration of capacity building grants, previously terminated by FC-XIV, initiated by the FC-XIIIth (Govt. of India, Ministry of Finance 2019). The Himalayan states are demanding special attention for evolving new paradigms of disaster management based on mitigation and capacity building that could balance economic interests and ecological conditions in the region. In recent years, the conditions of climate change have intensified the states' vulnerabilities, compelling them to advocate for a more comprehensive approach to disaster management. The IHR States specifically requested a special and separate climate fund in their memorandum to the Fifteenth FC. They also requested adequate funding for mitigation measures as a long-term strategy to counter the disasters induced by climate change. The demands of the IHR states highlight the urgent need for a comprehensive strategy to improve disaster management financing and operations. While many of these demands seem justified as they aim for better preparedness, response, and restoration in crises, they also reflect the multifaceted challenges in disaster management in the Indian Federation.

22.5 Response of the Union Government

The Indian Himalayan Region (IHR) raised a wide range of concerns and most of their concerns have been addressed by various finance commissions, particularly by the Fifteenth Finance Commission and some are still to be addressed. The response of the Union government can be categorized into three broad categories namely (i) demands for creating a disaster mitigation fund; (ii) amending the methodology for disaster funding, and (iii) inclusion of Himalayan-specific disasters in the notified list of disasters. The following section analyzes these three categories.

22.5.1 Union's Response to Disaster Mitigation Fund

The Fifteenth Finance Commission, while addressing the demands of the IHR for the creation of the mitigation fund, reviewed the existing structure and framework of disaster risk funding and found a major fallout. The existing funding was specifically oriented toward the response and missed out on enabling the wider approach of adaptation, reconstruction, preparedness, and recovery. The Thirteenth Finance Commission, highlighting the same issue, constituted the Capacity Building Fund (Govt. of India, Ministry of Finance 2010). While the CBF was not a mitigating fund as per the Disaster Management Act, it provided funds to states for preparedness. The CBF was later taken down by the Fourteenth Finance Commission, and also, no mitigation fund was recommended by the commission. Rather than constituting a dedicated mitigation fund, the commission suggested alternate mechanisms to carry out mitigation functions like the state districts should partner with private players to ensure funding for mitigation-related activities. The issue of mitigation funds was also addressed by the Supreme Court in 2016, when in the "Swaraj Abhiyan vs. Union of India & others" the court stated-

> Section 47 of the DM Act provides for the constitution of a National Disaster Mitigation Fund for projects exclusively for mitigation which, as mentioned earlier, means measures aimed at reducing, inter alia, the risk of a disaster or threatening disaster situation. Although the DM Act has been in force for more than 10 years, the National Disaster Mitigation Fund has not yet been constituted. There is, therefore, no provision for the mitigation of a disaster. … Risk assessment and risk management also appear to have little or no priority as far as the Union of India and the State Governments are concerned…. a National Disaster Mitigation Fund is required to be established. Unfortunately, no such Fund has been constituted to date. Accordingly, we direct the Union of India to establish a National Disaster Mitigation Fund within three months from today (Writ Petition (Civil) No. 857 of 2015).

Apart from the demands of IHR, the Fifteenth Finance Commission received views and opinions from the ministries and various other institutions of the central government regarding the creation of mitigation funds. As per the states, mitigation has emerged as a significant element of disaster management, and it has become very difficult for states to carry out mitigation measures due to insufficient funds. Thus, states want the central government to enable funding that can be solely used for mitigation measures. Along with the states, the NDMA has also argued for the creation of national and state disaster mitigation funds for the effective implementation of risk-reducing measures. The demand for the creation of NDMF has also been supported by the Ministry of Finance. The National Institute of Disaster Management has also argued in favor of establishing a separate funding mechanism that can be used to ensure research, capacity building, and spreading awareness. The FC-XV, along with input from various institutes and stakeholders, also sought expert opinion. For the same purpose, a study was commissioned under the "United Nations Development Program" in collaboration with NDMA for important policy suggestions. The study opined that disaster preparedness needs to be visible in measurable outcomes and the funds assigned should be used with accountability by the concerned authorities. The disasters that are funded by the government, as stated by the FC-XV, fall in the category of **"contingent liability."** Contingent liability comes into the picture in the case when an unforeseen probable event takes place. It is an obligation of the government, in case of a disaster, to deal with the unforeseen event. Contingent liabilities cannot reduce the risk of disaster as they only meet post-disaster requirements. After analyzing various aspects, the FC-XV argued in favor of creating a separate mitigation fund that can be used to avoid potential disasters to a certain level. Mitigation, as stated by the FC-XV, is "lessening or minimizing the adverse impacts of a hazardous event." Mitigation includes both structural and non-structural measures. Structural measures vary in terms of scale and intensity. They include both large projects like river embankments to control floods and small community-based actions and interventions to minimize the threats of disaster to a certain level. As per the FC-XV, the mitigation fund should not be utilized on large-scale projects, and the former should be accomplished through regular development plans. FC-XV stated "We are of the view that the mitigation fund created should be used for those local level and community-based interventions which reduce risks and promote environment-friendly settlements and livelihood practices" (XV Finance Commission, 2020, p. 235). The commission mainly focused on the community-led soft measures for which the mitigation fund can be used by the government. An indicative list of sanctioned mitigation activities is given in Annexure I.

These circumstances and background led the FC-XV to recommend the establishment of mitigation funds at both the national and state levels. A hybrid model of funding was created by the commission to cater to the response and mitigation requirements related to disasters. Instead of going for two separate funds, the FC-XV recommended the inclusion of the fund assigned for mitigation with the already existing national and State Disaster Response Fund. As a result, the "National Disaster Response and Mitigation Fund" and the "State Disaster Response and Mitigation Fund" came into existence.

A total sum of Rs. 68,463 crores was proposed by the commission for the NDRMF and Rs. 160,153 crores for the SDRMF for the period from 2021 to 2026. The share for mitigation and response-related measures was set at 20 percent and 80 percent, respectively (XV Finance Commission 2020, pp. 238–39). The existing NDRF was funded by the 'National Calamity Contingency Duty (NCCD)', but it has become insufficient to cater to the funding needs as most of the commodities are covered under the Goods and Service Tax. So, apart from NCCD, the commission also recommended an 'annual budgetary provision for NDRF'. To keep a check on the inflated and unnecessary demands by the states, the commission argued that central assistance through NDRF and NDMF should be based on a cost-sharing basis. For the initial demand of Rs. 250 crores, the state should contribute 10 percent of the amount, for the assistance of Rs. 500 crores states have to contribute 20 percent and if any demand exceeds Rs. 500 crores then the states have to contribute 25 percent of the total amount. If the SDMRF and NDRMF are insufficient to cater to the assistance needs of the central and state governments, then they should locate another channel of funding like crowd-funding, corporate social responsibility, and reconstruction board.

22.5.2 Region-Specific Mitigation Funds

Apart from the general funding mechanism, the commission also recommended four sectoral and region-specific amounts of Rs. 5950 crores out of the NDMF budget of Rs. 13,693 crores for the tenure from 2021 to 2026. The amount is allocated recognizing the regional variations in terms of their vulnerabilities like urban flooding, drought, and landslides that persist over some time. Long-term projects are required on the part of the government to address the region-specific vulnerabilities. Apart from these long-term projects, the commission also argued in favor of community-level interventions, which can also prove effective; thus, this specific amount is recommended under NDMF.

22.5.3 Managing Seismic and Landslide Risks in Ten Hill States

The Himalayan region is prone to constant seismic activities and landslides as well. The Himalayan states, including Himachal Pradesh, Uttarakhand, and all north-eastern states, as per the seismic map of India, fall under the zone IV and V of seismic activities (Jena et al. 2021). Being the youngest mountains in the world, the Himalayas are highly prone to earthquakes. While earthquakes pose a major threat, flooding, excessive rain, and development activities increase the risk of landslides in the reason. Over the years, landslides and earthquakes have caused damage to human resources and infrastructure as well. As a result, the commission pressed for effective mitigation plans for the Himalayan states to address the risk of landslides and earthquakes. To ensure the mitigative measures, the commission recommended a sum of Rs. 250 crores for Himachal Pradesh and Uttarakhand at Rs. 50 crore per annum. For all the north-eastern states, the amount allocated is Rs. 250 crores, which

Table 22.1 Allocation of funds for seismic and landslides risk in the Himalayan region

S. No.	States	Annual allocation	Total allocation (2021–2026)
1.	Himachal Pradesh	50	250
2.	Uttarakhand	50	250
3.	All North-Eastern States		250
Total			750

Source Main Report of Fifteenth Finance Commission 2021–26

can be increased if the allocated amount is utilized and exhausted before the five-year period ends (Table 22.1).

22.5.4 Mitigation Measures to Prevent Erosion

Coastal and River erosion is increasing year by year and has left an evident socio-economic impact on society. Therefore, the commission highlighted the need for mitigation measures to prevent erosion. Around 45 percent of India's coastline, as scientifically laid down in 2014, is severely prone to erosion. While coastal erosion is very alarming, the case of rivers is no different and they are also facing widespread soil erosion. There are two types of activities recommended by the commission: (i) mitigation measures to prevent erosion, which should be carried out under the NDMF and (ii) the resettlement of displaced people who are affected by erosion should be carried out under the NDRF. Floods caused by erosion should be addressed through large-scale projects, and they fall outside the purview of the Finance Commission. For community-level activities, the commission has recommended a sum of Rs. 1500 crores for the period from 2021 to 2026. States can avail of this fund by applying to the Ministry of Home Affairs and the National Disaster Management Authority. The states have to contribute 10 percent of the amount of the allocated fund.

22.5.5 Amending the Methodology for Disaster Funding

The role of the Fifteenth Finance Commission in the evolution of disaster finance mechanism is very significant, the commission departed from the expenditure-based approach followed by the preceding finance commission. The FC-XV advocated state-specific funding based on the "risks and vulnerabilities of the respective states." There was an urgent need on the part of the commission to amend the existing expenditure-based mechanism as it may enhance discordance in the allocation of funds to states as states that have lower and higher basis of expenditure lead to asymmetrical allocations of funds (Govt. of India, Ministry of Finance 2019). The commission, to address the same issue, introduced three new criteria along with the expenditure model for fund allocation. These criteria include the population, area, and risk profile of every state. Area and population shall be given 15 percent weightage separately. In the case of the Himalayan and North-Eastern states, an additional 11 percent weightage keeping in mind the requirement of infrastructure resilience as these areas are more prone and vulnerable to hazards.

Ranking the states concerning disasters is very important in making the funding mechanism more effective and efficient. Therefore, the "Disaster Risk Index'" has been prepared based on two significant parameters, including the "***probability of hazards***" and the "***extent of vulnerability***." There are various types of hazards, such as landslides, earthquakes, floods, and many more. Vulnerability refers to the weak capacities or ability of a state to respond and recover from a potential hazard, whether it is man-made or natural. There is no disaster database at the national level that can be used by the commission, and in such a situation, the commission has taken into consideration the risk exposure maps and national hazard zonation to assign a score for DRI. The commission assigned 70 percent weightage to the potential hazards and the remaining 30 percent to the vulnerability status of the state. The commission took into consideration four major hazards- drought, floods, earthquakes, and cyclones dividing them into three levels of scoring that are High (15 points), Medium (10 points), and Low (5 points). The commission used the Tendulkar Methodology while assessing the vulnerability status of the states based on the poverty line of states based on 2011–12 data. The vulnerability scale is divided into three groups, namely 10 points for states that have poverty below 13 percent, 20 points for states with poverty between 13 and 26 percent, and 30 points for states with poverty above 26 percent (Singh 2021).

22.5.6 Union's Response to the Himalayan-Specific Disasters

There is no clear definition of the types of disasters even in the Disaster Management Act. As a result, the Ministry of Home Affairs, in consultation with the earlier finance commission, formulated a list of naturally occurring disasters that fall under the purview of funding. But in the case of man-made disasters, there is no such list making them ineligible for funding under the NDRF and SDRF. This has led the states to make two demands- first, the list of naturally occurring disasters needs to be broadened, and it should also include certain state-specific disasters such as snake bites, heat waves, lighting, and many more; and secondly, man-made disasters should also be covered under the National Disaster Risk Fund. Addressing the demands of the states, the Fifteenth Finance Commission recommended that for the state-specific disaster, the states can utilize 10 percent of the funds available under the State Disaster Response Fund. These disasters should be local as defined by the state government and cannot be included in the list of naturally occurring disasters maintained by the Home Ministry. Concerning man-made disasters, the commission maintained that these disasters require enormous amounts of funds and cannot be funded through the

State Disaster Response Fund. However, the central government can provide funding to deal with man-made disasters through the National Disaster Response Fund.

Fourteenth Finance Commission received similar requests from the states to include state-specific disasters like a forest fire, snake bites, cold waves/heat waves, frost, erosion of river banks and sea coast, bamboo flowering, and road/rail/boat accidents in the list of disasters eligible for funding. These demands were supported by the Ministry of Home Affairs (MHA) and the National Disaster Management Authority (NDMA). States maintained that though these disasters have a local occurrence their impact is very deep. Therefore, states strongly demanded to provide funding from SRDF/NDRF for these disasters. Considering the persistent demands by the states supported by the MHA and NDMA, FC-XIV recommended that 10 percent of the SDRF can be utilized for state-specific disasters provided that the state notifies and includes these disasters approved by the State Disaster Management Authority (SDMA). Nevertheless, in the context of the man-made disasters, FC-XIV upheld the earlier stand of the FC-XIII and left it out of the scope of SDRF. However, it recommended that selected events be funded by NDRF (Reddy 2015).

Fifteenth Finance Commission has also received demands from the states to include state and location-specific disasters. FC-XV maintained that it would be difficult to quantify because of the difference in the severity of demanded disasters. Therefore, FC-XV maintained that the existing list of the notified disasters largely met the requirements of the states, and expressed no need to include region-specific disasters. However, FC-XV has initiated seven disaster-specific mitigation funds in which it has allocated Rs. 7500 crores for mitigation of fire and coastal and river bank erosion. It has allocated Rs. 5000 crore for strengthening the fire management services and Rs. 2500 crore for mitigation of coastal and river bank erosion (Singh 2021). Still, states are not satisfied with the list and continuously demanding to include additional state-specific disasters in the list.

22.6 Conclusion

India has a vast geography that runs from the Himalayas in the North and North East to the Indian and Arabian Oceans in the South and the Desert in the West. India is better explained as a sub-continent because of its geographical diversity. This is also evident in the report published by the government of India, which states that climate change will impact different geographical regions of India differently. The report published by the Ministry of Earth Science on "Assessment of Climate Change on Indian Region" in 2020 also divides India into four regions and assesses the possible impact of climate change specific to the region. This diversity of geography cannot be overlooked in the political discourse. The Indian Himalayan Region because of its geographical conditions faces a different set of vulnerability and hazards and demands a differential treatment. The 13 states and the Union territories have a different territorial identity. Based on the territorial theory of federalism these states have raised their concerns for a differential treatment in disaster risk funding and sought the response of the Union government accordingly. However, the analysis from the second to the fifteenth finance commission indicates that the Union government has a "one size fits all" approach while dealing with disaster management and hardly recognizes the differential demands of the Indian Himalayan Region. India's thrust for disaster resilience requires a specialized approach where a centralized list of disasters should have an add-on for the region-specific disasters and also rather than having five major disasters for calculating the Disaster Risk Index (DRI) there should be multiple sets of DRIs specific to the region-specific disasters. In addition, the methodology of the funding pattern is still based on past expenditures, and the hazard profile is used for the 30 percent funding only that needs to be improvised as soon as possible. In a nutshell, it can be argued that the Union government of India must transition from a mono-centric approach of disaster risk funding to a specialized approach leading to a better response to the concerns of the states located in the Indian Himalayan Region. In the absence of this transition, disaster resilience cannot be realized and the IHR will suffer due to the lack of a Himalayan-centric approach to Disaster risk funding.

References

Azad S, Nagarkoti Y (2024) Four forest staffers burnt alive in Binsar sanctuary. Times of India, 14 June 2024. Retrieved from https://timesofindia.indiatimes.com/city/dehradun/4-forest-staffers-burnt-alive-in-binsar-sanctuary/articleshow/110977653.cms

Bakshi PM (2001) The Constitution of India. Universal Law Publishing, Delhi

Barua A, Gulati V, Changakakati T, Bhaduri R, Dasgupta S, Vyas S, Khursheed A, Sanyal K, Esteves T, Murthy IK, Ravindranath NH, Sharma J (2021) How vulnerable are India's Himalayan Region states to climate change? Econ Political Weekly 56(11). Retrieved from https://www.epw.in/engage/article/how-vulnerable-are-indias-himalayan-region-states

Chakrabarti PGD (2009). Financing disaster management in India: a report for the Thirteenth Finance Commission. National Institute of Disaster Management. Retrieved from https://fincomindia.nic.in/asset/doc/commission-reports/13th-FC/reports/report23.pdf

Department of Science and Technology, Government of India (30 Nov 2018) Climate vulnerability assessment for the Indian Himalayan Region using a common framework. Retrieved from https://dst.gov.in/sites/default/files/IHCAP_Climatepercent20Vulnerabilitypercent20Assessment_30Nov2018_Final_aw.pdf

Dikshit RD (1997) Development in political geography: a century of progress. Sage Publication, New Delhi

Duchacek ID (1970) Comparative federalism: the territorial dimension of politics. Rinehart and Winston, New York

Government of Assam (n.d.) Flood & erosion problems. Water resources. Retrieved from https://waterresources.assam.gov.in/portlets/flood-erosion-problems

Government of India. Ministry of Earth Sciences (2024) Rajya Sabha Unstarred Question No. 666: India's land mass prone to earthquakes. Retrieved from https://www.moes.gov.in/sites/default/files/RS-English-666-08-02-2024.pdf

Government of India. Ministry of Tourism (2022) Rajya Sabha Unstarred Question No. 2534. Answered on March 24, 2022. Impact of Natural Disasters on Tourism. Retrieved from https://tourism.gov.in/sites/default/files/2022-03/usqpercent202534percent20forpercent2024032022.pdf

Government of Manipur (2020) DEMAND NO: 48—relief and disaster management. Retrieved from https://manipur.gov.in/wp-content/uploads/2020/02/dmd-48_2020-21.pdf

Government of Mizoram (2020) Mizoram State disaster management plan, 2020. Disaster Management & Rehabilitation Department. Retrieved from https://dmr.mizoram.gov.in/page/home

Govt. of India, Ministry of Finance (2004) Report of the Twelfth Finance Commission (2005–2010). New Delhi. Retrieved from https://fincomindia.nic.in/commission-reports-twelfth

Govt. of India, Ministry of Finance (2010) Report of the Thirteenth Finance Commission (2005–2010). New Delhi. Retrieved from https://fincomindia.nic.in/commission-reports-thirteenth

Govt. of India, Ministry of Finance (2015) Report of the Fourteenth Finance Commission. New Delhi. Retrieved from https://fincomindia.nic.in/commission-reports-fourteenth

Govt. of India. Ministry of Finance (2019) Report of the Fifteenth Finance Commission (2020–2021). New Delhi. Retrieved from https://fincomindia.nic.in/commission-reports-fifteenth

IPCC (2023) Climate Change 2023: synthesis report. Contribution of Working Groups I, II, and III to the Sixth Assessment Report of the Intergovernmental Panel on Climate Change [Core Writing Team, H. Lee and J. Romero (eds.)]. IPCC, Geneva, Switzerland, pp 35–115, https://doi.org/10.59327/IPCC/AR6-9789291691647

Jena R, Pradhan B, Naik SP, Beydoun G, Park HJ, lamri A, (2021) Earthquake vulnerability assessment for the Indian subcontinent using the Long Short-Term Memory model (LSTM). Int J Disaster Risk Reduction 66:102642. https://doi.org/10.1016/j.ijdrr.2021.102642

Mina U, Dimri AP, Farswan S (2023) Forest fires and climate attributes interact in the central Himalayas: an overview and assessment. Fire Ecol 19:14. https://doi.org/10.1186/s42408-023-00177-4

NITI Aayog (n.d.) Sustainable development in the Indian Himalayan Region. Retrieved from https://www.niti.gov.in/index.php/sustainable-development-indian-himalayan-region

Reddy YV (2015) Report of Fourteenth Finance Commission (2015–2020), vol I. Ministry of Finance, New Delhi

Reddy Y, Reddy G (2018) Indian fiscal federalism. OUP India, India

Salve NKP (1989) Second Report of the Ninth Finance Commission for (1990–95). Government of India Press, Delhi

Sharma S (2023) Finance commissions and disaster risk funding: a paradigm shift in methodology by Fifteenth Finance Commission. Indian J Public Adm 69(1):47–58. https://doi.org/10.1177/00195561221109092

Singh JS (2006) Sustainable development of the Indian Himalayan region: linking ecological and economic concerns. Current Sci 90(6):784–788. http://www.jstor.org/stable/24089189

Singh NK (2021) Report of fifteenth finance commission (2021–2026). Ministry of Finance, New Delhi

Supreme Court of India (2016) Swaraj Abhiyan vs. Union of India. Writ Petition (Civil) No. 857 of 2015

XV Finance Commission (2020) Finance commission in Covid times, vol I Main Report; Report for 2021–26. Retrieved from https://fincomindia.nic.in/asset/doc/commissionreports/XVFCpercent20VOLpercent20Ipercent20Mainpercent20Report.pdf

Chapter 23
Sustainable Governance Strategies for the Himalayan Ecosystem

Prabuddh Kumar Mishra and Suresh Chand Rai

Abstract The Himalayan ecosystem, a critical global natural resource, faces mounting challenges due to climate change, deforestation, and unsustainable development. Effective governance is essential to sustain this unique ecological region, which spans several countries in South Asia and serves as the water source for millions of people. This chapter explores the governance frameworks for supporting the Himalayan ecosystem, focusing on the integration of Sustainable Development Goals (SDGs) into regional environmental policies. Key SDGs, such as Climate Action (SDG 13), Life on Land (SDG 15), and Clean Water and Sanitation (SDG 6), are particularly relevant in addressing the complex environmental issues of the Himalayas. The chapter highlights current policies and strategies, including regional collaborations, national action plans, and community-based management approaches. It also assesses the challenges faced in policy implementation and the importance of transboundary cooperation between the Himalayan countries. Looking to the future, the chapter discusses strategies for adaptive governance, enhanced policy coordination, and the task of technology and local communities in ensuring the long-term sustainability of the ecosystem.

Keywords Himalayan ecosystem · Environmental governance · Transboundary governance Sustainable Development Goals (SDGs) · Climate action · Community-based management

P. K. Mishra
Department of Geography, Shivaji College, University of Delhi, Delhi, India
e-mail: prabuddh@shivaji.du.ac.in

S. C. Rai (✉)
Department of Geography, Delhi School of Economics, University of Delhi, Delhi, India
e-mail: raisc1958du@gmail.com

S. C. Rai (ed.), *Himalaya: Mountains of Destiny*,
https://doi.org/10.1007/978-3-031-94905-0_23

23.1 Introduction

The Himalaya, Asia's most prominent mountain range, is known as the "Abode of Gods." The Himalayan environment is extremely significant not just for its legendary significance but also because it controls the climate, acts as Asia's "Water Tower," and provides a range of ecological services to its people. The Himalayas are critical for global biodiversity, climate regulation, and the sustenance of nearly 1.9 billion people dependent on their resources (Molden et al. 2014a, b). The Himalaya mountain range regenerates the headwaters of five major river systems: Indus, Brahmaputra, Ganges, Mekong, and Yangtze, of which Indus, Brahmaputra, and Ganges are more related to the sustenance of lives in South Asia. These are among the largest, most heavily populated, and ecologically diverse river basins in the world. The Himalayan ecosystem is the stage for a series of productive systems, including high mountain tourism, forest product harvesting, grazing, farming, and relocating people moving out of urban centers. The immense importance of this part of the ecosystem acknowledges the policy, both at the regional and national levels. This mountain range, which stretches across eight countries, faces numerous ecological challenges, including climate change, deforestation, glacial retreat, and biodiversity loss. These challenges demand robust governance frameworks that balance ecological preservation with economic development. Effective governance involves multi-stakeholder cooperation, sustainable policy implementation, and community engagement to preserve this fragile ecosystem.

Sustainable development of the Himalayan region is dependent upon effective governance of environmental resources. Water, forests, wildlife, and agriculture are just a few of the region's many environmental resources. A complex web of institutions, protocols, decision-making processes, and well-established implementation methods are all part of the governance of these resources. In a broader sense, "environmental governance" refers to the several ways that formal and informal institutions work to manage and control the environment following different ecological, social, and cultural principles. In response to the various issues related to Himalayan environmental governance, national environmental policies and institutions are complex and changing, despite their enormous importance. In the Himalayan region, several state agencies have arisen, but they have difficulties in establishing links and coordinating at various levels, including local, sub-national, national, regional, and international. Regional and local governments have far less authority over planning and decision-making than do national and sub-national authorities. Sustainable natural resource management is frequently incompatible with this unequal vertical distribution of controlling power. There is a long history of Himalayan environmental governance including state agencies and local communities. Furthermore, there has been an increase in the activity of knowledge communities, private groups, and non-governmental organizations recently.

Environmental governance has additional challenges beyond the jurisdiction of national and sub-national bodies. One is that environmental organizations have not yet fully adapted to the complex geography of the Himalayas. As a result, there aren't enough context-specific and pertinent institutions in the region to establish connections between communities in river basins upstream and downstream. Another issue with the institution that hinders cooperation is the sectoral fragmentation of environmental governance. Climate change, ecological threats, and disaster risks are just a few of the contemporary problems that these initiatives seek to solve. Additional developments have pushed resource control toward market mechanisms, especially for the water, energy, and agricultural sectors.

Although many national and sub-national reforms aimed at decentralization and devolution have been successful on a local level, they cannot have a greater impact without corresponding adjustments to national and sub-national governments. They are essential for maintaining a suitable institutional structure and connections among different levels of government. The limitations are caused by inherent flaws in accountability and representation, failure to pay enough attention to power disparities, and poorly constructed policy frameworks. Uncertain lines of accountability and authority frequently impede effective devolution, and major policy changes are not implemented. There are still other issues with institutional efficacy and implementation. Limited cross-border knowledge sharing and recurring geopolitical standoffs in light of the region's growing global significance hinder cooperation among the governments of the Himalayan states. Local knowledge is rich and plentiful, but it is also underutilized in the processes that lead to policymaking. In a region that is experiencing fast socio-economic development, growing geopolitical tensions, and significant environmental threats, better political articulations are also desperately needed.

India's National Environmental Policy (NEP) 2006 contains special provisions for the protection of the mountains and Himalayas due to the importance of the Himalayan ecosystems (NEP 2006). As part of its National Action Plan on Climate Change (NAPCC), India recently unveiled a project named "National Mission for Sustaining the Himalayan Ecosystem NMSHE)," which has a special emphasis on the Himalayas. State and national governments must design and implement particular guidelines for sectors pertinent to the region's development to meet the goals. More attention has been paid to issues that are pertinent to the sustainable development of the Himalayan ecosystem, including resource degradation, urbanization, catastrophe risk reduction, pilgrimage and tourism, and raising awareness.

The policy interplays within countries are still unresolved and significantly impact the sustainable development in the Himalayan region. To ensure the well-being of the downstream communities of the Himalayan ecosystem, vastly improved governance is required. This chapter investigates a set of methods that could be employed to build sustainable governance for the Himalayan ecosystem. The significant impact of a good intervention implemented in the designated region is also exposed by applying a step, utilizing a typology that distinguishes physical, natural resource-based systems, and people. This approach can be widely applied to all community-based natural resource management conditions. Specifically, the importance of such an approach is also initiated for the question of the sustainable Himalaya governance initiative. We reflect on the relationships and discuss the broader implications of this governance for sustainability.

This chapter also provides a detailed overview of environmental governance in the Himalayas, but it also highlights new developments in environmental policy and practice and details the current status of governance in the area. Additionally, it highlights both opportunities and shortcomings in the areas of policymaking, regulatory frameworks and enforceability, and environmental governance's capacity to adjust to the growing effects of climate change. Its main argument is that the establishment of creative and regional institutional arrangements will be necessary to support shared policy initiatives, encourage intergovernmental dialogue, and facilitate cooperative trans-border community practices to manage the Himalayas' resources sustainably.

23.2 Environmental Policies and Governance in the Himalayan Region

Environmental policies and governance in the Himalayan region are critical due to the area's ecological fragility and its significant role in global climate regulation. The Himalayan ecosystem is governed through a combination of national policies and regional collaborations that address environmental conservation, climate adaptation, and sustainable resource management. International organizations and agreements emphasize the need for a coordinated approach to address environmental challenges in this region, recognizing the transboundary nature of climate change impacts and ecological degradation. Initiatives such as the *Hindu Kush Himalaya Assessment* and collaborations through *ICIMOD* highlight global efforts to promote sustainable mountain development and climate resilience (HIMAP 2019; ICIMOD 2019). Additionally, frameworks like the United Nations' Sustainable Development Goals (SDGs) provide a comprehensive roadmap, with SDG 13 (Climate Action) and SDG 15 (Life on Land) being particularly relevant. However, the global community acknowledges that effective governance in the Himalayas requires not only regional cooperation but also international support in terms of funding, technology transfer, and capacity building to ensure sustainable and equitable development (Shrestha et al. 2015). Effective governance involves multi-stakeholder cooperation, sustainable policy implementation, and community engagement to preserve this fragile ecosystem.

23.2.1 Transboundary Governance

Transboundary governance in the Himalayas is essential for addressing the region's shared environmental challenges, such as climate change, water resource management, and biodiversity conservation. The Himalayas span multiple countries making it imperative for these nations to work collaboratively. This geographical complexity makes environmental governance challenging. Issues like water resource-sharing, disaster risk management, and conservation policies require cooperation among these nations, which often have differing political and economic priorities (Adhikari and Taylor 2012). Regional institutions like the *International Centre for Integrated Mountain Development* (ICIMOD) play a significant role in facilitating transboundary cooperation by promoting data sharing, joint research, and policy dialogue aimed at sustainable mountain development (ICIMOD 2019). Additionally, initiatives such as the *Hindu Kush Himalaya Monitoring and Assessment Programme* (HIMAP) provide a scientific basis for joint decision-making and climate adaptation strategies (HIMAP 2019). However, effective transboundary governance remains challenging due to differing national interests, political tensions, and varying levels of economic development among these countries. Strengthening governance mechanisms through inclusive, multi-stakeholder approaches and aligning them with international frameworks like the Sustainable Development Goals (SDGs) can help address these challenges and foster greater regional stability and environmental protection (Sharma et al. 2019).

23.2.2 Frameworks for Collaborative Management

Successful transboundary governance models include the Indus Waters Treaty between India and Pakistan, which demonstrates the potential for collaboration despite geopolitical tensions. However, experts argue for more integrated approaches, such as regional environmental governance bodies, that facilitate knowledge exchange, resource-sharing, and unified climate action plans (Sharma et al. 2019).

23.2.3 Policy Frameworks and Environmental Legislation

Policy frameworks and environmental legislation in Himalayan countries are critical for conserving the region's unique ecosystems and ensuring sustainable development. Each country has established various laws and policies tailored to its environmental challenges. For example, India's *National Mission for Sustaining the Himalayan Ecosystem* (NMSHE) is part of the broader National Action Plan on Climate Change (NAPCC), focusing on climate resilience and ecological conservation (Government of India 2008). Nepal has implemented progressive measures, such as the *Forest Act of 1993*, which promotes community-based forest management, empowering local communities to sustainably manage forest resources (Ojha et al., 2016). Bhutan's environmental governance is guided by its constitutional mandate to maintain at least 60% forest cover, and its policy framework emphasizes the balance between development and environmental conservation, a principle rooted in its "Gross National Happiness" philosophy (Ura et al. 2012). These legislative measures are often supported by regional cooperation facilitated by organizations like

ICIMOD, which helps harmonize policies and address transboundary issues. Despite these frameworks, challenges such as enforcement gaps, resource constraints, and the impacts of climate change continue to strain environmental governance in the region (Singh 2020).

23.2.4 Legislative Gaps and Recommendations

Despite efforts, policy frameworks often fall short due to weak enforcement mechanisms, inadequate funding, and lack of coordination among various governmental and non-governmental agencies. Strengthening institutional capacities and investing in advanced monitoring technologies can bridge these gaps. Moreover, adaptive governance models that are flexible and responsive to changing environmental conditions are crucial (Blaikie and Muldavin 2004).

23.2.5 The Role of Science and Technology

Scientific advancements such as early warning systems are essential for monitoring glacial melt, tracking biodiversity, and predicting natural disasters. Remote sensing technology and Geographic Information Systems (GIS) provide critical data for policymakers. Collaborative research initiatives among academic institutions across the region have also led to a better understanding of climate impacts on the Himalayas (Pandey et al. 2017).

23.2.6 Data Sharing and Knowledge Dissemination

A major barrier to effective governance is the lack of data sharing among countries. Establishing a centralized data repository, accessible to all stakeholders, can enhance research and policy formulation. Initiatives like the South Asian Association for Regional Cooperation (SAARC) Disaster Management Centre have made progress, but more comprehensive efforts are needed (Shrestha and Aryal 2011).

23.2.7 Community-Centric Approaches

Local communities have lived in harmony with the Himalayan ecosystem for centuries, employing sustainable practices rooted in indigenous knowledge. Governance models that integrate this traditional wisdom can be highly effective. Community-based forest management and participatory watershed management are examples where local involvement has led to successful conservation outcomes (Agrawal and Gibson 1999). Empowering communities through education, training, and financial support is vital for sustainable governance. Micro-finance initiatives, eco-tourism projects, and community-led conservation programs can provide alternative livelihoods, reducing the pressure on natural resources. Additionally, gender-inclusive governance is necessary, as women in the Himalayas play a significant role in natural resource management (Sarkar and Pattanayak 2006).

23.2.8 Climate Change Adaptation and Mitigation

Climate change has accelerated glacial melting in the Himalayas, threatening water availability for millions. Effective governance requires integrating climate adaptation into development planning. Strategies like constructing artificial glaciers, rainwater harvesting, and improving irrigation efficiency can mitigate the water crisis (Bolch et al. 2012). The Himalayas are prone to natural disasters, including earthquakes, landslides, and floods. Risk reduction strategies should involve hazard mapping, community preparedness programs, and resilient infrastructure development. Governments must also invest in restoring natural ecosystems, such as wetlands and forests, which act as buffers against disasters (Liu et al. 2018).

23.2.9 International Efforts and Agreements

International agreements, such as the Paris Agreement, emphasize the importance of protecting fragile ecosystems like the Himalayas. The United Nations Sustainable Development Goals (SDGs) also provide a roadmap for balancing ecological preservation with economic development (United Nations 2015). The Hindu Kush Himalaya Monitoring and Assessment Programme is a key regional effort that addresses environmental and social challenges in the region. Such initiatives demonstrate the importance of collective action and the need for continuous funding and political commitment (Wester et al. 2019).

Governance for sustaining the Himalayan ecosystem is a complex yet imperative endeavor. It requires a holistic approach that integrates transboundary cooperation, stringent environmental policies, scientific advancements, and community engagement. As climate change accelerates, the urgency to implement effective governance frameworks becomes more pronounced. A sustainable future for the Himalayas hinges on collaborative efforts that prioritize ecological health, community well-being, and adaptive management.

23.3 Recent History of National Himalayas Environmental Policies

Natural resources in the Indian Himalayan region have long been controlled, appropriated, and managed by local customs, organizations, and practices. As the state has become more involved in modern society, natural resources are increasingly subject to centrally designed regulations and laws. In the Himalayas, environmental laws and regulations were first developed by central governments in its first five-year plan. Since the 1950s, central governments have had more power and involvement in formulating policies and planning for development, which has accelerated environmental governance in particular. After then, countries began drafting legislation pertaining to the environment and sustainable use of natural resources, particularly during the 1992 United Nations Conference on Environment and Development. The region has seen a sharp rise in the number of policy measures aimed at controlling the environment and common pool resources (CPRs), especially water, forests, and rangelands. In conservation policies and actions, the species-and-habitat-focused paradigm has mostly been supplanted by a participatory livelihood-based landscape approach (Sharma et al. 2010). India has formulated and evolved environmental policies aimed at sustainable management and conservation. This section outlines the recent history of these policies, highlighting their objectives, successes, and challenges.

23.3.1 National Action Plan on Climate Change (NAPCC) and the National Mission for Sustaining the Himalayan Ecosystem (NMSHE)

National Action Plan on Climate Change

The *National Action Plan on Climate Change* (NAPCC), launched in 2008, marked a milestone in India's climate policy. The plan comprises eight missions, each targeting specific sectors impacted by climate change. One of these is the *National Mission for Sustaining the Himalayan Ecosystem* (NMSHE), which focuses on the Indian Himalayan Region (IHR) (Government of India 2008).

National Mission for Sustaining the Himalayan Ecosystem (NMSHE)

The NMSHE seeks to understand the complex and diverse ecosystem of the Himalayas and devise sustainable strategies. The mission focuses on building an institutional and human capacity base to continuously assess and study the Himalayan environment. Establishing a research framework to study glacial retreat, biodiversity, and climate change impacts. The mission funds studies on glacial lake outburst floods (GLOFs) and other natural hazards to build scientific knowledge (Bhan et al. 2019). Recognizing the role of local communities, NMSHE supports participatory conservation efforts. This approach aims to incorporate indigenous knowledge into scientific frameworks. NMSHE emphasizes training and awareness programs for local communities and government officials, focusing on climate resilience and disaster preparedness.

Since 2015, NMSHE has emphasized creating robust data-sharing platforms among research institutions. Collaborations with universities and state governments have strengthened, enabling integrated studies on climate change impacts. In 2021, the mission renewed its focus on adaptive measures, such as promoting agroforestry and sustainable tourism (MoEFCC 2021).

23.3.2 Compensatory Afforestation Fund Act (2016) and Its Impact

The *Compensatory Afforestation Fund Act* (CAF Act) was enacted to ensure that funds collected from deforestation projects are efficiently used for afforestation and ecosystem restoration. States in the Indian Himalayas, such as Himachal Pradesh and Uttarakhand, have benefited from substantial funding under this Act to conserve forest cover and restore degraded land (Kumar 2017). The Act has led to large-scale afforestation projects and improved forest management. However, critics argue that the selection of afforestation sites has not always considered local ecological conditions, potentially threatening native biodiversity. Additionally, concerns about transparency and equitable fund distribution have been raised (Chakraborty 2020).

23.3.3 Forest Rights Act (2006) and Amendments

The *Forest Rights Act* (FRA) of 2006 is a landmark piece of legislation that recognizes the rights of forest-dwelling communities over land and forest resources. In the Himalayan states, the Act has been crucial in empowering indigenous populations and promoting community-based forest management. Despite its importance, the FRA has faced hurdles in implementation. In the Indian Himalayas, bureaucratic inefficiencies and a lack of awareness have hindered effective execution. Additionally, conservationists worry that the allocation of land rights may conflict with biodiversity preservation goals (Singh 2018). Recent amendments aim to simplify the claims process and enhance community participation, but progress has been slow.

23.3.4 Disaster Management and Resilience Policies

The Indian Himalayas are prone to a range of natural disasters, including earthquakes, landslides, and floods. The frequency and severity of these disasters have increased, largely due to climate change and unsustainable land use practices (Naithani 2021).

The *Disaster Management Act* of 2005 provides the framework for disaster preparedness and response in India. In the Himalayan context, this Act has been supplemented by state-level disaster management plans. For example, Uttarakhand and Sikkim have developed disaster risk reduction strategies focusing on early warning systems, community training, and resilient infrastructure development. In recent years, India has invested in technology-driven disaster management. The deployment of drones for landslide monitoring, real-time weather forecasting systems, and improved glacial lake monitoring are some advancements. The Indian National Centre for Ocean Information Services (INCOIS) and the India Meteorological Department (IMD) have collaborated to strengthen early warning mechanisms (Sharma et al. 2022).

23.3.5 Conservation of Biodiversity and Wildlife Policies

Project Snow Leopard was launched in 2009 as a flagship initiative to protect high-altitude wildlife, including snow leopards and their habitat. The project focuses on landscape-based conservation, scientific research, and community engagement. In 2020, a renewed focus on habitat corridors and human–wildlife conflict mitigation emerged, with additional funding allocated for research and conservation (Sinha 2020). The *Biodiversity Act* of 2002 established a legal framework for the conservation and sustainable use of biodiversity. In the Himalayas, state biodiversity boards have been set up to implement this Act, often working with local communities to document and preserve traditional ecological knowledge. However, limited financial resources and inadequate enforcement have posed challenges to effective biodiversity conservation (Joshi 2018).

23.3.6 Sustainable Development Policies: Tourism and Infrastructure

The Indian Himalayas are a popular tourist destination, but unregulated tourism has led to environmental degradation. In response, the Ministry of Tourism and state governments have developed eco-tourism policies that promote responsible travel practices. Guidelines include waste management, limiting the number of visitors to sensitive areas, and promoting homestays over large hotels (Thakur and Singh 2019). While infrastructure development, such as the construction of highways and hydropower projects, is critical for regional growth, it has environmental repercussions. The *Char Dham Highway Project* in Uttarakhand, for example, has faced scrutiny for causing deforestation and increasing landslide risks. The government has since revised environmental safeguards, emphasizing slope stabilization and eco-friendly construction practices (Pant 2021).

The recent history of environmental policies in the Indian Himalayas reflects a balancing act between development and conservation. While significant strides have been made in areas like climate resilience and community forestry, challenges such as policy enforcement, disaster preparedness, and sustainable tourism remain. Moving forward, a more integrated approach that prioritizes ecological health while supporting local communities will be crucial for the long-term sustainability of the Himalayas.

23.4 Environmental Governance and the Sustainable Development Goals

Encourage more regional cooperation between mountain areas and countries to advance a mountain-specific agenda for accomplishing the SDGs. The need for international cooperation in global or regional environmental governance is becoming more widely acknowledged. In the Himalayan region, this kind of cooperation must be founded on the exchange of data gathered via science, the creation of all-encompassing policies, and the acceptance of novel approaches that produce long-term, reciprocal benefits. By placing an emphasis on the distinctiveness of the HKH Mountains and its people and making sure that they are represented on higher decision-making platforms, it is necessary to renew the regional cooperation in institutions and processes that make decisions. Recognizing the requirement of providing adequate funding for the upkeep of mountain ecosystems is also crucial, as is creating incentive programs to promote both individual and group efforts toward sustainable use and conservation.

Adopted by the UN in 2015, the Sustainable Development Goals (SDGs) provide a worldwide framework for sustainable development that tackles environmental degradation, poverty, and inequality. The Himalayas are especially affected by several Sustainable Development Goals, including SDG 6 (Clean Water and Sanitation), SDG 13 (Climate Action), SDG 15 (Life on Land), and SDG 16 (Peace, Justice, and Strong Institutions) (United Nations 2015). Let's explore how environmental governance frameworks in the Himalayas are structured to meet these goals, the challenges they face, and potential pathways forward.

23.4.1 Governance Frameworks in the Himalayas

National and Regional Policies

Environmental governance in the Himalayas is complex due to its transboundary nature. Each country in the Himalayan region has distinct governance structures and policies, but there is increasing recognition of the need for a cooperative approach. India's *National Mission for Sustaining the Himalayan Ecosystem* (NMSHE) under the National Action Plan on Climate Change (NAPCC) addresses climate change impacts and promotes sustainable development (Government of India 2008). It focuses on research, monitoring, and community-based conservation efforts. Nepal has adopted policies such as the *Forest Act of 1993* and community forestry programs, empowering local communities to manage forest resources sustainably. This aligns with SDG 15, which emphasizes the protection of terrestrial ecosystems (Ojha et al. 2016). Conservation of the environment is emphasized in Bhutan's "Gross National Happiness" development concept. Since forests represent more than 70% of Bhutan's geographical area, deforestation and biodiversity loss have been lessened thanks to stringent environmental laws (Ura et al. 2012).

Regional Cooperation

Transboundary governance mechanisms, such as the *International Centre for Integrated Mountain Development* (ICIMOD), play a critical role in promoting regional cooperation. ICIMOD facilitates knowledge sharing and joint initiatives among Himalayan countries to address climate change and environmental degradation, supporting SDG 17 (Partnerships for the Goals) (ICIMOD 2019).

23.4.2 Strategies for Sustainable Environmental Governance

Community-Based Natural Resource Management (CBNRM)

CBNRM has been effective in promoting sustainable forest and water management. In Nepal, community forestry programs have successfully restored degraded forests while improving local livelihoods. In line with SDG 16 (Peace, Justice, and Strong Institutions), this model places a strong emphasis on local communities' involvement in decision-making processes (Chhetri et al. 2019).

Integrated Watershed Management

Given the interdependence of ecosystems in the Himalayas, integrated watershed management is crucial. It involves coordinated efforts to manage water, soil, and forest resources sustainably. Programs like India's *Catchment Area Treatment Plan* for major river basins have shown promise, but challenges remain in scaling up these initiatives (Gupta and Sah 2008).

Climate Adaptation and Disaster Risk Reduction

Strategies for disaster risk reduction (DRR) and climate adaptation are receiving more attention from governments in the Himalayan region. These include early warning systems for landslides and floods, climate-resilient infrastructure, and sustainable agricultural practices. Bhutan's *Climate-Resilient Villages* initiative, for example, integrates traditional knowledge with modern science to enhance community resilience (Wangdi et al. 2020).

Cross-border Collaboration

Environmental challenges in the Himalayas are not confined by national borders, necessitating cross-border collaboration. The *Hindu Kush Himalaya Monitoring and Assessment Programme* (HIMAP) is a collaborative effort that provides scientific assessments and policy recommendations to address climate change and environmental degradation (HIMAP 2019).

23.4.3 Relevance to the Sustainable Development Goals

The Himalayas are critical for regional water security, supplying water to billions. Effective governance ensures sustainable water management, addressing both supply and quality issues. Integrated water resource management practices, as promoted by ICIMOD, support SDG 6 by balancing human and ecological needs (ICIMOD 2019). Climate action in the Himalayas is crucial for global climate stability. National policies, such as India's *NAPCC* and Nepal's *National Climate Change Policy*, focus on climate mitigation and adaptation measures, but implementation remains a challenge. Strengthening governance structures to enforce climate policies can accelerate progress toward SDG 13 (Sharma et al. 2019). Forest conservation and biodiversity protection are central to SDG 15. The Himalayas' rich biodiversity requires comprehensive governance mechanisms to curb habitat loss and promote reforestation. Programs like Bhutan's *National Biodiversity Strategies and Action Plan* emphasize the importance of preserving genetic diversity and ecosystem services (Ura et al. 2012). Effective environmental governance relies on strong institutions and inclusive decision-making. Addressing environmental conflicts, such as disputes over water rights or forest resources, is essential for fostering peace and justice. Community-based governance models in Nepal have demonstrated the importance of empowering local institutions to manage natural resources sustainably (Chhetri et al. 2019).

23.5 Conclusion

Effective governance and the incorporation of sustainable development ideas into regional policy are essential to the Himalayan ecosystem's sustainability. Deforestation, unsustainable resource exploitation, and climate change all present difficulties that need a multifaceted strategy that acknowledges the region's distinct transboundary character while incorporating global frameworks like the SDGs. Environmental governance in the Himalayas is integral to achieving the Sustainable Development Goals. The region's unique environmental challenges, from climate change to biodiversity loss, require coordinated, multi-stakeholder approaches that emphasize sustainability and resilience. While significant progress has been made, particularly in community-based resource management and climate adaptation, continued efforts are needed to strengthen governance frameworks and foster cross-border collaboration. The Himalayas' future depends on a holistic approach that integrates environmental, social, and economic considerations, ensuring that this vital region continues to support life for generations to come. The chapter highlights how crucial it is for the Himalayan nations to work together, stressing the necessity of unified and flexible policies that support both domestic and international environmental objectives. Moving forward, the focus must shift toward strengthening governance frameworks, enhancing local community participation, and utilizing technology to monitor and manage ecological changes. By prioritizing the health of the Himalayan ecosystem, these strategies can ensure the continued provision of essential resources like water and biodiversity for millions of people while supporting the achievement of the SDGs.

References

Adhikari B, Taylor K (2012) Transboundary water governance: adapting to climate change impacts in the Hindu Kush-Himalaya. Climate Policy 12(2):195–209

Agrawal A, Gibson CC (1999) Enchantment and disenchantment: the role of community in natural resource conservation. World Dev 27(4):629–649

Bhan U, Arya A, Singh R (2019) Climate change and the Indian Himalayan region: policy and governance perspectives. J Mt Sci 16(3):345–360

Blaikie P, Muldavin J (2004) Upstream, downstream, China, India: the politics of environment in the Himalayan region. Ann Assoc Am Geogr 94(3):520–548

Bolch T, Kulkarni A, Kääb A, Huggel C, Paul F, Cogley JG, Stoffel M et al (2012) The state and fate of Himalayan glaciers. Science 336(6079):310–314

Chakraborty R (2020) Evaluating the effectiveness of the Compensatory Afforestation Fund Act in India. Environ Policy Law 50(2):112–125

Chhetri BBK, Johnsen FH, Konoshima M, Yoshimoto A (2019) Community forestry in the hills of Nepal: determinants of user participation in forest governance. Forest Policy Econ 50:164–171

Government of India (2008) National action plan on climate change. Prime Minister's Council on Climate Change

Government of India (2010) National Mission for Sustaining the Himalayan Ecosystem (NMSHE). Ministry of Environment, Forest and Climate Change

Gupta A, Sah MP (2008) Impact of anthropogenic activities on the Himalayan environment. Curr Sci 94(1):40–46

HIMAP (2019) The Hindu Kush Himalaya Assessment: mountains, climate change, sustainability, and people. ICIMOD, Kathmandu, Nepal

ICIMOD (2019) Annual report 2019. International Centre for Integrated Mountain Development, Kathmandu, Nepal

Joshi K (2018) Challenges in implementing the Biodiversity Act in the Indian Himalayas. Biodivers Conserv 27(6):1473–1489

Kumar S (2017) The impact of the Compensatory Afforestation Fund Act on forest management in India. Forestry J India 43(1):25–36

Liu J, Zhang Z, Xu X, Kuang W, Zhou W, Zhang S, Li J et al (2018) Spatial patterns and driving forces of land use change in the Himalayas. Land Use Policy 71:429–440

Ministry of Environment and Forests, Government of India (2006) National Environment Policy

MoEFCC (Ministry of Environment, Forest and Climate Change) (2021) Annual report on the National Mission for Sustaining the Himalayan Ecosystem. Government of India

Molden D, Sharma E, Eriksson M (2014a) The Himalayan ecosystem: an overview of environmental change. Mt Res Dev 34(4):315–324

Molden D, Shrestha AB, Vaidya RA (2014b) Water resource management in the Hindu Kush Himalayas. Mt Res Dev 34(2):101–123

Naithani A (2021) Landslides and floods in Uttarakhand: causes and mitigation strategies. Nat Hazard Rev 22(4):506–518

Ojha H, Ghimire S, Pain A (2016) The role of community forestry in achieving the sustainable development goals in Nepal. Forest Policy Econ 63:31–39

Pandey R, Joshi S, Rajeev K (2017) Climate change adaptation in the Himalayas: governance, institutions, and policy framework. Environ Sci Policy 77:17–26

Pant M (2021) Environmental impact assessment of the Char Dham Highway Project. Environ Eng Manag J 20(5):903–915

Sarkar S, Pattanayak SK (2006) Empowering women through microfinance: evidence from the Himalayas. J Environ Manage 79(1):95–113

Sharma E, Molden D, Rahman A, Khatiwada YR (2019) The Hindu Kush Himalaya Assessment: key findings and actions for sustainable mountain development. Mt Res Dev 39(2):1–8

Sharma P, Gupta V, Mehta R (2022) Advancements in disaster management in the Indian Himalayas. J Geosci 10(1):120–135

Shrestha AB, Aryal R (2011) Climate change in the Hindu Kush-Himalayas and its impacts on water and ecosystems. Mt Res Dev 31(3):201–207

Singh A (2018) The forest rights act and its impact on the Indian Himalayan communities. J Forest Policy 29(3):134–148

Singh SP (2020) Environmental challenges in the Himalayas: policy implications and conservation strategies. Mt Res Dev 40(1):91–99

Singh SP, Thadani R (2015) Ecosystem services and Himalayan forest ecosystems. J Mt Sci 12(4):764–776

Sinha R (2020) Project snow leopard: a decade of conservation efforts. Wildlife Conservation J 7(4):134–148

Ura K, Alkire S, Zangmo T, Wangdi K (2012) A short guide to Gross National Happiness Index. Centre for Bhutan Studies

Wester P, Mishra A, Mukherji A, Shrestha AB (2019) The Hindu Kush Himalaya Assessment: mountains, climate change, sustainability, and people. Springer International Publishing, Cham

MIX
Papier aus verantwortungsvollen Quellen
Paper from responsible sources
FSC® C105338

If you have any concerns about our products,
you can contact us on
ProductSafety@springernature.com

In case Publisher is established outside the EU,
the EU authorized representative is:
Springer Nature Customer Service Center GmbH
Europaplatz 3, 69115 Heidelberg, Germany

Printed by Libri Plureos GmbH
in Hamburg, Germany